HVAC/R

TERMINOLOGY

A Quick Reference Guide

BY

RICHARD WIRTZ

Wirtz, Richard
HVAC/R Terminology: A Quick Reference Guide / by Richard Wirtz
ISBN: 1-930044-22-4

Cover photo: © ESCO Press
Editor: Julie Stelzer
Cover Designer: Scott Wills

esco press

Mount Prospect, IL 60056

Printed in the United States of America
1 2 3 4 5 6 7 8 9 10

ISBN: 1-930044-22-4
UPC: 65188110019-4

This book is dedicated to the wonderful and dedicated people of the Heating, Air Conditioning, Refrigeration

ACKNOWLEDGMENTS

It is not possible to write a document of this size and magnitude without the helpful assistance of other people and organizations. As with any book of this type, it is a team effort. They helped me gather vital information that is critical to the successful completion of this book.

Disclaimer:

This book is meant as an industry reference guide. Terms contained within may deviate from standard dictionary definitions as they are meant to be applicable within the scope of the HVACR industry.

Table of Contents

-A-

A - symbol used for the measurement of current flow (amperage).

Abs - acrylonitrilebutadiene styrene.

Ac - alternating current.

Acb - air circuit breaker.

Acu - air conditioning unit.

Acfm - the actual volumetric flow rate of air-vapor mixture. Unit is cubic feet per minute.

Ach - air changes per hour.

ACR - air conditioning and refrigeration.

ADC - analog to digital converter.

ADP - apparatus dew point.

Aec - annual energy cost.

AEV - automatic expansion valve.

AFCI - arc fault circuit interrupter

AFUE - annual fuel utilization efficiency.

Ag - silver.

Ahp - air horsepower.

AHU - air handling unit.

AIF - Acoustic Insulation Factor.

Ail - amber indicator light.

Al - aluminum.

ALU - arithmetic logic unit.

Am - ammeter.

Amb - ambient.

Amp - ampere.

Ann - annunciator.

Arm - armature.

As - ammeter switch.

ASCII Code - American Standard Code for Information Interchange.

ATC - automatic temperature compensation.

Atm - atmosphere.

Auto - automatic.

Aux - auxiliary.

AWG - American Wire Gauge.

AXV - automatic expansion valve.

"A" cock - a manual gas valve placed in the main gas line to shut off flow of gas to the piece of equipment. A "B" cock fitting is tapped into the side of the "A" cock for the pilot gas shut off valve. The "a" cock valve shall be installed so that the "B" cock fitting is on the upstream side of the gas flow.

A-coil - component on the low pressure side of the refrigeration system where liquid evaporates and absorbs heat into the evaporator. So called because the physical shape of the coil resembles the letter "A".

A-weighted - a term referring to the single number logarithmic summation of the 8 octave bands that have been adjusted to account for response of human ear to sound pressure level.

Abort - premature termination of a program during execution.

Abrasion - a scrape or other damage on an object's surface. A general wearing away of a surface by constant scratching, usually due to the presence of foreign matter such as grit, or metallic particles in a lubricant. It may also cause a break down of a material such as the tooth surfaces of gears. Lack of proper lubrication may result in abrasion.

Abrasion resistance - the ability of a material or cable to resist surface wear.

ABS pipe - acrylonitrilebutadiene styrene plastic pipe that is used for water, drains, waste, and venting (DWV).

Absolute - a chemical substance relatively free of impurities.

Absolute filtration rating - the diameter of the largest hard spherical particle that will pass through a filter under specific test conditions. This is an indication of the largest opening in the filter elements.

Absolute humidity - the total amount of moisture that is present in one cubic foot of air, as indicated by grains of moisture per cubic foot.

Absolute pressure - pressure measured from a perfect vacuum as a base. The common gauge expresses a pressure in pounds per square inch called gauge pressure. When the gauge is open to the atmosphere, it reads zero. The sum of gauge pressure (psig) plus atmospheric pressure (14.696 lb per square inch). Atmospheric pressure is 14.7 psi at sea level. It varies with location and atmospheric conditions. It is accurately indicated by a barometer. Absolute pressure can only be zero in a perfect vacuum.

Absolute temperature - the measurement of temperature from absolute zero (-459.7° F). The temperature as read on the Fahrenheit scale plus 460 degrees.

Absolute viscosity - a term used interchangeably with viscosity to distinguish it from either kinematic viscosity or commercial viscosity. Absolute viscosity is the ratio of shear stress to shear rate. It is a fluid's internal resistance to flow. The common unit of absolute viscosity is the poise. Absolute viscosity divided by fluid density equals kinematic viscosity. It is occasionally referred to as dynamic viscosity. Absolute viscosity and kinematic viscosity are expressed in fundamental units. Commercial viscosity such as Saybolt viscosity is expressed in arbitrary units of time, usually seconds.

Absolute zero - the temperature at which molecular motion theoretically stops. This temperature is considered to be -459.7° F and -273° C. At absolute zero there is considered to be a complete absence of heat.

Absorb - to soak up. The incorporation of a liquid into a solid substance, as by capillary, osmotic, solvent, or chemical action.

Absorbent - substance that has the ability to pick up, take up, or hold some other substance. One example would be the salt solution used to attract water in an absorption water chiller.

Absorbent filter - a filter medium that holds contaminant by mechanical means.

Absorber - the component, in an absorption cycle, where the heat laden refrigerant vapor is re-absorbed into the solution.

Absorber plate - the section of the solar collector responsible for gathering the solar radiation from the sun.

Absorber shell - the lower shell of a lithium bromide absorption chiller, containing the absorber and the evaporator, where water vapor is absorbed by the lithium bromide solution in the absorption cycle.

Absorber, sound - a device, panel, or material specially designed to absorb sound energy. Such devices are usually constructed of porous materials composed of organic or mineral fibers.

Absorption - (1) a process whereby a material extracts one or more substances present in an atmosphere or mixtures of gases or liquids. Accompanied by physical change, chemical change, or both, of the material. (2) a type of chiller machine that relies on heat energy to cool a liquid, such as water, and then uses the cooled liquid to cool something else.

Absorption coefficient - the absorption coefficient of a material or sound absorbing device is the ratio of the sound absorbed to the sound incident on the material or device. The sound absorbed by a material or device is usually taken as the sound energy incident on the surface minus the sound energy reflected.

Absorption refrigerator - a system in which the refrigerant, as it is absorbed in another liquid, maintains the pressure difference needed for successful operation of the system.

Absorption, solar - the ability of a solar collector to soak up light and therefore capture the heat of the sun. Absorption (or absorptance) is measured as a percentage of the total radiation available.

Absorption system - one of the most common types of cooling systems associated with solar radiation.

Absorptivity - the ratio of radiant energy absorbed by an actual surface at a given temperature to that absorbed by a black body at the same temperature.

AC electrolytic capacitor - this capacitor can house a large amount of capacitance in a small case size, is designed for short term use only, and used as the starting capacitor on single phase motors.

Accelerate - to add to the speed or velocity; or to hasten progress of development.

Accelerated aging - a test performed on materials that simulate or duplicate long time environmental conditions in a relatively short period of time.

Accelerated depreciation - the declining balance and sum of the years' digits methods which allocate a greater portion of depreciation to the early years of the life of the asset.

Acceleration time - when a motor is used to accelerate a rotating load, acceleration time is the response time of the motor to bring the load from zero to normal running speed.

Accelerator - a chemical additive that hastens a chemical reaction under specific conditions.

Acceptable air quality - air in which there is no known contaminants at harmful concentrations and with which a substantial majority (usually 80%) of the people exposed do not express dissatisfaction.

Acceptable thermal environment - an environment which at least 80% of the occupants would find thermally acceptable.

Acceptor - a dopant material such as boron that has fewer outer shell electrons than required in an otherwise balanced crystal structure, providing a hole that can accept a free electron.

Acceptor ion - an atom in a doped semiconductor that accepts an electron or gives up a hole.

Access - that which enables a device, appliance or equipment to be reached by ready access or by means that first requires the removal or movement of a panel, door or similar obstruction.

Access door - a door, cover or lid that can be moved or opened for inspection or servicing of internal components.

Access fitting - a valve or fitting that provides or creates an opening into a closed system.

Access time - the time it takes a computer to produce a bit of information from its memory section, also called "read time", or the time it takes a computer to store information in its memory section. Also called "write time."

Access to - that which enables a device, appliance or equipment to be reached by ready access or by a means that first requires the removal or movement of a panel, door or similar obstruction.

Accessible - easily approached, removed or exposed. Not permanently concealed.

Accessible, equipment – admitting close approach: not guarded by locked doors, elevation, or other effective means.

Accessible hermetic - the assembly of the motor and compressor inside a single bolted housing. Also referred to as a semi-hermetic compressor.

Accessible, readily - having direct access without the need of removing any panel, door, or similar covering of the item described.

Accessible, wiring methods - capable of being removed or exposed without damaging the building structure or finish, or not permanently closed by the structure or finish of the building.

Accessories - those items that add to the effectiveness or convenience of a system, but are not essential to its operation.

Accumulation test - test used to establish the relieving capacity of boiler safety valves.

Accumulator, electronic - the register within a computer where the results of all arithmetic and logical operations are placed.

Accumulator, refrigeration - a storage tank that receives any liquid refrigerant from the evaporator and prevents it from flowing into the suction line. It allows small amounts of liquid refrigerant to boil away before entering the compressor. Sometimes used to store excess refrigerant in heat pump systems during the winter cycle.

Accuracy - the precision of a reading in relation to the actual measurement. The extent to which the measured value of a quantity agrees with the accepted, consensus, or true value of that quantity. The accuracy of a measurement is its "closeness" to a defined true or reference value. For example, a measurement that is uncertain by ±1% is 99% accurate.

Acetone - a liquid solvent used in acetylene storage tanks.

Acetylene - fuel gas used for soldering, brazing, or cutting. Molecular weight is 26.038 and specific volume is 14.7 cubic feet/lb. Chemical formula is C_2H_2. Flammability limits are 2.75 to 100% by volume.

Acfm - the actual volumetric flow rate of air vapor mixture. Units are cubic feet per minute.

Acid - either an inorganic or organic compound (1) that reacts with metals to yield hydrogen; (2) that reacts with a base to form a salt; (3) that dissociates in water to yield hydrogen or hydronium ions; (4) that has a pH of less than 7.0; (5) that neutralizes bases or alkaline media. All acids contain hydrogen and turn litmus paper red. They are corrosive to human tissue and are to be handled with care.

Acid aerosol - acidic liquid or solid particles that are small enough to become airborne. High concentrations of acid aerosols can be irritating to the lungs and have been associated with some respiratory diseases, such as asthma.

Acid condition - the condition in which refrigerant or oil in the system are mixed with acidic fluids.

Acid test - a test that provides a positive indication whether the concentration of acid level in the oil being tested is either safe or unsafe.

Acid test ratio - a calculation of a firm's liquidity position; that is, the ratio of its quick assets (those that can be easily converted to cash) to current liabilities.

Acidity - water is a chemical combination of hydrogen (H) and oxygen (O) that is represented by the formula H_2O. It may also be treated as a chemical combination of hydrogen ions (H) and hydroxyl ions (OH). If there is a greater number of hydrogen ions than hydroxyl ions as a result of the chemical action of impurities or solutes, the solution is acid. A greater number of hydroxyl ions results in an alkaline solution. The degree of acidity or alkalinity of a substance is known as the hydrogen-ion concentration and is called the pH value. A pH value of 7.0 indicates neutral water; a value less than 7.0, acidity; and a value greater than 7.0, alkalinity.

Acme lock - a sheet metal lock made by turning the edges of adjoining sheets 180 degrees in opposite directions, hooking them together and locking them tight with a grooving tool.

Acoustical - pertaining to sound.

Acoustical duct lining - ducts with a lining designed to control or absorb sound and prevent transmission of sound from one room to another.

Acoustical treatment - the application or use of any sound absorbers, building materials, or structures and construction techniques for the purposes of controlling noise and improving the acoustical environmental.

ACR tubing - an air conditioning/refrigeration grade tubing. This tubing is a type "L" copper that is very clean, dry and has both ends sealed to prevent contaminants from entering.

Acrolein - a warning agent often used with methyl chloride to call attention to the escape of refrigerant. The material has a compelling, pungent odor, and causes irritation of the throat and eyes. Acrolein reacts with sulfur dioxide to form a sludge.

Across the line - a method of motor starting that connects the motor directly to the supply line on starting or running. Also known as full voltage starting.

Acrylonitrile butadiene styrene (ABS) - material used to made plastic pipe and fittings used for drain, waste, and vent piping.

Activated alumina - a chemical in the form of aluminum oxide (Al_2O_3) that is found as a desiccant in filter driers. Operates by adsorption of water molecules.

Activated carbon - a specially processed carbon that is used in filter driers or used to filter and/or clean air.

Activated charcoal - material at is either powered, granular, or pelleted carbon, coal or coconut shells that are capable of absorbing odors, gases, and vapors because of their fine pores.

Activator - a chemical additive used to initiate the chemical reaction in a specific chemical mixture.

Active region - an area of transistor operation between cut-off and saturation.

Active recovery - a self contained (independent) method of recovering refrigerant from mechanical refrigeration systems in which the gauge manifold is connected to the suction and the discharge service valves. Refrigerant is removed from the system by a recovery unit and stored in a recovery cylinder.

Active solar system - solar system that use the forced movement of a fluid to transfer heat from the collector to the heated space. Systems that use electrical and/or mechanical devices to help collect, store, and distribute the sun's energy.

Actual air delivery in SCFM - The standard cubic feet per minute of compressed air actually delivered at the air compressor's discharge port.

Actual body rating - see valve body rating.

Actual evaporation - the total quantity of water evaporated from any temperature feed water (in pounds) to steam at the working pressure and temperature.

Actual torque - the actual torque required to drive a machine. It is the torque measured at the input shaft of a machine being driven.

Actuator - a controlled motor, relay or solenoid in which the electrical energy is converted into a rotary, linear, or switching action. An actuator can effect a change in the controlled variable by operating the final control elements a number of times. Valves and dampers are examples of mechanisms that can be controlled by actuators. Also see proportional, spring return, and two-position actuators.

Acute angle - an angle less than 90 degrees.

Adapted person - a person who has occupied a space for more than 15 seconds. According to ASHRAE, such a person is considered to be at least partially adapted to odors in the space. The 15 second definition for adaptation does not imply that all persons will accept the surroundings in this time frame.

Adapter - special fittings used to quickly and easily connect different parts for different applications.

ADC - an analog to digital converter. An electronic device that converts analog signals to an equivalent digital form, in either a binary code or a binary-coded decimal code. When used for dynamic waveforms, the sampling rate must be high to prevent aliasing errors from occurring.

Add-a-relay - relays that are especially designed to satisfy the remote mounting requirements of bypassing a furnace for control when adding air conditioning.

Addendum - the radial distance from the pitch circle to the top of the gear tooth.

Addendum - modification or change made to the construction documents (plans or specifications) during the building process.

Additive - a compound that enhances some property of, or imparts some new property to, the base fluid. In some hydraulic fluid formulations, the additive volume may constitute as much as 20% of the final composition. The more important types of additives include anti-oxidants, anti-wear additives, corrosion inhibitors, viscosity index improvers, foam suppressants and fuel stabilizers.

Add-on heat pump - a special application of a heat pump where the unit is installed to work in conjunction with an existing gas, oil, or electric furnace.

Address - numeric code that precedes a controller's command to the controlled device. The controlled device will respond only to commands preceded by its assigned address code number.

Address bus - a unidirectional bus over which digital information appears to identify either a particular memory location or a particular I/O device.

Address selector - switches on the controlled device that are used to assign it to an address code number.

Adhesion - the property of a lubricant that causes it to cling or adhere to a solid surface.

Adhesion filter - this type of filter is used by most window air conditioners. It removes up to 90% of the dirt until it gets loaded. Adhesive filters can be of the permanent or throw-away type.

Adhesive - a substance capable of holding materials together by surface attachment.

Adhesive failure - rupture of an adhesive bond, such that the plane of separation appears to be at the adhesive-adhered interface.

Adhesive solvent - an adhesive having a volatile organic liquid as a vehicle. Also see solvent cement.

Adhesive wear - is often referred to as galling, scuffing, scoring, or seizing. It happens when sliding surfaces contact one another, causing fragments to be pulled from one surface and to adhere to the other.

Adiabatic - describes a system or thermodynamic process in which virtually no gain or loss of heat is allowed to occur with (to or from) the surroundings.

Adiabatic compression - the ability to compress a gas without removing or adding heat.

Adiabatic expansion - when steam is expanded in such a manner that there is no heat flow into or away from the steam and all the heat energy lost by the steam is converted into work, the process is called "adiabatic expansion." Steam expanding behind the piston of a steam engine after the point of cut-off approaches adiabatic expansion.

Adiabatic process - a process in which there is neither a loss nor gain in the total heat content. A constant enthalpy process.

Adiabatic saturation - theoretical achievement of 100% moisture saturation of air without gaining or losing heat energy.

Adjacent conductor - any conductor next to another conductor either in the same multi-conductor cable layer or in adjacent layers.

Adjustable - to bring to a more satisfactory state or to regulate.

Adjustable auto-transformer - a transformer in which parts of one winding are common to both the primary and the secondary circuits. The output voltage is adjustable from zero to line voltage.

Adjustable differential - a means of changing the difference between the control cut-in and cut-out points.

Adjustable wrench - an adjustable open end wrench has a sliding jaw that moves by an adjusting screw. This wrench should not be used in place of an open end wrench. It is used when an odd size nut or bolt needs to be worked on.

Adjusted dry bulb temperature - the average of the air temperature and the mean radiant temperature at a given location. The adjusted dry bulb temperature is approximately equivalent to operative temperature at air motions less than 0.4 m/s (80fpm) when the mean radiant temperature is less than 50° C (120° F).

Adjusting entries - entries required to adjust the accounts to their correct balances at the end of an accounting period.

Adjustment - the procedure required to produce the exact setting, response or effect desired.

Adjustment factor - a multiplier or penalty that is applied to conductors when there are more than three current carrying conductors in a raceway or cable. An adjustment factor is also referred to as a derating factor.

Administrative controls - a number of measures used to reduce worker exposure, including work practices, labeling and warning devices, training, environmental monitoring, assignment scheduling, housekeeping, maintenance, and management.

Admixture - a material added before or during concrete mixing. Used as a water repellent, coloring agent, or as a retarder or accelerator for curing time.

Adsorb - to attract and retain gas or liquid molecules on the surface of another material.

Adsorbent - a substance that has the property to hold in suspension molecules of fluids without causing a chemical or physical change. Certain solid materials such as silica gel, activated carbon and activated alumina have this property.

Adsorbent filter - a filter medium intended to hold soluble and insoluble contaminants on its surface by molecular adhesion.

Adsorption - the action, associated with surface adherence, of a material in extracting one or more substances present in an atmosphere or mixture of gases and liquids, unaccompanied by physical or chemical change.

Adsorptive filtration - the attraction to, and retention of particles in, a filter medium by electrostatic forces, or by molecular attraction between the particle and the medium.

Advertising - is the act of attracting public attention to a product or business. Advertising is usually understood to mean public attention you have paid to attract.

Aeration - the exposure of a substance to air circulation.

Aerodynamic noise - In duct systems, aerodynamic noise is caused by instabilities of the air flow through elbows, dampers, branch wyes, pressure reduction devices, silencers and other duct components.

Aerosol - an assemblage of small particles, solid or liquid, suspended in air. The diameters of the particles may vary from 100 microns down to 0.01 micron or less. Examples would be dust, fog, and smoke.

Affinity - an attractive force between two substances or particles that causes them to enter into and remain in chemical combination. Refrigeration oil has an affinity for moisture.

Aftercooler - heat exchanger used for the reduction of heat in gas of a compressor. Also used in the form of an economizer or thermal exchange unit.

Agent - any substance, force, radiation, organism, or influence affecting the body. The effects may be beneficial or injurious.
Aggregate - inert material mixed with cement and water to produce concrete.

Aging - the effect on materials of exposure or exposing to an environment for an interval of time.

Agitator - a device that is used to cause motion in a confined fluid.

Air - the elastic, invisible mixture of gases that surrounds the earth and composed of 78% nitrogen, 21% oxygen, and 1% rare gases including carbon dioxide, krypton, neon, argon, ozone, helium and ammonia. Also called atmosphere. One Btu will raise the temperature of 55 cubic feet of air 1° F. Also see ambient, atomizing, combustion, compressed, excess, induced, preheated, primary, seal, secondary, and ventilation air.

Air acetylene - a mixture of air and acetylene gas that when ignited is used for soldering, brazing, and other applications.

Air, ambient - the air surrounding an object.

Air-atomizing oil burner - a burner for firing oil in which the oil is atomized by compressed air, which is forced into and through one or more streams of oil breaking the oil into a fine spray..

Air balancing meter - an instrument that reads directly from exhaust or supply grilles in ceilings, floors or walls. It reads out in cubic feet per minute.

Air barrier - the membrane of the flexible duct which actually contains the air volume.

Air borne - carried in or transmitted by air. Certain diseases such as influenza are air borne.

Air borne sound - sound which reaches the point of interest by radiation through the air.

Air bound - a condition in which a bubble or other pocket of air is present in a pipeline or item of equipment and, by its presence, prevents or reduces the desired flow or movement of the liquid or gas in the pipeline or equipment.

Air blast - forced air circulation.

Air break - an inverted opening placed in the chimney of a gas furnace to prevent back pressure from outside wind from reaching the furnace flame or pilot.

Air change(s) - the number of times in an hour the air in a room is changed either by mechanical means or by the infiltration of outside air leaking into the room through cracks around the windows or doors. In residential air conditioning, one air change per hour is normally sufficient to dissipate undesirable odors and to keep the air within the structure from becoming stale. This is not to be confused with the re-circulation of air within the structure due to the operation of a furnace or evaporator blower. Air within the structure may re-circulate three or four times per hour on the heating cycle and five to seven times per hour on the cooling cycle, depending on the heat loss in winter and gain in summer in relation to the volume of air within the structure.

Air circulating - air for cooling, heating, or ventilation distributed to habitable spaces.

Air circulation - natural or imparted circulation of air.

Air cleaner - a device designed for the purpose of removing air-borne impurities, such as dust, gas, vapor, fumes, and smoke. Air cleaners include air washers, air filters, electrostatic precipitators and charcoal filters. Styles are either permanent or throw-away.

Air cleaning - an IAQ control strategy to remove various airborne particulates and/or gases from the air. The three types of air cleaning most commonly used are particulate filtration, electrostatic precipitation, and gas sorption.

Air coil - a coil used with some types of heat pumps that may be used as either an evaporator or condenser.

Air, combustion - the air necessary to provide for the complete combustion of fuel and usually consisting of primary air, secondary air, and excess air.

Air, compressed - air having a pressure greater than atmospheric pressure.

Air compressor - a device to compress air and used to activate pneumatic controls.

Air conditioner - any device or equipment that is used to control the temperature, humidity, cleanliness, and movement of air in a conditioned space.

Air conditioner, room - encased assembly designed as a unit for mounting in a window, through a wall, or as a console. It is designed for free delivery of conditioned air to an enclosed space without ducts.

Air conditioner, unitary - an evaporator, compressor, and condenser combination where one or more assemblies are designed to be assembled together to form one system.

Air conditioner, window - self-contained room conditioner arranged to be supported in or connected with a window opening, circulating outside air over the high side coil (condenser) and room air over the low side coil (evaporator).

Air conditioning - the simultaneous control of all or at least the first three, of the following factors affecting the physical and chemical conditions of the atmosphere within a structure: temperature, humidity, motion, distribution, dust, bacteria, odors, toxic gases and ionization --- most of which affect in greater or lesser degree human health or comfort.

Air conditioning, comfort - the process of treating air so as to control simultaneously its temperature, humidity, cleanliness and distribution to meet the comfort requirements of the occupants of the conditioned space.

Air conditioning, industrial - air conditioning for other uses than comfort.

Air conditioning, summer - comfort air conditioning carried out primarily when outside temperature and humidity are above those to be maintained in the conditioned space.

Air conditioning, winter - heating, humidification, air distribution, and air cleaning, where outside temperatures are below the inside or room temperature.

Air conditioning unit - a device designed for the treatment of air. It consists of a means for ventilation, air circulation, air cleaning, and heat transfer with a control to maintain the temperature within the prescribed limits.

Air connector - a passageway or duct for transferring air between an air duct or plenum and an air terminal unit or an air inlet or air outlet. A category of flexible duct not meeting the requirements of an Air Duct per UL 181 Standard (not tested for flame penetration, puncture and impact) and having limitations on use, length and location as defined by NFPA 90A and 90B. Air connectors are identified by a "round shape" listing label of the listing agency.

Air control damper - used in central fan systems to control the mixture of air admitted to the system. Installed in the fresh air intake, return air and exhaust ducts.

Air cooled condenser - the piece of equipment that the heat of compression is transferred from the condensing coils to the surrounding air. This process may be either by natural convection currents or by a fan or blower.

Air cooler - a device or mechanism designed to lower the temperature of the air passing through it. May or may not remove latent heat.

Air cooler, dry - removes sensible heat from the dehydrated air whenever it leaves the dehydrator at an elevated temperature.

Air core inductor - a coil wound on an insulated core that does not have a metal core.

Air core solenoid - solenoid that has a hollow core instead of a solid core.

Air cushion tank - a closed tank, generally located above a boiler and connected to a hydronic system in such a manner that when the system is initially filled with water, air is trapped within the tank. When the water in the system is heated it expands and compresses the air trapped within the air cushion tank, thus providing space for the extra volume of water without creating excessive pressure. Also called an expansion tank.

Air cycle air conditioner - any system that removes heat from the air and transfers this heat to the air.

Air diffuser - a circular, square, or rectangular air distribution outlet, generally located in the ceiling and comprised of deflecting members discharging supply air in various directions and planes, and arranged to promote mixing of primary air with secondary air.

Air, dilution - the air that enters the relief opening of a draft hood or draft diverter, or the air that enters another opening, such as a barometric damper, in an appliance flue or venting system.

Air distribution system - any system of ducts, plenums and air handling equipment that circulates air within a space or spaces and includes systems made up of one or more air-handling units. These systems may include air connectors, air duct fittings, dampers, plenums, and fan assemblies.

Air, dry - air without water vapor.

Air duct - the network of pipes and fittings that are used to control airflow to and from the conditioned space. A category of flexible duct tested and classified as to the Surface Burning Characteristics in accordance with the UL 181 Standard and identified by a "rectangular shape" listing label of the listing agency.

Air duct covering - a material such as an adhesive, insulation, banding, a film coat, or a jacket used to cover the outside surface of an air duct, fan casing, or duct plenum.

Air duct lining - a material such as an adhesive, insulation, a coating or film used to line the inside surface of an air duct, fan casing, or duct plenum.

Air economizer - a ducting arrangement and automatic control system that allows a cooling supply fan system to supply outside air to reduce or eliminate the need for mechanical cooling during cool weather.

Air exchange - the process of respiration or breathing. Inhalation and exhalation of air into and out of the lungs.

Air exchange rate - (1) the number of times that the ventilation system replaces the air within a room or area within the building. (2) the number of times that the outdoor air replaces the volume of air in a building per unit of time, typically expressed as air changes per hour.

Air exhaust - air being removed from any space or piece of equipment and conveyed directly to the atmosphere by means of openings or ducts.

Air film - interior and exterior air surface film coefficients for winter design conditions. A stagnant air flow that exists on construction surfaces and reduces the flow of heat through the construction.

Air film resistance - the thermal resistance of the air film along a wall, ceiling, or floor surface.

Air filter - non-mechanical air cleaning devices to include panel type frames filled with treated media. Such filters will be of the throw-away or cleanable types.

Air flow - the distribution or movement of air.

Air flow switch - any controller that is used to prove proper air movement by measuring pressure or air velocity. Also known as a sail switch. Also see lock-out and running interlocks.

Air friction chart - a chart used to determine proper round duct sizes for proper air flow.

Air fuel ratio - the ratio of air supply flow rate to the fuel supply flow rate, measured in the same units under the same conditions. The optimum air/fuel ratio is the minimum ratio that will provide complete combustion of the fuel with enough excess air to maintain a stable flame. Also see stoichiometric air fuel ratio.

Air gap - the space between the magnetic poles or between rotating and stationary assemblies in a motor or generator.

Air gas ratio - the ratio of combustion air supply flow rate to the fuel gas supply flow rate.

Air handler - a fan blower, heat transfer coil, filter, and housing parts of the system. Air handlers do not have provisions for heating or combustion.

Air handling unit (AHU) - mechanical equipment used to heat, cool, humidity, dehumidify and/or pressurize an airstream before it is distributed to a zone or conditioned space.

Air heat exchanger - a device used to exchange heat between air and another medium at different temperature levels, such as air-to-air, air-to-water, or air-to-refrigerant.

Air heater - an indirect-fired appliance intended to supply heated air for space heating and other purposes, but not intended for permanent installations.

Air horsepower - the power output developed by a fan in moving a given air rate against a given resistance. Unit is horsepower and symbol is ahp.

Air in system - air that has not been removed completely from the refrigeration system.

Air infiltration - the leakage of air into structures through cracks, crevices, windows, doors, and other openings, caused by wind pressure and/or temperature difference.

Air infiltration retarder - fabric or film material or system designed and installed primarily to reduce the movement of air through a wall or into the thermal insulation.

Air inlet - any opening through which air is removed from a space and returned to an air distribution system.

Air inlet, water tower - opening in a cooling tower through which air enters. Sometimes referred to as the louvered face on induce draft towers.

Air inlet valve - a door or opening that lets either outside or inside air enter the plenum assembly.

Air instrument - oil free and moisture free compressed air used to operate pneumatic control devices.

Air lateral - pipe that supplies the air to the ice cans contained in the freezing tank.

Air main - pipe that carries air to the air laterals supplying ice cans contained in the freezing tanks.

Air, make-up - air that is provided to replace air being exhausted from a space or building. This air may or may not be filtered.

Air mixing - consist of dampers regulating the flow of two separate air supplies to each zone, one heated and one cooled. A zone thermostat usually regulates the amount of air delivered to the zone. Occasionally, a third supply consisting of unconditioned, or "neutral" air, is also used.

Air mixing valve - a valve that opens ducts to blend hot and cold air to get desired output temperature. Sometimes called the blend door.

Air motion switch - used to sense suction and/or discharge pressure across a fan or coil.
Air movement, active - air moved by means of a mechanical device such as a fan or blower.

Air movement, passive - air moved by the difference in the weight of the air due to the temperature difference (change in specific gravity) of the air.

Air of the coil temperature - the actual temperature of air leaving the coil.

Air, outdoor - air taken from outdoors and not previously circulated through the system.

Air outlet - any opening through which air is delivered to a conditioned space from an air distribution system.

Air outlet valve - a door that directs either through the heater core or through the evaporator.

Air, outside - external air. Atmosphere exterior to the refrigerated or heated space. Ambient air.

Air over - motors that are intended for fans or blower service. These motors must be located in the air steam to provide motor cooling.

Air plant - compressed air used in industrial plants for air operated tools. Not usually used for instruments or breathing air. In some cases, may not contact any process materials.

Air pollutants - any particulate matter, gas or odor, including but not limited to, smoke, dust, soot, vapor, mist or irritating, noxious or obnoxious acids, alkalies, fumes or gases or any combination of these. Uncombined water vapor is not an air pollutant.

Air pollution - the presence in the out-of-doors atmosphere of one or more air pollutants in such quantities as to be, or tend to be, injurious to human, plant or animal life, or to property, or as to interfere unreasonably with the enjoyment of life and property.

Air pressure control - a switch used to detect air pressure drop across the coil in an outdoor heat pump unit due to ice buildup.

Air pressure drop test - when air pressure of up to 100 pounds per square inch is maintained for a specified length of time to determine if a product or system is leak free.

Air, primary - the air necessary for combustion that is mixed with the gas in a burner before ignition takes place.

Air purger - a device that separates air bubbles from the moving system fluid and ejects them from the system.

Air rate, water tower - mass flow of dry air per square foot of cross sectional area in the tower's heat transfer region per hour. Unit is pounds per square foot per hour. Symbol is G.

Air, recirculated - return air passed through the conditioning unit before being again supplied to the conditioned space.

Air register - (1) a air distribution outlet designed to deliver air in a concentrated stream into the occupied space. A register usually has a damper vane to regulate air flow out of the register. (2) a burner mounting that may admit secondary air to the combustion chamber.

Air, reheated - in an air conditioning system the final step in treatment, in the event the temperature is too low.

Air return - the air that is returned from the conditioned or refrigerated space.

Air rich - the air/fuel ratio supplied to a furnace that provides more air than the optimum ratio.

Air saturated - the mixture of dry air and saturated water vapor all at the same dry bulb temperature.

Air, secondary - the air surrounding a flame that is necessary to complete combustion.

Air sensor - a device that registers changes in air conditions such as pressure, velocity, temperature, or moisture content.

Air sensing thermostat - a thermostat in which the sensing element is located in a conditioned space or outside ambient air.

Air separator - a device that separates air from water in a system containing a movable fluid.

Air shutter - an adjustable device for varying the size of the inlet or inlets regulating primary and/or secondary air. An air shutter may be automatically or manually operated.

Air side - any part of a system that has to do with air movement.

Air side pressurization - the pressure on the air side of a diaphragm type expansion tank before fluid enters the tank. This pressure is adjusted by adding or releasing air through the Schrader valve on the tank.

Air source heat pump - heat pumps that use air as the source of heat while operating in the heating mode and as a heat sink while operating in the cooling mode.

Air, specific heat - the quantity of heat absorbed by a unit weight of air per unit temperature rise. The two most common specific heat valves in this industry are air at .24 and water at 1.

Air spill over - the refrigerating effect formed by the cold air from the freezing compartment in the refrigerator spilling over, or flowing into the normal storage area of the refrigerator.

Air, standard - air having a temperature of 70° F, a relative humidity of 36%, an atmospheric pressure of 14.696 psi, a specific volume of 13.33 ft3, and a specific density of 0.075 lb/ft3.

Air standard conditions - conditions at which Btu/h ratings for summer air conditioning equipment is rated at 95° F dry bulb and 75° F wet bulb at the condenser inlet and 80° F dry bulb and 67° F wet bulb at the evaporator inlet.

Air, supply - that air delivered to the conditioned space and used for ventilation, heating, cooling, humidification, or dehumidification.

Air supported structure - a structural and mechanical system that is constructed of high strength fabric or film and achieves its shape, stability, and support by pre-tensioning with internal air pressure; air structures may be used for temporary applications.

Air temperature stratification - the natural tendency for higher temperature air to rise to the upper portion of a room or interior space while cooler air settles near the floor.

Air terminal unit - an appliance receiving, conditioning, and/or delivering air supplied through an air distribution system.

Air throw - the distance an air stream travels from an outlet to a position where its motion has been reduced to a velocity of 50 feet per minute.

Air-to-air heat pump - an air source heat pump system that is ultimately used to heat and/or cool air.

Air-to-liquid heat pump - an air source heat pump system that is ultimately used to heat or cool a liquid.

Air transfer opening - an opening designed to allow the movement of environmental air between two different spaces or conditions.

Air travel, water tower - distance that air travels in its passage through the fill. Measured vertically on counterflow towers and horizontally on crossflow towers.

Air tunnel - a refrigerated tunnel with rapid air circulation through which the product to be frozen or passed.

Air type collector - a solar collector that uses the air as a transfer fluid.

Air valve - a device used to reduce high pressure or high velocity air streams to a lower pressure for discharge into a conditioned space.

Air velocity - speed of the air vapor mixture through a specific region or space. Unit is feet per minute. Symbol is V.

Air vent - a valve, either manual or automatic, used to remove air from the highest point of a coil or piping assembly.

Air vent riser - a vertical length of pipe at the highest point in the system to gather and retain air in the system as the liquid flows past the riser. The air vent is located at the top of the riser.

Air, ventilation - the quantity of supply air drawn from outdoors needed to maintain the desired amount of oxygen and the quality of air within a designated space.

Air washer - a device used to clean air that may increase or decrease the relative humidity.

Air water system - HVAC system which distributes heated or chilled air and heated or chilled water to condition the air in a space.

Airway - the passageway through which air enters the body and goes to the lungs.

Ak - area factor of an outlet or inlet, measured in square feet, is a flow factor determined from the discharge or intake velocity.

Alarm - an audible or visible signal indicating a non-standard or abnormal condition.

Alarm circuit - an electrical circuit that includes a bell, horn, or similar device to signal desired conditions.

Alcohol brine - a solution of water and alcohol that remains a liquid at or below 32° F.

AL/CU - terminal marking on switches and receptacles rated 30 amperes or greater suitable for use with aluminum, copper, and copper clad aluminum conductors. If not marked, it is suitable for copper conductors only.

Aldehyde - a class of highly reactive organic compounds that can be produced during incomplete combustion of a fuel gas. They have a pungent distinct odor.

Algae - a low form of plant life, found floating free in water.

Algorithm - a rule or procedure for solving a recurrent mathematical problem.

Alignment - being brought into a straight line.

Alkali - broadly, any compound having highly basic properties, i.e., one that readily ionizes in aqueous solution to yield OH ions, with a pH above 7, and turns litmus paper blue. Alkalies are hydroxides of an alkali metal and belong to group IA of the periodic table (Li, Na, K, Rb, Cs, Fr). Ammonium salts may also be called alkaline. Alkalies are caustic. Their high concentrations of OH ions in solution destroy tissue by abstracting fluids or by corrosive deoxidation. Alkali burns are to be treated by quickly washing the afflicted area with copious amounts of water. Common commercial alkalies are sodium carbonate (soda ash), NaOH, lye, potash, caustic soda, KOH, water glass, and bicarbonate of soda.

Alkyl benzene - an organic lubricant that is made from the raw chemicals propylene, a colorless hydrocarbon gas, and benzene, a colorless liquid hydrocarbon. Used often when incorporating CFC and HCFC based refrigerant blends. Some HCFC based blends are soluble in a mixture of mineral oil and alkyl benzene up to 20% concentration of mineral oil.

All air system - HVAC system which distributes heated or chilled air from a central plant through ducts to conditioned zones as opposed to systems which use air and another fluid or water.

Allen head screw - a screw with a recessed head designed to be turned with a hex shaped wrench.

Allen wrench - hexagonal (6 point) tip used to fit socket headed screws or set screws. Can either be straight, 90°, T-handled, folding, or socket.

Alley pipe - the main bank of pipe headers located inside the limits of the structure. Also known as a pipe bridge when outside the limits of the structure.

Alloy - a combination of two or more metals to form a new or different metal.

Alloy 11 – a compensating alloy used in conjunction with pure copper as the negative leg to form extension wire for type R and S platinum-platinum rhodium thermocouples.

Alloy steel - a steel that owes its distinctive properties to elements other than carbon.

Allowable load - the load that induces the maximum permissible unit stress at a critical section of a structural member.

Allowance - minimum clearance between mating parts.

Alnico - an alloy of aluminum, nickel, iron, and cobalt used to make permanent magnets.

Alphanumeric - a numbering system containing numbers and letters.

Alpha particle - a positively charged atomic particle that is ejected in certain radioactive transformations.

Alteration - a change in a mechanical system that involves an extension, addition or change to the arrangement, type or purpose of the original installation.

Alternating current (AC) - a flow of electricity that reaches maximum in one direction, decrease to zero, then reverses itself and reaches maximum in the opposite direction. This cycle is repeated continuously. This can be single phase, two phase, three phase and poly phase. The number of such cycles per second is the frequency in hertz (Hz). The average value of voltage during any cycle is zero. In a 60 cycle current, the direction of flow reverses every 120th of a second. Its advantage over direct or unidirectional current is that its voltage can be stepped up by transformers to the high values that reduce transmission costs.

Alternating current resistance - the resistance offered by any circuit to the flow of alternating current.

Alternation - one half of a sine wave. There is a positive and a negative alternation for each alternating current sine wave.

Alternative refrigerant - one of the newer refrigerants that are replacing the traditional CFC refrigerants that have been used for many years. Many of these refrigerants have very low ozone depletion and global warming indices. Some are completely chlorine free.

Alternator - an alternating current machine that changes mechanical energy into electrical energy.

Altitude correction - the change in atmospheric pressure at higher altitudes has the same effect on a vapor pressure bellows as does a barometric change. The effects of altitudes are greater and of a permanent nature so that adjustments may be made in the control settings to correct for the effect of altitude.

Alumel - an aluminum nickel alloy used in the negative leg of a type K thermocouple. Trade name of Hoskins Manufacturing Company.

Aluminum - a lightweight but relatively strong metal. Often alloyed with copper to increase hardness and strength.

Ambient - surrounding.

Ambient air - generally, the dry bulb air temperature surrounding or in the vicinity of an object or person.

Ambient compensated - designed so that varying temperatures of air at the control do not affect the control setting.

Ambient conditions - surrounding conditions or conditions on all sides.

Ambient noise - the quiet-state noise level in a room or space, that is a composition of sounds from many external sources both near and far, over which one individually has no control.

Ambient sensor - a sensor that measures the temperature of the outside air temperature.

Ambient switch - an open on temperature fall switch that keeps the compressor from coming on if the ambient temperature is below a preset value.

Ambient temperature - the temperature of the air or other medium immediately surrounding a object or device. This may be considered to be the temperature of the air entering the air cooled condenser. In order to check the temperature of the air entering the condenser, hold a thermometer in the entering air stream to the condenser. Do not let the thermometer touch anything while the temperature is being taken, and be sure the bulb is not exposed to sun rays. Since the operating conditions of an air cooled system, such as discharge pressure, suction pressure and ampere draw are influenced by ambient temperature, it is important to know the ambient temperature when checking a system for correct operation.

Ambient wet bulb temperature (water tower) - the wet bulb temperature of the air encompassing a cooling tower, not including any temperature contribution by the tower itself. Generally measured upwind of a tower, in a number of locations sufficient to account for all extraneous sources of heat.

American National Acme Thread - a form of thread with a 29° angle that is used for feed screws, jacks, and vises.

American national standard thread - a group of three main series of threads, consisting of national fine, national course, and national pipe.

American Standard Code for Information Interchange (ASCII) - a standard, eight level code developed by an ANSI subcommittee, having seven code bits for data and an eighth bit for parity. Developed to simplify and standardize the intercommunication link between computers, communication circuits, and digital output equipment, such as a printer or video.

American standard pipe thread - type of screw thread commonly used on pipe and fittings to assure a tight seal.

American wire gauge - a standard used in the determination of the physical size of a conductor determined by its circular mil area. Usually expressed as AWG. As the diameter gets smaller, the number gets bigger, e.g., AWG #14 = 0.0641 in. while AWG #12 = 0.0808 in. There are 40 electrical conductor sizes from #36 through 0000. Starting from the smallest AWG size, each successive size is approximately 1.26 times larger than the previous AWG size. Conductor sizes through 0000 are shown using the AWG designation. Conductors larger than 0000 are shown in circular mils area. Previously referred to as the Brown and Sharpe gauge.

Ammeter - an electric meter calibrated and used to measure the amperes that are present in a circuit. Ammeter styles are in-line or clamp-on.

Ammonia - a chemical combination of nitrogen and hydrogen (NH_3). Identified as R-717. Boiling point at atmospheric pressure is -28.2° F, molecular weight is 17.031 and specific volume is 22.6 cubic feet/lb. Color code is black.

Amorphous - the condition of a solid in which the atoms are not arranged in an orderly pattern; not crystalline.

Amortisseur - winding named for the set of type "A" squirrel cage bars in a synchronous motor rotor.

Amp - abbreviation for ampere

Ampacity - total amperes expected at maximum load times 1.25. The current in amperes a conductor can carry continuously under the condition of use without exceeding its temperature rating as determined by the National Electric Code.

Amperage - the electron or current flow of one coulomb per second past a given point in a circuit.

Amperage interrupting capacity - the point beyond which a circuit protection device can no longer interrupt current flow.

Amperage relay - relay that operates on the amperage (current flow) that is generated by the run winding. Another common name is a current relay.

Ampere - the standard unit of electrical current. Equivalent to 6.28 quintillion electrons flowing through one ohm of resistance pushed by one volt of electrical energy in one second. Symbol is "I" for intensity of current.

Ampere fuse rating - the current capacity of a fuse. When a fuse is subjected to a current above its ampere rating, it will open the circuit after a predetermined period of time.

Ampere hour - a capacity rating of storage batteries that refers to the maximum current that can flow from the battery during a specified time frame.

Ampere squared seconds - the measure of heat energy developed within a circuit during the fuse's clearing. It can be expressed as "melting I2T" or the sum of them as "clearing I2T". "I" stands foe effective let-through current (RMS), that is squared, and "T" stands for the time of opening, in seconds.

Ampere turns - the term used to measure magnetic force. Represents product of amperes times number of turns in coil of electromagnet. Base unit of magnetomotiveforce.

Amplification - the increasing in value of an input signal by an amplifying device. There is voltage, current, and power amplification.

Amplification factor - ratio of the output of an active electron device to its input.

Amplification site - areas in buildings where biological growth develops and grows due to moisture, nutrition, and lack of light.

Amplifier - a device used to increase the magnitude of a small input signal to proportions sufficient to perform some desirable function. Also see flame signal amplifier.

Amplifier saturation - the point at which an increased input signal no longer increases the amplifier output.

Amplitude - the maximum deviation from a baseline of a varying sine wave. A measurement of the distance from the highest to the lowest excursion of motion, as in the case of mechanical body oscillation or the peak-to-peak swing of an electrical waveform.

Amplitude modulation (AM) - a method of transmitter modulation achieved by varying the strength of the RF carrier at an audio signal rate.

Amps - a measurement of the electrical current flowing in an electrical circuit.

Analytical ferrography - the magnetic precipitation and subsequent analysis of wear debris from a fluid sample. This approach involves passing a volume of fluid over a chemically treated microscope slide which is supported over a magnetic field. Permanent magnets are arranged in such a way as to create varying field strength over the length of the substrate. This varying strength causes debris to precipitate in a distribution with respect to size and mass over the ferrogram. Once rinsed and fixed to the substrate, this debris deposit serves as an excellence media for optical analysis of the composite wear particulates

Analyze - to separate into parts so as to find out their nature, purpose and use. To examine the parts of a whole carefully and in detail.

Analyzer - a device in the high side of an absorption system for increasing the concentration of refrigerant in the vapor entering the rectifier or condenser.

Analog - being continuous or having a continuous range of values. Data that is displayed by continuously variable physical quantity, rather than discrete steps.

Analog bar graph - some digital multimeters and clamp-ons provide this type of display to present the signal as it would appear on an analog meter. This is especially useful for viewing trends.

Analog control - a device or control loop that generates and responds to modulating signals. The data in analog signals is represented by a continuous varying magnitude.

Analog control signal - an electrical voltage or current that varies over a predetermined range, and typically used to control a modulating device.

Analog inputs and outputs - an input or output of variable value that is provided to a control system. Usually a variable voltage, current or pressure.

Analog meter - a meter that uses a mechanical magnetized needle movement across a graduated scale to indicate the measured value.

Analog tachometer - a tachometer that provides a continuous indication of speed of some rotary device.

Usually an output voltage is directly proportional to the speed of the rotating device and a voltmeter, calibrated in RPM, is the output indicator.

Analog-to-digital converter - an electronic device used to interface analog sensor signals with a digital controller. The ADC converts the magnitude of the input signal into a binary word.

Analogy - an explanation of something by comparing its point with something else that is usually more familiar than the idea or object being explained.

Analyzer - a device used to check the condition of electrical components.

Anchor - rigid support that keeps a pipe from translation or rotational movement at one point along the piping system. This anchor prevents transmission of forces and moments (thermal, shock, vibration) between both sides of the pipe.

Anchor bolts - hold down points for equipment made from mechanical steel machine bolts, steel rods, or other devices and put in place before the concrete embedment is poured.

Anemometer, deflecting vane - the deflecting vane anemometer consists of a pivoted vane enclosed in a case. Air exerts a pressure upon the vane as it passes through the instrument from an upstream to a downstream opening. The movement of the vane is resisted by a hairspring and a dampening magnet. The instrument gives instantaneous readings of directional velocities on an indicating scale. The instrument is supplied with various types of remote connected and direct connected measuring tips or jets. This instrument is commonly known and referred to as a velometer and has an accuracy of" 2% of full scale. The instrument may be used for measurements of air velocity through both supply and return air terminals using the proper jet and the proper air terminal factor (effective area) for the air flow calculation. The instrument may also be used for measuring some lower velocities where the instrument case itself is placed in the air stream.

Anemometer, hot wire - an instrument where the operation of this instrument depends on the fact that the resistance of a heated wire will change with its temperature. Incorporated with the instrument is a probe that is provided with a special type of wire element that is energized by electrical current. As air flows over the element in the probe, the temperature of the element is changed from that which exists in still air and the resistance change is indicated as a velocity on the indicating scale of the instrument. Some instruments are also provided with temperature scales that can be utilized simply by setting the proper selector button. Static pressure can be measured if the proper cap is placed over the probe.

Anemometer, rotating vane - an instrument where the propeller or rotating vane consists of a light wind driven wheel connected through a gear train to a set of recording dials that read the linear feet of air passing through the wheel in a measured length of time. The instrument is made in various sizes: 3", 4" and 5". Each instrument requires individual calibration. At low velocities, the friction drag of the mechanism is considerable. In order to compensate for this, a gear train that over speeds is commonly used. For this reason, the correction is often additive at the lower range and subtractive at the upper range. Most of these instruments are not sensitive enough to use below 200 fpm. Their useful range is from 200 to 2000 fpm. The required instrument test accuracy minimum is 1% of indicated value (using a corrective chart).

Anemometer, bridled vane - this instrument utilizes displacement of an air vane to read directly in fpm. Face area is approximately 3" in diameter. It can be used to measure the air flow from grilles by determining the average velocity over the face or may be used with other types of diffusers with a special attachment that directs the jet of air through the instrument. It may also be used with an extension handle to reach high sidewall outlets and is provided with a scale lock that locks the indicator in place once the reading is made. The required instrument test accuracy minimum is 3%. Also known as a fluorite meter.

Aneroid barometer - an instrument used to measure atmospheric pressure. This barometer uses a closed bellows linked to a needle and can be easily carried from place to place.

Angle - the figure formed when two lines come together at some point.

Angle of lead or lag - the angle between the applied voltage (EMF) and current flow (I) in an AC circuit in degrees. In an inductive (L) circuit, voltage (E) leads current (I); in a capacitive circuit, current (I) leads voltage (E).

Angle iron - a structural shape whose section is a right angle.

Angle valve - a type of globe valve design, having pipe openings at right angles to each other. Usually, one opening is on the horizontal plane and one is on the vertical plane.

Angstrom unit (A) - a unit of measurement of a wavelength of light and other electromagnetic radiation, equal to one ten-thousandth (1/10,000) of a micron.

Anhydride - a compound derived from another compound (e.g., an acid) by removing the elements that compose water, i.e., hydrogen and oxygen.

Anhydrous - devoid or without water. Describes a substance in which no water is present in the form of a hydrate or as water of crystallization.

Anhydrous ammonia - ammonia that is free of water especially water of crystallization.

Anhydrous calcium sulphate - a dry chemical made of calcium, sulphur and oxygen (CaSo4).

Animal dander - tiny scales of animal skin.

Anion - a negative ion. An ion that moves toward the anode in electrolysis.

Anneal - to subject to high heat with subsequent cooling. Relative to copper, the act of heating to just below its melting point and then allowing it to cool slowly. This usually lowers the tensile strength while improving the flex life and its flexibility characteristics.

Annealed tubing - annealed copper tubing is also simply called soft copper tubing. It has been heat treated to make it soft, easy to bend and easy to flare. Over time, or as a result of pounding, rough handling or bending, the tubing also tends to harden. Because annealed tubing is so flexible, it needs support from clamps or brackets. Soft copper tubing is used in home refrigeration and air conditioning and in some commercial installations. It is available in 25, 50, and 100 foot rolls. It can be flared, soft soldered, or brazed.

Annual fuel utilization efficiency (AFUE) - ratio of annual output energy to annual input energy, that includes any non-heating season pilot input loss, and for gas or oil fired furnaces or boilers, does not include electrical energy.

Annunciator panel - a device that indicates an off-standard or abnormal condition by visual and/or audible signals.

Annunciator light - a light that is illuminated at the time a point exceeds a preset alarm limits. An individual annunciator light can be provided for each input variable being alarm scanned, or a common alarm annunciator light can be utilized to indicate trouble somewhere in the system. These lights can also be used to indicate a status.

Anode - the positive terminal or electrode of an electronic device such as a solid state diode. The P region of a P-N junction diode.

Antagonism - when the effect of one chemical or material counteracts (works against) the effect of another.

Anthracite coal - anthracite coal is black, hard and brittle, with a very shiny appearance. It is more difficult to ignite than bituminous and burns with a short blur-yellow flame. It does not produce much smoke when burning. Semi-anthracite coal would exhibit properties between bituminous and anthracite. Anthracite coal has a moisture content of 2.80%, volatile matter 1.16%, fixed carbon 88.21% and ash at 7.83%. The ultimate analysis would should a hydrogen content of 1.89%, carbon 84.36%, nitrogen 0.63%, oxygen 4.40% and a calorific value of 13,298 Btu/lb.

Anti-flooding device - a primary control that causes the flow of fuel to be shut off upon a rise in fuel level or upon receiving excess fuel, and which operates before the hazardous discharge of fuel can occur.

Anti-foam agent - one of two types of additives used to reduce foaming in petroleum products. Silicone oil to break up large surface bubbles and various kinds of polymers that decrease the amount of small bubbles entrained in the oils.

Anti-friction bearing - the name given to all bearings in general, but more specifically to the ball or roller-type bearings that are also known as a rolling element bearing.

Anti-oxidant - a compounding ingredient added to a substance that is designed to retard possible degradation from contact with oxygen, particularly in processing at or exposures to high temperature.

Anti-ozonant - a substance that slows down material degradation due to ozone reaction.

Anti-short cycling device - a control that prevents a load from stopping and starting in rapid succession. This device is used mainly to prevent compressors from being restarted within a certain time after it has been stopped.

Anti-static additive - an additive that increases the conductivity of a hydrocarbon fuel to hasten the dissipation of electrostatic charges during high speed dispensing, thereby reducing the fire/explosion hazard.

Anticipating control - a control that is artificially forced to cut in or out before it would otherwise, therefore starting cooling before needed or stopping the heating before the control point is reached. This tends to reduce the temperature fluctuation or override of the control.

Anticipation - a method of reducing the operating differential of the system by adding a small resistive heater inside the thermostat to raise the internal temperature of the thermostat faster than the surrounding room temperature. This causes the thermostat to shut off the heating equipment and start the cooling equipment sooner than it would if affected only by the room temperature.

Anticipator - a small heater placed inside the thermostat housing used to adjust or "anticipate" the thermostat operation to produce a closer temperature differential than the mechanical capability of the control. A heating thermostat anticipator is wired in series with the thermostat contacts while a cooling thermostat cooling anticipator is wired in parallel with the thermostat contacts.

Antifreeze - materials such as salt, alcohol, or glycol products that are added to water to lower the fusion point.

Antifreeze protection - feature in some differential thermostats that turns the pump on and circulates warm water through the solar collector when its temperature is near freezing. This avoids damage to the collector.

Antimicrobial - agent that kills microbial growth.

Aperture - a service connection used to make access into a sealed refrigeration system. Two common types are the clamp-on piercing valve and the schrader valve.

Aperture disc - a thin, flat, circular plate of any material with an opening or hole (orifice) of pre-determined diameter to restrict the area viewed through it.

Apparatus dew point - the temperature of the air leaving an air conditioning (evaporator) coil if all of the air has been affected by the coil temperature. Also known as the effective coil temperature.

Apparent power - the power apparently available for use in an AC circuit that contains a reactive load. It is the product of the measured voltage and current in a circuit containing reactance expressed in volt-amperes. It then is multiplied by the power factor to obtain the true or actual power available. A circuit in which voltage and current reach their peak values at different times in a cycle.

Appliance - (1) a current consuming device that utilizes energy or raw materials to product light, sound, heat, power, refrigeration, or air conditioning to do work. (2) Any device that contains and uses a Class I or Class II substance as a refrigerant and that is used for household or commercial purposes, including any air conditioner, refrigerator, chiller, or freezer.

Appliance, automatically controlled - a piece of equipment equipped with an automatic burner ignition and safety shut-off device and other automatic devices that accomplish complete turn-on and shut off of the gas to the main burner or burners and graduate the gas supply to the burner or burners, but does not affect complete shut off of the gas.

Appliance, building heating - a fuel burning or electric boiler operating at not over 50 psig pressure, a central furnace, or a heater intended primarily for heating spaces having a volume exceeding 25,000 cubic feet.

Appliance, categorized vent diameter - the minimum vent diameter/area permissible for category I appliances to maintain a non-positive vent static pressure when tested in accordance with a nationally recognized standard.

Appliance, fan assisted combustion - an appliance equipped with an integral mechanical means to either draw (induced draft) or force (forced draft) products of combustion through the combustion chamber or heat exchanger.

Appliance, residential heating - fuel burning or electric heating appliances, not including high pressure steam boilers, for heating building spaces having a volume not exceeding 25,000 cubic feet and other heat producing appliances of the type mainly used in residences but that might be used in other buildings, such as cooking stoves and ranges, clothes dryers, fireplace stoves, domestic incinerators, laundry stoves, water heaters, and heat pumps.

Appliance, stationary - an appliance that is not easily moved from one place to another in normal use. An example is a refrigerator.

Appliance, vented - an appliance designed and installed in such a manner that all of the products of combustion are conveyed directly from the appliance to the outside atmosphere through an approved chimney or vent system.

Appliances, nonresidential - These appliances are divided into four categories as designated by their flue gas temperature. (1) low heat appliance. These are commercial, industrial, or institutional appliances needing a chimney capable of withstanding a continuous flue gas temperature not exceeding 1000° F. (2) 1400° F appliance. These are commercial, industrial, or institutional appliances needing a chimney capable of withstanding a continuous flue gas temperature not exceeding 1400° F. (3) Medium heat appliance. These are commercial, industrial, or institutional appliances needing a chimney capable of withstanding a continuous flue gas temperature not exceeding 1800° F. (4) High heat appliance. These are commercial, industrial, or institutional appliances needing a chimney capable of withstanding a continuous flue gas temperature exceeding 1800° F.

Applied voltage - the amount of voltage connected to a circuit or device.

Apprentice - a beginner in a trade, usually serving several years as a helper to a journeyman.

Approved - acceptable to an authority or regulatory agency having jurisdiction.

Approved agency - an established and recognized agency that is approved by the code official and regularly engaged in conducting tests or furnishing inspection services.

Approved Equipment Testing Organization - any organization that has applied for and received EPA approval.

Approach - difference between the cold water temperature and either the ambient or entering wet bulb temperature. In a cooling water tower application, the difference between the wet bulb temperature of the entering air and the temperature of the leaving water.

Approach temperature - the difference in temperature between the refrigerant and the leaving water in a chilled water system. In an evaporative cooling device, the difference between the average temperature of the circulating water leaving the device, and the average wet bulb temperature of the entering air.

Apron - piece of horizontal trim under the sill of the interior casing of a window.

Apron wall - that portion of a skeleton wall below the sill of a window.

Aquadag tube - an electron tube sensitive to ultraviolet radiation. When saturated with ultraviolet radiation, this tube delivers less than one microwatt to the flame signal amplifier.

Aquastat - a switching device that is sensitive to heat, located on a hot water line, coil, tank or boiler. Controls the heating or temperature of the water, turning the fuel burning equipment on or shutting it off when it reaches a preset or given temperature.

Aqueous - describes a water based solution or suspension. Frequently describes a gaseous compound dissolved in water.

Aquifer - the water being underground formation that serves to supply water to geothermal heat pump systems. The water is contained in soil, sand, gravel, or rock and its depth varies along with its geographical location.

Arc - a portion of a circle.

Arc, electrical - a luminous bridge formed in a gap between two conductors or terminals when they are separated. A flash caused by the ionizing of a gas or a vapor. A spark.

Arc-fault circuit interrupter - a device intended to provide protection from the effects of arc faults by recognizing characteristics unique to arcing and by functioning to de-energize a circuit when an arcing fault is detected.

Arc gap protector - a device that conducts at high voltage levels to bypass an electronic network, thus protecting the network against damage from the voltage interference.

Arc weld - to weld by electric arc. The work is usually the positive terminal.

Architectural drawings - the plans show the dimensions and finished structure. Architectural drawings show walls, roofs, partitions, floor plans, mechanical equipment, elevations, and sections. From these drawings you should have a good idea of what the structure will look like when it is completed.

Arcing - the forming of an electric arc across two physically separated contacts.

Arcing time - the amount of time from the instant the fuse link has melted until the over current is interrupted, or cleared.

Area - square measure found by multiplying length times width, or (for a circle) the pi value of 3.1416 by the radius squared.

Area, core - the total plane area of the portion of a grille, face, or register, bounded by a line tangent to the outer edges of the outer openings through which air can pass.

Area, free - the total minimum area of the openings in an air inlet or outlet through which air can pass.

Area of a circle - the measurement of the surface within a circle. To find the area of a circle, multiply the product of the radius times the radius by Pi (3.1416). Commonly written as $A = \pi r2$.

Argon - an inert, colorless, and harmless gas used instead of air in sealed spaces between panes of glass in insulating glass units to increase insulation. Argon is less conductive to heat than air.

ARI conditions - operating parameters that are 95° F dry bulb outdoor ambient. Indoor conditions are 80° F dry bulb and 67° F wet bulb. These conditions are used to verify the operating performance of air conditioning equipment.

Arithmetic logic unit - a section of a microprocessor that performs mathematic and logic routines.

Armature - the moving element in an electro-mechanical device, such as the movable iron core in a solenoid, the rotating part of the generator or motor, the movable part of a relay, or the spring-mounted iron portion of a bell or buzzer.

Armature reaction - the effect in direct current generators and motors in which current flow through the armature conductors reacts with the main magnetic field and increases sparking between the brushes and commutator.

Armor - a covering or wrapping of metal that is used to protect interior conductors.

Armor clamp - a fitting used to attach armored cable to a junction box or other device.

Armored cable - a flexible metal protective covering enclosing electrical conductors.

Aromatic oil - a refrigeration mineral oil in which the alkylbenzene lubricant group originates.

Arrestor, lighting rod - a device used to protect buildings, including electrical devices, from damage by lightning.

Arrestor, spark - wire screen secured to the top of an incinerator to confine sparks and other products of burning. Also designed to restrict the entry of birds, leaves and water to the chimney stack.

Artificial fuels - man-made fuels, including all manufactured and by-product fuels.

Artificial pulse - cardiac pulse that is felt on a victim while chest compressions are being performed.

Artificial weathering - the exposure of materials to cyclic laboratory conditions involving changes in temperature, relative humidity, and ultraviolet radiant energy, with or without direct water spray, in an attempt to produce changes in the material similar to those observed after long-term continuous outdoor exposure. The laboratory exposure conditions are usually intensified beyond those encountered in actual outdoor exposure in an attempt to achieve an accelerated effect.

Artist cap - a fabricated hood, usually of sheet metal, placed on top of a stack to prevent down drafts or erratic drafts because of the directional wind effects.

As built plans - when required, a modified set of working drawings that are prepared for a construction project that includes all variances from the original working drawings that occurred during the project construction.

Asbestos - a fire-resistant, heat resistant mineral. A silicate of calcium and magnesium used in fireproofing and insulation.

ASCII - American Standard Code for Information Interchange. A seven or eight bit code used to represent alphanumeric characters. It is standard code used for communications between data processing systems and associated equipment.

Ash - a measure of the amount of inorganic material in lubricating oil. The amount of ash is determined by burning the oil and weighing the residue. Results are expressed as percent by weight. The noncombustible matter that remains after a fuel is burned.

ASME Boiler code - standard specifications issued by the American Society of Mechanical Engineers for the construction of boilers.

Aspect ratio - in air distribution outlets, the ratio of the length of the core opening of a grille, face, or register to the width. In rectangular ducts it is the ratio of the width to the depth. In buildings it is the ratio of the length to the width.

Asperities - microscopic projections on metal surfaces resulting from normal surface finishing processes. Interference between opposing asperities in sliding or rolling applications is a source of friction, and can lead to metal welding and scoring. Ideally, the lubricating film between two moving surfaces should be thicker than the combined height of the opposing asperities.

Aspirating burner - a power burner that uses venturi action to pull the gas into the air stream. (Also see venturi and zero governor burner.)

Aspirating psychrometer - an instrument that draws a sample of air through it for humidity measurement purposes.

Aspiration - induction of air into the primary air stream.

Aspiration hazard - the danger of drawing material into the lungs, leading to an inflammatory response that can be fatal.

Aspirator hole - a small hole drilled in the side of the accumulator dip tube. It permits small quantities of oil to enter the outlet tube and be drawn back to the compressor.

Aspirator mixer - a gas/air proportioning device that causes the flow of combustion air to induce the proper amount of gas into the air stream. It is used with low pressure air and zero gas pressure. Also called a suction type mixer. Also see zero governor burner.

Assembly - portion of an envelope component represented by an arrangement and connection of building construction materials with a specific thermal transmittance or thermal conductance.

Atm - Atmosphere. A unit of pressure equal to the average pressure that air exerts at sea level. 1 atm = 1.013 x 105 N/m2, or 14.7 lb/in2, or 760 mm Hg or 101 kPa. Generally used in connection with high pressures.

Atmosphere - the invisible mixture of gases (air) surrounding the earth. A unit of pressure defined as the pressure of 760 millimeters of mercury at sea level and at 0° C (32° F), that is equal to 14.696 pounds per square inch (psi), or 29.921 inches of mercury.

Atmospheric burner - a gas burner relying on atmospheric pressure to bring in combustion air. It simply permits the air required for combustion to enter the combustion chamber, or it draws air into the combustion chamber from the surrounding the burner. Two methods are used to draw air into the combustion space.

Natural draft burner - the natural draft produced by the products of combustion moving up through the chimney, or stack, draws air in at the bottom of the burner.

Inspirating burner - the fuel gas is allowed to escape at high velocity into the throat of a venturi tube, creating a lower static pressure that draws in the air. Also called an injection or venturi mixing burner.

A typical atmospheric burner draws in part of the combustion air by venturi action and the rest by a natural draft.

Atmospheric dust spot efficiency - measurement of a device's ability to remove atmospheric air from test air.

Atmospheric ozone - as the ultraviolet rays from the sun reach the earth they combine with smog and pollution and cause atmospheric ozone. This ozone is harmful and it must not be confused with the stratospheric ozone layer.

Atmospheric pressure - the pressure exerted on the earth's surface by the weight of the air and moisture above it. A unit of pressure equal to about 14.696 psi, or 29.92 inches Hg mercury, at 59° F at sea level. Pressure is measured in pounds per square inch. (Also called barometric pressure.)

Atmospheric regulator - see zero governor.

Atmospheric water tower- refers to the movement of air through a cooling tower purely by natural means, or by the aspirating effect of water flow.

Atom - the smallest particle of an element that can exist alone or in combination. Every atom is made up of a positively charged nucleus and a set of negatively charged electrons that revolve around the nucleus. The nucleus is made up of positively charged protons and neutrons that have no charge. Atoms link together to form molecules.

Atomic excitation - the application of some form of external energy to an atom.

Atomic number - the number of electrons or protons in the center or nucleus of an atom.

Atomic weight - the weight of an atom approximately equal to the number of protons and neutrons in the nucleus of an atom.

Atomization - making a liquid into a mist-like spray.

Atomize - to reduce a liquid into a multitude of tiny droplets or a fine spray.

Atomizer - a device that breaks liquid fuel into tiny droplets. It may require the assistance of an atomizing medium such as steam or air. Also see mechanical atomizer.

Atomizing air - the air supplied to an air-atomizing oil burner (usually about 10%) that is used to break the oil stream into tiny droplets. The atomizing air is also later used for combustion.

Atomizing humidifier - a type of humidifier that discharges tiny water droplets into the air, which evaporate very rapidly into the duct air steam or directly into the conditioned space.

Atomizing medium - a supplementary fluid, such as steam or air that assists in breaking oil into a finely divided state.

Atomizing oil burner - an oil burner that delivers fuel oil to the combustion zone in the form of tiny droplets. It uses some form of nozzle or orifice to spray the oil into the combustion chamber. An atomizing oil burner is classified by the medium used to break up the oil into fine particles.

Air-atomizing oil burner - compressed air.

Steam-atomizing oil burner - steam at pressures greater than 25 pounds per square inch (psig).

Mechanical-atomizing oil burner - no medium; the oil is pumped at high pressures (50 to 300 psig) through a suitable burner nozzle orifice that breaks it into a fine mist.

Synonymous with pressure atomizing oil burner. (Also see pressure atomizing gun type burner.)

Air and steam atomizing oil burners can be further classified by the location at which the atomization takes place.

External mix oil burner - the high velocity air or steam strikes the oil after it has left the nozzle.

Internal mix oil burner - the high velocity air or steam strikes jets of incoming oil in a mixing chamber, and the oil is discharged from the nozzle in completely atomized form.

Attachment - any accessory that increases the capabilities of an instrument.

Attachment plug - a device that is inserted into a receptacle and establishes the connection between the conductors of the attached flexible cord and conductors attached permanently to the receptacle.

Attenuation - the sound reduction process in which sound energy is absorbed or diminished in intensity as the result of energy conversion from sound to motion or heat.

Attentuator - a device that decreases the amount of signal voltage or current.

Attic - that part of a structure immediately below the roof that is unoccupied, unconditioned and located outside of the building envelope. Attics are usually closer to outdoor temperature than conditioned space temperature. Many times HVAC equipment is located here and must be accessed for proper service.

Attic fan - an exhaust fan to discharge air near the top of a building while cooler air is forced (drawn) in at a lower level.

Attic, habitable - is an attic that has a stairway as a means of access and egress and in which the ceiling area at a height of seven and one-third (7 1/3) feet above the attic floor is not less than one-third (1/3) the area of the floor next below.

Attractant - a type of salt solution, usually lithium-bromide, used as an absorbent in absorption refrigeration.

Audio frequency - pertaining to that band of frequency that is audible to the human ear. Usually 20-20,000 cycles per second.

Audit - an examination made by independent accountants in order to express an opinion as to the fairness of financial statements.

Auger - the screw like center part of an ice flaker that scrapes ice from a cylindrical evaporator.

Augers - screw-like devices used to compress refrigerant gas in a screw compressor.

Authority - the adjustment on a receiver controller that determines the effect of the reset signal from a secondary transmitter as a percentage of the primary signal from the primary transmitter.

Authority having jurisdiction (AHJ) - the organization, office, or individual responsible for approving equipment, an installation, or a procedure. The authority having jurisdiction has the responsibility for making interpretations of the rules, for deciding upon the approval of equipment or materials and for granting the special permission when required.

Authorized employee - a person who locks and tags out machines or equipment in order to perform service or maintenance.

Auto - short for automatic.

Auto-ignition temperature - the minimum temperature at which a substance ignites without application of a flame or spark. Do not heat materials to greater than 80% of this temperature.

Auto zero - an automatic internal correction for offsets and/or drift at zero voltage input.

Automatic - self-regulating or self-acting. Capable of responding to certain predetermined conditions such as temperature, pressure, or humidity.

Automatic air vents - air vents that can automatically eject air as it accumulates within a closed system.

Automatic blowdown system - system programmed to control frequency and duration of continuous blowdown operations of a boiler.

Automatic burner - a burner that starts and stops automatically.

Automatic changeover thermostat - a thermostat that selects either heating or cooling, depending on room temperature and the heating and cooling set points.

Automatic combination gas valve - a valve for gas furnaces that incorporates a manual shut-off control, an internal 100% safety pilot shut-off, gas valve, and pressure regulator.

Automatic control - a system that reacts to a change or unbalance in one of its variables by adjusting the other variables to restore the system to the desired balance. For example, a system for control of air-fuel ratio can be set up so that a change in the setting of the combustion air valve results in a corresponding change in the fuel input.

Automatic control valve - an electrically operated valve that contains a valve body and a valve actuator or motor. A signal from some remote point can energize the actuator or motor to open or close the valve, or to proportion the rate of flow through the valve. Also see constant, level, firing rate, modulating, motorized, and safety shut-off valves.

Automatic defrost - system of removing ice or frost from evaporators automatically.

Automatic expansion valve (AEV or AXV) - pressure controlled valve that reduces high pressure liquid refrigerant to a low pressure liquid/vapor mixture refrigerant. This type of metering device is usually used on constant load applications and maintains a constant evaporator pressure. When the load increases on the system, this valve tends to close down instead of opening up more.

Automatic feed water valve - a valve that controls the high and low water level in the boiler, opening and closing automatically.

Automatic fire alarm system - a manual fire alarm system containing automatic detecting device(s) that actuates a fire alarm signal.

Automatic fire check - a device for stopping the progress of a flame front in burner mixture lines (flashback) and for automatically shutting off the fuel-air mixture.

Automatic fire door - a fire door or other opening protective constructed and arranged so that, if open, it shall close when subjected to:

1. A predetermined temperature.
2. A predetermined rate of temperature rise, or
3. Smoke or other products of combustion.

Automatic flue damper - a device used between the vent system and the draft diverter on a heating furnace to prevent heat escaping through the vent system during the "off" cycle.

Automatic gas pilot device - a gas pilot incorporating a device, that acts to automatically shut off the gas supply to the appliance burner if the pilot flame is extinguished.

Automatic gas shut-off device - a device constructed so that the attainment of a water temperature in a hot water supply system in excess of some predetermined limit acts in such a way as to cause the gas to the system to be shut off.

Automatic ice cube maker - refrigerating mechanism designed to produce ice cubes in quantity automatically.

Automatic ignition - a system in which a burner is ignited directly, without manual intervention by an automatically supervised ignition spark or pilot.

Automatic non-return valve - valve located on the steam line closest to the shell of the boiler that turns the boiler on and off line automatically. This valve also protects the system in the event of a large steam leak on any boiler.

Automatic pilot - a device that is designed to shut off the gas supply automatically to the burner(s) in the event the source of ignition fails. This device may interrupt gas flow to the main burner, the pilot, or both.

Automatic refrigerating system - one that regulates itself to maintain a definite set of conditions by means of automatic controls and values usually responsive to temperature or pressure.

Automatic reset - a compensating feature incorporated into certain control instruments. It automatically resets the instrument control point when the controlled medium (air or water, etc., under the instrument's control) deviates from the instrument set point. This shift in control point returns the controlled medium to the set point condition. The reset function is continuous during the time period the deviation exists.

Automatic safety shutoff valve - see safety shutoff valve.

Automatic sprinkler - a device, connected to a water supply system, that opens automatically at a predetermined fixed temperature and disburses a spray of water.

Automatic system - a system that regulates itself and requires no manual operations. A system in which the burner is cycled completely automatically under control of an automatic temperature or pressure controller.

Automatic valve - an automatic device consisting essentially of a gas valve and operator that opens and closes the gas supply to the main burner(s) as required during normal operation of an appliance. It may be operated mechanically, electrically, or by pressure of the gas acting on a flexible diaphragm; the pressure on the diaphragm being controlled by an electrically operated pilot valve.

Automatic variable pitch fan - a propeller type fan whose hub incorporates a mechanism that enables the fan blades to be re-pitched simultaneously and automatically. They are used on cooling towers and air cooled heat exchangers to trim capacity and/or conserve energy.

Automatically ignited burner - a burner in which fuel to the main burner is automatically turned on and ignited.

Autoranging - the capability of an instrument to automatically select the most appropriate range in response to the applied stimulus.

Autotransformer - a transformer in which both the primary and secondary coils have turns in common. Step up or step down of voltage is accomplished by taps on the common windings. Output of the secondary depends on whether all or part of the split coil (primary or secondary) is connected to the load.

Auxiliary contacts - extra switches added to magnetic contactors or starters that operate simultaneously with the load contacts. Such switches are used to energize secondary circuits.

Auxiliary device - devices in automatic control systems that are not automatic controls such as air compressors, transformers, thermometers, etc.

Auxiliary drain pan - a complete drain pan that is larger than the outside dimensions of the unit that contains the evaporator coil. Used to prevent water damage to surfaces under the unit in the event of blockage of the condensate drain.

Auxiliary drive - belt, chain, or gear drive used with a gear motor or speed reducer to provide additional speed reduction and/or torque multiplication.

Auxiliary heat (heat pump) - heating provided by equipment other than the heat pump when the heat loss exceeds the balance point of the structure. When this auxiliary heat is electric, auxiliary heaters may be called strip heaters. These strip heaters may be staged by outdoor thermostats.

Auxiliary heat (solar) - a back-up heating system that is used when there is not enough solar radiation to operate the primary solar system. These back-up systems can be gas, oil, or electric.

Auxiliary limit - one of two high limit contacts connected in series with an automatic gas valve. Required on downflow and horizontal furnace applications.

Auxiliary loads - any additional heating load served by a hydronic space heating system.

Auxiliary potentiometer - a potentiometer mounted in or on a modulating motor, that is used to control an external device in response to movements of the motor.

Auxiliary switch - a device that may be added to a damper or valve motor to provide electric switching action at a predetermined point in the motor travel. It is sometimes called an end switch because of its customary location on the motor.

Available floor area - the floor area in a room that can be used to release heat as part of a floor heating system.

Available head - the difference in pressure that can be used to circulate water in a system. The difference that may be used to overcome friction within the system.

Available heat - the quantity of heat released in a combustion chamber that is available for useful purposes. It is the total quantity of heat released minus both the dry flue gas loss (sensible) and the moisture loss (latent).

Avalanche - the breakdown of a diode resulting when the reverse voltage reaches a value sufficient to ruin the diode.

Average floor velocity - the average speed of a fluid particle across a specified cross section in a piping component.

Average house - construct type that has insulated walls and ceilings, vapor barriers, loose doors, windows and fireplace damper. In regards to infiltration and building construction standards the air changes per hour is considered to be one.

Average voltage - an AC voltage or current value based on the average of all instantaneous values for one alternation. To convert from average voltage to peak voltage use the formula:

Epk = 1.57 · Eav. To convert from average voltage to effective voltage use the formula Eeff = 1.11 · Eav.

Average responding - a measurement that uses the DC average of the rectified AC which is scaled to indicate RMS value. In the case of a pure sine wave, the DC is multiplied by 1.1 to indicate the RMS value.

Averaging control - control achieved when two or more control sensing elements are paired together to give one signal in response to the average condition sensed by the two. Such elements are often used to avoid over control due to sudden changes at any single location.

Averaging relay - used where the application requires operation of a final control device or setting a controller, by the average signal from two or more controllers.

Averaging thermostat element - a thermostat sensing element that responds to the average temperature sensed at many different points in a duct or process medium.

Aviation snips - hand tools used to cut sheet metal straight or in a circle.

Avogadro's Law - equal volumes of gases at the same temperature and pressure contain the same number of particles.

Avoided cost - incremental cost for the electric utility to generate or purchase electricity that si avoided through the purchase of power from a cogeneration facility.

AWG - American Wire Gauge. A system of designating wire size (diameters). The smaller the numbers, the larger the gauge of the wire.

Awl - a pointed instrument used for making or piercing small holes.

Axial center - that point on a shaft or other rotating member about which all other points rotate. The axial centerline is a line drawn through and connecting all the axial center points within the member. The line between the north and south poles of the earth about which the earth rotates is called its axis.

Axial fan - a fan that creates an air movement parallel to its motor shaft. Term used to describe propeller type fan blades.

Axial flow - stream of air whose motion is approximately symmetrical along a line, although some spreading and drop or rise can occur from diffusion and buoyancy effects.

Axial thrust load - the external loading or force (as in the push of a fan blade) acting lengthwise along the shaft or in the direction of the shaft.

Axis - the centerline of a geometric figure.

Axis of rotation - the axis of rotation (spin axis) is that straight line about which a body rotates.

Azeotrope - a blend of two or more components whose equilibrium vapor phase and liquid phase composition are the same at a given pressure and having constant maximum and minimum boiling points.

Azeotropic blend - two or more refrigerants mixed together that will have only one boiling and/or condensing point for each system pressure. Negligible fractionation or temperature glide will occur.

Azeotropic mixture - a constant boiling point mixture. A mixture of two liquids that boils at constant composition -- the vapor's composition is the same as the liquids. When the mixture boils, at first the vapor has a higher proportion of one component than is present in the liquid, so this proportion in the liquid falls over time. Eventually, maximum and minimum points are reached, at which the two liquids distill together with no change in composition. An azeotrope composition depends on pressure. An example of an azeotropic mixture would be R-502 where the mixture consists of 48.8% R-22 and 51.2% R-115. The refrigerants do not combine chemically, yet azeotropic mixture provides refrigerant characteristics desired.

Azimuth - the angular distance between the true south and the point on the horizon that is directly below the sun.

-B-

Bas - building automation system.

Bat - battery.

Bat chg - battery charger.

BC - back connected.

BEI - biological exposure indexes.

BHCC - Better Heating-Cooling Council.

Bhp - brake horsepower.

Bi - backward inclined.

Bkr - breaker.

Bpcdh74 - break point cooling degree-hours base 74°F.

Bphdd65 - break point heating degree-days base 65°F.

Bri - building related illness.

BSI - British Standards Institute.

Btu - British thermal unit.

Btuh - British thermal unit heat.

Btu/hr -Btu per hour.

BV - back view.

"B" cock - a manual pilot line gas valve tapped into the upstream side of the "A" cock valve. Some of these valves have a means of adjusting the pilot gas pressure under a screw cover.

B-vent - a venting system for gas stoves or other gas appliances using approved piping through the roof or into a chimney. The inner lining is made of aluminum and the exterior is made of galvanized metal.

Babbitt - a soft, white, non-ferrous alloy for bearings, mostly of tin with small amounts of copper, tin and antimony.

Back draft - down draft often leads to back draft. When poor draft is unable to carry hot gases from the heating appliance, the result is that gases force themselves back into the boiler room through dampers and fire doors; and in extreme cases, back draft can result in considerable damage.

Back EMF - the amount of voltage produced across the start and common terminals or connections of a single phase motor. Also referred to as counter electromotive force.

Backfill - earth placed against the outside foundation walls to fill the void made from the digging of the foundation.

Backfire - a condition that causes the brazing torch flame to go out with a loud cracking sound. This occurs because the burning speed of the fuel gas (inward) exceeds the flow velocity of the fuel gas (outward). Also known as "extinction pop."

Back pressure - pressure against which a fluid is flowing, resulting from friction in lines, restrictions in pipes, valves, pressure in vessel to which fluid is flowing, hydrostatic head, or other impediment that causes resistance to fluid flow.

Back pressure, refrigeration - the pressure in the low side of a refrigerating system. Also called suction or low side pressure.

Back pressure regulator - an automatic valve in the suction line that prevents the suction pressure from dropping below the desired set point.

Back seat port - a port entering the valve body behind the back seat. Since no pressure can reach this port when the valve is back seated, the port can be used for attaching test gauges or making other connections to the system while under pressure.

Back seated - term describing the turning of the stem counter-clockwise until it stops at the wide open position. It is in this position that the valve usually closes off the gauge connection fitting or port. For normal operation, the valve is always in a back seated position.

Back-to-back - a method of connecting two tubes or semiconductor devices, either in series or parallel, but connected in opposite directions with respect to each other. Such a parallel connection is also called an inverse-parallel circuit.

Back vent - the portion of the stack that continues upward and eventually joins back into the main stack above the highest fixture. Also known as a continuous vent.

Backflow - reverse flow in a water system from the normal or intended direction. Use of a backflow preventer check valve is recommended.

Backflow preventer - a piping component that contains the functional equivalent of two check valves and a vent port. It prevents any liquid within a hydronic system as in the case of a boiler or humidifier, from flowing backwards and contaminating a fresh water source.

Background noise - sounds other than the wanted signal. In room acoustics, the irreducible noise level measured in the absence of any building occupants.

Backup power - electric energy available from or to an electric utility during an unscheduled outage to replace energy ordinarily generated by the facility or the utility. Frequently referred to as stand-by power.

Backward inclined fan - the tips of the blades are inclined away from the direction of rotation. Used in commercial/industrial heavy duty heating and cooling systems that require heavy duty construction, non-overloading characteristics and stable air delivery. These blowers operate at higher efficiencies than forward curved blowers. Not as quiet as forward curved blowers because they operate at higher speeds. Can be used in systems up to 3" static pressure. Smaller diameter wheels are supplied with flat blades; larger diameter wheels are supplied with air foil blades to improve efficiency.

Bad debt - an uncollectible account receivable.

Baffle - a metal or refractory surface used to direct the flow of air or flue gases or to minimize the effects of flame, heat, or flue gases.

Baffle, sound - a shielding structure or partition used to increase the effective length of a sound transmission path between two locations. Such structures often are constructed or surfaced with sound absorbing materials and are frequently used to seal open plenums above ceilings or below floors.

Bag assembly - closely related filter media parts grouped together in a single package to remove dust particles from the air stream.

Bag filter - filters intended to remove fine dust particles that cause surface soiling. These filters are constructed of a special blend of paper and fiber in the shape of a bag. This shape creates an extensive filter surface area, which results in low air velocities across the filter. Standard as well as extra length bags are typically offered, the latter being used where larger than normal air quantities are involved.

Bag filter house - see bag assembly.

Bake out - a process to remove or reduce chemical components in building furnishings, such as panels, carpeting and furniture. They are heated to higher temperatures than normal occupied space temperature in order to force the chemicals to vaporize.

Balance fitting - a pipe fitting or valve designed so that its resistance to flow may be varied. These are used to balance the pressure drop in parallel circuits.

Balance point - the outdoor temperature at which the output of the heat pump in a specific application is equal to the heat requirements of the structure. At this point a backup or auxiliary heat source automatically comes on to assist the heat pump in maintaining the desired temperature setting on the thermostat.

Balance point, initial - the outdoor ambient temperature at which the heating capacity of the heat pump only, balances the heat loss of the conditioned space.

Balance point, second - the outdoor temperature at which the heating capacity of the heat pump plus the auxiliary heat capacity balances the heat loss of the conditioned space.

Balance pressure ventilating system - a system where the mechanical fresh air supply is balanced against the mechanical exhaust to maintain a constant air pressure in the conditioned space.

Balanced draft - when the intake damper is automatically controlled by the pressure in the furnace. Furnace pressure is maintained slightly below atmospheric pressure.

Balanced heat recovery - this occurs when internal heat gain equals recovered heat and no external heat is introduced to the conditioned space. Maintaining balance may require raising the temperature of the recovered heat.

Balanced port TXV - a thermostatic expansion valve that will meter refrigerant at the same rate as when the condensers head pressure is low. This valve compensates for a wide range of head pressure changes and ensures more uniform operation of the valve.

Balanced valve - a valve so designed that the fluid pressure exerted on the valve discs is equalized, or balanced, thereby reducing the force required by the operator to open and close the valve.

Balancing - the process of measuring and adjusting the air circulation in an air conditioning system to assure proper temperatures in each section of the conditioned space. Also a procedure used to repair motor rotors and impellers that are out of round due to wear.

Balancing ell - see balance fitting.

Balancing relay - a relay with an armature pivoted between two electromagnetic coils. The armature controls the position of the wiper on a feedback potentiometer in the electrical circuit. If the currents through the two coils differ, the armature moves toward the coil with the most current (strongest electromagnetic force). As the armature moves, the potentiometer wiper changes the resistance in series with both coils until the currents are equal. The armature stops when the circuit is balanced. Besides the electro-mechanical type, there is also a solid state type that performs the same functions electronically. A balancing relay is used in a Honeywell modutrol motor to position it. Also called a mouse trap.

Balancing valve - manually controlled valve used to increase resistance and reduce water flow through a given branch circuit.

Ball bearing - an anti-friction device that is used to allow free turning and support of the rotating member of the device. It consists of an outer and inner ring with races and has steel balls sandwiched between the rings. Used where higher load capacity is required or periodic lubrication is impractical.

Ball bearing motor - type of motor that uses ball bearings to provide low friction contact between the rotating motor shaft and the motor itself.

Ball check valve - a device consisting of a ball and orifice. Pressure on one side of the device will seat the ball across the orifice and stop flow. Pressure in the opposite direction will shift the ball from the seated position permitting flow. Has a large pressure drop across it.

Ball valve - a straight through valve that creates very little pressure drop or resistance to flow.

Ball valve bleed - an arrangement that varies the exhaust stream by positioning a small ball in a shaped opening. This increases output pressure linearity and produces a significant amount of internal feedback.

Ball peen hammer - a 12 or 16 ounce hammer used for hammering and the balled end is used for rounding rivets and metal. Also known as a machinist's hammer.

Ballast - a special type of resistor whose resistance increases rapidly with an increase in current through it. It then limits the current flow through it and a load (such as a fluorescent lamp) connected to it.

Balloon - circle with extension line used to label an item on a pipe support detail or isometric drawing.

Balloon type gasket - a flexible refrigerator door gasket having a large cross section area.

Balustrade - a protective or decorative railing. A row of balusters topped by a rail.

Band gap energy - the amount of energy (in electron volts) required to free an outer shell electron from its orbit about the nucleus to a free state, and thus to promote it from the valence level to the conduction level.

Bandwidth - (1) the range of frequencies within which performance remains within specified limits. (2) amount of charge in analog input required to cause a controlled device to move from 0 to 100%.

Bar - a unit of pressure. One bar equals .98690 atmospheres or 14.504 pounds per square inch.

Bar chart - a representation of the operating sequence of a control or controller using bars shaded to show the times when various circuits are energized.

Bar magnet - a magnet with a pole at one end of a straight metal bar.

Bar slip - rigid, S shaped sheet metal strip used to connect sheet metal duct sections.

Barometer - an instrument used to measure atmospheric pressure. It may be calibrated in inches or millimeters of mercury. Two types of barometers are mercury column and aneroid.

Barometric draft regulator - a balanced damper device attached to a chimney, vent connector, breeching, or flue gas manifold to protect combustion equipment by controlling chimney draft. A double acting barometric draft regulator is one whose balancing damper is free to move in either direction to protect combustion equipment from both excessive draft and back draft.

Barometric pressure - see atmospheric pressure.

Barrier - an insulator that separates electrical terminals on a relay, contactor, compressor terminals or any other electrical device.

Barrier energy - the energy given up by an electron in penetrating the cell barrier or a measure of the electrostatic potential of the barrier.

Base - the main supporting frame or structure of an assembly, excluding the legs.

Base (chemical) - a substance that (1) liberates OH ions when dissolved in water; (2) liberates negative ions in a solvent; (3) receives a hydrogenation from a strong acid to form a weaker acid; (4) gives up two electrons to an acid. Bases have a pH value less than 7 and turns litmus paper blue.

Base (electronic) - the semi-conductor region between the collector and emitter of a transistor. The base controls the current flow through the collector-emitter circuit.

Base flood elevation - a reference point, determined in accordance with the building code, based on the depth or peak elevation of flooding, including wave height, that has a 1 percent (100 year flood) or greater chance of occurring in any given year.

Base load - the minimum amount of electric power that a utility must supply in a 24-hour period. Utilities typically operate their most efficient system or device to produce as much useful energy as was consumed in its manufacture and construction.

Base metal - the metal that you are going to weld, braze or cut.

Base plate - metal plate attached to equipment, vessels, or other components to provide a transition piece from equipment to foundation. Usually attached to the concrete foundation by anchor bolts embedded in concrete.

Base speed - the speed with which a shunt or compound motor runs when it is supplied with it rated voltage.

Base, thermostat - is that portion of the thermostat that contains the internal wiring of the thermostat. This is the portion of the thermostat that contains the sensing device and contacts for activating the heating and/or cooling system.

Baseboard - a terminal unit resembling the room base trim of a house. These units are the most popular terminal unit for residential systems.

Baseboard system - an electric or hot water system that is used for whole house, spot, or individual room heating. They are economical to install and can be controlled by individual room thermostats. The units are thin and mounted on an outside wall just above the carpet or flooring. The heaters are natural draft and as the air enters near the bottom it passes over the heating element, is heated, and rises in the room. Since there is a difference in the density of the air, the cool air falls and sets up the natural draft convection air current.

Baseboard tee - a special fitting that resembles a 90 degree elbow with an additional threaded tapping. This fitting is often used to mount an air vent at the outlet end of a finned tube baseboard element.

Basecoat - the first coat applied to a surface (prime coat for painting).

Baseline data - measurements taken from a refrigeration system for converting it to the use of alternative refrigerants. The data should include temperatures and pressures at the evaporator, temperatures and pressures at the condenser, compressor suction and discharge, expansion valve superheat, and compressor amps.

Basement - that part of the building that is wholly or partly below ground level.

Baseplate - heavy steel plate placed on concrete providing support for columns.

Basic charge - the amount of factory installed refrigerant in the condensing unit, evaporator or line set. Some factory pre-charged units have the exact amount of refrigerant to match components with no further adjustment at the time of field installation. Others have excess refrigerant in one or more components (usually the condenser) to allow for final full charge adjustment and compensate for line length changes. Each of the components has only a small amount of refrigerant for the purpose of maintaining the integrity of the system and is referred to as a holding charge.

Basic refrigeration control - device that starts, stops, regulates and/or protects the refrigeration system and its components.

Basic refrigeration cycle - a process in which a circulating refrigerant absorbs unwanted heat from a place where it is not wanted and transfers it to a place where it is unobjectionable.

Basin - see collection basin or distribution basin.

Basin curb - top level of the cold water basin retaining wall; usually the datum from which pumping head and various elevations of the tower are measured.

Batch type furnace - a furnace that is shut down periodically to remove the old charge and add new charge; as distinguished from a continuous furnace.

Bath - a liquid solution used for cleaning, plating, or maintaining a specified temperature.

Bathroom - a room containing a bathtub, shower, spa, or similar bathing fixture.

Batten - narrow wood strips used to cover joists. Also, wood strip used to secure adjoining boards.

Batter - slope of the exposed face of a retaining wall.

Battery - electricity producing cells, connected in series, that use interaction of metals and chemicals to create direct current electrical flow.

Battery capacity - the ability of a battery to produce a given current over a period of time measured in ampere hours. A rated 100 ampere-hour battery will theoretically produce one ampere for 100 hours or two amperes for 50 hours, etc.

Battery resistance - the internal resistance between plates and electrolyte in a cell or battery.

Baud - a unit of signaling speed. Speed in bauds is the number of signaling elements per second (bits/sec). While baud can be expressed in bits per second when each signal event represents only one bit, some codes permit each signal event to represent other than one bit baud.

Baudelot cooler - a heat exchanger in which water flows by gravity over the outside of tubes or plates.

Bay, construction - a regularly repeated spatial element defined by beams or ribs and their support.

Bay, cooling tower - the area between adjacent transverse and longitudinal framing bents of a cooling tower.

Bay window - a composite of three windows, usually made up of a large center unit and two flanking units at 30 or 45 degree angles to the wall. A bay projects from the wall of the structure.

Bead, carpentry - narrow projecting molding with a rounded surface.

Bead, plastering - metal strip imbedded in plaster at the projecting corner of a wall. It serves as a guide in plastering and helps resist abrasion.

Bead, welding - is a narrow layer or layers of metal deposited on the base metal as the electrode melts.

Beaded tube end - the rounded exposed end of a boiler rolled tube when the tube metal is formed over against the sheet in which the tube is rolled.

Beam - a structural member whose prime function is to carry transverse loads, as a joists, rafters, and girders.

Beam compass - a compass used to draw or make large circles and arcs.

Beam loading - the application of a load to a pipe between two points of support, usually expressed in pounds and the distance between the centers of the supports.

Bearing - a low friction device for supporting and aligning a moving part. Basic types are: sleeve, ball, and unit.

Bearing plate - steel plate placed under one end of a beam or truss for load distribution.

Bearing wall - a wall supporting any vertical load in addition to its own weight.

Becquerel - the SI unit for the rate of radioactive decay in a sample of material. One Becquerel (Bq) equals one disintegration per second. Also see Curie.

Bel - a measurement unit of gain that is equivalent to a 10:1 ratio of power levels. Symbol is "B".

Bell and spigot - pipe joint formed with sections of cast iron pipe with a wide opening (bell) at one end and a narrow end (spigot) at the other, fitted by chalking with oakum and lead.

Bell end - the enlarged portion of a pipe that resembles the socket portion of a fitting and that is intended to be used to make a joint by inserting a piece of pipe into it. Joining may be accomplished by solvent cements, adhesives, or mechanical means.

Bell transformer - a small transformer used to change the line voltage for operation of door bells and chimes.

Bellows - a metallic accordion-like box that can be compressed mechanically or with a fluid pressure (like a spring), and that will return to its normal shape when the pressure is released.

Bellows seal - a method of sealing the valve stem. The ends of the sealing material are fastened to the bonnet and to the stem. The seal then expands and contracts the stem level.

Bellows type thermostat - a thermostat that uses a bellows that expand and contract in relation to temperature changes that open or close an electrical set of contacts.

Belly band mount motor - electric motor mounted with a strap around the motor secured with brackets on the strap.

Below floor tube and plate system - a radiant floor heating system in which the tubing and metal heat transfer plates are attached to the underside of a subfloor.

Below grade - walls or that portion of walls that are entirely below the finished grade or which extend two feet or less above the finished grade.

Belt - a rubber like, continuous loop placed around two or more pulleys to transfer rotary motion from the drive pulley to the driven pulley.

Belt driven - driver and driven, as motor and compressor, equipped with suitable sheaves or pulleys, and connected by one or more belts to operate at a speed ratio established by the relative diameters of the pulleys.

Belt tension gauge - a tool used to check the tension of a belt on belt and pulley assemblies.

Bench grinder - a rotating machine that takes off unwanted material by running an abrasive wheel against the work. It may also be used to redress chisels and sharpen tools.

Bench mark - point of reference from which measurements are made.

Bend supports - preformed elbows or clips that allow flexible polymer tubing to be bent to a relatively small radius without kinking.

Bending spring - a coil spring that is mounted on the inside or outside of a tube to keep it from collapsing while bending.

Bent - a transverse or longitudinal line of structural framework in a water tower composed of columns, girts, ties, and diagonal bracing members.

Bent tube boiler - see water tube boiler.

Bentonite - a clay material used in well drilling to seal the walls of the bore hole to reduce the loss of drilling mud into the earth.

Bernoulli's theorem - in a stream of liquid, the sum of elevation head, pressure head, and velocity remains constant along any line of flow, provided no work is done by or upon the liquid in the course of its flow, and decreases in proportion to energy lost in flow.

Bevel - that angle that is formed by a line or a surface that is not at a right angle to another line or surface.

Bezel - a sloping or slanting surface, as the cutting edge of a chisel.

BHP - brake horsepower. A measure of the output power of a motor.

Bias - a DC voltage applied to the base of a transistor to preset its operating point.

Bi-directional data bus - a data bus in which digital information can be transferred in either direction. With reference to a micro- computer, the bi-directional data path by which data is transferred between the CPU, memory, and input-output devices.

Bidirectional filter drier - a liquid line filter drier that is designed for heat pump applications where refrigerant can flow through the drier in either direction.

Bidirectional thermostatic expansion valve - a thermostatic expansion valve designed for heat pump applications where a built in check valve permits refrigerant to flow through the valve unimpeded while the system is operating in the reverse cycle or mode.

Billing charges (utility) - a charge for the use of a unit of electricity. See demand charge, and energy (consumption charge) charge.

Billing units - unit of electricity for which a charge is assessed. See watt, kilowatt, and kilowatt/hour.

Bimetal element - two metals each having different rates of expansion for each change in temperature. For a typical room thermostat the materials are brass and invar, with brass having the greatest coefficient of expansion. Other materials could be steel and copper. These two metals are fused together in such a manner as to form one piece. When they are heated or cooled, one will expand at a faster rate than the other, thus causing a bending of the two metals. This action can be used to open or close electrical contacts or other suitable movement.

Bimetal overload - a simple thermal overload that breaks the power supplying a small motor directly. When it reaches a high enough temperature, the bimetal opens by warping. When the overload cools, the bimetal will warp back and close the circuit.

Bimetal relay - an overload device that opens a set of contacts by a temperature that corresponds to the current draw of a load.

Bimetal strip - an element that is constructed of two dissimilar metals with unequal expansion rates that are welded together. The two metals have different coefficients of expansion and bend with changes in temperature, thereby opening or closing the contact points. This type of element is used commonly in room thermostats but can be used in insertion controllers as well.

Bimetal thermostat - thermostat that operates on the principle that different metals expand and contract at different rates at difference temperatures. A bimetal strip in the device is used to open and close the thermostats contact points.

Bimetallic - in a bimetallic device, two dissimilar metals (usually steel and copper) are fastened together. They respond differently to temperature changes. These different responses cause the assembly to distort and move contacts.

Bimetallic strip - see bimetal strip.

Bin, ice storage - a container in which ice can be stored.

Bin switch - this switch senses when the ice makers bin is full and turns of the refrigeration system.

Bin thermostat - senses the level of ice in the ice storage (harvest) bin.

Bin weather data - tables published by the Air Conditioning Contractor's Association showing the number of hours per year the local ambient temperatures will be at. Shown in 5 degree increments.

Binary - having two elements, parts or divisions.

Binary code - having only two conditions or values. No intermediate units like "on" or "off", "minimum" or "maximum", or "open" or "closed". These states are usually given the symbols 0 and 1.

Binary control - a device or control loop that can only recognize or respond to the change of state of a signal from one value to another. Examples would be on-off, high-low, yes-no, 1-0 etc. Binary devices are also called two position devices. Having two elements, parts, or divisions.

Binary counter - an interconnection of flip-flops having a single input so arranged to permit binary counting. Each time a pulse appears at the input the counter changes state and tabulates the number of input pulses for readout in binary form. It has 2n possible counts, where n is the number of flip-flops.

Binary input - data or signal indicating one of two conditions, for example, the operating status of a fan as either on or off.

Binary point sensor - whose output indicates that either a monitored characteristic exists or it is zero. For example, an air vane in a fan stream indicates that the fan is either on or off.

Binder - usually a spirally served tape or thread used for holding assembled cable components in place awaiting further manufacturing operations.

Binding post - a threaded terminal for connecting wire to electric equipment. Also known as a stud.

Bioaerosols - biological contaminants, including microbials and fungal spores that become part of the air stream in the building.

Biocide - chemicals used to kill living organisms. In facility air quality management the three most common are disinfectants, sterilizers, and sanitizers. Any biocide used to clean a facility must be approved by the EPA for the application.

Biological contaminants - agents derived from or that are living organisms (e.g., viruses, bacteria, fungi, and mammal and bird antigens) that can be inhaled and can cause many types of health effects including allergic reactions, respiratory disorders, hypersensitivity diseases, and infectious diseases. Also referred to as micro biologicals or microbials.

Biological exposure indexes - a system to determine the amount of a material absorbed into the human body by measuring it or its metabolic products in tissue, fluid, or exhaled air.

Biologicals - biological contaminants. Agents derived from or that are living organisms such as bacteria, fungi and viruses.

Bipolar transistors - dual poled transistors used by operational amplifiers for input.

Bitumen - is the brown or black viscous residue from the vacuum distillation of crude petroleum. It also occurs in nature as asphalt "lakes" and "tar sands." It consists of high molecular weight hydrocarbons and minor amounts of sulfur and nitrogen compounds. Also called asphalt or tar.

Bituminous coal - one of the two coals of great commercial value. It is soft, black in color, and often layered with alternate bands of dull shading. Bituminous coal ignites easily and burns with a smoky yellow flame. Metallurgical coal is bituminous coal. The properties of bituminous coal would shows a moisture content of 3.24%, volatile matter 27.13, fixed carbon 62.52%, and a ash content of 7.11%. The ultimate analysis would show a sulfur content of 0.95%, hydrogen 5.24%, carbon 78.00%, nitrogen 1.23%, oxygen 7.47% and a calorific value of 13,919 Btu/lb.

Black line print - a photocopy or a drawing from a computer (CAD).

Black oxides - contaminants that form by the combination of oxygen and copper during the brazing process.

Blade screwdriver - a tool used to tighten or loosen slotted head items.

Blank ports - the ports on a manifold gauge set that are used to secure the free ends of the refrigerant hoses. This prevents dirt from entering the hoses.

Blast freezer - a low temperature evaporator that uses a fan to force air rapidly over the evaporator surface.

Blast furnace gas - a gas of low Btu content recovered from a blast furnace as a by-product and used as a fuel.

Bleed - a process of slowly reducing the pressure of liquid or gas from a system or cylinder by slightly opening a valve.

Bleed off - the continuous or intermittent wasting of a small fraction of the circulating water to a cooling tower to prevent the build up and concentration of scale-forming chemicals in the water.

Bleed port - valve located on the oil pump (fuel oil unit) for bleeding air from the system. Bleeding of the air from the lines is only needed on a one pipe application.

Bleed resistor - a fixed resistor that is installed to permit built up electrons to flow from one side of a capacitor to slowly and safely migrate to the other side of the capacitor until the two sides are equal in value.

Bleed valve - a valve with a small opening inside that permits a minimum fluid flow when the valve is closed.

Bleeder - (1) an intentional leak, usually used to reduce pressure. (2) a pipe attached to the condenser to siphon off liquid flow that is parallel to main refrigerant flow.

Bleeder resistor - a resistor that is normally soldered in place, and wired in parallel with the capacitor in order to discharge it in the off cycle.

Bleeding - intentional leakage (usually of gas). Leakage that might occur during shut-down in a safety shut-off valve "block and bleed" arrangement. Gas bleeds off to the atmosphere through a normally open vent valve. When the safety shut-off valve(s) is open, the vent or bleed valve closes and "bleeding" stops. Also see double block and bleed.

Blended refrigerants - refrigerants that are created by mixing two or more components together to achieve a desired result.

Blind flange - solid platelike fitting used to seal the end of a flanged end pipeline (also known as a dead end).

Blind hole - one that does not penetrate completely through a component.

Blind rivet - a permanent metal fastener used to join two pieces of sheet metal with a strong connection that will not loosen with vibration.

Block and bleed - see double block and bleed.

Block diagram - a diagram used to show how the parts of a system fit together.

Blocked suction - a method of cylinder unloading. The suction line passage to a cylinder in a reciprocating compressor is blocked, thus causing that cylinder to stop pumping.

Blow back - the drop in pressure in the boiler that occurs after the safety valve has opened.

Blow down (boiler) - the discharged material from relief valves.

Blow down tank - a ASTM coded tank vented to the atmosphere that protects sewer lines from boiler pressure and high temperature when blowing down.

Blow down valve - found on the boiler blowdown line at the lowest part of the water side of the boiler.

Blow down, water tower - a system in a cooling tower whereby some of the circulating water is bled off and replaced with fresh water to dilute the sediment, or control concentrations of salts in the sump.

Blow through unit - a term applied to central fan systems that are arranged so that coils, sprays, and other air treating elements are on the discharge side of the fan.

Blow, throw - the distance an air stream travels from an outlet to a position at which air motion along the axis is reduced to a velocity of 50 feet per minute.

Blower - motor driven, forward curve centrifugal type fan that delivers air to the desired space under pressure. May be either belt or direct drive.

Blower door - a large powerful variable speed fan mounted in a doorway that blows air into (pressurizes) or sucks air out (depressurizes) a house. It's used to test for air leakage in a house. The size and complexity of the fan varies, but all blower doors have adjustable frames around the fan so that they can fit snugly into most doorways.

Blower relay - an electro-magnetic device that energizes the blower motor during the cooling cycle and continuous fan operation mode.

Blower section - the motor and centrifugal type fan that circulates air through the furnace or fan coil unit for treatment of heating, cooling, or ventilation.

Blueprint - a photographic print in white on a bright blue ground, used especially for copying maps, mechanical drawings, and architectural drawings or plans.

Board foot - a unit of cubic content, used in measuring lumber. It is equal in volume to a piece 1 foot square and 1 inch thick.

Body, nominal - the nominal pressure rating of the valve body (exclusive of packing, disc, etc.) Expressed in psig. This rating is often cast on the valve body. The function of this rating is to provide a convenient method of classifying the valve by pressure class for identification purposes. A valve of specified body material and specified nominal valve body rating often has universally accepted construction characteristics, such as wall thickness, flange dimensions, etc., determined by specifications issued by societies such as the A.S.A. Note that the nominal valve body rating is not the same as actual valve body rating

Body, valve - the correlation between safe permissible flowing medium pressure and flowing medium temperature (body only). Each nominal valve body rating has definite corresponding permissible pressures at various temperatures. Thus the permissible pressure is usually dependent on the temperature. The maximum pressure and temperature may be limited by packing and disc material.

Body rating - see valve body rating.

BoHP - boiler horsepower.

Boil - to be heated to a point where bubbles of vapor rise and break at the surface. To evaporate.

Boiler - a closed vessel in which a liquid is heated or vaporized. Used to supply hot water or steam for heating, processing, or power purposes. If steam is generated, it is a steam boiler. If the temperature of the water is raised without boiling, it is classified as a hot water boiler. Also see cast iron sectional, dry-back, firebox, fire tube, high pressure, high temperature, low pressure, packaged, scotch, water tube, and wetback boilers.

Boiler capacity - the rate of heat output in Btu/h measured at the boiler outlet, at the design inlet and outlet conditions and rated fuel/energy input.

Boiler, cast iron sectional - a boiler consisting of individual hollow cast iron sections serving as water jackets surrounding the furnace area.

Boiler drain - a valve that connects to a hose thread and can be placed at any point in a system requiring drainage.

Boiler efficiency - the ratio of the heat absorbed per pound of fuel fired, to the heat of complete combustion of one pound of fuel.

Boiler, electric - boiler which uses electric resistance heating elements to heat water. Has no flue or chimney.

Boiler feed water valve - another name for a pressure reducing valve that automatically allows water to flow into a hydronic system if the system pressure drops below a preset value.

Boiler fire tube - boiler system with heating tubes located within the shell. The use of tubes submerged in boiling water increases the area of contact between the water and the fire gasses passing through the tubes.

Boiler heating surface - the area of the heat transmitting surfaces in contact with the water (or steam) in the boiler on one side and the fire or hot gases on the other.

Boiler, high pressure - boiler furnishing steam at pressures of 15 pounds per square inch gauge or higher (205Kpa) or for heating water to a temperature in excess of 250° F or at a pressure in excess of 160 psig.

Boiler horsepower - the equivalent of the heat required to evaporate 34.5 pounds of water per hour from a temperature of 212° F (100° C) into dry, saturated steam at the same temperature. Equivalent to (970.3 x 34.5) 33,475 Btu/h, 9.803 kilowatts, 139.5 square feet EDR (equivalent direct radiation) for steam and 223 square feet EDR for water.

Boiler, hot water - a low-pressure hot water boiler having a volume exceeding 120 gallons, or a heat input exceeding 200,000 Btu/hr, or an operating temperature exceeding 200° F that provides hot water to be used outside the boiler. A boiler furnishing hot water at pressures not more than 30 pounds per square inch gauge.

Boiler jacket - the outer sheet metal or polymer shell of a boiler.

Boiler lay-up - removing a boiler from service for an extended period of time. A boiler can be laid up wet or dry.

Boiler, low pressure - a boiler furnishing steam at pressures that are not more than 15 pounds per square inch gauge or for furnishing water at a maximum pressure of 160 psig at 250° F.

Boiler rating - the guaranteed output of a boiler in Btu's per hour, or in square feet of radiation, as determined in a test laboratory such as the Institute of Boiler and Radiator Manufacturers (IBR) or the American Society of Mechanical Engineers (ASME).

Boiler reset control - a control action and/or device that automatically changes the upper temperature limit of a boiler based on outdoor temperature.

Boiler room log - a data sheet used to record pressures, temperatures, and other operating conditions of a boiler on a continuous basis.

Boiler rotation - a control technique that varies the firing order of two or more boilers in a multiple boiler system with the goal of equalizing the operating hours of each boiler.

Boiler, shell - a vessel containing water with a portion of the shell exposed to a source of heat.

Boiler short cycling - an undesirable condition during which the burner of a boiler operates for a very short period of time. It leads to a high number of on/off cycles, flue gas condensation and premature wear of components

Boiler shut down - a sequence of operations completed when taking a boiler off line.

Boiler start up - a sequence of operations completed when preparing a steam boiler for service.

Boiler, supplementary - a boiler designed to burn one type of fuel (gas, oil, or solid) that is intended for supplementing a boiler burning another type of fuel (gas, oil, or solid) by means of a common heat transfer medium.

Boiler tubes - used to carry water or heat and gases of combustion. May be straight or bent tubes.

Boiler vent - line coming off the highest part of the steam side of the boiler that is used to vent air from the boiler when filling with water and when warming the boiler. Also used to prevent a vacuum from forming when taking the boiler off line or when a boiler is cooled or drained. Also known as an air cock.

Boiler, waste heat - boiler that uses heat generated for another purpose that otherwise would be released to the atmosphere.

Boiler water tube - the reverse of the water tube boiler. Hot gasses are in contact with the outside surface of the tubes and the water is inside.

Boilers in battery - two or more boilers connected to a common steam header.

Boiling - see boil.

Boiling point - the temperature at which the vapor pressure of a liquid is equal to the surrounding atmospheric pressure so that the liquid becomes a vapor; i.e., the temperature at which the upward pressure of the molecules escaping from the surface equals the downward pressure of the atmosphere. Flammable materials with low boiling points generally present special fire hazards.

Boiling temperature - temperature at which a fluid changes from a liquid to a gas.

Boiling water reactor - in the boiling water reactor, the heat produced in the nuclear fuels results in the boiling of the reactor coolant. In this design, the steam is created directly in the rector and then used to turn a turbine.

Boil-out - the evaporation of water that occurs when the storage water is heated to the boiling point and solar heating continues.

Bolt - a fastener that is used with a nut and normally used for fastening heavy metal parts together.

Bolt circle - a circular center line on a drawing, containing the centers of holes about a common center.

Bolt extractor - a threaded tool used to remove a broken bolt or screw.

Bond - (1) the continuity of an electric connection across a joint or otherwise separated conductors. (2) when referring to bricklaying and masonry chimneys, that connection between brick, stone, or other masonry units formed by lapping them upon one another in carrying up the work, thereby forming an inseparable unit or mass.

Bonded core - an oversized type of refrigerant filter drier with a high moisture absorbing capacity used in removing contaminants from the system.

Bonding - a safety practice. The interconnecting of two objects (tanks, cylinders, etc.) with bonding clamps and bare wire. This will equalize the electrical potential between the objects and help prevent static sparks or discharge that could ignite flammable materials.

Bonding jumper - a reliable conductor between metal parts that establishes electrical continuity.

Bonding jumper, circuit - the connection between portions of a conductor in a circuit to maintain required Ampacity of the circuit.

Bonding jumper, equipment - the connection between two or more portions of the equipment grounding conductor.

Bonding jumper, main - the connection between the grounded circuit conductor and the equipment grounding conductor at the service.

Bonnet - the part of the furnace casing that forms a plenum chamber from where supply ducts receive conditioned air. Also called supply plenum.

Bonnet switch - a N/O temperature activated fan switch that senses bonnet temperature and closes when the temperature climbs to approximately 130° F.

Bonnet and body connections - in valve construction, the method to physical connect the operating mechanism of a valve to the body. The five different designs are threaded, union, bolted, welded and pressure sealed.

Booster - common term applied to the use of a compressor when used as the first stage in a cascade refrigerating system.

Booster compressor - the compressor on a parallel compressor system that is dedicated to the coldest evaporators.

Booster heater - a heater used to raise the temperature of oil from that required for pumping to that required for atomization.

Booster pumps - additional pumps required on long or higher lines to maintain sufficient flow and pressure. These pumps could be used for water, refrigerant, or air.

Boot - a piece of ductwork that is used to connect ducts with registers.

Bore, cylinder - the inside diameter of a cylindrical hole.

Bore, machine - to enlarge a hole with a boring bar or tool in a lathe, drill press, or boring mill.

Bore hole - holes made with a rotary drill rig and bit without a well casing. Bore holes are used for vertical ground loop systems.

Bore scope - device that allows a person to examine sections of air handling equipment visually without the need to make large inspection holes.

Borings - taking sample cylinders at varying depths of subsurface material at a proposed building site to determine the character of bearing material.

Boron - a chemical element, atomic number 5, semi-metallic in nature, and used as a dopant to make p-silicon.

Boss - a projection on a casting or forging.

Bottled gas - liquefied petroleum (LP) gas usually considered to be propane.

Bottoming cycle - cogeneration facility in which the energy input to the system is first applied to another thermal energy process. The rejected heat that emerges from the process is then used for power production.

Bottoming tap - is used first to cut threads to the bottom of a hole. A bottom tap is used when threads must be cut to the bottom of a blind hole. For most threading cases, the thread is started with the tapered tap, then with the plug tap and finally with a bottoming tap.

Boundary layer - a thin layer of fluid that moves slowly along the inside wall of a pipe. The thicker the boundary layer is, the poorer the heat transfer between the bulk of the fluid and the tube wall.

Bourdon tube - a tube that responds to pressure changes. The tube, with elliptical cross section, is shaped into an arc or spiral with one end attached to an indicating, recording, or controlling device. A pressure within the tube makes it less elliptical and more nearly circular, thus actuating the attached device.

Bowden cable - a tube containing a wire that is used to regulate a valve or control from a remote point.

Box - an enclosed panel for connecting conductors or mounting devices.

Box header boiler - see water tube boiler.

Box lock - see government lock.

Box wrench - generally used where nuts and bolts cannot be reached by a socket wrench. Box wrenches are generally 6 or 12 point and give the Technician a strong and safe grip. These wrenches come in straight, offset, or double offset. Most all box wrenches are double ended. Both ends may have the same size with one end offset, or may be the same pattern and different sizes.

Boyle's Law - a law of physics where the volume of a gas varies as the pressure varies, if the temperature remains the same. If pressure is doubled on quantity of gas, volume becomes one half. If volume becomes doubled, gas has its pressure reduced by one half. $P1V1=P2V2$

Bracket - another term for the end shield or end bell of a motor.

Braid - a fibrous or metallic group of filaments interwoven in cylindrical form to form a covering over one or more wires.

Brake horsepower - the actual power output of a motor, turbine, or engine. The power to drive a machine at full load. Does not include the power lost in the transmission of force due to friction and heat.

Brake winding - the winding in a electric motor that holds the armature against a brake shoe to hold the motor in its open position after the motor limit switch opens. When both the motor winding and the brake winding are energized, the strength of the motor winding overcomes the brake winding.

Brake wire - wire used in mobile home, travel and truck trailers to supply current to the electrical braking system.

Branch circuit appliance - a branch circuit supplying energy to one or more outlets to which appliances are to be connected. Such circuits are to have no permanently connected lighting fixtures not a part of an appliance.

Branch circuit center - a central supply, from which the branch circuits extend.

Branch circuit, electrical - a secondary circuit leading from the main electrical circuit to a load or a group of loads.

Branch circuit, general purpose - a branch circuit that supplies a number of outlets for lighting and appliances.

Branch circuit, individual - a branch circuit that supplies only one utilization equipment. Also called a dedicated circuit.

Branch circuit, multi-wire - a branch circuit consisting of two or more ungrounded conductors having a potential difference between them, and a grounded conductor having equal potential difference between it and each ungrounded conductor of the circuit and that is connected to the neutral or grounded conductor of the system.

Branch circuit, oil - a secondary circuit leading from a main oil circulating loop to a burner or group of burners.

Branch circuit, piping - that portion of the piping system that connects a terminal unit to the circuit.

Branch circuit, pneumatic - air lines, used in a pneumatic control system, that transmit the varying air pressure, produced by a pneumatic controller. This changing air pressure positions actuators, such as damper motors, control valves or similar devices.

Branch circuit selection current - branch circuit selection current is the value in amperes to be used instead of the rated-load current in determining the ratings of motor branch-circuit conductors, disconnecting means, controllers, and branch-circuit short circuit and ground fault protective devices whenever the running overload protective device permits a sustained current greater than the specified percentage of the rated load current. The value of branch circuit selection current will always be equal to or greater than the marked rated load current.

Branch connection - a pipeline that runs away from the main or run of a pipeline.

Branch current - the current through a parallel branch.

Branch duct - a smaller duct section or run connected to the main trunk line. Branch ducts normally feed one section of the conditioned space. Consideration must be given to sizing these ducts with regards to velocity values to keep noise down in the conditioned space.

Branch lines - air lines in a pneumatic control system in which a varying air pressure is maintained by the action of controllers to position valves, dampers or similar devices.

Branch resistance - the total resistance of a parallel branch.

Branch soil pipe - a pipe that receives the discharge of toilets and urinals.

Branch voltage - the voltage across a parallel branch.

Branch waste pipe - a horizontal or vertical pipe that receives the discharge of small fixtures and conveys it to the stack.

Brass - an alloy of copper and zinc.

Braze welding - a weld produced by heat and a filler metal without the help of capillary action.

Brazed joint - a joint obtained by the joining of metal parts with metals or alloys that melt at a temperature above 1,000° F but lower than the melting temperature of the parts to be joined.

Brazing - a method of joining metals with nonferrous filler (without iron) material or alloy at a temperature higher than 840° F (427° C) and lower than the melting point of the base metal. The filler material is distributed between the closely fitted surfaces of the joint by capillary attraction.

Brazing filler metal - a metal or alloy used to fill the space between the two parts being joined.

Brazing sheet - brazing filler material in sheet form.

Brazing temperature - the temperature to which the base metal is heated for brazing.

Break - electrical discontinuity in the circuit generally resulting from the operation of a switch or circuit breaker.

Break even cost - the cost of a photovoltaic system (in dollars) per kilowatt of generating capacity) at which the cost of the electricity it produces exactly equals the price of electricity from a competing source.

Break even temperature - the outdoor temperature at which the total heat losses from conditioned spaces equal the internally generated heat gains.

Break lines - used on blueprints when you cannot see the whole object. Or they are used to show that, to save space, the full length of the object has not been shown. When the break on the drawing is long, long break lines are used. Short breaks are used when the break is short, such as across a beam.

Breakdown torque - the maximum torque that a motor will develop under increasing load conditions without an abrupt drop in speed and power.

Breakdown voltage - that voltage at which an insulator or dielectric ruptures.

Breaker - a device usually used in a breaker panel that is capable of being used as a disconnect switch and an overload for the circuit it is supplying power to. Can either be thermal or current activated.

Breaker panel - an electric panel that houses breakers used to distribute power to circuits in the structure.

Breaker points - metal contacts that open and close a circuit at timed intervals.

Breaker strip - a strip of wood or plastic used to cover the joint between the outside case and the inside liner of a refrigerator.

Breakout - a term used to define a wire or group of wires in a multi-conductor configuration that terminate somewhere other than at the end of the configuration.

Breakout noise - the transmission or radiation of noise from some part of the duct system to an occupied space in the building.

Breathing zone - area of a room in which occupants breathe as they stand, sit, or lie down.

Breeching - a passageway to conduct flue gasses from the furnace or boiler to the chimney. Also called the flue pipe or vent connector. Connections must be airtight with end joints made by welding, lapping and bolting, or with companion flanges. Clean out openings should be provided, with expansion joints used in breechings of long runs.

Bridge - the air apace between the earth and the buried ground loops. Bridging reduces the efficiency of the ground loop system because of reduced contact area.

Bridge circuit - an electrical network arranged so that the voltage at a given point in the circuit can be zeroed by adjusting the electrical parameters (resistance, inductance, and capacitance) in the branches until the currents balance.

Bridge contacts - those contacts on contactors and motor starters that are moveable. When the device is energized the bridge contacts change position to make or break an electrical circuit. May or may not be field replaceable.

Bridge rectifier - a device constructed with four diodes that converts both positive and negative cycles of AC voltage into DC voltage. The bridge rectifier is one type of full wave rectifier.

Bridging - system of bracing between floor beams to distribute floor loads or joists between joists to stiffen floor.

Brine - a generic term for a solution of water and anti-freeze such as propylene glycol, ethylene glycol, or calcium chlorine to prevent the freezing of water below 32° F.

Brinelling - permanent deformation of the bearing surfaces where the roller (or balls) contacts the races. Brinelling results from excessive load or impact on stationary bearings. It is a form of mechanical damage in which metal is displaced or upset without attrition.

Brinell - a method of testing the hardness of metals.

Bring-up time - the time required to raise a cold furnace to operating temperature.

British thermal unit (Btu) - the means to define and measure heat energy. One Btu will raise the temperature of one pound (approximately one pint) of water 1° F from 62 to 63. One Btu will also raise the temperature of approximately 55 cubic feet of air from 70 to 71° F. This is equivalent to the heat content of a standard wooden kitchen match completely burned. This rate of heat transfer, whether heating or cooling, is expressed in Btu's per hour (Btu/hr). Heating and cooling equipment capacity is rated on this basis. One Btu = 1055 Joules or 778.169 foot/pound.

Bronze - an alloy of eight or nine parts of copper and one part of tin.

Brown and Sharpe - the former name of the American Wire Gauge system of designation of wire sizes.

Brown coat - in three coat plastering, the second coat of plaster troweled over the scratch coat. This provides a base for the white finish coat.

Brown-out - low line voltage that can cause misoperation and possible damage of equipment.

Brush - an instrument used to apply paste soldering flux to items to be soldered.

Brush holder - a device used in motors and generators to hold the brushes in position on the commutator.

Brushes - sliding contact made of carbon and graphite that make connections to a rotating armature in a generator or motor.

Btu - abbreviation for British thermal unit. The quantity of heat needed to raise the temperature of 1 pound of water 1° F.

Btu meter - a device that indicates to an operator how many Btu's are being generated by a unit.

Btu per hour - the hourly rate of heat transfer expressed in Btu's. This can be either a heat loss or a heat gain. It can also be either the heat producing or the heat removal capacity of the equipment. Expressed as Btu/h.

Bubble memory - bits are stored as the presence (1) or absence (0) of microscopic magnetic cylinders (which resemble bubbles when viewed on end through a microscope), called domains in a thin film of magnetic garnet or orthoferrik. Bubble memories provide high capacity read/write, non-volatile data storage.

Bubble point - the saturated liquid pressure for zeotropic blends, as listed on a temperature pressure chart. The bubble point indicates the two pressures for a given temperature in which the first bubble of vapor appears in a liquid refrigerant.

Buck - door frame placed in a wall or partition to which the door moldings are attached. A completely fabricated steel door frame set in a wall or partition to receive the door.

Bucket brigade - a colloquial term for the electronic self-checking circuitry in the flame signal amplifier of some flame safeguard controls. The term refers to the periodic discharging "dumping" of one capacitor in the circuit into another. If this dumping stops, due to a failure in the flame detection system, the control shuts down the burner.

Bucket trap (inverted) - a float trap with an open float. The float or bucket is open at the bottom. When the air or steam in the bucket has been replaced by condensation, the bucket loses its buoyancy and when it sinks it opens a valve to permit the condensate to be pushed into the return.

Bucket trap (open) - the bucket (float) is open at the top. Water surrounding the bucket keeps it floating and the pin is pressed against the seat. Condensate from the system drains into the bucket. When enough has drained into it so that the bucket loses its buoyancy, it sinks and pulls the pin off its seat and steam pressure forces the condensate out of the trap.

Buff - to polish or finish on a buffing wheel composed of fabric with abrasive powders.

Buffer - a temporary data storage device used to compensate for a difference in data flow rate or event times, when transmitting data from one device to another.

Buffer resistor - the resistor in a voltage regulator across which changes of voltage occur while maintaining the output voltage constant.

Buffer tank - an insulated, water filled storage tank that forms the link between a heat source and a hydronic distribution system. This tank allows the rate of heat input from the heat source to be different than the rate of heat output to the distribution system.

Buffer zone - an area within the home between the conditioned zones and the outside. Thus it normally is not conditioned. These include attics, attached garages, crawlspaces, basements, and enclosed porches.

Building - a separate and distinct structure occupied or intended for supporting or sheltering any occupancy..

Building automation system - a centralized control system that organizes into manageable segments the many tasks of operating a building. These segments include gathering data, analyzing data, and taking corrective action. Also known as a facility management system.

Building drain - the soil or waste stacks feed into this horizontal drain that carries the wastes from the stacks to the building sewer outside the building. The building drain extends just three feet outside the building.

Building envelope - the exterior surfaces of a building --- specifically, walls, roof and floor (if over unheated basement, slab, or crawlspace).

Building heat load - the rate at which heat must be added to a building to maintain interior comfort.

Building official - the official authorized to act in behalf of a jurisdiction code enforcement agency or its authorized representative.

Building piping - the fuel piping from the point where it enters the building or foundation to and including the outlets and upstream to the gas meter or Lp gas or fuel oil tank outlet.

Building plans - the final drawings from which a building is built. Also known as working drawings.

Building service equipment - the mechanical, electrical and elevator equipment, including piping, wiring, fixtures and other accessories that provide sanitation, lighting, heating, ventilation, fire fighting and transportation facilities essential for the habitable occupancy of the building or structure for its designated use and occupancy.

Building sewer - at a point three feet outside the building, the building drain becomes a building sewer that connects to a municipal or private system.

Building trap - a trap placed in the buildings sewer line.

Building water main - the portion of the water supply system that begins at the municipal or private source and extends horizontally to the furthest riser.

Building wire - wire used for light and power in permanent installations utilizing 600 volts or less. Usually in an enclosure and which will not be exposed to outdoor environments.

Built-up system - an HVAC system in which the designer selects components individually on the basis of system specifications and operating characteristics as opposed to a factory assembled (packaged) system.

Bulb and capillary sensor - temperature sensor bulb which contains a thermally sensitive fluid which expands or retracts through the capillary.

Bulb pressure - in a thermostatic expansion valve, the only pressure that pushes to open the valve. As the evaporator outlet temperature increases, the bulb pressure would increase pushing the valve more toward the open position.

Bulb, sensing - that part of a sealed fluid device that reacts to temperature to be measured, or which will control a mechanism.

Bunch strand - any number of conductor strands twisted together in one direction with the same lay length.

Bulkhead - a structure on the roof of a building covering a water tank or service equipment.

Bull dog snips - hand tin snips used to cut 24 to 16 gauge metal. They can be used to cut straight or medium sized outside circles and curves.

Bull head tee - the installation of a pipe tee in such a way that water enters (or leaves) the tee at both ends of the run (the straight through section of the tee) and leaves (or enters) through the side connection only. A tee having a branch that is larger than the run.

Bunker - in commercial installations, the space in which ice or cooling element is installed.

Bunker "C" - term for heavy oil.

Bunsen type burner - a gas burner in which combustion air is injected into the burner by the gas jet emerging from the gas orifice, and air is premixed with the gas supply within the burner body before the gas burns on the burner port.

Burden voltage - the voltage drop across the input terminals of a current measuring device, caused by internal shunt resistance. Burden voltage contributes measurement error, and should be as low as possible.

Buried cable - cable that lies below the grade surface. Also called direct burial cable.

Burner - a device that introduces fuel and air properly mixed and in the correct proportions, into a combustion chamber where the mixture is burned and the products of combustion removed. Burner types are injection, atmospheric, power, and yellow flame. Also see atmospheric, atomizing oil, automatically ignited, automatic, combination, inshot, long flame, main, manually ignited, multi-port, nozzle mixing, partial premix, power, premix, raw gas, rotary oil, semi-automatic, upshot, and warm-up burners.

Burner controller - an automatic temperature or pressure sensing device connected in the input circuit of a flame safeguard control. It indirectly acts to turn a burner on or off in order to maintain pressure, air temperature, water temperature, or liquid level within the desired limits.

Burner, fan - centrifugal-type fan, located inside the burner assembly, that supplies primary air for mixing with the fuel.

Burner flexibility - the degree at which a burner can operate with reasonable characteristics with a variety of fuel gases and/or variation in input rate (gas pressure).

Burner head - that end of a burner beyond the nozzle or ports.

Burner input control valve - see firing rate valve.

Burner motor - the motor which drives the blower or fan that delivers air at proper velocity into the combustion chamber. With a oil burner assembly the burner motor also operates the fuel pump unit.

Burner refractory - see refractory.

Burning in suspension - the combustion of a fuel when burned in air without support. Flame impingement does not happen in this situation.

Burning rate - the time it takes a specified sized sample of solid material (e.g., 1 in. by 1 in.) to burn a designated distance. The rate is given in units of distance/time.

Burning speed - the speed that the flame travels through the gas/air mixture.

Burning velocity - see flame propagation rate.

Burnish - to finish or polish by pressure upon a smooth rolling or sliding tool.

Burnout - accidental passage of high voltage through an electrical circuit or device that causes damage.

Burr - a fine, feathered or jagged edge on the inside of materials resulting from punching or cutting.

Burst strength - the internal pressure required to break a pipe or fitting. This pressure will vary with the rate of built-up of the pressure and the time during which the pressure is held.

Bus - a path over which digital information is transferred, from any of several sources to any of several destinations. Only one transfer of information can take place at any one time. While such transfer of information is taking place, all other sources that are used are tied to the bus must be disabled.

Bus bar - a primary power distribution source that is connected to a main power source. Usually a heavy, rigid metallic bar or strap that carries a large current and makes a common connection between several circuits. Bus bars are usually un-insulated and located where the service enters the building.

Bus, bi-directional - a twisted pair of wires over which two way communications are transmitted between the controller and controlled devices.

Bushing - a usually removable lining or sleeve of metal or other material that is inserted or screwed into an opening to limit its size, resist wear or erosion, serve as a guide, or protect wires from abrasion and possible short circuit. An internally and externally threaded plug for connecting a pipe or fitting with another of different size.

Busway - an enclosed system used for power transmission that is voltage and current rated.

Butane - a gaseous hydrocarbon fuel; a mixture of easily liquefiable hydrocarbon gases, principally butane. Butane is one of the components of raw natural gas, and is also refined from petroleum refining processes. Chemical formula is C_4H_{10}, maximum burning speed of 33 inches per second, ignition temperature of 800° F, molecular weight is 58.124, specific volume of 6.5 cubic feet/lb, boiling temperature of 10.89° F, and a specific gravity of 2.0.

Butanes - organic compounds that have four carbon atoms in the molecule, all bonded by single bonds.

Butt weld - a circumferential weld in pipe fusing the abutting pipe walls completely from inside wall to outside wall.

Butterfly damper - a single blade damper, pivoted about its center. The blade somewhat resembles the opened wings of a butterfly.

Butterfly valve - a throttling valve for a pipe, consisting of a rotating hinged plate that somewhat resembles the opening wings of a butterfly; often used as a firing rate valve. This type of valve does not close tightly, so a separate safety shut off valve must be used with it for final close off.

Button clip - when two or more pieces of sheet metal are to be joined by slipping together, the union can be made tight by the use of a tool that forces a deep indentation into all pieces of metal at one time. These indentations are generally spaced on about 18 inch centers. Also known as Hickey punched.

Bx cable - a flexible armored electric cable used for wiring buildings and electrical and electronic equipment. It consists of a flexible metal cable (strip wound) containing insulated wires.

Bypass - the pipe or duct, usually controlled by a valve or damper, for conveying a fluid around an element of a system.

Bypass, hot gas - a connection from the discharge directly to the suction side of the compressor. Sometimes used as a means of capacity control.

Bypass humidifier - certain humidifiers that use the difference in pressure between the supply and return ducts to create air movement across the humidifiers wetted media.

Bypass line - tubing that permits refrigerant flow to bypass the expansion valve when the coil is operating as a condenser in a heat pump installation.

Bypass line, boiler - a pipe that passes around a control, heater or steam trap. Used so that a heating plant can be operated while equipment is being serviced or repaired.

Bypass, relief - a connection from the discharge, directly to the suction side of the compressor, made either internally or externally. A port that opens at a set point and relieves abnormally high discharge pressure to the suction side of the compressor.

Bypass, switch - a function of a timing device whereby the circuit across a control mechanism is de-energized until a desired state has been achieved, e.g., until a desired temperature or pressure has been reached.

Bypass valve - small diameter pipe connected from inlet to outlet side of valve to allow line contents to pass to the other side of disk or gate to equalize pressure for ease of operation.

-C-

C - (1) thermal conductance, (2) Centigrade, (3) Celsius.

CAA - Clean Air Act.

Cap - (1) capacity (2) capacitor.

Cat - catalog.

Cc - (1) closing coil, (2) compressor contactor.

CCA - canonical correlation analysis.

CCU - central control unit.

Ccw - counter clockwise.

Cdh74 - cooling-degree hours base 74EF.

CEMF - counter electromotive force.

Cf - cubic feet. The standard measurement for natural gas.

CFC - chlorofluorocarbon.

Cfh - cubic feet per hour.

Cfm - cubic feet per minute.

CFR - Code of Federal Regulations.

Ckt - circuit.

Cl - chlorine

Clf - cooling load factor.

CMOS - complementary metal oxide semiconductor.

Cmp - Coupled ocean atmosphere computer model

CO - carbon monoxide.

CO_2 - Carbon dioxide.

Com - common.

Cond - (1) conductor (2) condition.

Conn - connection.

Cop - coefficient of performance.

Cp - centipoise.

Cpvc - chlorinated polyvinyl chloride.

Cr - control relay.

CRT - cathode ray tube.

Cs - capacitor start motor.

Cscr - acronym for capacitor start capacitor run motor.

Csir - capacitor start induction run.

Csr - capacitor start/run motor.

Cv - capacity index.

Cw - clockwise.

Cw/ccw - reversible.

CWA - Clean Water Act.

Cwr - chilled water return

Cws - chilled water supply

"C" factor - the amount of heat, in Btu's, conducted through one square foot of a material of a specified thickness, per hour, per degree temperature difference.

Cabinet - an enclosure designed for flush or surface mounting used to house the branch circuit connections and protective devices.

Cabinet thermostat - one of two thermostats in a flex tray ice maker that, if opened, interrupts the water fill cycle.

Cable - an arrangement of conductors in a protective covering. It is arranged so that the conductors can be used separately or in groups. Small cable is called stranded wire or cord.

Cable joint - a connection of two or more lengths of cable. Often called a splice.

Cad cell - see cadmium sulfide cell.

Cadd - computer aided drafting/design. Sometimes referred to as CAD.

Cadmium - a chemical element, atomic number 48. Cadmium is used in making certain types of solar cells.

Cadmium alloys - brazing alloys containing cadmium, that is a toxic metal when molten. It emits highly poisonous cadmium oxide fumes that can cause illness or death.

Cadmium plated - parts that have a thin corrosion resistant covering of cadmium metal.

Cadmium sulfide cell (cad cell) - a photo conductive device that responds to the visible light emitted by an oil flame. In the presence of visible light, the electrical resistance of the cadmium sulfide becomes very low and the cell conducts current. The cad cell is used in flame detectors for oil burner systems.

Cage valve - a single seated valve containing a flame or box (cage) inside which the plug slides thus opening the ports.

Cages - enclosures around ladders at a certain elevation to prevent personnel from falling.

Calcium - a soft, silver white metallic element found combined with limestone, chalk and marble, etc. An element that is often dissolved in water causing hardness.

Calcium chloride ($CaCl_2$) - a granular form chemical used as a drier to absorb moisture from refrigerants. This material is soluble in water, and in the presence of large quantities of moisture may dissolve and plug up the drier or even pass into the system beyond the drier.

Calcium sulfate - a chemical compound ($CaSo_4$) that is used as a drying agent or desiccant in liquid line driers.

Calibrate - to re-position indicators as required to obtain accurate measurements.

Calibration - to systematically adjust the graduations so that the actual output matches the desired output.

Calibration point - the set point at which control is calibrated. Normal calibration is 9 psig in branch line when ambient temperature equals set point.

Calipers - instrument for measuring both inside and outside diameters.

Call for heat - when the controlled temperature or pressure drops below the operating set point, the burner controller contacts make to initiate burner operation.

Callback - a return trip to a piece of equipment that was not properly serviced the first time.

Calorie - two different calorie units are used by scientists. The calorie used by medical science is a small heat unit. It equals the quantity of heat required to raise the temperature of one gram of water from 4° C to 5° C. The calorie used by engineering science is a large heat unit. It is equal to the amount of heat required to raise the temperature of one kilogram of water 1° C. In the SI system it is recommended that the joule unit of energy be used in place of the calorie.

Calorimeter - a device used to measure quantities of heat or to determine specific heats.

Cam - a mechanical component that is oblong, giving a reciprocating motion when rotated.

Candela - the International System (SI) unit of luminous intensity. The term has evolved from considering a standard candle, similar to a plumbers candle, as the basis of evaluating the intensity of other light sources. Sometimes the term candlepower is used to describe the relative intensity of a source.

Candle - a unit of luminous intensity or brightness.

Candlepower - the measure of the intensity of light in a given direction.

Candling (candle lighting) - the deterioration of the pilot to an unsatisfactory condition, characterized by a lazy, yellow flame; usually caused by a high ambient temperature or an excessively large pilot burner orifice.

Canonical correlation analysis - a forecasting tool that predicts developing patterns of temperature and precipitation based upon the patterns of global sea surface temperature, and U.S. surface temperature and precipitation, from the most recent, four non-overlapping seasons.

Cant strip - beveled strip placed in the 90 degree angle between the roof and an abutting wall to avoid a sharp bend in the roofing material. A strip placed under the lowest row of tiles on a roof to give it the same slope as the rows above it.

Cantilever - projecting beam or member supported at only one end.

Canvas collar - a flexible connector (used to be made of canvas but now plastic coated material) between the air handler and the main and/or return duct runs to lessen vibration and noise transmission of the equipment through the ducts.

Cap - a fitting that seals off the end of a pipe or tubing.

Cap end - the end enclosure of a vessel. Can be dished elliptical, or flanged and bolted.

Cap screw - set screws that have heads.

Cap seam - a Dutch Fold capped with a cap strip.

Cap strip - a strip of metal that has had its two edges turned 180 degrees inward and on the same side of the metal strip. It is used to form a drive seam.

Capability - maximum load that a generating unit, generating station, or other electrical apparatus can carry under specified conditions for a given period of time without exceeding approved limits of temperature or stress.

Capacitance - property of non-conductor (condenser or capacitor) that permits the storage of electrical energy in an electrostatic field. The inherent property of an electric circuit that opposes a change in voltage. Measured in microfarads.

Capacitive coupling - electrical interaction between two conductors caused by the capacitance between them.

Capacitive circuit - a circuit that contains at least one capacitor.

Capacitive load - a leading load; a load that is predominantly capacitive, so that the alternating current leads the alternating voltage, i.e., the voltage does not change direction until after the corresponding current does.

Capacitive reactance - the opposition or resistance to an alternating current as a result of capacitance; expressed in ohms.

Capacitor - an electrical storage device that consists of two conductive metal plates with a dielectric insulator between the plates. They are used to "boost" voltage for a single phase wound motor. The two types of capacitors used primarily in the HVAC/R industry are the starting (electrolytic) and running (oil filled). Capacitors are rated in microfarads. Also, an electrical condenser used for power factor correction.

Capacitor analyzer - an instrument that determines whether a capacitor is open or shorted whether it is within its microfarad ratings and if it is operating at the correct power factor percentage.

Capacitor Laws - there are five laws that govern the application of capacitors in relation to substitution.

First: the voltage rating of any capacitor must be equal to, or greater than, the capacitor being replaced.

Second: when replacing a run capacitor, the microfarad capacitance tolerance shall not vary more than +/- 10% of the original capacitor.

Third: when replacing a start capacitor, the microfarad capacitance tolerance shall not vary more than +20% - 0% of the original capacitor.

Fourth: when wiring capacitors in series, the following rule applies.

$$Ct = \frac{C1 \times C2}{C1 + C2}$$

Fifth: when wiring capacitors in parallel, the following rule applies.

Ct = C1 + C2........

Capacitor start motor - a modified version of the split phase motor, employing a capacitor in series with its starting winding, to produce a phase displacement for starting.

Capacitor start/run motor - a high starting torque and good running efficiency motor. It utilizes a start capacitor to increase starting torque and a run capacitor to increase running efficiency. This type of motor must be used in single phase refrigeration systems that use a thermostatic expansion valve as the metering device.

Capacitor start/induction run motor - type of motor commonly used in situations where the motor must start under a heavy load, but once started can operate with just the run winding. Usually contains one start capacitor with means to remove it from the electrical circuit once the motor is up and running and a run capacitor to improve motor efficiency. Abbreviated CSR.

Capacity - (1) a measure of the ability of air conditioning equipment to remove or add heat. Usually given in terms of Btu per hour or watts (metric). (2) The ability to hold, contain, or absorb. (3) the rated load of a motor, apparatus or device, or the maximum load of which it is capable under existing service conditions.

Capacity charge - see demand charge.

Capacity classifications of stokers - stokers are classified according to their coal-feeding rates. Although some residential applications still use stokers, the primary thrust of stoker application is in commercial and industrial areas. The following classification has been made by the U.S. Department of Commerce, in cooperation with the Stoker Manufacturers Association:

Class 1: capacity under 60 lb of coal per hour
Class 2: capacity 60 to 100 lb of coal per hour
Class 3: capacity 100 to 300 lb of coal per hour
Class 4: capacity 300 to 1200 lb of coal per hour
Class 5: capacity 1200 lb of coal per hour and over

Capacity, condensing unit - the refrigerating effect in Btu per hour produced by the difference in total enthalpy between refrigerant liquid leaving the unit and the total enthalpy of the refrigerant vapor entering the unit. Generally measured in tons or Btu per hr.

Capacity control - term used for various methods of control intended to limit the compressor from cycling on and off.

Capacity control actuator - a device that responds to variations in the suction pressure and proportions a fluid pressure to a mechanism on the cylinder liner that loads or unloads the compressor cylinders.

Capacity, cooler refrigerating, net - the rate of heat removal from a fluid flowing through a cooler (air, water, brine, etc) at stated conditions; difference in specific enthalpies of the cooling fluid entering and leaving the cooler, the latent heat of fusion and the subcooling heat of the ice (frost) must be added in determining the net cooler refrigerating capacity.

Capacity credits - value included in the utility's rate for purchasing energy, based on the savings accrued through the reduction or postponement of new generation capacity that results from purchasing power from cogenerators.

Capacity, expansion valve - the refrigerating effect in Btu per hour, or tons, each of 12,000 Btu per hour produced by the evaporation of refrigerant passed by the valve under specified conditions.

Capacity factor - the output of a generating plant for a specified period of time, say a year, divided by the output if the plant had operated continuously at full rated capacity for the same period.

Capacity, heat - the amount of heat necessary to raise the temperature of a given mass one degree. Numerically, the mass multiplied by the specific heat.

Capacity, ice making - actual productive ability of a system making ice. This is less than the rated (ice making) capacity as some refrigeration is used in cooling the water to freezing point, cooling the ice below the freezing point, and overcoming heat leakage.

Capacity, ice melting - refrigeration equal to the latent heat of fusion of a stated weight of ice at 144 Btu per hr.

Capacity, ice melting equivalent - the amount of heat absorbed by one pound of ice at 32° F in liquefying to water at 32° F, or 144 Btu.

Capacity index (Cv) - the quantity of water, in gallons per minute at 60° F, that flows through a given valve with a pressure drop of one psi. The flow of any fluid through the same valve can be calculated from the Cv. Also called flow coefficient.

Capacity modulation - a method of varying the capacity of a compressor.

Capacity modulation, compressor - a mechanism to control the capacity of a compressor by rendering one or more cylinders ineffective.

Capacity modulation, cylinder - the method used for modulating the capacity of a compressor cylinder by interrupting the normal action of the suction valve. Usually activated by a device controlled by electricity or oil pressure.

Capacity modulation, multi-step compressor - a compressor capacity modulation system arranged to reduce capacity in two or more consecutive steps.

Capacity modulation, single step compressor - a compressor capacity modulation system arranged to reduce capacity in one step. Usually reduces capacity to one-half of full capacity.

Capacity, peak - output of a machine or system for a short time under an extreme load.

Capacity reducer - in a compressor, a device such as a clearance pocket, movable cylinder head, or suction bypass by which compressor capacity can be adjusted without otherwise changing the operating conditions.

Capacity, refrigerating - the rate of heat removed from a medium or space to be cooled, at stated conditions. The term refrigerating effect is used to denote heat transfer to or from the refrigerant itself in a refrigeration system, whereas the term refrigeration capacity is used to denote the rate of heat removal from a medium or space to be cooled and usually measured in Btu/hr or tons/24 hr.

Capacity, refrigerating, gross - the total rate of heat removal from all sources by the evaporator of a refrigerating system at stated conditions. It is numerically equal to the system refrigerating effect.

Capacity, refrigerating, net - the remaining rate of heat removal from all sources by the evaporator of a refrigerating system, at stated conditions, after deducting internal and external heat transfers to the evaporator that occur before distribution of the refrigerating medium and after its return.

Capacity, refrigerating, useful - the refrigerating capacity available for the specific ultimate cooling function for which the system was designed.

Capacity, refrigerating, volumetric - the refrigerating capacity of a system per unit volume of refrigerant circulated at the compressor suction.

Capacity, useful latent, dehumidifying - the available refrigerating capacity of an air conditioner for removing latent heat from the space to be conditioned.

Capacity, useful sensible - the available refrigerating capacity of an air conditioner for removing sensible heat from the space to be conditioned.

Capacity, water tower - the amount of water (gpm) that a cooling tower will cool through a specified range, at a specified approach and wet bulb temperature.

Capillary attraction - the attraction of a liquid material between two pieces of material such as two pieces of copper or copper and brass. For instance, in a joint made up of copper tubing and a brass fitting, the solder filler material has a greater attraction to the copper and brass than to itself and is drawn into the space between them.

Capillary tube - a type of refrigerant metering device. Usually consists of several feet of tubing having a small inside diameter. Selection criteria are based on length and inside diameter. Friction of liquid refrigerant and bubbles of vaporized refrigerant within the tube serve to restrict flow so that correct high and low side pressures are maintained while the compressor is operating. A capillary tube refrigerant control allows high and low side pressures to balance during the off cycle. Also small diameter tubing used to connect temperature control bulbs to control mechanisms.

Capillary tube cleaner - a service tool used to apply up to 15,000 psi of pressure to force and clear any obstruction out of the interior of the tube.

Carbon - a non-metallic element #6 on the Periodic Table. Diamonds and graphite are pure forms of carbon. Carbon is a constituent of all organic compounds. It also occurs in combined form in many inorganic substances; i.e., carbon dioxide, limestone, etc.one of the elements used in refrigeration.

Carbon brushes - material connected to slip rings on rotors to create resistance.

Carbon dioxide (CO_2) - (dry ice) a heavy, colorless gas produced by the combustion and decomposition of organic substances and as a byproduct of many chemical processes. CO_2 will not burn and is relatively nontoxic and unreactive. A normal component of outside air at 300 to 400 parts per million. CO_2 is exhaled by humans at rates typically varying from 0.007 to 0.033 cubic feet per minute and it is considered a contaminant at certain levels in occupied spaces. High concentrations, especially in confined spaces, can create hazardous oxygen-deficient environments that can cause asphyxiation. Useful as a fire extinguishing agent because it blocks the oxygen and smothers a fire. It is 1.5 times as heavy as air, molecular weight of 44.01, specific volume of 8.76 and has a sublimation temperature of -109.2°F. Also used as a refrigerant. Refrigerant number is R-744.

Carbon filter - an air filter using activated carbon as an air cleansing agent.

Carbon monoxide (CO) - a deadly, colorless, odorless, flammable, and very toxic gas produced by the incomplete combustion of carbon compounds and as a byproduct of many chemical processes. While carbon monoxide itself is odorless, it is accompanied often by the aldehydes formed which have a distinctive acrid odor. A chemical asphyxiant; it reduces the blood's ability to carry oxygen. Hemoglobin absorbs CO two hundred times more readily than it does oxygen. It has a molecular weight of 28.01, specific volume of 13.7 cubic feet/lb. and a boiling temperature of -312.7° F.

Carbon monoxide air free - Air-free measurement of CO takes account of the amount of excess air by incorporating an adjustment to the as-measured ppm value, thus simulating air-free (oxygen-free) conditions in the combustion gases. To do this, a reading of oxygen (O_2) or carbon dioxide (CO_2) percentage is taken from the combustion gases along with the as-measured CO reading. This can be done with a meter having the capability of measuring CO and O_2 or CO_2 percentage, or it can be done with two different meters, one measuring CO ppm and one measuring O_2 or CO_2 percentage. If air-free CO is determined with a single meter, the meter will have an integral electronic chip that will calculate the air-free level from as-measured CO ppm and O_2 percentage. If two meters are used, the equations below can be used to determine the air-free level of CO in a combustion gas sample.

For natural gas or propane, using as-measured CO ppm and O_2 percentage:
For propane, using measured CO ppm and CO_2 percentage:
For natural gas, using measured CO ppm and CO_2 percentage:

Where:
COAFppm = Carbon monoxide, air-free ppm
COppm = As measured combustion gas carbon monoxide ppm
O_2 = Percentage of oxygen in combustion gas, as a percentage
CO_2 = Percentage of carbon dioxide in combustion gas, as a percentage

Carbon monoxide indicator - a test instrument used to check flue gas samples of fossil fuel units to determine percent of carbon monoxide present.

Carbon pile - a series of carbon discs held loosely together by an insulated holder, generally found in a battery load tester. The tighter the discs are pressed together, the less resistance to current flow.

Carbon residue - coked material remaining after an oil has been exposed to high temperatures under controlled conditions.

Carbon steel - by common custom, steel is considered to be carbon steel when no minimum content is specified or required for aluminum, boron, chromium, cobalt, columbium, molybdenum, nickel, titanium, tungsten, vanadium, or zirconium, or any other element added to obtain a desired alloying effect; when the specified minimum for copper does not exceed 0.40 percent; or when the maximum content specified for any of the following elements does not exceed the percentages noted: manganese 1.65, silicon 0.60, copper 0.60.

Carbon tetrachloride - a colorless, non-flammable and very toxic liquid used as a solvent. It is very toxic and should never be allowed to touch skin and fumes must not be inhaled.

Carbonization - formation of carbonaceous deposits that may be produced by decomposition of lubricating oil or other organic materials.

Carburize - to heat a low carbon steel to approximately 2000°F in contact with material that adds carbon to the surface of the steel, and to cool slowly in preparation for heat treatment.

Carburizing flame - the flame (also called a reducing flame) that results from supplying excess acetylene.

Cargo, chilled - cargo maintained at an assigned temperature above its freezing point.

Cargo, refrigerated - cargo maintained at an assigned temperature by means of mechanical refrigeration.

Carnot cycle - a sequence of operations forming the reversible working cycle of an ideal heat engine of maximum thermal efficiency. It consists of isothermal expansion, adiabatic expansion, isothermal compression and adiabatic compression to the initial state.

Carrene 7 - a chemical combination of carbon, chlorine and fluorine. A refrigerant in Group 1. Referred to as R-500.

Carry over - (1) the spillage of water, from a boiler, into steam lines. (2) the cross-over of flame from one burner to another.

Carryover ports or slots - ports or slots in a gas burner head which are designed to carry the flame from one burner to the next to improve burner ignition.

Carrying capacity - the ability of a conductor to carry a current, expressed in amperes.

Cartridge fuse - a large ampere capacity fuse consisting of a current responsive element inside a fuse tube with terminals on both ends that fit into special holders.

CAS number - an assigned number used to identify a material. CAS stands for Chemical Abstracts Service, a Columbus, Ohio based organization that indexes information published in chemical abstracts by the American Chemical Society and provides index guides by which information about particular substances may be located in the abstracts when needed. CAS numbers identify specific chemicals and are assigned sequentially. The number is useful because it is a concise, unique means of material identification.

Cascade system - an arrangement in which two or more refrigerating systems that are used in series. It uses the cooling coil of one machine to cool the condenser of another machine. This type of system produces ultra-low temperatures.

Case - in a commercial installation, an individual cooling system. A grocery store will typically have a number of cases operating at different temperatures.

Case hardened - heat treating ferrous metals (iron) so surface layer is harder than the interior or core.

Casement window - a type of window with a side hinged sash that opens like a door. This type of window is the best for catching breezes and providing cross ventilation.

Casing - (1) an enclosure forming the outside of an appliance, no parts of which are likely to be subjected to intense heat. (2) the exposed trim molding, framing, or lining around a door or window. (3) the large diameter steel pipe used in a well hole.

Casing ground - a ground wire attached to the casing of the equipment to prevent the possibility of a shock resulting from a ground fault.

Casing loss - heat loss through the walls of a heater or furnace to the space surrounding the appliance.

Cast brass - cast in sand molds and is used for water valve housings and boat propellers.

Cast copper - copper that has been shaped with the use of a mold.

Cast iron - iron melted and poured into molds that usually has a rough texture. Cast iron has a gray, grainy appearance. It is a ferrous metal and can be brazed or fusion welded with an oxyacetylene torch.

Cast iron sectional boiler - a boiler composed of a number of separate sections fastened together, resulting in a large heat transfer area. It can be fired by an upshot or inshot gas burner, or by a pressure-atomizing oil burner.

Casting - making an object by pouring molten metal into a mold.

Catalyst - a substance that modifies (slows, or more often quickens) a chemical reaction without being consumed in the reaction.

Catalytic combustion stove - a stove that contains a cell-like structure consisting of a substance, washcoat, and catalyst producing a chemical reaction causing pollutants to be burned at much lower temperature.

Catastrophic failure - sudden, unexpected failure of a machine resulting in considerable cost and downtime.

Categories - classes or divisions in a system of classification.

Categorized vent diameter - the minimum vent diameter permissible for Category I appliances to maintain a non-positive vent static pressure when tested in accordance with nationally recognized standards.

Cathode - the negative terminal of an electrical device. The N region of a P-N junction diode. Electrons leave the device at this terminal.

Cathode-bias resistor - resistor connected from the cathode of a tube to ground, across which bias is developed.

Cathode ray tube (CRT) - a large vacuum tube in which electrons emitted from a cathode are formed into a narrow beam and accelerated so that upon striking a phosphorescent screen, a visible pattern of light energy is produced.

Cathodic protection - a method of preventing oxidation (rusting) of exposed metal structures such as bridges by imposing between the structure and the ground a small electrical voltage that opposes the flow of electrons, and is greater than the voltage that is present during oxidation.

Cation - a positively charged ion, that is attracted to the cathode during electrolysis.

Caulk - the act of filling cracks and seams with caulking compound.

Caustic embrittlement - this refers to the cracking of boiler steel occurring while the boiler is in service but not subjected to excessive pressure or temperature. Such failures are attributed to the boiler water's being too caustic, that is, too alkaline.

Cauterizing flame - a flame that results from too much acetylene.

Cavitation - formation of an air or vapor pocket (or bubble) due to lowering of pressure in a liquid, often as a result of a solid body, such as a piston or propeller, moving through the liquid; also, the pitting or wearing away of a solid surface as a result of the collapse of a vapor bubble. Cavitation can occur in a hydraulic system as a result of low fluid levels that draw air into the system, producing tiny bubbles that expand explosively at the pump outlet, causing metal erosion and eventual pump destruction. Cavitation is very undesirable in circulators.

Cavitation erosion - a material-damaging process which occurs as a result of vaporous cavitation. Cavitation refers to the occurrence or formation of gas or vapor filled pockets in flowing liquids due to the hydrodynamic generation of low pressure (below atmospheric pressure). This damage results from the hammering action when cavitation bubbles implode in the flow stream. Ultra-high pressures caused by the collapse of the vapor bubbles produce deformation, material failure and, finally, erosion of the surfaces.

Cavity wall - a wall built of masonry units or of plain concrete, or a combination of these materials, arranged to provide an air space within the wall, and in which the inner and outer parts of the wall are tied together with metal ties.

CC - a marking on wire connectors and soldering lugs indicating they are suitable for use with copper clad aluminum conductors only.

CC/CU - a marking on wire connectors and soldering lugs indicating they are suitable for use with copper or copper clad aluminum conductors only.

Cds cell - acronym for a photo resistance cell made of cadmium sulfide. This solid state device changes its electrical resistance when exposed to light. Used on oil burners as a safety burner shutdown device.

Ceiling damper - a device installed to limit radiant heat transfer through an outlet or air inlet opening in the ceiling of a floor/roof-ceiling assembly having not less than a 1 hour fire resistance rating. Such a device is described in the construction details for such tested floor/roof-ceiling assemblies.

Ceiling diffuser - any air diffuser located in the ceiling used to provide a horizontal distribution pattern of air.

Ceiling effect - the tendency of an air steam moving along a wall or ceiling surface to remain in contact with the surface. Also called wall effect.

Ceiling outlet - a round, square, rectangular, or linear air diffuser located in the ceiling that provides a horizontal distribution pattern of primary and secondary air over the occupied zone and induces low velocity secondary air motion through the occupied zone.

Ceiling plenum - space below the flooring and above the suspended ceiling that accommodates the mechanical and electrical equipment and that is used as part of the air distribution system. The space is kept under a negative pressure.

Cell barrier - a very thin region of static electric charge along the interface of the positive and negative layers in a photovoltaic cell. The barrier inhibits the movement of electrons from one layer to the other, so that higher energy electrons from one side diffuse preferentially through it in one direction, creating a current, and thus a voltage across the cell. Also called the depletion zone or the cell junction.

Cell, electric - a single unit that produces voltage or current by converting chemical or radiant energy into electrical energy. Individual cells wired in series make up a battery. Also see cadmium sulfide and lead sulfide cells.

Cell junction - the area of immediate contact between the two layers (positive and negative) of a photovoltaic cell. The junction lies in the center of the cell barrier or depletion zone.

Cell, water tower - the smallest tower subdivision that can function as an independent unit with regard to air and water flow. It is bounded by either exterior walls or partition walls. Each cell may have one or more fans and one or more distribution systems.

Cellar - that portion of a building that is partly or completely below grade and having at least one half its height below grade.

Cellulose media - a filter material made from plant fibers. Because cellulose is a natural material, its fibers are rough in texture and vary in size and shape. Compared to synthetic media, these characteristics create a higher restriction to the flow of fluids.

Celsius - German language word for Centigrade originated by Andrew Celsius. The metric system temperature scale. A temperature measuring system in which 0° equals the freezing point of water and 100° equals the boiling point.

Center distance - when applied to speed reducers, it is the distance between the centerlines of the input (high speed) shaft and the output (low speed) shaft. Shaft centerlines may be parallel or at right angles to each other. Center distance is often used to designate a single reduction worm gear speed reducer, such as 1.33", 1.75".

Center drill - a special drill to produce bearing holes in the ends of a work place to be mounted between centers. Also called a "combined drill and countersink".

Center lines - center lines on blueprints showing the center of the objects in the building that can be divided into two equal halves. They are not used for the entire building but are used to show the center of columns, equipment and fixtures. This line is light in weight and alternates long and short dashes.

Center of glass - all of the glass area of a window except that within 2.5" from the edge of the glass. Used in measuring and calculating glazing performance such as R-values and U-values.

Center punch - a tool that has a 60° to 90° point and is generally used for starting detailed drilling work.

Center tap - a connection at the electrical center of a winding, or mid-way between the electrical ends of a resistor or other portion of a circuit. Commonly used in the secondary of a transformer with 240 volts on the primary to provide two 120 volt branch circuits from the secondary.

Center to center - the measurement to the center of one fitting to the center of another fitting.

Centigrade - a temperature scale that is divided into 100° increments; where water boils at 100° and water freezes at 0° abbreviation is "C". Also see Celsius scale.

Centimeter - metric unit of linear measurement that equals 0.3937 inches. One hundredth part of a meter.

Centipoise - the metric measurement of the viscosity of a liquid. 1 centipoise = 0.01 poise.

Centistokes - a unit of kinematic viscosity. 1 centistoke = 0.01 stoke.

Central control - a system whereby all data analysis and control actions are performed at a single point.

Central cooling - same as central heating except that cooling (heat removal) is supplied instead of heating. This could be accomplished with a chilled water distribution system supply and return for air conditioning.

Central dearating device - a device that separates air bubbles from the moving system fluid and ejects them from the system.

Central fan system - a mechanical indirect system of heating, ventilating, or air conditioning consisting of a central plant where air is heated and/or conditioned and then by fans or blowers circulated through a system of distributing ducts.

Central furnace - a self-contained, indirect fired furnace designed to supply heated air through ducts to spaces remote from or adjacent to the furnace location.

Central heater - a stationary, indirect fired, vented appliance in one of three classes: boilers, central furnaces, floor furnaces, recessed furnaces, and floor mounted unit heaters connected to a duct system.

Central heating - supply of thermal energy from a central plant to multiple points of end use, usually by steam or hot water or for space and/or service hot water heating. Central heating includes large scale plants serving university campuses, medical centers, or military installations and central building systems serving multiple zones. Also called district heating plants.

Central power - the generation of electricity in large power plants with distribution through a network of transmission lines (grid) for sale to a number of users.

Central station - central location of condensing unit with either wet or air cooled condenser. Evaporator located as needed and connected to the central condensing unit.

Central system - an HVAC system which conditions air at a central plant and distributes it to zones using one or more fans and a duct system.

Centralized computerized controller - energy control device, centrally located, that makes control decisions based on operating data, programmed information, and stored data. Can be used to optimize energy consumption of many devices throughout a building.

Centrifugal - moving or tending to move away from the center.

Centrifugal atomizing oil burner - a burner in which oil is thrown by centrifugal force from a rotating cup into an air stream. Also see rotary oil burner.

Centrifugal compressor - compressor that compresses gaseous refrigerants by centrifugal force. It is not positive displacement but it is similar to a blower. A compressor used for large refrigeration systems.

Centrifugal blowers - air moving devices in which the air flow is perpendicular to the shaft on which the wheel is mounted. The wheel is mounted in a scroll-type housing that is necessary to develop the rated pressures. The four classes of centrifugal blowers are determined by wheel blade position with respect to the direction of rotation.

Centrifugal fan - a fan rotor or wheel within a scroll type housing and including driving mechanism supports for either belt or direct drive.

Centrifugal force - the force that an object moving along a circular path exerts on the body constraining the object and that acts outwardly away from the center of rotation.

Centrifugal pump - a fan-shaped impeller rotating in a circular housing, pushing liquid towards a discharge opening. Usually used where large flow of liquid at relatively low pressure (head or lift) is desired.

Centrifugal switch - a switch that is wired in series with the start capacitor and/or start winding. When the motor reaches approximately 75% of its total RPM, the device opens and disconnects either the start capacitor or start winding from the circuit. The switch then closes when the motor slows to approximately 50% of total RPM.

Centrifugal type distributor - a refrigerant distributor located at the inlet of the evaporator. A swirling action is created within the device to evenly distribute the refrigerant to each of the evaporator circuits.

Ceramic - pertaining to a product made from inorganic, nonmetallic materials fused or fired at high temperatures. Used as an insulation in cables when extremely high temperatures are to be encountered.

Ceramic ignitor - electric ignition system used to create ignition of the gas air mixture in the combustion chamber.

Ceramic radiants - clay insulators that act as heat radiators when they themselves are heated.

Certification - process of establishing that a technician has the required knowledge of refrigerant recovery techniques and Environment Protection Agency regulations prohibiting the venting of refrigerants to the atmosphere. EPA certification is required by passing an EPA approved test and is maintained under section 608 of the Clean Air Act of 1990.

Certified refrigerant recycling or recovery equipment - equipment certified by an approved equipment testing organization to meet EPA standards.

CFM - air quantity in cubic feet per minute.

CFR - Code of Federal Regulations. A collection of the regulations established by law.

Chain (traveling) grate stoker - a cross-feed stoker that is used with larger capacity boilers because of the ability to feed coal at a faster rate than other stokers.

Chain tongs - used when installing very large pipe in place of a pipe wrench. A chain takes the place of the upper hook jaw in the common pipe wrench, making it easier to handle.

Chain wrench - a type of pipe wrench used for working on larger diameter pipes. This wrench is adjustable and can work easier in confined areas and on round, square or irregular shaped objects. Also referred to as chain tongs.

Chamfer - a narrow inclined surface along the intersection of two surfaces. A bevel or slope.

Change of air - introduction of new air to a conditioned space, measured by the number of complete enclosed space changes per unit time. Usually per hour.

Change of state - the process that a substance (either a compound or element) goes through when changing from one physical state to another. An example of this would be when water becomes ice or water becomes steam it goes through a change of state.

Change plate - holds the post in place and allows the dies to recede to make the taper on the thread. The change plate must be between the two standard marks when starting a thread.

Changeover - the process of switching an air conditioning system from heating to cooling, or vice versa.

Channel - a conductive path through a semiconductor between the source and the drain.

Characterized guide - a valve guide with notches or V's cut into them to determine the valve flow characteristics. Also known as a skirted guide.

Charge - a load; the maximum or necessary quantity that a container or apparatus is built to accommodate; e.g., fuel in a furnace. In electricity, the electrical energy stored in a capacitor or battery or held on an insulated object. The amount of refrigerant in a system or to put refrigerant into a system.

Charge, basic - amount of refrigerant placed in a refrigerating unit.

Charge, holding - evaporators and replacement compressors are shipped with a holding charge to prevent the entrance of air and moisture. This charge may be either nitrogen or refrigerant.

Charge, operating - the total amount of refrigerant required by a system for correct operation.

Charge, test - an amount of gas forced into a refrigerating system for leak testing purposes.

Charging, electrical - the storing of electrical energy.

Charging, refrigerant - the procedure of adding refrigerant to the system.

Charging board - a specially designed panel or cabinet fitted with gauges, valves and refrigerant cylinders used for charging refrigerant and oil into the refrigeration system.

Charging cylinder - a calibrated container in which a predetermined weight of refrigerant may be charged or dispensed into a refrigeration system. Commonly used on small systems.

Charging hose - a flexible, plastic or rubber hose, capable of withstanding high system pressures and is used to transfer and move refrigerant from one component to another. Charging hoses are designed for a working pressure of 500 psi and an average breaking or bursting pressure of 2,000 psi.

Charging manifold - specifically designed for technicians to charge, evacuate, and recover refrigerant into and out of refrigerating and air conditioning systems.

Charging rate - the amount of amperage flowing from the generator or alternator to charge a battery.

Charging scale - a scale used to weigh refrigerant when charging a refrigeration or air conditioning system.

Charging station - is a piece of equipment is used to charge a unit, an appliance, or an automobile's air conditioning unit. The station includes a vacuum pump, a charging cylinder, manifold gauges, and other useful charging instruments.

Charles' Law - gas law that explains the relationship between pressure and volume. It states: "at a constant pressure, the volume of a confined gas varies directly as the absolute temperature; at a constant volume, the pressure varies directly as the absolute temperature." Also see appendix.

Chase - (1) to cut threads with an external cutting tool. (2) An opening for piping, ductwork, and wiring in a multi-story construction.

Chaser dies - steel tools for cutting or finishing screw threads.

Chassis - a sheet metal box, frame, or single plate on which the components of a device are mounted; the assembled frame and parts.

Chatter - the rapid opening and closing of a relay or solenoid due to a malfunction. Also see relay chatter.

Cheater - pipe extension used to increase leverage of a wrench. Not recommended for safety reasons.

Check model - model fabricated after all the design work has been completed. It is simply a three dimensional representation of the paper design.

Check valve - a flow control device that permits fluid flow only in one direction.

Checking relay - a relay in the self-checking circuit of some flame safeguard controls that makes and breaks periodically ("checking" or "pecking" action) when the flame signal is strong enough. While checking, it transfers energy to hold in the flame relay. It stops "checking" if the flame signal becomes too weak or disappears.

Cheek - the side of a hammer head.

Chelating agent - a substance that can remove heavy metal toxins such as lead, mercury, or arsenic from the body by complexing them and allowing their excretion in urine.

Chemical cartridge respirator - a respirator using various chemical substances to purify various chemical contaminative gases and vapors. Approved for concentrations no more than 10 times the threshold limit value (TLV) for a half face piece and 100 times the TLV for a full-face piece, provided the contaminant has warning properties such as odor or irritation near the TLV.

Chemical cell - a source of DC electricity that converts chemical energy into electrical energy.

Chemical compound - a chemical compound is formed when two or more chemical elements combine into a new substance.

Chemical concentration - the amount of a specific chemical in the boiler water.

Chemical energy - the energy in the fuel that converts to heat energy during the combustion process.

Chemical energy content - the theoretical maximum energy content of a fuel based on its chemical composition.

Chemical instability - the tendency of substances to break down into other, possibly corrosive substances.

Chemical resistance, piping - (1) the effect of specific chemicals on the properties of plastic piping with respect to concentration, temperature and time of exposure. (2) The ability of a specific plastic pipe to render service for a useful period in the transport of a specific chemical at a specified concentration and temperature.

Chemical sensitivity - evidence that suggest that some people can develop health problems characterized by effects such as dizziness, eye and throat irritation, chest tightness, and nasal congestion that appear whenever they are exposed to certain chemicals. People may react to even trace amounts of chemicals to which have been become "sensitized".

Chill - to harden the outer surface of cast iron by quick cooling, as in a metal mold.

Chill factor - calculated number based on temperature and wind velocity.

Chilled water system - (1) a closed loop circulating water system normally operates with a design supply water temperature of 40 to 55° F, usually 44 to 45° F, and at a pressure of up to 120 psi. Anti-freeze or brine solutions may be used for applications that require temperatures below 40° F or for coil freeze protection. (2) System in which a mechanical refrigerating system at a central location cools water that is then piped to various parts of the building.

Chiller - an air conditioning system that circulates chilled water to various cooling coils in an installation.

Chiller barrel - the evaporator component of a chiller, consisting of a multiple arrangement of pipes submerged in a liquid refrigerant. Heat is removed from the water as it flows through the chiller barrel.

Chiller, dry expansion - a device designed to utilize the evaporation of refrigerant within a tube bundle, the refrigerant being metered to the inside of the tubes by means of one or more thermostatic expansion valves. The liquid to be chilled passes through the chiller between the shell and tubes.

Chiller, flooded - a device designed to utilize the evaporation of refrigerant on the outside surface of a tube bundle; the refrigerant level being maintained by a float valve. The liquid to be chilled passes through the tubes of the chiller.

Chiller purge unit - a system that removes air and non-condensables from a low pressure chiller.

Chiller sequencing - an energy conserving control strategy which determines the most efficient combination of chillers required to operate to meet the cooling load of the space conditions.

Chilling - ooling) the lowering of the temperature of a substance by the removal of heat in the temperature range above freezing.

Chilling room - room where products are cooled prior to cold storage.

Chimney - a primarily vertical shaft or structure enclosing one or more flues for conveying flue gases to the outside atmosphere. Can be one of the following types:

1. Factory-built chimney - a chimney composed of listed and labeled factory built components assembled in the field in accordance with the terms of listing to form the completed chimney.
2. Masonry chimney - a field constructed chimney of solid masonry units, bricks, stones, listed masonry chimney units, or reinforced Portland cement concrete, lined with suitable chimney flue liners.
3. Metal chimney - a field constructed chimney of metal.

Chimney cap - a protective covering or housing for the top of a chimney intended to prevent the entry of rain, snow, animals, and birds, and to prevent downdrafts.

Chimney connector - a pipe that connects a fuel-burning appliance to the chimney. Also called a smoke pipe.

Chimney effect - the tendency of air or gas in a duct or other vertical passage to rise when heated due to its lower density compared with that of the surrounding air or gas. In buildings, the tendency toward displacement, caused by the difference in temperature, of internal heated air by unheated outside air due to the difference in density of outside and inside air.

Chimney flue - flue gas passageway in a chimney.

Chip - to cut away metal with a cold chisel.

Chips - excess pieces broken off, fragments.

Chisel - a hand tool with a cutting edge at one end and is used to cut when struck on the blunt end with a hammer.

Chlorinated polyvinyl chloride (CPVC) - type of plastic pipe and fittings made from chlorinated polyvinyl chloride with colorants, filters, plasticizers, stabilizers, lubricants added that can be used with temperatures up to 180° F at 100 psig.

Chlorine - a greenish-yellow gas with a pungent, suffocating odor. It is highly toxic and corrosive gas that is irritating to the eyes and mucous membranes. Although noncombustible, it is a strong oxidizer that supports combustion of other organic materials. It is an extremely reactive and explosive gas. Chlorine reacts with many common substances such as acetylene, ammonia, hydrogen, ether, fuel gas, hydrocarbons, turpentine, and finely divided metals. Has a molecular weight of 70.906, specific volume of 5.4 cubic feet/lb., and a boiling temperature of -29.3° F.

Chlorine monoxide - a molecule made of one oxygen atom and one chlorine atom. It's found in the stratosphere when ozone depletion is taking place. By measuring chlorine monoxide, scientists can determine the degree of ozone depletion.

Chlorofluorocarbons (CFC) - any of several compounds made up of carbon, chlorine, and fluorine. CFC's were used as aerosol propellants and refrigerants until they were found to be harmful to the earth's protective ozone layer.

Choke - an inductor designed to present an impedance to AC current, or to be used as the current filter of a DC power supply.

Choke coil - an inductor coil used to block the flow of AC current and pass DC current.

Choke tube - a throttling device used to maintain correct pressure difference between the high and low side of the refrigeration system. Capillary tubes are sometimes called choke tubes.

Chopper - a mechanical or electronic device that switches on and off at a regular rate to interrupt a DC voltage or current and thus convert it to AC.

Chuck - a mechanism for holding a rotating tool or work piece.

Circuit - a tubing, piping or electrical wire installation that permits flow from the energy source back to the energy source.

Circuit (electrical) - an electron path that completes a loop. Circuits generally consist of a power source, conductors, a load, and a switch to control the current flow. An electric current flows through the circuit when the path is complete and when a potential difference is applied.

Circuit feeder - wiring used to connect the service panel and subdistribution box.

Circuit breaker - a safety device in the service panel that protects the wiring, and is designed to open the circuit automatically on a predetermined overload of current and close a circuit by manual means without injury to itself when properly applied within its rating. There are thermal and magnetic types.

Circuit - general purpose - conductors (or cable) that supply electricity to lights and outlets.

Circuit ground - the portion of a circuit used as a common potential point.

Circuit, individual - conductors (or cable) that supply electricity to just one place, such as a furnace or air conditioner.

Circuit main - the portion of the main in a multiple circuit system that carries only a part of the total capacity of the system.

Circuit, parallel - an arrangement of electrical devices in which all positive terminals are joined to one conductor and all negative terminals to other conductor. Is considered to be a current divider circuit where the voltage is the same across all load circuits.

Circuit, pilot - secondary circuit used to control a main circuit or a device in the main circuit.

Circuit protection devices - fuses and circuit breakers.

Circuit, refrigerating - the assembly of refrigerant containing parts and their connections used in a refrigerating cycle.

Circuit, series - electrical path in which electricity must pass through all loads and then back to the energy source. Is considered to be a voltage divider circuit where the current is the same through all parts of the circuit.

Circuit switching - an on-off control function. Conduction occurs during the on time with no conduction during the off time.

Circular mil - measuring units for the area of an electrical conductor, obtained by squaring the diameter of a wire in mils. One circular mil equals 0.001 inch in diameter. Used to express the wire cross sectional area.

Circular platform - steel grating or solid plate on a vessel or tower that provides an access area at different elevations to manways, valves, instruments, piping, and other areas.

Circulating loop - the main loop in which oil is circulated from the storage tanks to the branch circuits and then back to the storage tanks.

Circulating lubrication - a system of lubrication in which the lubricant, after having passed through a bearing or group of bearings, is recirculated by means of a pump.

Circulating water rate - quantity of hot water entering the cooling tower.

Circulation - the free flow of fluids.

Circulator - a motor driven device used to mechanically circulate water in a system. Also called a pump.

Circumference - the perimeter of a circle.

Circumference of a circle - the measurement around the perimeter of a circle. To find the circumference, multiply Pi (3.1416) by the diameter or 2π x radius. Commonly written as πd.

Circumscribe - to draw a line around.

Cistern - a holding tank or vessel for holding rainwater.

Clamshells - sections of the heat exchanger. Burners fit into special openings into the bottom of the clamshells.

Clamp circuit - also called clamping circuit. Any circuit designed to hold (clamp) voltage or current to a specific level.

Clamp-on ammeter - a current measuring device with jaws that are placed around a single current carrying conductor. The clamp-on ammeter measures current by sensing the strength of the magnetic field around the conductor.

Class I atmospheres –

Group A; acetylene.

Group B; butadiene, ethylene oxide, hydrogen, propylene oxide, manufactured gases containing more than 30% hydrogen by volume.

Group C; acetaldehyde, cyclopropane, diethyl ether, ethylene.

Group D; acetone, acrylonitrile, ammonia, benzene, butane, ethanol, ethylene, dichloride, gasoline, hexane, isoprene, methane (natural gas), methanol, naphtha, propane, propylene, styrene, toluene, vinyl acetate, vinyl chloride, and xylene.

Class I refrigerants - those fully halogenated chlorofluorocarbon liquid refrigerants that have a boiling point greater than 68° F and are normally packaged in drums. No flame propagation.

Class 1 stoker - these are primarily for residential heating and designed for quiet, automatic operation. These stokers are usually underfeed types and are typically screw-feed. Class 1 stokers feed coal to the furnace intermittently, in accordance with temperature or pressure demands. A special control is needed to ensure stoker operation in order to maintain a fire during periods when no heat is required.

Class II atmospheres - Group E; aluminum, magnesium and other metal dusts with similar characteristics.

Group F; carbon black, coke or coal dust.

Group G; flour, starch or grain dust.

Class II refrigerants - those refrigerants and halons requiring a minimum cylinder service pressure not exceeding 260 psig. Moderately flammable.

Class 2 and 3 stokers - these are usually of the screw-feed type, without auxiliary plungers or other means of distributing the coal. They are used extensively for heating plants in apartment buildings, hotels, and industrial plants. They are the underfeed type and are available in both the hopper and bin-feed type. These units are also built in a plunger-feed type with an electric motor, steam, or hydraulic cylinder coal-feed drive. Class 2 and 3 stokers are available for burning all types of anthracite, bituminous, and lignite coals. The tuyere and retort design varies according to the fuel and load conditions. Stationary grates are used on bituminous models, and the clinkers formed from the ash accumulate on the grates surrounding the retort. Class 2 and 3 anthracite stokers are equipped with moving grates that discharge the ash into a pit below the grate. This ash pit may be located on one side or both sides of the grate and, in some installations, is of sufficient capacity to hold the ash for several weeks of operation.

Class III atmospheres - easily ignitable fibers, such as rayon, cotton, sisal, hemp, cocoa fiber, oakum, excelsior and other materials of similar nature.

Class III refrigerants - those high pressure refrigerants requiring a minimum cylinder service pressure exceeding 250 psig. Highly flammable.

Class IV refrigerants - include refrigerants that are flammable or mixtures that could become flammable in the event of a leak. These refrigerant cylinders have a red band or red top.

Class 4 stokers - these stokers vary in details of design, and several methods of feeding coal are practiced. Underfeed stokers are widely used, although overfeed types are used in the larger sizes. Bin-feed and hopper models are available in underfeed or overfeed types.

Class 5 stokers - these are made with spreader, underfeed, chain grate, traveling grate, or vibrating grate types. Various subgroups reflect the type of grate and method of ash discharge.

Class 10 or 20 - refers to time needed for a bimetallic overload relay to trip (disconnect) the motor from the load. Class 10 takes 10 seconds at 6 times the rated trip current. Class 20 takes 20 seconds at 6 times the rated trip current.

Class A motor insulation system - a Class A insulation system is a system utilizing materials having a preferred temperature index of 105 and operating at such temperature rises above stated ambient temperature as the equipment standard specifies based on experience or accepted test data. This system may alternatively contain materials of any class, provided that experience or a recognized system test procedure for the equipment has demonstrated equivalent life expectancy. The preferred temperature classification for a Class A insulation system is 105°C.

Class B motor insulation system - a Class B insulation system is a system utilizing materials having a preferred temperature index of 130 and operating at such temperature rises above stated ambient temperature as the equipment standard specifies based on experience or accepted test data. This system may alternatively contain materials of any class, provided that experience or a recognized system test procedure for the equipment has demonstrated equivalent life expectancy. The preferred temperature classification for a Class insulation system is 130°C.

Class CC fuses - 660 volt, 200,000 ampere interrupting rating, branch circuit fuses with overall dimensions of 13/32" x 1.5". Their design incorporates rejection feature that allows them to be inserted into rejection fuse holders and fuse blocks that reject all lower voltage, lower interrupting rating 13/32" x 1.5" fuses. They are available from 1/10 amp through 30 amps.

Class F motor insulation system - a Class F insulation system is a system utilizing materials having a preferred temperature index of 155 and operating at such temperature rises above stated ambient temperature as the equipment standard specifies based on experience or accepted test data. This system may alternatively contain materials of any class, provided that experience or a recognized system test procedure for the equipment has demonstrated equivalent life expectancy. The preferred temperature classification for a Class insulation system is 155° C.

Class G fuses - 300 volt, 100,000 ampere interrupting rating branch circuit fuses that are size rejecting to eliminate over fusing. The fuse diameter is 13/32" while the length varies form 1 5/16" to 2 1/4". These are available in ratings from 1 amp through 60 amps.

Class H fuses - 250 volt and 600 volt, 10,000 ampere interrupting rating branch circuit fuses that may be renewable or non-renewable. These are available in ampere ratings of 1 amp through 600 amps.

Class H motor insulation system - a Class H insulation system is a system utilizing materials having a preferred temperature index of 180 and operating at such temperature rises above stated ambient temperature as the equipment standard specifies based on experience or accepted test data. This system may alternatively contain materials of any class, provided that experience or a recognized system test procedure for the equipment has demonstrated equivalent life expectancy. The preferred temperature classification for a Class insulation system is 180° C.

Class J fuses - these fuses are rated to interrupt 200,000 amperes AC. They are labeled as "current-limiting", are rated for 600 volts AC, and are not interchangeable with other classes.

Class K fuses - these are fuses listed as K-1, K-5, or K-9 fuses. Each subclass has designated I2T and IP maximums. These are dimensionally the same as Class H fuses, and they can have interrupting ratings of 50,000, 100,000, or 200,000 amps. These fuses are current-limiting, however, they are not marked "current limiting" on their label since they do not have a rejection feature.

Class L fuses - these fuses are rated for 601 through 6000 amperes, and are rated to interrupt 200,000 amperes AC. They are labeled "current limiting" and are rated for 600 volts AC. They are intended to be bolted into their mountings and are not normally used in clips. Some Class L fuses have designed in time-delay features for all purpose use.

Class R fuses - these are high performance fuses rated 1/10-600 amps in 250 volt and 600 volt ratings. All are marked "current limiting" on their label and all have a 200,000 amp interrupting rating. They have identical outline dimensions with the Class H fuses but have a rejection feature that prevents the user from mounting a fuse of lesser capabilities (lower interrupting capacity) when used with special Class R fuses will fit into either rejection or non-rejection clips.

Class T fuses - a UL classification of fuses in 300 volt and 600 volt ratings from 1 amp through 1200 amps. They are physically very small and can be applied where space is at a premium. They are fast acting fuses, with an interrupting rating of 200,000 amps RMS.

Class of fire - the five classes of fires are Class A, started from wood, paper, or other combustible materials containing carbon; Class B, started from oil, grease, or flammable liquids; Class C, started from electrically energized devices; Class D, started from combustible metals; and Class K, started from grease from commercial cooking equipment.

Class of refrigerating systems - for ASHRAE Standard 34 --- number designation and safety classification of refrigerants provides an alpha-numeric classification system whereby all refrigerants are placed in one of six categories based on flammability and toxicity. An "A" designation indicates low toxicity; "B" higher. Numbers 1, 2 and 3 indicate increasing flammability. Thus A1 is no flame propagation, low toxicity refrigerant; B3 indicates higher flammable, higher toxicity refrigerant. This system is used in the ASHRAE Standard 15 --- Safety Code for Mechanical Refrigeration.

Classes of fuses - the industry has developed basic physical specifications and electrical performance requirements for fuses with voltage ratings of 600 volts or less. These are known as standards. If a type of fuse meets the requirements of a standard, it can fall into that class. Typical classes are K, RK1, RK5, G, L, H, T, CC, and J.

Clean Air Act - Public Law PL 91-604, 40 CFR 50-80. EPA has jurisdiction. Effective December 31, 1970, and significantly amended several times, most recently in 1990. The regulatory vehicle that sets limitations and monitors airborne pollution hazardous to public health or natural resources. The EPA sets national ambient air quality standards. Enforcement and issuance of discharge permits are carried out by the states and are called state implementation plans. The Clean Air Act is directed toward by-products discharged into the air from stationary sources (i.e., factories) and mobile sources (i.e., automobiles) rather than use and assessment of specific chemicals.

Clean out (**chimney**) - fitting located at the base of a masonry chimney that allows for cleaning out those items that would otherwise clog up the passageway for flue gasses to escape. It is located at a point lower than where the vent connector enters the chimney.

Clean out (**plumbing**) - fitting that allows its cover to come off so that the pipes may be cleaned if they are clogged.

Clean room - a facility or enclosure in which air content and other conditions (such as temperature, humidity, and pressure) are controlled and maintained at a specific level by special facilities and operating processes and by trained personnel. A specially designed and conditioned space where air cleanliness is critical to the room's function.

Cleanliness level - a measure of relative freedom from contaminants.

Clearance - the distance by which one part clears another part.

Clearance bearing - a journal bearing in which the radius of the bearing surface is greater than the radius of the journal surface.

Clearance, **compressor** - space in compressor cylinder not occupied by piston at the end of the compression stroke, or volume of gas remaining in cylinder at same point. Measured in percentage of piston displacement.

Clearance, **furnace** - the minimum distance through air measured between the heat producing surface of the mechanical appliance, device or equipment and the surface of the combustible material or assembly.

Clearance pocket - a small space in the compressor cylinder from which compressed gas is not completely expelled. This space is called the compressor clearance space or pocket. For effective operation, compressors are designed to have as small as clearance space as possible.

Clearance space - the space in the cylinder area not occupied by the piston at the end of the compression stroke, or volume of gas remaining in cylinder at the same point. Measured in percentage of piston displacement.

Clearance vapor - in reciprocating compressors, the refrigerant trapped between the valve plate and the piston, the clearance volume, when the piston is at top dead center.

Clearance volume - the volume at the top of the stroke in a compressor cylinder between the top of the piston and the valve plate.

Clearing time - the total time between the beginning of the over current and the final opening of the circuit at rated voltage by an over current protective device. Clearing time is the total of the melting time and the arcing time.

Cleat - a ladder crosspiece of rectangular cross section placed on edge upon which a person may step while ascending or descending a ladder.

Cleats - used to fasten sheet metal to other materials. Usually formed from 2 inch wide strips that are folded 180 degrees at one end for fastening under a 180 degree fold along the edge of the sheet metal being anchored. The cleat is usually nailed to other materials and the remaining metal is bent back over the nail heads.

Climate - the sum total of weather that takes place over a period of years.

Climate control - a space in which an ideal climate is maintained by some devices.

Climatology - a 30 year average of daily and hourly meteorological observations. The average of this 30 year mean, typically for a three month calendar period, is the base for the seasonal outlooks. Changes from the seasonal norm are discussed in the long range forecasts.

Clo value - a numerical representation of a clothing ensemble's thermal resistance. 1 Clo = 0.155m2 K/W (0.88 ft2hF/Btu.

Clock timer - a time delay device that uses an electric clock to measure the time delay.

Clock - a pulse generator that controls the timing of clocked logic devices and regulates the speed at which such devices operate. It serves to synchronize all operations in a digital system.

Clock input - that terminal on a flip-flop whose condition or change of condition controls the admission of data into a flip-flop through the synchronous inputs, and thereby controls the output state of the flip-flop. The clock signal performs two functions: (1) it permits data signals to enter the flip-flop, and (2) after entry, it directs the flip-flop to change state accordingly.

Clock thermostat - a thermostat with a built-in clock that automatically re-adjusts the control point at a predetermined time at night and restores the daytime setting at another predetermined time in the morning.

Clockwise - from left to right in a circular motion. The direction that the hands on a clock move.

Close coupled - bolting two or more inline circulators together end-to-end, in the same flow direction, to increase head.

Close nipple - a short section of pipe threaded along its entire length. Also known as an all-thread nipple.

Close off panels - sheet metal panels fastened between casing sides and coils, filter racks, damper assemblies, etc., to prevent air leakage around the equipment.

Close off pressure - the maximum allowable pressure difference to which a valve may be subjected while fully closed.

Close-off rating - the maximum allowable pressure drop to which a valve may be subjected while fully closed. It is a function of the power available from the valve actuator for holding the valve closed, and is independent of the valve body rating.

Closed circuit, electrical - an electrical circuit that forms a complete path so that electrical current can flow through it.

Closed circuit, oil system - a system in which oil may be pumped completely through the circulating loop and back into the oil storage tank.

Closed container - container sealed by means of a lid or other device so that neither liquid nor vapor will escape from it at ordinary temperatures.

Closed cycle - any cycle in which the refrigerant is used over and over again.

Closed delta - a four wire connection that uses three transformers to provide 120 volt single phase, 240 volt three phase, and 208 volt (wild leg) power.

Closed impeller - see impeller.

Closed loop - (1) an automatic control system that reflects the results of the controller, back to the controller, so that it can again take corrective action if the set point is not being maintained. (2) a piping system that is sealed at all points from the atmosphere.

Closed loop system - geothermal heat pump system in which the fluid that facilitates the heat exchange between the refrigerant and the earth is contained within a sealed piping arrangement.

Closed loop heat pump - a heat pump system that reuses the same heat transfer fluid, which is inside plastic pipe that is buried within the earth, or within a lake or pond used for the heat source.

Closed piping system - piping system which has no openings to the atmosphere. A boiler or refrigeration system would be considered closed loop or closed piped.

Closed system - a refrigeration system in which the refrigerant supply does not deplete. As the refrigerant flows through the system, it undergoes a series of changes and, when the cycle is complete, the refrigerant has been returned to its original state.

Closed vapor compression cycle - this is the most common type of refrigeration system used for both HVAC and industrial processes. This type of refrigeration cycle can either be a single stage, compound, multistage, or cascade.

Closely spaced tees - two tees that are installed immediately adjacent to each other to form a primary/ secondary connection between two piping circuits.

Clothes dryer - a device used to dry wet laundry by means of heat. Dryer classifications are as follows:

Type I - factory built package, multiple production. Primarily used in family living environments. Usually the smallest unit physically and in function output.

Type II - factory built package, multiple productions. Used in businesses intended for public use and access. Not designed for use in individual family living environments.

Cloud point - the temperature at which waxy crystals in an oil or fuel form a cloudy appearance. This temperature is usually 20 degrees higher than the pour point.

Clutch, armature - the part of the automotive compressor magnetic clutch that is attracted by the magnetic field and causes the compressor to be turned on by the belt drive when the magnetic field is energized.

Clutch, field - the coils of wire through which the current flows to create the magnetic force that engages the magnetic clutch.

Clutch, magnetic - clutch built into an automobile compressor flywheel, operated magnetically, that allows a pulley to revolve without driving compressor when the refrigerating effect is not required.

Clutch head - a screwdriver used with screws for sheet metal and trim where "looks" are needed. The tip of the screwdriver is very strong and stays in the screw opening with only a small amount of effort.

Clutch rotor - that part of the clutch that is driven by the belt whether the clutch is engaged or not.

CMOS logic - logic circuitry that uses complementary metal oxide semiconductors.

CO_2 indicator - an instrument used to indicate the percentage of carbon dioxide in flue gases.

CO/ALR - terminal markings on switches and receptacles rated 15 and 20 amperes that are suitable for use with aluminum, copper, and copper clad aluminum conductors. If not marked, they are suitable for copper or copper clad conductors only.

Coal feeder - controls the flow of coal entering the pulverizer.

Coal gas - fuel gas produced by the distillation of coal.

Coal gate - used to control the depth of coal entering the boiler furnace on chain grate stokers.

Coal ram - distributes coal evenly into the center retort on underfeed stokers and forces the coal up to the top where it is burned.

Coal scale - measures and records the amount of coal fed to stoker fired or pulverized coal fired boilers.

Coaxial - a cable configuration having two cylindrical conductors with coincidental axes, such as, conductor with a tubular shield surrounding the conductor and insulated from it.

Cock - a manually operated faucet or valve for shutting off or throttling down the flow of a liquid or gas, usually by means of a tapered plug.

Code - the local, state, or national rules that govern safe installation and service of systems and equipment for the purpose of safety of the public and trade personnel. Codes relate to minimum requirements for the installation work. They tell an installer which products are approved for certain applications within the codes body jurisdiction, as well as how these products must be installed.

Code conversion - the changing of the bit grouping for a character in one code into the corresponding bit grouping in another group.

Code installation - a refrigeration or air conditioning installation which conforms to the local or national code for safe and efficient installations.

Code letter - provided on alternating motors nameplates that indicates the type of bars used in the rotor.

Code of federal regulations (CFR) - regulations that are generated and published by a branch of the U.S. government. They must be adhered to for any undertaking regulated by the Federal Government.

Coefficient of conductivity - the measure of the relative rate at which different materials conduct heat. Copper is a good conductor of heat and therefore, has a high coefficient of conductivity.

Coefficient of coupling (K) - a decimal value that indicates the amount of magnetic coupling between coils.

Coefficient of discharge - for an air diffuser, the ratio of net area or effective area at vena contracta of an orificed air stream to the free area of the opening.

Coefficient of expansion - increase in unit length, area, or volume for one degree rise in temperature. The coefficient of expansion is not constant and varies with changes in temperature. For linear expansion it is expressed as the change in length of one unit of length of a substance having one degree rise in temperature.

Coefficient of friction - the number obtained by dividing the friction force resisting motion between two bodies by the normal force pressing the bodies together.

Coefficient of heat transmission, overall - the amount of heat (Btu) transmitted from air to air in 1 hr/ft2 of the wall, floor, roof, or ceiling for a difference in temperature of 1° between the air on the inside and outside of the wall, floor, roof, or ceiling. Symbol is "U".

Coefficient of performance (COP) - a term used to measure the efficiency of a heat pump system. It is defined as the heat output of a heat pump or electric elements, divided by the heating value of power consumed in watts at standard test conditions of 17° and 47° F. COP's are established through laboratory testing, using standard procedures developed by the Air Conditioning and Refrigeration Institute. Application of COP's are limited to a heat pumps steady state performance in the heating mode only. It is calculated by dividing the systems total heating capacity (including circulating fan heat but excluding supplementary resistance heat) by the total electrical input (watts) x 3.413.

$$\text{COP} = \frac{\text{Rate of useful energy output in Btu\textbackslash h}}{\text{Rate of energy input in watt x 3.414 Btu/watt}}$$

Cogeneration - primary source of energy that is also used to produce a secondary source of energy. An example would be the use of waste heat from an electrical energy generation system to heat a building. To qualify as a cogeneration facility in the United States, a plant must meet certain energy efficiency standards.

Coil - any heating or cooling element made of pipe or tubing connected in series. Wire arranged in a helically or spirally shape that creates a magnetic field when current passes through it.

Coil deck - insulated horizontal partition between refrigerated space and bunker.

Coil, direct expansion - evaporator coil that allows or permits liquid refrigerant to evaporate inside this coil.

Coil, expansion - an evaporator constructed of pipe or tubing.

Coil, holding - that part of a magnetic starter or relay that causes the device to operate when energized.

Coil process line - the line imposed on a psychrometric chart defining the efficiency of a cooling coil, which is the sensible, latent, and total heat removal capacity of the unit.

Coil, refrigerant - an evaporator or condenser made up of tubing either with or without extended surfaces such as fins.

Coil, solenoid - an electrical winding having an open core. Used for operating relays, contactors, valves, and motor starters.

Cold - the relative absence of heat.

Cold anticipator - a component part of the space thermostat that causes the system to begin operation in the cooling mode shortly before the thermostat's contacts actually call for cooling. This helps to reduce the variation of temperature in the space. The cold anticipator is a fixed resistor that is wired in parallel with the cooling contacts on the thermostat. Also known as a cooling anticipator.

Cold bending - bending process for pipes with diameters (up to 42 in.) and walls (up to 3/4 in.). Wrinkling, excessive thinning, and ovality can be avoided with proper equipment and fixtures.

Cold contacts - contacts in a stack relay that are in the closed position when the flue pipe is cold. They would then open up when the flue pipe is heated to an appropriate temperature.

Cold control - slang for the line voltage thermostat used to regulate the interior temperature of a refrigerator.

Cold deck - the ductwork supplying cooling air in an HVAC dual duct system with mixing boxes at the discharge locations.

Cold feed - to allow water to enter into a closed system, as with a boiler.

Cold joint - a soldered joint made with insufficient heat.

Cold junction - that part of a thermoelectric element that is attached to a load, through which electricity generated by the thermocouple is conducted to the load. The electricity generated by a thermocouple increases as the temperature difference between the hot and cold junctions increases.

Cold load - spring reading (a function of amount of compression) when pipe suspended from a spring hanger is at room temperature. Cold load is higher than hot load if the thermal movement of the pipe is up and vice versa.

Cold plate evaporator - with this evaporator, the tubing is attached to a metal plate. Then the metal plate wraps around the freezer compartment. This type is often used in systems that do not have forced air evaporators.

Cold rolled steel - open hearth or Bessemer steel containing .12% to .20% carbon that has been rolled while cold to produce a smooth, quite accurate stock.
Cold storage - a trade or process of preserving perishables on a large scale by refrigeration.

Cold trap - a device to help trap moisture during the evacuation of a refrigeration system.

Cold wall - refrigerator construction that has the inner lining of refrigerator serving as the cooling surface.

Cold water temperature, water tower - temperature of the water leaving the collection basin, exclusive of any temperature effects incurred by the addition of make-up and/or the removal of blowdown.

Collapse - occurs when a magnetic field suddenly changes from its maximum value to a zero value.

Collar - a round flange or ring fitted on a shaft to prevent sliding.

Collar beam - horizontal tie beam connecting rafters considerably above the plate.

Collection basin - vessel below and integral with the tower where water is transiently collected and directed to the sump or pump suction line.

Collector (electrical) - either the high potential electrode of an n-p-n transistor or the low potential electrode of a p-n-p transistor.

Collector (solar) - a panel like item used to collect and absorb solar energy.

Collector base junction - the P-N junction formed by the collector and base regions.

Collector efficiency - the ratio of the weight of dust collected to the total weight of dust entering the dust collector.

Collector ring - metal rings mounted on a rotating part of an electric machine (such as a generator) so that stationary brushes, riding on the rings can conduct electrical current in or out of the rotating part.

Collar beam - horizontal tie beam connecting rafters above the plate.

Color code - a color system for circuit identification by use of solid colors, tracers, braids, surface printing, etc. Or a system of colored bands on electrical components used to indicate the components electrical value.

Color, fuel oil - new, fresh #2 fuel oil is normally champagne-straw colored but is dyed red to differentiate it from taxable over the road diesel fuel. Fuel oil tends to discolor (darkens) when exposed to light, so color is not a good indicator for high quality fuel.

Color metric detection device - a device for detecting the presence of a particular substance, such as carbon monoxide, in which the presence of that substance will cause a color change in a material in the detector.

Column - vertical load carrying member of structural frames.

Combination burner - a burner by which more than one fuel can be burned, either separately or simultaneously. Also called a multi-fuel burner. A dual fuel burner uses one prime fuel, but can switch to a standby fuel under peak load conditions. A gas oil burner is capable of burning either gas or oil some types can burn both at the same time.

Combination circuit - a circuit that has one portion connected in series with the voltage source and another part connected in parallel. Sometimes referred to as a complex circuit.

Combination duct system - a duct system constructed with two or more materials. One example could be ductboard and galvanized sheet metal. Another could be

Combination fan and limit control - a temperature operated control that regulates the operation of the blower assembly and provides safety protection against overheating of the furnace heat exchanger.

Combination gas valve - a main line gas control that includes the main valve, pilot valve, and pressure regulator in a body assembly.

Combination heating cooling thermostat - a temperature activated control that is installed when adding an air conditioning system to an forced air furnace system.

Combination pliers - designed to permit jaws to be opened for gripping larger diameters and also to provide cutting for wires and cotter pins by use of a side-cutting jaw.

Combination snips - these sheet metal snips are used for straight cutting of metal, ranging from 30 to 26 gauge. They can cut medium sized outside curves or circles. They are the most common sheet metal snips used in the HVAC industry.

Combined collector - a photovoltaic device or module that provides useful heat energy in addition to electricity.

Combustible - a term used by the NFPA, DOT, and others to classify certain liquids that will burn on a basis of flash points. Both NFPA and DOT generally define combustible liquids as having a flash point of 100° F (38° C) or higher. Non-liquid substances such as wood and paper are classified as ordinary combustibles by the NFPA. OSHA defines combustible liquid within the hazard communication law as any liquid having a flash point at or above 100° F (38° C), but below 200° F (93.3° C). Also any mixture having components with flash points of 200° F (93.3° C), or higher, the total volumes of that make up 99% or more of the total volume of the mixture.

Combustible assembly - walls, floors, ceilings, or other assemblies constructed of one or more components materials that are not defined as noncombustible.

Combustible constituent - that part of a fuel that will burn, releasing energy.

Combustible gas - a gas or vapor considered to be flammable when exposed to a spark or flame.

Combustible liquids - any liquids having a flash point at or above 100° F (38° C), and are divided into the following classifications. Combustible liquids differ from flammable liquids in that the flash point of all combustible liquids is at or above 100° F (38° C). The range of flash point dictates the class of combustible liquid. The flash point range of 100° F to 140° F for Class II liquids was based on a possible indoor ambient temperature exceeding 100° F (38° C). Only a moderate degree of heating would be required to bring the liquid to its flash point in this type of condition. Class III liquids, that have flash points higher than 140° F (60° C) would require a significant heat source besides ambient temperature conditions to reach its flash point.

Class II - liquids having flash points at or above 100° F (38° C) and below 140° F (60° C).
Class IIIA - liquids having flash points at or above 140° F (60° C) and below 200° F (93° C).
Class IIIB - liquids having flash points at or above 200° F (93° C).

Combustible material - a material that will ignite and burn. As pertaining to materials adjacent to or in contact with heat-producing appliances, chimney connectors, steam and hot water pipes, and warm air ducts, it is material made of or surfaced with wood, compressed paper, plant fibers, or other material that will ignite and burn. Such material shall be considered as combustible even though flame-proofed, fire- retardant treated, or plastered.

Combustion - the act or process of burning; the rapid oxidation of fuel accompanied by the production of heat and light. For combustion to take place, the fuel must be heated to its ignition temperature and brought into contact with oxygen. Complete combustion requires and adequate supply of oxygen. Also see complete, incomplete, and perfect combustion.

Combustion air - the air required for complete and satisfactory combustion of the fuel; does not include the air used for atomization. Also called main air or secondary air. The actual amount of air supplied is usually expressed as a percentage of the theoretical amount required for complete combustion.

Combustion air switch - see airflow switch

Combustion analysis - the measurement of the efficiency of a given flame, based on such factors as air-fuel ratio, stack temperature, CO_2 and O_2.

Combustion analyzer - an instrument used to measure percentages of CO_2, O_2, excess air, N_2, SO_2, CO, stack temperature, smoke and combustion efficiency,

Combustion appliance - a fuel burning (oil, gas, coal, or wood) device such as a range, furnace, or water heater.

Combustion chamber - the portion of the heating or process equipment where fuel is burned. Also called the firebox.

Combustion control - a control that regulates the fuel to air ratio supplied to the burner.

Combustion control valve - see firing rate valve.

Combustion detector - see flame detector.

Combustion efficiency - the efficiency of a heat source in converting the chemical energy content of its fuel into heat, based on measurements of exhaust stack gas temperature and carbon dioxide content.

Combustion, perfect - the complete burning of a fuel, with no ash or smoke resulting. Excess air and O_2 for this condition would be zero and the percentage of CO_2 would be maximum for that particular fuel. Not practical to achieve in the field.

Combustion products - constituents in the exhaust stream resulting from the combustion of flue gas in air, including the inert gases like nitrogen, argon, water vapor, and ash, but excluding excess oxygen.

Combustion rate - the Btu liberation of fuel (gas) per cubic foot of combustion space per hour.

Combustion safeguard - see flame safeguard control

Combustion space - that space in a boiler or furnace available for the fuel to complete its combustion. Combustion flame must burn in suspension.

Combustion tests - a group of checks made on a furnace or boiler to determine that efficient combustion is taking place. Sample tests are smoke, draft, stack temperature, CO_2 and O_2.

Comfort air circulation - the adjustment of a residential heating system to provide comfortably uniform conditions from floor to ceiling and from room to room the whole season through.

Comfort air conditioning - the process of treating air to simultaneously control its temperature, humidity content, cleanliness and distribution to meet the comfort requirements of the occupants in the conditioned space. A temperature level of 70° F to 74° F with 40% to 60% relative humidity is generally accepted as being ideal. Nature provides very few hours when temperature and humidity are simultaneously within these ranges. In the winter it is normally necessary to heat the air and add humidity to achieve comfort. Normal summer operation is to lower the temperature and humidity levels.

Comfort chart - chart used in air conditioning to show the dry bulb temperature and humidity for human comfort conditions.

Comfort cooler - a system used to reduce the temperature in the living space in homes. These systems are not complete air conditioners as they do not provide complete control of heating, humidifying, dehumidification, and air circulation.

Comfort cooling - refrigeration for comfort as opposed to refrigeration for storage or manufacture.

Comfort line - a line on the comfort chart showing a relationship between the effective temperature and the percentage of adults feeling comfortable.

Comfort range - a range of conditions, especially temperature and humidity, that provide a comfortable indoor environment.

Comfort zone - the area on a psychrometric chart that show conditions of temperature, humidity, and sometimes air movement, in which 50% or more adults feel comfortable.

Commercial - applications performed at businesses or institutions such as office buildings, hospitals, apartments and universities.

Commercial buildings - buildings such as stores, shops, restaurants, motels, and large apartment buildings.

Commercial refrigeration - refrigeration appliances used in the retail food and cold storage warehouse sectors. Retail food includes the refrigeration equipment found in supermarkets, convenience stores, restaurants, and other food service establishments. Cold storage includes the equipment used to store meat, produce, dairy products, and other perishable goods. All of the equipment contains large refrigeration charges, typically over 75 pounds.

Commercial refrigeration system - a reach-in or service refrigerator of commercial size with or without the means of cooling.

Commissioning - start up of a building that includes testing and adjusting HVAC, electrical, plumbing, and other systems to assure proper functioning and adherence to design criteria. Commissioning also includes the instruction of building representatives in the use of these building systems.

Common - a terminal, connection, leadwire, circuit, or other part of an electrical network that is shared by other parts of the network. The neutral side of a 115 volt system.

Common name - a designation for a material other than its chemical name, such as code name, code number, trade name, brand name, or generic name.

Common pipe segment - the piping path in a multiple branch system that is in common with branch circuits and through which the full system flow passes.

Common terminal - terminal on a motor that is connected to both the start and the run windings and/or a junction terminal.

Common vent - a point where two appliances are connected to one venting system.

Commutation - the process of applying direct current of the proper polarity at the proper time to the rotor windings of a DC motor through split rings and brushes.

Commutator - that part of the armature of an electric motor to which the coils of the motor are attached. It consists of segmented sections, electrically insulated from the rotor and from each other. Brushes then ride on the commutator, thereby connecting the armature coils to the power source.

Commutator bars - the metal current carrying members of a commutator which make contact with the brushes.

Companion angle joints - a transverse joint that is formed when angle frames are bolted, riveted or welded to duct ends and are then bolted together with gaskets (or caulked) to prevent air leakage.

Companion flange - any flange suited to connect with a flange valve or fitting or with another flange to form a joint.

Comparator - a linear circuit used for comparing two input signals and provides a digital output in accordance to the input signals.

Compartment - an enclosed subdivision within a ship. Also relates to a room.

Compartment, freezer - the freezer compartment in a household refrigerator is the compartment designed for short-term storage of food temperature below 32° F (0° C); in a household combination refrigerator-freezer it is that compartment(s) designed for storage of foods at temperature of 8° F (-13.3° C); and in a household freezer is that compartment, or compartments, designed for extended storage of frozen foods at a recommended rating temperature of 0° F (-17.8° C) and having inherent capacity for freezing of food.

Compatible fittings - specially designed male-female unions.

Compensated control - a method of control in which the setting of one controller is automatically changed when the conditions at another control point change. The change is made in accordance with a definite schedule. The term more commonly used is reset. For example, in a system using hot water for heating, the water temperature may be reset from 100° F up to 180° F as the outside air temperature ranges from 60° F down to 0° F.

Compensation - a correction to keep a controller from drifting from its control point with changes in load or to help the controller shift its response with special changes.

Complete circuit - a continuous path for electrons to flow from a higher potential to a lower potential.

Complete combustion - burning with stoichiometric proportions of air and fuel plus a moderate amount of excess air to ensure that all of the fuel is burned.

Compliant scroll compressor - term used to describe the yielding mating scrolls inside a scroll compressor. They will give or separate from each other in the present of liquid refrigerant.

Complex circuit - another term for a combination circuit.

Comply - to act in accord with a rule, standard, or law.

Component - one (or more) of the parts of a whole.

Component arrangement diagram - the portion of a label diagram that illustrates the physical location of the electrical components in the unit.

Component isolation - the use of valves on all piping connections to a given device that allows that device to be isolated for service if necessary.

Composite tubing - a term often used to describe tubing made of both metal and polymer materials. PEX-AL-PEX tubing is an example of composite tubing.

Composition disc - a washer made from fiber, leather, neoprene or other material that will withstand change in pressure or temperature.

Composition wall - a wall built of a combination of two or more masonry units of different materials bonded together, one forming the back-up and the other the facing elements.

Compound - the chemical combination of two or more elements to make an entirely different material combined together chemically. A distinct substance formed by the chemical combination of two or more elements in definite proportions.

Compound gauge - an instrument for measuring pressures both above and below atmospheric pressure. Typically this is a gauge to measure suction pressure.

Compound wound motor - a direct current motor that has two separate field windings. One, usually the predominating field, connected in parallel with the armature circuit, and the other connected in series with the armature circuit.

Compound refrigerating systems - a system that has several compressors or compressor cylinders in series. The system is used to pump low pressure vapors to condensing pressures.

Compound retard gauge - this special compound gauge allows an accurate reading within a certain range, between zero and 100 psig.

Compressed air - air at a pressure of five or more pounds per square inch above atmospheric pressure.

Compressed air atomizing oil burner - see atomizing oil burner.

Compressed gas - a gas or mixture of gases in a closed container having an absolute pressure exceeding 40 psi at 70° F (21.2° C); or having an absolute pressure exceeding 104 psi at 130°F (54.4° C) regardless of the pressure at 70° F (21.2° C) or a liquid having a vapor pressure exceeding 40 psi at 100° F (38° C) as determined by ASTM D-323-72.

Compressibility - the change in volume of a unit volume of a fluid when subjected to a unit change of pressure.

Compressible flow - flow of high pressure gas or air that undergoes a pressure drop sufficient to result in a significant reduction of its density.

Compressible fluids - fluids, usually gases, which can be compressed into smaller volumes when pressure is applied to them.

Compression - the term used to denote increase of pressure on a fluid by using mechanical energy.

Compression, compound - compression by stages in two or more cylinders.

Compression, dual - split suction valving arrangement on compressor for carrying two suction pressures.

Compression, efficiency - ratio of work required to compress, adiabatically and reversibly, all the vapor delivered by a compressor (per stage) to the actual work delivered to the vapor by the piston or blades of the compressor.

Compression fitting - pipe and tubing fittings used for applications that involve relatively low pressure. Also called a union type fitting.

Compression gauge - instrument used to measure positive pressures (pressures above atmospheric pressures) only. These gauges are usually calibrated from 0 to 300 pounds per square inch of gauge (psig).

Compression, multistage - compression in more than one step. Usually the discharge of one compressor goes into the suction of another compressor.

Compression ratio - ratio of the volume of the clearance space to the total volume of the cylinder. This is taken from the bottom dead center to that at top dead center. In refrigeration it is also used as the ratio of the absolute low side pressure to the absolute high side pressure.

Compression, single stage - compression in one stage.

Compression strength - the ability of an object or material to resist being crushed.

Compression system - a refrigerating system in which the pressure imposing element is mechanically operated. Distinguished from an absorption system that uses no compressor.

Compression tank - a closed tank containing air that can be compressed to provide additional room for expanding water.

Compression, wet - a system of refrigeration in which some liquid refrigerant is mixed with vapor entering the compressor so as to cause the discharge vapors from compressor to be saturated rather than superheated.

Compressive stress - occurs when two forces of equal intensity act from opposite directions, pushing toward the center of an object. Fire tubes in a firetube boiler are subject to compressive stress.

Compressor - the pump of a refrigerating mechanism that draws a vacuum or low pressure on cooling side of refrigerant cycle and squeezes or compresses the gas into the high pressure or condensing side of the cycle. Three main types are reciprocating, centrifugal and rotary.

Compressor, **booster** - a compressor for very low pressures, usually discharging into the suction line of another compressor.

Compressor, **centrifugal** - a non-positive displacement compressor that depends for pressure rise, at least in part, on centrifugal effect.

Compressor, **compound** - a compressor having two or more cylinders wherein the discharge of one cylinder is fed to another cylinder for greater compression.

Compressor crankcase - the internal part of the compressor that houses the crankshaft and lubricating unit.

Compressor crankshaft seals - prevents air from entering the compressor, and oil and refrigerant from escaping.

Compressor cylinder - the portion of a reciprocating compressor in which the piston is housed. The valve plate is located on top of the cylinder.

Compressor displacement - volume, in cubic inches, represented by the area of the compressor piston head or heads multiplied by the length of the stroke.

Compressor, **double-acting** - one that has two compression strokes per revolution of the crankshaft per cylinder --- i.e., both faces of the faces of the piston are working faces.

Compressor, **double suction** - split suction; valving arrangement on compressors for carrying two suction pressures.

Compressor, **external drive** - see compressor, open type.

Compressor head - the component part of the compressor that houses the compressor valves and the valve plate. The compressor head has both high and low pressure compartments, which separate the suction and discharge refrigerant.

Compressor, **hermetic** - a mechanical compressor combination consisting of a compressor and a driving motor, both of which are enclosed in the same housing, with no external shaft seals, the motor operating in the refrigerant atmosphere.

Compressor, **horizontal** - compressor with horizontal cylinder, or, in small sizes, with horizontal crankshaft.

Compressor, **multi stage** - compressor having two or more compressive stages. Discharge from each stage is the intake pressure of the next in series.

Compressor oil cooler - one or more piping systems used for cooling the crankcase oil.

Compressor, **open type** - a compressor in which the crankshaft extends through the crankcase and is driven by an external motor.

Compressor, **positive displacement** - a refrigerant compressor in which an increase of refrigerant gas or vapor pressure is attained by changing the internal volume of the compression chamber.

Compressor, **reciprocating** - a compressor that uses a piston and cylinder mechanism to provide pumping action. A positive displacement type compressor.

Compressor, **reciprocating open** - a reciprocating compressor coupled externally to an open motor or drive source.

Compressor, **reciprocating hermetic** - a sealed, reciprocating compressor and motor combination with no external coupling.

Compressor, **refrigerant** - that component of a refrigerating system that increases the pressure of a compressible refrigerant fluid, and simultaneously reduces its volume, while moving the fluid through the device.

Compressor, **rotary** - a positive displacement compressor in which a rotating cylinder revolves on an axis that is offset from the cylinder centerline. This system uses vanes, eccentric mechanisms, or other rotating devices to provide pumping action.

Compressor, **rotary lobe** - device used to produce pressure for gas and air by the use of lobed rotors.

Compressor, **screw** - the spiral flutes of one screw acts as a cylindrical pathway for the refrigerant to be moved and compressed by the mating screw as they are rotated at high speed.

Compressor, **single stage** - a compressor having only one compression step between the high and the low side pressure.

Compressor shaft seal - a leak proof seal between the crankshaft and compressor body.

Compressor seal - the device that prevents leakage between the crankshaft and the compressor body on open type compressors.

Compressor, semi-hermetic - a hermetic refrigerant compressor whose housing is sealed by one or more gasketed joints and is provided with means of access for servicing internal parts in the field.

Compressor, vertical - compressor with vertical cylinder, or, in small sizes, with vertical crankshaft.

Compressor unit - a condensing unit without the condenser and liquid receiver.

Compressor valve - a reed type valve found inside the compressor. It lets the low pressure gas coolant to enter the compressor and discharges the high pressure gas from the compressor.

Concave - a curved depression in an object.

Concealed - rendered inaccessible by the structure or finish of the building. Wires in concealed raceways are considered to be concealed, even though they may become accessible by withdrawing them.

Concealed location - a location that cannot be accessed without damaging permanent parts of the building structure or finish surface. Spaces above, below or behind readily removable panels or doors shall not be considered as concealed.

Concentrating collector - a device that uses the reflective surfaces to concentrate the sun's rays onto a smaller area, where they are absorbed and converted to heat energy.

Concentration - the quantity of one constituent dispersed in a defined amount of another. For example, the amount of gas or gases of interest present in a space. Usually expressed in percent of total volume, parts per million (ppm) or percent of the lower explosive limit (lel).

Concentrator - a photovoltaic array that includes an optical component such as a lens or focusing mirror to direct incident sunlight onto a solar cell of smaller area.

Concentrator, air conditioning - that part of an absorption chiller where the dilute salt solution is boiled to release the water.

Concentric - having a common center.

Concrete - a mixture of Portland cement, sand, gravel and water.

Concrete drawings - drawings showing wall thickness, column location, concrete or block walls, floor and wall penetrations, openings and drains, embedments, channels, and in place anchor bolts.

Concrete thin slab - a layer of concrete in the range of one to two inches thick placed over hydronic tubing to create a radiant floor panel.

Condensable - a gas that can be easily converted to liquid form, usually by lowering the temperature and/or increasing pressure. Steam, which changes state as the temperature drops, is a common example.

Condensate - the liquid that separates from a gas (including flue gases) due to a reduction in temperature. In a steam heating system, the water formed by cooling steam as in a radiator. The capacity of traps, pumps, etc., is sometimes expressed in pounds of condensate they will handle per hour.

Condensate drain pan - container located under the evaporator coil to catch the condensate or defrost water and has provisions for a drain tube. The condensate pan should collect water and direct it to a drain without stagnation. Also called drain pan.

Condensate drain line - piping arrangement used to carry condensate from the condensate drain pan to an appropriate point of disposal.

Condensate drain trap - installed in the condensate drain line when the drain is at a negative pressure to ensure proper drainage of the condensate.

Condensate line - the refrigeration line that extends from the outlet of the condenser to the inlet of the receiver. If you do not have a receiver, then this line would be the liquid line. The condensate line is usually sized larger than the liquid line to accommodate liquid refrigerant flowing out of the condenser while refrigerant vapor is being exchanged.

Condensate pump - a device used to remove fluid condensate that collects beneath an evaporator coil.

Condensate tank - where condensed steam (water) is stored before it is delivered back to the open feedwater heater by the condensate pump.

Condensate water drain - a u-shaped channel or pan placed underneath an evaporator coil to catch the condensed water vapor as it drops from the coil. Provisions are made in this pan to connect a condensate drain fitting.

Condensation - (1) liquid or droplets that form when a gas or vapor is cooled below its dew point. (2) the result of steam giving off its heat in radiation to the air, liquid or other cooling medium and returning it to a liquid state that is returned to the boiler as feed water. (3) condensation is used to indicate the water vapor present in the flue gases caused by the combustion of hydrogen.

Condensation point - the temperature at which the removal of sensible heat (superheat) from a vapor or gas at a given pressure will begin a change of state to a liquid form.

Condensation resistance factor - an indication of a window's ability to resist condensation. The higher the value the better.

Condense - the action of changing a gas or vapor to a liquid.

Condenser - (1) a heat exchanger where the heat absorbed by the refrigerant in the evaporator is rejected. This is the place where the latent heat of condensation takes place. Heat flows from the refrigerant to the air (passing across the condenser) causing the gas to change back to a liquid without a change in temperature or pressure. Condensing temperature must be higher than ambient air. (2) An electrical device that stores an electrical charge.

Condenser, air cooled - a heat exchanger that transfers heat to the surrounding air.

Condenser, atmospheric - a condenser cooled with water that is exposed to the atmospheric.

Condenser, barometric - a condenser that operates under a vacuum.

Condenser coil - in mechanical refrigeration, a section of coiled tubing where gas refrigerant is cooled below its boiling point.

Condenser comb - comb-like device, metal or plastic, that is used to straighten the metal fins on condensers or evaporators.

Condenser, direct contact - equipment used to condense steam exhausted from a turbine. Can be the barometric or low level jet type with parallel flow or counterflow design options.

Condenser, double pipe - two concentric tubes in which the vapor in the outer tube is condensed by water flowing through the inner tube.

Condenser, evaporative - a condenser in which part of the heat rejection may be accomplished by raising the temperature of an air stream passing over a heat exchange surface and the remainder by evaporation of water sprayed or otherwise distributed over the heat exchange surface.

Condenser fan - forced air device used to move air through an air cooled condenser.

Condenser flooding - a method of maintaining a correct head pressure by adding liquid refrigerant to the condenser from a receiver to increase the head pressure.

Condenser heat rejection - the total amount of heat removed from a system by the condenser.

Condenser heat rejection factor - the amount of heat absorbed in the evaporator as a percentage of the condenser heat rejection.

Condenser, jet - type of heat exchanger that condenses vapor and also removes noncondensable gases without additional equipment.

Condenser, open shell and tube - one in which the water passes in a film over the inner surfaces of the tubes, that are open to the atmosphere.

Condenser saturation temperature - the temperature at which a vapor changes to a liquid at a specific pressure. Also referred to as condensing temperature.

Condenser, shell and coil - a condensing coil enclosed in a cylinder filled with cooling water.

Condenser, shell and tube - one in which the water passes over the inner surfaces of the tubes, that are open to the atmosphere.

Condenser, surface - large tubular device used mainly for condensing steam to even out the flow.

Condenser temperature split - condensing temperature of the refrigerant vapor less the dry bulb temperature of the air flowing over the condenser coil. For air cooled condensers (<10 SEER) using R-22 as the refrigerant, the normal condenser temperature split is 30°F to 32°F. This temperature split will change as the equipment gets more and more efficient.

Condenser, water cooled - heat exchanger that is designed to transfer heat from hot gaseous refrigerant to water.

Condenser, water pump - a forced water moving device used to move water through a condenser.

Condenser water reset - energy conserving control strategy which reduces the temperature of condenser water to the lowest economical temperature on the basis of the ambient wet bulb temperature and the load being handled by the chiller.

Condensing - the process of vapor turning to liquid upon being cooled.

Condensing furnace - a high efficiency, usually above 85% and above, gas or oil forced air appliance that allows the condensation of the products of combustion thereby extracting the latent heat of vaporization. This condensate then gets drained away to a convenient drain. Combustion air intake and exhaust vent pipes are usually PVC schedule 40 plastic pipe.

Condensing medium - a substance, such as air or water, that is used to remove heat from the condensing coil to change the refrigerant vapor to a liquid.

Condensing pressure - the pressure inside a condenser at which refrigerant vapor gives up its latent heat of vaporization and becomes a liquid. This varies with the condensing temperature.

Condensing temperature - the temperature inside a condenser at which refrigerant vapor gives up its latent heat of vaporization and becomes a liquid. If high pressure R-22 refrigerant vapor enters the condenser at 270 psig, it will change to a liquid at the corresponding temperature of 122° F due to the heat removed from the vapor by the lower temperature of the air entering the condenser. The higher the temperature of air entering the condenser, the higher the discharge temperature will be, the higher the condensing temperature will be, and the higher the current will be. The condensing temperature should normally not exceed ambient temperature plus 30° F with a clean coil and air cooled condenser of average size.

Condensing turbine - a steam turbine with an exhaust steam pressure below atmospheric pressure, such that steam is directly and completely condensed.

Condensing unit - the component on the high side of the refrigeration cycle with consists of a compressor, condenser coil, and condenser fan and motor. The unit is usually electrically wired with the required electrical components such as contactor or starter, starting relay, overload protection and start and run capacitors on single phase units. Some units may be equipped with high and low pressure switches.

Condensing unit service valves - shut off hand valves mounted on the condensing unit to enable the service person to install and/ or service unit.

Conditioned air - air that ha been heated, cooled, humidified, or dehumidified to maintain an interior space within the comfort zone. Also referred to as tempered air.

Conditioned space - that part of a building that is controlled in terms of temperature and humidity for the comfort of occupants.

Conductance, electrical - the ability of an object to permit the flow of electrons. The opposite of resistance. Conductance is a ratio of the current flow to the potential difference causing the current flow.

Conductance, surface film - the time rate of heat flow per unit area under steady conditions between a surface and the ambient fluid for a unit temperature difference between the surface and the fluid. In English units its value is usually expressed in Btu per (hour) (square foot) (Fahrenheit degree) temperature difference between surface and fluid.

Conductance, thermal - the amount of heat (Btu) transmitted from surface to surface in one hour through 1 ft2 of a material or construction for the thickness or type under construction for a difference in temperature of 1° F between the two surfaces. Symbol is "C"

Conduction - the transfer of heat from particle to particle without movement of the object itself The flow of heat through a body having unequal temperatures this flow of heat is from a point of higher to a point of lower temperature.

Conduction band - energy level at which electrons are not bound to (orbiting) a specific atomic nucleus, but are free to wander among the atoms. The outermost energy level of an atom.

Conduction level - the point at which an amount of voltage or current will cause a device to conduct. Energy level at which electrons are not bound to (orbiting) a specific atomic nucleus, but are free to wander among the atoms.

Conduction, thermal - the process of heat transfer through a material medium in which kinetic energy is transmitted by particles of the material from particle to particle without movement of the material.

Conductive losses - heat lost from a building by contact with outside air or heat lost by leakage and by infiltration.

Conductivity - the ability or ease of heat to flow through a material. Copper is considered to be one of the better conductors of heat and insulating materials are extremely poor conductors.

Conductivity, electrical - a term used in describing the capability of a material to carry an electrical charge. Usually expressed as a percentage of copper conductivity - copper being one hundred percent (100%). Conductivity is expressed for a standard configuration of conductor.

Conductivity, thermal - the amount of heat (Btu) transmitted in one hour through 1 ft2 of a homogeneous material 1" thick for a difference in temperature of 1°F between the two surfaces of the material. Symbol is "K." Materials are considered homogeneous when the value of "K" is not affected by variation in thickness or in size of sample within the range normally used in construction.

Conductivity circuit - a flame proving circuit that operates on an alternating current flame signal and utilizes the electrical conduction of a flame, as distinguished from rectification. Also see conductivity flame rod system and flame conductivity.

Conductivity flame rod system - a flame rod detection system based on the ability of the flame to conduct a current; conducts the same amount of current in either direction, as distinguished from flame rectification.

Conductivity values - indication of the Btu per square foot of area for each Fahrenheit degree of temperature difference per hour for each inch of thickness of the material.

Conductor - a substance or body that is capable of transmitting electricity or heat. A substance that allows electrons to flow freely through it. An object that has good conductance.

Conductor, bare - a conductor having no covering or electrical insulation whatsoever.

Conductor, covered - a conductor encased within material of composition or thickness that is not recognized by the NEC as electrical insulation.

Conductor, insulated - a conductor encased within material of composition or thickness that is recognized by the NEC as electrical insulation.

Conductor, thermal - a material that readily transmits heat by means of conduction.

Conduit - a tube or trough for carrying and protecting electrical wires. Can be rigid, flexible, metallic or nonmetallic. Conduit provides the strength needed to run electrical lines throughout an industrial installation.

Conduit body - a separate portion of a conduit or tubing system that provides access through a removable cover to the interior of the system at a junction of two or more sections of the system or at a terminal point of the system.

Conduit fittings - materials used to connect a conduit system of electrical wiring.

Conduit run - a continuous electrical duct between two points or devices.

Confined space - (1) a room or space that contains less than 50 cubic feet per 1,000 Btu/h of the aggregate input rating of all appliances installed within that space. (2) a space large enough and so configured that an employee can physically enter and perform assigned work, has limited or restricted means for entry and exit, and is not designed for continuous employee occupancy.

Congeal - to come together and lose viscosity.

Conical - shaped like a cone.

Connected load - the electrical power that would be needed if every load connected to the system were drawing power at the same time.

Connected load, boiler - the total load in Btu/h attached to the boiler. It is the sum of the outputs of all terminal units and all heat to be supplied by the boiler for process applications.

Connecting rod - that part of a compressor mechanism that connects the piston to the crankshaft, carrying the back and forth motion of the crankshaft levers to the pistons, causing the pistons to move back and forth. One end of the connecting rod is connected to a wrist pin that is connected to the piston. The other end of the connecting rod is connected to the crankshaft.

Connection - joining of two or more parts; a union.

Connection diagram - a circuit diagram that shows the connections of an installation or its component parts. It may cover internal or external connections. Referred to as a pictorial wiring diagram.

Connection, parallel - systems where flow is divided among two or more pathways from a common starting point.

Connection, series - system where the flow must pass through each component. Flow through the first then enters the second and then the third etc. If any of the components are broke or plugged, flow will stop at that point.

Connector, pressure - a device that establishes a connection between two or more conductors or between one or more conductors and a terminal by means of mechanical pressure and without the use of solder.

Conradson test - a test that indicates the percentage of carbon residue resulting from the burning of a given fuel oil. Also known as a carbon residue test.

Consequent pole motor - a multispeed alternating current motor that changes the motor speed by changing the number of its stator pole.

Conservation of energy - the principle that assumes that energy can neither be created or destroyed.

Console - a total unit or system of controls located in one area and enclosed. A window air conditioner is a console air conditioner.

Consolidation formation - materials below the surface of the earth that contain very little water for use in geothermal heat pump systems. Examples of consolidated formations include granite, sandstone, and limestone.

Constant load hanger - device that consists of a lever mechanism in a housing and a coil spring. The movement of piping, within certain limits, will not change the spring forces holding up the piping; thus no additional forces are introduced to the piping being supported.

Constant-level valve - an automatic control valve for maintaining, within a reservoir, a constant level of fuel for delivery to the burner.

Constant pressure circulator - a circulator controlled by a variable speed drive so that it provides a constant differential pressure at all flow rates.

Constant pressure valve - an automatic expansion valve that holds the pressure at a constant level regardless of the load.

Constant speed motor - a motor whose speed is constant or substantially constant over the normal range of loads.

Constant temperature valve - a control valve that maintains a set, desired fluid pressure.

Constant volume air system - those systems that are designed to provide a constant airflow and to vary the air temperature to meet heating and cooling needs. The percentage of outdoor air may be held constant, but is often controlled either manually and/or automatically to vary with outdoor temperature and humidity. Controls may include a minimum setting that should allow the system to meet ventilation guidelines for outdoor air quantities under design conditions.

Constant volume variable temperature - a system control which varies the temperature and humidity of the air stream while maintaining air flow at a constant volume.

Constantan - copper nickel alloy used in a J-type thermocouple.

Constrictor - a tube or orifice used to restrict flow of a gas or a liquid.

Construction documents - all of the written, graphic and pictorial documents prepared or assembled for describing the design, location and physical characteristics of the elements of the project necessary for obtaining a building permit. The construction drawings shall be drawn to an appropriate scale.

Consumed - used up; wasted; destroyed, as by a fire.

Consumption - the maximum amount of gas per unit of time, usually expressed in cubic feet per hour, or Btu per hour, required for the operation of the appliance or appliances supplied.

Contact, auxiliary - a pilot or auxiliary contact that is operated whenever the main contact or contacts are closed, such as the holding circuit contact in a magnetic starter. Also an interlock contact.

Contact bounce - when a mechanical switch is turned on, the contacts are forced together. This causes the contacts to open and close several times before making firm contact.

Contact chatter - see contact bounce.

Contact points - two movable points or areas that complete a circuit when pressed together. These points are usually made of tungsten, platinum, or silver.

Contact rating - the electrical power or current handling capability of a switch or relay contacts. Usually measured in amperes, volts, or horsepower, either resistive or inductive.

Contact section - section of a solder-melting type overload relay that is connected in series with the coil of the motor starter.

Contacts - conducing devices that may open or close to complete or interrupt a circuit.

Contactor - an electrical device for making or breaking load carrying contacts by a pilot circuit through a magnetic coil.

Contaminant - dirt, moisture, or other substance foreign to the refrigerant or oil in a system. These foreign materials can cause unwanted chemical reactions, component failure and have a negative effect on system operation, life or reliability.

Contaminant failure - any loss of performance due to the presence of contamination. Two basic types of contamination failure are: perceptible -- gradual loss of efficiency or performance, and catastrophic -- dramatic, unexpected failure.

Continuity - unbroken, continuous path or line through which electricity may flow.

Continuity check - a test to see if a circuit is an open or closed path.

Continuity equation - an equation expressing the conservation of mass in a steady-flow system.

Continuity principle - at a given time, the same quantity of air is flowing in every section of a duct.

Continuity testing - using an ohmmeter to check for open circuits.

Continuous air circulation - the adjustment of a blower assembly for continuous operation, thereby providing comfortably uniform conditions from floor to ceiling and from room to room. It is good practice to maintain air circulation within the design standards of the equipment manufacturer and to obtain the highest level of circulation possible without causing excessive drafts and noise. This practice is considered mandatory with an electronic or mass media type air cleaner.

Continuous blow down - the expulsion of water continuously from steam lines.

Continuous check relay - see dynamic self check.

Continuous combustion - this process burns fuel in a sustained manner as long as fuel and air are continuously fed to the combustion zone and the fuel/air mixture is within the flammability limits. Continuous combustion is more common than pulse combustion and is used in most fuel burning equipment.

Continuous component check - see dynamic self check.

Continuous cycle absorption system - a system that has a continuous flow of energy input.

Continuous duty - the operation of a circuit or device under a constant load without exceeding certain temperature limits for an indefinite period of time.

Continuous flow rating - the rate at which a water heating device can supply heated water at a given temperature on a sustained basis. Often expressed in gallons per hour or gallons per minute.

Continuous furnace - a furnace operated on an uninterrupted cycle, in which the charge is being constantly added to, moved through, and removed from the furnace; as distinguished from a batch furnace.

Continuous ignition - ignition by an energy source that is continuously maintained throughout the time the burner is in service, whether the main burner is firing or not.

Continuous load - a load where the maximum current is expected to continue for three hours or more.

Continuous pilot - a pilot that burns without turndown throughout the entire time the burner assembly is in service, whether the main burner is firing or not. The pilot ignites fuel that may leak into the combustion chamber when the burner is off, thus preventing a hazardous condition at start-up. "Continuous pilot" is Underwriters Laboratories Inc. term for the old term "standing pilot."

Continuous run timer - a defrost timer connection (pigtail to terminal1, terminal 3 to neutral) that permits the motor to operate on a continuous basis.

Contour - (1) the outline or shape of an object. (2) on a land map denoting elevations, a line connecting points with the same elevation.

Contract drawings - the final drawings from which a building is built. Also called building plans and/or working drawings.

Contract modifications - after the agreements have been signed, any additions, deletions, or modifications of the work to be done are accomplished by change order, supplement instruction, and field order. They can be issued at any time during the contract period.

Contraction - shrinkage resulting from cold temperatures.

Contrahelical - a term meaning the application of two or more layers of spirally twisted, served, or wrapped materials where each successive layer is wrapped in the opposite direction to the preceding layer.

Control - a device that directly operates a system to regulate the fuel, air, water, or electrical supply to the controlled equipment. It may be automatic, semi-automatic, or manual.

Control agent - the medium that is manipulated by the control system to cause a change in the controlled medium. For example, suppose a heating coil through which steam is flowing is used to heat a room. The room thermostat is placed so that it measures temperature (controlled variable) in the room air (controlled medium), and operates a valve that regulates the flow (manipulated variable) of the steam (control agent) through the heating coil. Heat from the coil is thus furnished to the room.

Control algorithm - the operating instructions or logic that a controller follows.

Control, anticipating - one which, by artificial means, is activated sooner than it would be without such means, to produce a smaller differential of the controlled property.

Control box enclosure types - The National Electrical Manufacturers Association (NEMA) has established standard types of enclosures for control equipment. They are:

a. NEMA Type 1 - General Purpose: Intended primarily to prevent accidental contact with control apparatus. It is suitable for general purpose applications indoors, under normal atmospheric conditions. Although it serves as a protection against dust, it is not dust-proof.

b. NEMA Type 3 - Dusttight, Raintight, and Sleet Resistance: Intended for outdoor use, and for protection against wind blown dust and water. It has a watertight conduit entrance, mounting means external to the box, and provisions for locking. Although it is sleet resistance, it is not sleet proof.

c. NEMA Type 3R - This is similar to Type 3, except it also meets UL requirements for being rainproof. When properly installed, rain cannot enter at a level higher than the lowest live part.

d. NEMA Type 4 - Watertight and dust-tight: This enclosure is designed to exclude water. It must pas a hose test for water, and a 24 hour salt spray test for corrosion. This enclosure may be used outdoors on a cooling tower or other appropriate piece of equipment. It is usually a gasketed enclosure of cast iron or stainless steel.

e. NEMA Type 4X: Similar to Type 4, except it must pass a 200 hour salt spray test for corrosion. It is usually a gasketed enclosure of fiber reinforced polyester.

f. NEMA Type 6 - Submersible, Watertight, Dust-tight and Sleet Resistance: Intended for use where occasional submersion may be encountered. Must protect equipment against a static head of water of 6 feet for 30 minutes.

g. NEMA Type 12 – Dust-tight and Drip-tight: Enclosure intended for indoor use. It provides protection against fibers, flyings, lint, dust, dirt, and light splashing.

h. NEMA Type 7 - Hazardous Locations - Class I Air Break: This enclosure is intended for use indoors in locations defined by the National Electrical Code for Class I, Division I, Groups A, B, C or D hazardous locations.

i. NEMA Type 9 - Hazardous Locations - Class II Air Break: Intended for use indoors in areas defined as Class II, Division 1, Groups E, F or G hazardous locations.

Control bus - a set of signals that regulate the operation of a microcomputer system, including I/O devices and memory. They function much like "traffic" signals or commands. They may also originate in the I/O devices, generally to transfer to or receive signals from the CPU.

Control building - centrally located structure that houses board mounted instrumentation, control panels, and monitoring equipment necessary for the operation of the central plant. Often provides laboratory, kitchen, locker room, work, and limited office space for control room workers.

Control cable - a multi conductor cable made for operation in control or signal circuits, usually flexible, relatively small in size, and with relatively small current ratings.

Control circuit - a circuit that controls some load in the entire control system, whether it is a relay or contactor coil or a major load.

Control, compressor - see motor control.

Control contacts - in an oil pressure failure switch, conducting parts of a relay that acts as switches and provides power to the heater of the timer.

Control, defrosting - a device to automatically defrost the evaporator. It may operate by means of a clock, door cycling mechanism, or during the "off" portion of the refrigerating cycle.

Control, dew point - method of moisture control which uses a dry bulb temperature sensor downstream from a cooling coil to maintain discharge air at saturation by controlling the discharge temperature.

Control differential - (1) the intended difference in the sensed control variable from where the controller output is turned on to when it is turned off. For example, if a thermostat contact is expected to open at 72° F and close at 68° F the control difference is 4° F. (2) The characteristic of a two position transfer function that creates a delay between changes in the output signal to reduce the possibility of the output signal repeatedly cycling on and off whenever the control point is near the set point.

Control, discharge air - a method of zone control in which the controller receives its input signal from a temperature sensor in the discharge air stream.

Control, floating - method of control in which the controller produces a gradual continuous action in the controlled device.

Control fuse - the fuse wired in series located in the control circuit that protects the transformer. A blown control fuse is an indication of a short circuit in the control circuit.

Control, high pressure - a pressure responsive device (usually an electric switch) actuated directly by the refrigerant pressure on the high side of the refrigeration system.

Control hunt - a condition of extreme instability, in which the controlled condition changes rhythmically from a value much larger than that desired to one much smaller, as the controller alternately over corrects in each direction.

Control, integral - a method of control in which the controller positions the controlled device to maintain the set point without error upon a change in load.

Control joint - a groove penetrating part way through a concrete slab for the purpose of forcing a crack to occur along a predetermined path.

Control loop - a group of interconnected control devices that operate collectively to maintain a single process at its desired set point.

Control, low pressure - a pressure responsive device (usually an electric switch) actuated directly by the refrigerant pressure on the low side of the refrigeration system.

Control medium - the agent, manipulated by the controlled device that serves as the transport system of heating or cooling. Examples would be chilled water, hot water, steam etc.

Control module - contains the printed circuit card, motor, wiring, harness, and switches to receive signals from inputs, including several thermistors, interpreting them and reacting properly to allow the cuber to operate.

Control, motor - a temperature or pressure operated device used to control the running of the motor.

Control panel - board used for mounting instruments and controls for remote control operation of a process.

Control piping - all piping, valves, and fittings used to interconnect air, gas, or hydraulically operated control apparatus or instrument transmitters and receivers.

Control point - the actual value of the controlled variable (temperature, pressure or humidity) being maintained by a controller at any given time. Not necessarily the same as the control set point.

Control, pressure motor - a high or low pressure controller that is connected into the electrical circuit and used to start and stop the motor when there is a need for refrigeration or for safety purposes.

Control, proportional - a method of control in which the amount of corrective action applied to a controlled device is proportional to the amount of load sensed by the controller.

Control, proportional-integral-derivative (PID) - a control algorithm that enhances the PI control algorithm by adding a component that is proportional to the rate of change (derivative) of the deviation (error) of the controlled variable. This serves to anticipate future errors under current conditions.

Control range - the difference between the maximum and minimum points of an adjustable control.

Control, refrigerant - a device used to regulate the flow of liquid refrigerant into the evaporator such as a capillary tube, expansion valves, high and low side float valves.

Control relay - a switch or switches (contacts) together with a magnetic coil that controls the switch operation.

Control, return air - zone control under which controller receives its input signal from a temperature sensor in the room.

Control state - the current condition of the output of an on/off controller (e.g., the "on" or the "off" state).

Control station - pipe loop arrangement consisting of control valve, various fittings, and isolating (block) valves to regulate the flow, pressure, or quantity of material at a predetermined location.

Control system - the entire system, including the controlled system and the automatic control equipment. See controlled system.

Control, temperature - a thermostatic device that automatically stops and starts the motor. The operation of which is based on temperature changes.

Control transformer - used to change the value of line voltage to the value needed for the control circuit.

Control valve - valve that regulates the flow or pressure of a medium that affects a controlled process. Control valves are operated by remote signals from independent devices using any of a number of control media such as pneumatic, electric, or electrohydraulic.

Control voltage - the voltage used in the electrical circuit that contains the low current devices such as the thermostats.

Control zone - the section of a furnace in which temperature is controlled by the throttling action of a single valve. Also see valve zone.

Controlled evaporator pressure - a controlled system that maintains a definite pressure or range of pressures in the evaporator.

Controlled device - a device in a control system that responds to the signals sent by a controller to start, stop or otherwise vary the operation of conditioning equipment.

Controlled medium - the substance (usually air, water, or steam) whose characteristics (such as temperature, pressure, flow rate, volume, level, or concentration) are being controlled.

Controlled space - the volume of the controlled medium. For example, a room in which the air temperature is being controlled.

Controlled system - the system made up of all equipment in which the controlled variable exists, but which does not include the automatic control equipment.

Controlled variable - that quantity or condition of a controlled medium that is measured and controlled. For example, temperature, pressure, flow rate, volume, level, or concentration.

Controller - a device that senses and measures changes in the temperature, pressure or moisture content and indirectly acts to maintain the controlled variable within preset limits. A system of switches, relays, and instrumentation used to regulate voltage, current, speed, and other predetermined actions of an electrical machine or group of machines.

Controller, dual pressure - a controller consisting of two pressure bellows that operate a switch within a single enclosed case. When applied to a compressor control, one bellows functions on a change in suction pressure, the other bellows on a change in discharge pressure.

Controller feedback - the change in the controller's output in response to a measured change in the controlled variable transmitted back to the controller input for evaluation.

Controller, humidity - a controller sensitive to changes in humidity.

Controller, pressure - an automatic control device that is actuated and responsive to pressure.

Controller, temperature - a controller sensitive to changes in temperature.

Controls - devices designed to regulate the gas, air, water, or electrical supply to a gas appliance. These may be either manual or automatic.

Convection - this is the transfer of heat energy in liquids and gases from one location to another by the motion of molecules that carry heat. Convection is natural when the movement of the liquid or gas is due to the difference in the weight of hot and cold liquid or gas. Convection is forced when fans, jets or pumps are used to move the liquid or gas.

Convection, forced - the transfer of heat resulting from the forced movement of liquid or gas by means of a fan or pump.

Convection loss - heat loss through the walls and roof of a building, through movement of heated air.

Convection, natural - the circulation of a gas or liquid due to the difference in density resulting from temperature differences.

Convection superheater - a component located in a boiler and receives heat from convection currents.

Convector - a concealed radiator. An enclosed heating unit located (with enclosures) either within, adjacent to, or exterior to the room or space to be heated, but transferring heat to the room or space mainly by the process of convection. A shielded heating unit is also termed a convector. If the heating unit is located exterior to the room or space to be heated, the heat is transferred through one or more ducts or pipes.

Convenience outlet, gas - a permanently mounted, hand operated device providing a means for connecting and disconnecting an appliance or an appliance connector to the gas supply piping. The device includes an integral, manually operated gas valve with a nondisplaceable valve member so that disconnection can be accomplished only when the manually operated gas valve is in the closed position.

Conventional current flow - current flow that is assumed to be in one direction from high charge concentration (+) to low charge concentration (-).

Conversion burner - a unit consisting of a burner and its controls utilizing gaseous fuel for installation in an appliance originally utilizing another fuel.

Conversion efficiency (cell) - the ratio of the electric energy produced by a solar cell under full sun conditions to the energy from sunlight incident upon the cell.

Conversion factors - force and power may be expressed in more than one way. A horsepower is equivalent to 33,000 foot pounds of work per minute, 746 watts, or 2,546 Btu per hour. These values can be used for changing horsepower into foot pounds, BTU or watts, etc, etc.

Converter - a heat exchange unit designed to transfer heat from one distributing system to another. These may be either steam to water or water-to-water units. They are usually the shell and tube design.

Converter, electrical - a device used in relation to power supplies that changes DC to a higher or lower DC voltage.

Convex - a rounded surface on an object.

Coordinates - the position or locations of points on the X, Y, and Z planes.

Cooled space - enclosed space within a building that is cooled by a cooling system whose sensible capacity exceeds 5 Btu/h (ft2) or is capable of maintaining a space temperature of 90° F or less at design cooling conditions.

Cooler - a heat exchanger that removes heat from a substance.

Cooling anticipator - a nonadjustable heating resistor that is placed in parallel with the cooling circuit inside the thermostat. It produces a false heat in the thermostat during the equipment off cycle, to cause a shorter off cycle, to maintain a more even temperature inside the space.

Cooling coil - any section of coiled tubing (usually an evaporator coil) used to lower the temperature of the surrounding area. Also see evaporator coil.

Cooling degree days - for each degree that the average temperature is above 65° F produces one cooling degree day.

Cooling degree hours - unit, based upon temperature difference and time, used in estimating cooling loads of residences in summer. For any hour when the dry-bulb temperature is greater than a reference temperature (base 65° F for this standard) there are as many cooling degree hours as degrees Fahrenheit difference in temperature between the average hourly temperature and the reference temperature.

Cooling leg - a length of un-insulated pipe through which the condensate flows to a trap and that has sufficient surface to permit the condensate to dissipate enough heat to prevent flashing when the trap opens. In the case of a thermostatic trap, a cooling leg may be necessary to permit the condensate to drop a sufficient amount in temperature to permit the trap to open.

Cooling mode - the operating phase of a system that removes heat and/or moisture from a conditioned space.

Cooling run factor - the % of time that the unit is expected to operate to handle the cooling load during the warmest month of the year.

Cooling system, direct expansion - a cooling and dehumidification device that cools air or other fluids by the evaporation of mechanically compressed gas in an evaporator coil. A condensing coil then removes this transferred heat to a different space.

Cooling system, evaporative - housed in a cabinet containing a pump, distribution tubes, water pads, and a blower. The pump supplies water to the distribution tubes that carry the water to pads on three sides of the cabinet. The blower draws outdoor air through the moist water pads. Some of the water in the pads absorbs heat from the air and evaporates. The air cannot be re-circulated. This system works most effectively in relatively dry climates.

Cooling system, multiple stage - a cooling system that increases its capacity by stages in response to cooling demand.

Cooling tower - found on many water cooled systems, the piece of equipment that cools the water returning from the condenser. Cooling towers can ideally cool return water to a temperature within 7 degrees of the wet bulb temperature of the air flowing through the tower.

Cooling tower, forced draft - a cooling tower in which the flow of air is created by one or more fans discharging air into the fan.

Cooling tower, induced draft - a cooling tower in which the flow of air is created by one or more fans drawing the saturated air out of the tower.

Cooling tower, natural draft - a cooling tower in which the flow of air depends upon natural air currents or a breeze. Generally applied where the spray water is relatively hot and will cause some convection currents.

Cooling water - water used for condensation of refrigerant; condenser water.

Coping - a protective cap, top or cover of wall parapet, pilaster, or chimney.

Copper - a metal that conducts electricity and is widely used in piping systems. It is recognizable by it brownish-red color. Symbol is Cu.

Copper clad aluminum conductors – conductors drawn from a copper clad aluminum rod with the copper metallurgically bonded to an aluminum core. The copper forms a minimum of 10% of the cross sectional area of a solid conductor or each strand of a stranded conductor.

Copper losses - the heat losses in electrical machines due to the resistances of the copper wire used for windings; also called I2R losses.

Copper plating - a film of copper deposited by electrical, immersion, or other means on the surface of another material such as iron or steel; in refrigeration usually on compressor walls, pistons, discharge valves, and shaft seals.

Copper strip corrosion test - a qualitative measure of the tendency of a petroleum based product to corrode pure copper.

Copper tubing - copper tubing is the most widely used tubing type in the HVAC/R industry. It has a high thermal conductivity. Copper is also easy to solder, braze, bend and flare and has a high resistance to corrosion. It takes a coolant that is caustic as ammonia to corrode copper. Copper cannot be used in systems where ammonia is the coolant. Copper tubing is classified in terms of wall thickness, temper and type.

Coproduction - conversion of energy from a fuel (possibly including solid or other wastes) into shaft power (which may be used to generate electricity) and a second or additional useful form of arrangement of conversion to shaft power and either process or space heating. Cogeneration is a form of coproduction.

Corbel - a projection, each stepped progressively further forward in height, used to support an overhanging member above.

Core, air - coil of wire not having a metal core.

Core area - the total plane area of the portion of the grille, face, or register, bounded by a line tangent to the outer edges of the outer openings through which air can pass.

Core, cables - a term used to denote a component or assembly of components over which other materials are applied such as additional components, shield, or armor.

Core loss - refers to the conversion of electrical energy to heat energy in the core material of a magnetic device.

Core, magnetic - iron or steel materials used to make laminated internal sections of electromagnets. The magnetic center of a magnetic field.

Core saturation - when the atoms of a metal core material are aligned in the same pattern so that no more magnetic lines of force can be developed.

Core solder - this is a solder wire or bar with flux as the core.

Cornice - the exterior trim of a structure at the meting of the roof and wall.

Corona - a discharge due to ionization of a gas (usually air) due to a potential gradient exceeding a certain critical value.

Correction factor - a multiplier or penalty that is applied to conductors when the ambient temperature is greater than 30° C (86° F). These multipliers are found in section 310 of the National Electric Code.

Corrective action - the built in action of a controller to correct drift. The action taken by a controller to keep the control point as close as possible to the set point.

Corresponding pressure - the pressure at which both the water and steam and liquid refrigerant and refrigerant gas (vapor) are the same temperature

Corrosion - the process or result of a material being eaten or worn away, usually by chemical attack or electrochemical reaction. It is a transformation process in which the metal passes from its elemental form to a combined (or compound) form.

Corrosion inhibitor - additive for protecting lubricated metal surfaces against chemical attack by water or other contaminants. There are several types of corrosion inhibitors. Polar compounds wet the metal surface preferentially, protecting it with a film of oil. Other compounds may absorb water by incorporating it in a water-in-oil emulsion so that only the oil touches the metal surfaces. Another type of corrosion inhibitor combines chemically with the metal to present a non-reactive surface.

Corrosive - a chemical that causes visible destruction of or irreversible alterations in living tissue by chemical action at the site of contact; or in the case of leakage from its packaging, a liquid that has a severe corrosion rate on steel. A solid or liquid waste that exhibits a "characteristic of corrosivity," as defined by the Resources Conservation and Recovery Act (RCRA), may be regulated by EPA as a hazardous waste.

Cotter pin - a split pin used as a fastener, usually to prevent a nut from unscrewing.

Coulomb - the quantity of electricity transferred by electric current of one ampere in one second. This unit of measurement equals 6.25 x 1018 electrons.

Counter - a device capable of changing states in a specified sequence upon receiving appropriate input signals. The output of the counter indicates the number of pulses that have been applied. A counter is made from flip-flop and some gates. The output of all flip-flops is accessible to indicate the exact count at all times.

Counter electromotive force (CEMF) - "generator action" that takes place in motors when voltage is induced into the rotor conductors that opposes the source voltage. The voltage produced by self-inductance.

Counter flow - in a heat exchanger, the direction of flow of the cooling (or heating) liquid is opposite to that of the cooled (or heated) liquid. A forced air furnace that has its air flow downward as opposed to upward.

Counter flashing - sheet metal strip in the form of an inverted L, built into a wall to overlap the flashing and make the roof watertight.

Counter weight - a dead weight added to a compressor crankshaft to balance the crankshaft.

Counterbore - to enlarge a hole a given amount.

Counterclockwise - from right to left in a circular motion.

Counterflow - flow in the opposite direction. The opposing direction of fluids. The coldest portion of one meeting the one with the warmest portion of the other.

Counterflow arrangement - a piping arrangement that causes two fluids to flow in opposite directions as they pass through a heat exchanger. This is generally desirable for maximum heat exchange.

Counterflow condenser - a water cooled condenser in which the cooling water first contacts the condenser near its coolest point (outlet) and is tapped off near it warmest point (inlet).

Counterflow, fluid - the opposing direction of flow of fluids where the coldest portion of one meeting with the warmest portion of the other.

Counterflow furnace - a furnace that has the air stream in a downward direction. A downflow application. Same as a downflow furnace.

Counterflow spiral - a shape for a radiant floor tubing circuit that is intended to evenly distribute heat within a room.

Counterflow, water tower - a tower in which the air flow direction through the fill is countercurrent to that of the falling water.

Countersink - to bevel a hole to receive a flat or fillister head fastener.

Coupled ocean atmosphere computer model (CMP) - a tool that is a mix of forecast and measured, persistent sea temperatures. CMP skill is estimated from 45 years of simulations using the atmospheric portion of the model coupled with specified sea surface temperatures.

Coupling - a sleeve (female fitting) used to join the ends of two lengths of pipe or tubing. Also a device connecting a motor drive shaft to the driven shaft in a direct drive system.

Coupling, electrical - (1) the method of connecting two circuits together. Direct coupling circuits are connected with a wire while indirect coupling or AC coupled circuits are connected through a capacitor or transformer. (2) The amount of mutual inductance between coils.

Coupling, quick disconnect - a coupling which can quickly join or separate lines. These couplings may or may not have check valves built into them. Examples of these disconnects would be an pneumatic air hose or a refrigeration manifold gauge set.

Course - a layer of masonry units, shingles, tiles, etc., running horizontally.

Covalent bond - atoms joined together that share electrons to form a stable molecule.

Cover - the cover protects the internal wiring of a thermostat base from dust, lint and other harmful substances. It also contains the temperature setting lever, system switching levers and the temperature indicator.

Cover plate - this sheet of glass or transparent plastic sits above the absorber in a flat plate collector.

Crackage - joint in a structure that permits movement of a gas or vapor through it, even under a small pressure difference.

Cracked - state in which a valve is opened slightly by turning the stem about one or two turns.

Cracking - the process of breaking down hydrocarbons by heat and pressure into lighter hydrocarbons of lower molecular weight. For example, the breaking down of petroleum into gasoline.

Cracking a valve - opening of a valve a very small amount.

Cradle mounted motor - a motor with a mounting cradle that fits the motor end housing on each end and is held down with a bracket.

Crankcase - the housing for the compressor crankshaft and a sump for the compressor oil.

Crankcase heater - a heating device fastened to the crankcase or lower portion of the compressor housing intended to keep the oil in the compressor crankcase at a higher temperature than the rest of the system to reduce the migration of the refrigerant to the coldest part of the refrigeration system.

Crankcase pressure regulator - an automatic valve installed in the suction line that prevents the pressure in the crankcase from exceeding the set point.

Crankshaft - the shaft in a compressor that is driven by a motor, and rotates to provide movement to the pistons.

Crankshaft seal - leak proof joint between the crankshaft and the compressor body.

Crankshaft throw - the distance between the centerline of the main bearing journal and the centerline of the crank pin or eccentric.

Crawl space - shallow space between the first tier of beams and the ground (no basement).

Creep - the time-dependent part of strain resulting from stress, that is, the dimensional change caused by the application of load over and above the elastic deformation and with respect to time.

Creosote - a mixture of unburned organic material found in the smoke from a wood burning appliance.

Crest - the top edge of two adjoining threads.

Cricket - small false roof designed to throw off water from behind an obstacle.

Crimp - to squeeze or contract. Also a collapse or kink in a section of tubing often at a bend in the tubing.

Crimp termination - a wire termination that is applied by physical pressure of terminal to wire.

Crisper - a drawer or compartment in a refrigerator designed to provide high humidity along with low temperature to keep vegetables, especially leafy vegetables, cold and crisp.

Criteria pollutant - standards set forth by the EPA for the following pollutants in outdoor air: nitrogen dioxide, sulfur dioxide, ozone, lead, carbon monoxide, and total suspended particles.

Critical bias line - a graph of the gate current required to fire an SCR at given values of anode voltage.

Critical charge - term used to describe the amount of refrigerant in air conditioning or refrigeration systems that are equipped with a capillary tube metering device. All of the refrigerant is moving through the system, so the amount of refrigerant in the system must be within a very specific range to ensure proper system operation.

Critical point - of a substance, state point at which liquid and vapor have identical properties.

Critical pressure - condition of refrigerant at which liquid and gas have the same properties or the point at which the density of water and steam is the same..

Critical pressure drop - fluid flow through a valve increases with increased pressure drop until a critical value is reached. Any drop in excess of this value can cause noise and wear.

Critical pressure ratio - the ratio of the pressure at the minimum area in a converging-diverging nozzle to the inlet pressure. The velocity at this section of the nozzle is sonic under certain conditions.

Critical temperature - the highest temperature at which vapor and liquid can have and still be condensable by the application of pressure.

Critical temperature, electronics - the temperature at which a doped semiconductor will lose its intrinsic properties.

Critical time override - an electromechanical device that bypasses the normal operation of a load management system during times that are extremely busy for building operation.

Critical time programmer - a clock that programs loads to bypass their normal cycling schedule during critical times.

Critical velocity - the velocity above which fluid flow is turbulent.

Critical vibration - vibration that is noticeable and harmful to structure or machinery.

Crocus cloth - used as a sandpaper with a smooth surface.

Cross - a four way connector fitting for pipe or tubing.

Cross braking - when addition support is desired in sheet metal ductwork (or sheet metal panels), the sheet may be placed diagonally in a press brake and slightly crimped, then turned and similarly crimped between the two other opposite points, forming a criss-cross in the flat surface of the sheet.

Cross charged - a sealed container containing two fluids that together create a desired pressure temperature curve.

Cross connected - two pipes or systems of flow connected to each other, usually to provide an equalization or interplay of pressures. For plumbing systems the direct connection between a water supply and a system carrying unsafe water.

Cross drum boiler - see water-tube boiler.

Cross flow, water tower - air flow direction through the fill is essentially perpendicular to that of the falling water.

Cross linked polyethylene - polyethylene in which long molecular chains have been bonded together to form a stronger and more temperature resistant material. The acronym PEX is often used to represent cross linked polyethylene. Primarily used in panel radiant heating situations.

Cross liquid charge bulb - a type of charge in the sensing bulb of the TXV that has different characteristics from the system refrigerant. This is designed to help prevent liquid refrigerant from flooding to the compressor at start-up.

Cross over bridge - a piping segment that allows water to flow from a supply piping main to a return piping main. It usually contains a pair of closely spaced tees or a heat emitter.

Cross over igniter - a slotted length of metal that acts as a bridge between burners.

Cross reference - a book that lists the mechanical or electrical twin of a part made by the same or a different manufacturers.

Cross sectional area - the area of the cut surface of an object cut at right angles to the length of the object.

Cross, straight - fitting in which all outlets are of the same diameter.

Cross "T" - used on connections on a water column for inspection of steam and water lines to ensure they are clean and clear.

Cross talk - electrical interference between two adjacent insulated conductors whereby a signal in one of the conductors will be picked up by the adjacent conductor.

Cross vapor charge bulb - similar to the vapor charge bulb but contains a fluid different from the system refrigerant. This is a special type of charge that produces a different pressure/temperature relationship under different conditions.

Crown - the raised contour, as on the surface of a pulley.

Crude oil - unrefined oil in its natural state as it comes from the ground. Also called petroleum or crude oil.

Crushed hem or fold - a straight edged sheet that has been turned back 180 degrees, and the fold flattened.

Cryogenic fluid - a substance that exists as a liquid or gas at ultra-low temperatures of -250° F (157° C) or lower.

Cryogenics - the science of refrigeration that deals with producing temperatures of -250° F (157° C) or lower.

Crystal - a structure formed by the covalent bonding together of atoms.

Crystal lattice - network of linked atoms that makes up the physical structure of a crystal.

Crystallization - when a salt solution becomes too concentrated and part of the solution turns to salt.

CU - a marking on wire connectors and soldering lugs indicating they are suitable for use with copper conductors only.

Cube of fan speed law - the formula that determines the increase in horsepower required to increase the fan speed. If we wanted to double the speed of a motor the horsepower required would increase eight fold. For example, if you had a 2 horsepower fan operating at 1800 rpm and you wished to double the speed to 3600 would require a 4 horsepower motor. (8 x .5 = 4)

Cuber analyzer - a device used for checking the control module and probes in an ice making machine. Thermistor probes signals are substituted by operating switches in the analyzer, that send all combinations of signals to the card. Thus, a defective component can be identified by the process of elimination. This can also be used to dry cycle an electronic cuber for demonstration or schools.

Cubic foot of gas - the amount of gas that would occupy one cubic foot (0.3M) when at a temperature of 60° F (16° C), saturated with water vapor and under a pressure equivalent to that of 29.921 inches mercury column.

Cubic feet per minute (CFM) - the volume of air, expressed as cubic feet per minute, passing through a piece of equipment.

Cubic measurement - the measure of volume, found by multiplying height, width, and length.

Cumulative compound motor - this is a compound motor where the magnetic field of the series coils aids the magnetic field of the shunt coil.

Cumulative compressor run timer - this defrost timer connection (pigtail to terminal 2, timer contact between terminals 1 and 4) allows the timer motor to operate only when the compressor is in operation and the thermostat is closed.

Cupro nickel - a material used in heat exchangers. This material is made with an alloy of copper for the high corrosion resistance for acid cleaning.

Curb, roof - raised sections formed at the time the basic roof foundations are installed. They are intended as raised bases for equipment foundations and should be flashed between the roof and the equipment

Curie - a unit used to describe the rate decay of a radioactive substance. One Curie (Ci) equals 37 billion disintegrations per second. 1 Ci = 3.7×10^{10}

Curing - maintaining the humidity and temperature of freshly placed concrete to assure hydration and proper hardening of the concrete.

Current - transfer of electrical energy in conductor by means of electrons changing position. It is usually expressed in amperes. Symbol is (I) for "intensity" of current.

Current carrying capacity - the current a conductor of a given size is capable of carrying safety without exceeding its own insulation and jacket temperature limitations.

Current control ratio - as applied to relays, the contact, or load current divided by the pull-in current.

Current limit relay - a relay that protects a circuit from overload by opening its contacts and breaking the circuit when the current flow reaches a predetermined high.

Current limiting fuse - a fuse operation relating to short-circuits only. When a fuse operates in its current-limiting range, it will clear a short-circuit in less than 2 cycle. Also, it will limit the instantaneous peak let through current to a value substantially less than that obtainable in the same circuit if that fuse were replaced with a solid conductor of equal impedance.

Current limiting resistor - a component of an oil pressure failure switch that is center tapped to allow either 240 volts to be connected in series with the heater, or for 120 volts to be connected to the center tap.

Current overload - an overload that opens a set of contacts on high current draw and allows them to close when the current draw has decreased. It usually is a pilot duty device.

Current penetration - the depth a current of a given frequency will penetrate into the surface of a conductor carrying the current.

Current rating - the amount of current flow a device is designed to withstand.

Current regulation - the process of maintaining the output current of a power source constant under varying conditions.

Current relay - a normally open relay that is opened or closed by the starting current of an electric motor. The relay allows the start capacitor and start windings to drop in or out of the starting circuit.

Current transformer - a transformer used to extend the range of an AC ammeter.

Curtain type damper - a damper composed of flexible material, that moves in a vertical plane as it is rolled.

Curtain wall - a non-bearing enclosure wall not supported at each story.

Curved line - a line of which no part is straight.

Curved tooth file - a file that is used on aluminum and steel sheets. Most are designed to cut on the forward stroke.

Cushion (hydraulic) - a deceleration device sometimes built into the ends of a cylinder to restrict the flow of fluid at the outlet port.

Cut away - an illustration of a component in which the outer walls or housing are partially removed to permit viewing of internal parts.

Cut-in - the temperature or pressure value that closes a control circuit.

Cut-off - immediate shutdown of a system after loss of the main burner flame, with no attempt to recycle or to relight the burner.

Cut-out - (1) temperature or pressure value that opens a control circuit. (2) box or enclosure that contains fuse holders and usually also a switch.

Cut-through resistance - the ability of a material to withstand mechanical pressure, usually a sharp edge or prescribed radius, without separation.

Cutting attachment - this is a separate cutting attachment attached to the cutting torch.

Cutting fluid - any fluid applied to a cutting tool to assist in the cutting operation by cooling, lubricating or other means.

Cutting oxygen lever - a lever actuated valve on an oxygen acetylene cutting attachment, used to supply the stream of oxygen needed for cutting.

Cutting tip - this is the part of the oxygen cutting torch from which the oxyacetylene gas is released.

Cwle - the rotation of a motor when viewed from the lead end.

Cwse - the rotation of a motor when viewed from the shaft end.

Cycle, control - a sequence of operations under automatic control intended to maintain the desired conditions at all times.

Cycle efficiency - the net efficiency of a heat source accounting for heat losses during off cycles. Cycle efficiency decrease as run fraction decreases

Cycle, electrical - the voltage generated by the rotation of a conductor through a magnetic field from a zero reference, in a positive direction, back to zero, then in a negative direction and back to zero. One complete cycle equals 360° of rotation.

Cycle, system - series of events or operations that have a tendency to repeat in the same order.

Cycles - segment of AC signal where current travels from zero to a positive peak down through zero to a negative peak and returns to the same amplitude as at the start.

Cycles per second - the number of complete alternating current waved in one second of current flow.

Cycling - (1) continuous oscillation occurring without periodic stimuli. (2) A situation in a closed loop system where the controller's response (sensitivity) to an input change causes this instability.

Cycling rate - the number of complete cycles that the system goes through in one hour. One complete cycle includes both "on" and "off" times.

Cyclo - a refrigerant having carbon atoms bound to each other in a ringed arrangement. ASHRAE includes a C at the beginning of the refrigerant number. Example: R-C318.

Cyclone separator - separates water droplets from steam by centrifugal force and by changing directions.

Cylinder - (1) a device that converts fluid power into linear mechanical force and motion. This usually consists of movable elements such as a piston and piston rod, plunger or ram, operating within a cylindrical bore. (2) A closed container for fluids.

Cylinder head - the top to the cylinder on the high pressure side of the compressor. The part that encloses the compression end of the compression cylinder.

Cylinder, hydraulic - a part of the valving mechanism that is used in a capacity control actuator system.

Cylinder, refrigerant - cylinder in which refrigerant is purchased and dispensed. Color code painted on cylinder indicates type of refrigerant cylinder contains.

Cylinder spring - the spring located between the cylinder head and discharge valve assembly of a reciprocating compressor to insure protection against damage from slugs of liquid refrigerant or oil. The entire discharge valve assembly will "lift" against the spring and pass the slug, then return to its normal operation.

Cylinder unloading - a method of providing capacity control by causing a cylinder in a reciprocating compressor to stop pumping.

Cylindrical commutator - a commutator with its contact surfaces that are parallel to the rotor shaft.

-D-

Da - direct acting

Db - dry bulb.

DB - decibel

Dba - sound level reading on the a-weighted scale of a sound meter.

Dac - digital to analog converter.

Dc - direct current.

DDC - direct digital control.

Dia - diameter.

Diag - diagram.

Diff - differential.

Disc - disconnect.

Dm - demand meter.

Dmm - digital multi meter

Dms - Drill Manufacturer's Standard (DMS)

Dpc - differential pressure control.

DPDT - double pole double throw.

DPST - double pole single throw.

DS - disconnect switch.

Dsv - discharge service valve.

DTS - defrost termination switch.

Dwg - drawing.

DWV - drain waste and vent.

Dx - direct expansion.

D type flip-flop - D stands for data. A flip-flop whose output is a function of the input that appeared one clock pulse earlier. For example, if a logic 1 appeared at the input, the output after the next clock pulse will be a logic 1.

D'arsonval meter - the internal portion of a meter that has a stationary permanent magnet, an electromagnetic coil, and a pointer that moves in direct proportion to the current flow through the coil.

Daily range - the range of outdoor ambient temperatures encountered in an area over a 24 hour period. The temperatures are taken at the design high and low temperature for the area.

Daily range, high - a range of outdoor ambient design temperatures both high and low that exceeds 25° F. A low humidity area.

Daily range, low - a range of outdoor ambient design temperatures both high and low that is less than 15° F. A high humidity area.

Daily range, medium - a range of outdoor ambient design temperatures both high and low that is in the 15° and 25° F range. A medium humidity area.

Dalton's Law of Partial Pressure - where the vapor pressure exerted on a container by a mixture of gases is equal to the sum of the individual vapor pressures of the gases contained in mixture.

Damp location - partially protected locations under canopies, marquees, roofed open porches, and like locations, and interior locations subject to moderate degrees of moisture, such as some basements, some barns, and some cold storage warehouses.

Damper - a device for introducing a variable resistance for regulating the volumetric flow of air through an air outlet or inlet. Also see butterfly, curtain, flap, louver, slide type, opposed and parallel blade.

Damper, fire - a damper placed in a duct or at an outlet to prevent or retard the emission of smoke and/or heat into a room or space during a fire. Such devices are also designed to prevent or retard the spread of fire throughout the structure, thus confining it to a specific area.

Damper, flue gas - a damper located on the downstream side of the combustion chamber of a fuel-burning appliance, usually in a flue passage of the appliance or in the chimney or vent connector.

Damper motor - a motor that, through mechanical linkage, opens or closes a damper. Also see firing rate motor.

Damper vent device, automatic - a device intended for installation in the venting system, in the outlet of or downstream of the appliance draft hood, of an individual automatically operated fuel gas burning appliance and that is designed to open the venting system automatically when the appliance is in operation and to close off the venting system automatically when the appliance is in a standby or shutdown condition.

Damper, volume - a device that will restrict, retard, or direct the flow of air in any duct, or the products of combustion of heat producing equipment, vent connector, vent or chimney.

Dampening - the reduction of noise, pulsation, or vibration.

Data hold - the ability to freeze the display on any given measurement. Useful in applications where the instrument is not easily read while a measurement is being made.

Data plate - an equipment identification label, usually including such items as brand name, model, serial, voltage, and amperage.

Daughters (progeny) - the new radio nuclides formed by the radioactive decay of a parent radio nuclide.

Davit - small crane at the top of a column (tower) used to lift equipment.

Daylight transmittance - the percentage of visible light that glazing transmits through a window. A standard clear dual pane has a daylight transmittance of 82%.

Day-night thermostat - a thermostat that can be indexed to control at one temperature during day operation and another temperature at night.

DbA - sound level reading (below 55db sound pressure level, but frequently used to gauge levels to 85 db) on the a-weighted scale of a sound meter. This a-weighted adjusts response of the meter to approximate that of the human ear.

DbB - a filtering system that has characteristics that roughly match the response characteristics of the human ear at sound levels between 55 and 85 db.

DbC - a filtering system that has characteristics that roughly match the response characteristics of the human ear at sound levels above 85 db. The filtering system in this case is flat with frequency.

Dc generator - a rotating machine that produces a form of direct- current (DC) voltage.

Dc power supply - an element of an electronic system consisting of three components; a full wave rectifier, a filter, and a voltage regulator.

DDC - direct digital control. A computer based means of programming the operation of operating controls.

De - equivalent diameter in inches. $\sqrt{A \times 4 / \pi}$ Where A = area in square inches.

Dead - portion of an electrical system having no voltage or charge.

Dead air space - a sealed layer of dry, motionless air having an insulating value.

Dead band - a temperature band between the on and off of an output in the on off control action. No heating or cooling takes place. Band occurs between the times when the heat is turned off upon rising temperature and turned on upon falling temperature.

Dead end line/system - the end of a pipe line that does not lead back to an oil storage tank, so the oil in that end of the line cannot be re-circulated. A system that does not contain a return line to an oil storage tank, so the oil cannot be re-circulated in a closed circuit.

Dead front - without exposed live parts.

Dead heading - an undesirable operating condition in which a circulator is running, but all flow through it is blocked.

Dead leg - a grounded phase of a three phase delta transformer.

Dead load - the weight of a structure itself, including the weight of fixtures or equipment permanently attached to it.

Dead short - connections that permits an unrestricted flow of electrons.

Dead space - the short distance between a burner port and the base of the flame.

Dead time - any time period of delay between the action of the controller and the resulting action of the actuator.

Dead weight tester - used to test a pressure gauge so that it can be recalibrated.

Deaerator - a separator that removes air from the system fluid through the application of bubble dynamics.

Deaerating feedwater heater - type of open feedwater heater equipped with a vent condenser.

Deaeration - the act of removing air from substances like a boiler's water supply.

Debug - the process of removing a problem or trouble from a system or piece of equipment.

Decay rate - the rate at which the sound pressure level in an enclosed space decreases after the sound source has stopped. In is measured in decibels per second.

Decibel (db) - unit used for measuring the relative loudness of sounds. One decibel is equal to approximate difference of loudness ordinarily detectable by the human ear, the range of which is about 130 decibels on scale beginning with one for faintest audible sound.

Decimal equivalent - a fraction converted to the number of tenths or hundredths that the fraction is equal to.

Deck (coil deck) - insulated horizontal partition between refrigerated space and the evaporator space.

Declination of sun - the angle above or below the equatorial plane. It is plus if north of the plane, and minus if below. Celestial objects are located by declination.

Decomposition - process of chemical change, breaking up of structures, spoilage.

Dedendum - distance from pitch circle to bottom of tooth space.

Dedicated - set apart or committed to a definite use. Addressed to a specific task, job or circuit.

Dedicated circuit (electrical) - an electrical circuit that supplies power to one appliance only. Fused for that one circuit only.

Dedicated geothermal well - a system where water is drawn from the top of the well and is returned to the bottom of the well. The well is dedicated to supply water for the use of the heat pump only.

De-energize - to stop the electron flow to an electric device.

Deep vacuum - a vacuum reading approaching 30" of mercury.

Deep well submersible pump - a centrifugal pump in which a number of impeller assemblies, in a housing, are mounted on a shaft directly coupled to a submersible motor. The entire assembly is located at the bottom of the well. Power is brought to the motor by a waterproof cable.

Default value - a value that will automatically be selected by the controller if a specific value is not manually chosen.

Deficiency of air - a supply of air that is inadequate for complete combustion of a fuel. This is the same as excess of fuel. Also see fuel rich.

Definite purpose contactor - similar in construction to a general purpose contactor. However, they are lighter duty and often considered to be a throw away product when compared to an equivalent NEMA or IEC contactor. Used in HVAC, data and food processing equipment.

Definite purpose motor - a motor designed for use on a particular application, with operating characteristics and mechanical construction required by the application. Examples are sump pump motor or an oil burner motor.

Deflect - to swerve; to bend or turn to one side.

Deflection - (1) the departure of an indicator or pointer from the zero reading on the scale of an instrument. (2) the deformation of a structural member as a result of loads acting on it.

Deflection temperature - the temperature at which a specimen will deflect a given distance at a given load under prescribed conditions of the test.

Defrost - to melt or do away with frost and ice.

Defrost, automatic - an automatic defrost system is one in which the defrost cycle is automatically initiated and terminated, with automatic resumption of normal refrigeration at the conclusion of the defrost action. The defrost water is disposed of automatically.

Defrost condensate - the condensate or water from a defrost application of a heat pump or refrigeration system.

Defrost control - a device that senses the accumulation of frost on an evaporator and/or initiates and terminates the defrost cycle. This control causes the system to reverse the refrigerant flow to remove the frost and/or ice from the coil.

Defrost cycle - refrigerating cycle in which the evaporator frost and ice accumulation is melted.

Defrost, manual - one in which defrosting of the refrigerated surface is accomplished by natural or manual means with manual initiation and manual termination of the overall defrost operation.

Defrost mode - the portion of the operation of the system in which the frost and/or ice on the evaporator is removed.

Defrost termination thermostat - a temperate sensing device that initiates and terminates the defrost cycle at preset temperatures.

Defrost thermostat - a temperature sensing device that initiates and terminates the defrost cycle at predetermined temperatures.

Defrost timer - a device connected into electrical circuit that shuts unit off long enough to permit ice and frost accumulation on evaporator to melt.

Defrost valve - a four way reversing valve used in air to air heat pumps.

Defrosting - the process of removing frost accumulation from evaporators.

Defrosting, hot gas - the use of high pressure refrigerant gas used to defrost the low side of the refrigerant system. May be started by time, temperature or a combination of the two.

Defrosting type evaporator - an evaporator operating at such temperatures that ice and frost on surface melts during off part of the operating cycle.

Degradation - refers to a reduced level of service at which a system is operated.

Degas - removing air from a liquid, usually by ultrasonic and/or vacuum methods.

Degreasing solvent - a solution or solvent that is used to remove oil or grease from parts.

Degree (electrical) - the 360th part of the angle subtended at the axis of the machine by two consecutive field poles of like polarity. One mechanical degree is thus equal to as many electrical degrees as there are pairs of poles in the machine.

Degree (physics) - a unit of temperature measurement.

Degree day - a unit that represents 1° of difference from a given point in average outdoor temperature of one day and is often used in estimating fuel requirements for a building. Degree days are based on average temperature over a 24 hour period. As an example, if an average temperature for a day is 50° F the number of degree days for that day would be equal to 65° F minus 50° F or 15 degree days. Degree days are useful when calculating requirements for heating purposes.

Degree of superheat - the difference between the boiling point of the refrigerant and the actual temperature above the boiling point.

Dehumidification - the condensation of water vapor from air by cooling below the dew point or removal of water vapor from air by chemical or physical methods.

Dehumidifier - a device used to remove moisture from air in an enclosed space. Can either be an air cooler or washer used for lowering the moisture content of the air passing through it or an absorption or adsorption device for removing moisture from the air.

Dehumidifier, surface - an air conditioning unit, designed primarily for cooling and dehumidifying air through the action of passing the air over wet cooling coils.

Dehumidify - the ability to remove water vapor from the atmosphere. To remove water or liquid from stored goods.

Dehumidifying effect - the difference between the moisture contents, in pounds per hour, of the entering and leaving air, multiplied by 1.060.

Dehydrate - to remove water or water vapor; to become dry; to lose water. Liquid water, hygroscopic water, and water of crystallization or water of hydration are included.

Dehydrated oil - a lubricant that has the water removed to an acceptable level.

Dehydration - the removal of water vapor from air by the use of absorbing materials or the removal of water from stored goods.

Dehydrator - a refrigeration accessory that removes water from liquid refrigerant. Most commonly called a drier or filter drier.

Dehydrator/receiver - a small tank that serves as liquid refrigerant reservoir and also contains a desiccant to remove moisture. Used on most automobile air conditioning installations.

De-ice control - a device used to operate a refrigerating system in such a way as to provide melting of the accumulated ice or frost.

Delay - the period of time before an event, or the period of time of an event.

Delay-on-break - a function of a timing device whereby delay is initiated when the switch is opened or the circuit is de-energized. When set time has elapsed, load may energize immediately on normal demand. If abnormal conditions are sensed, load will de-energize and delay is initiated.

Delay-on-make - a function of a timing device whereby delay is initiated when the switch is closed or when the circuit is energized.

Delayed ignition - an undesirable condition where there is a delay in the ignition of a combustible mixture of gas and air that has accumulated in the combustion chamber.

Delayed mixing - a process in which the fuel and air leave the burner nozzle unmixed and, thereafter, mix relatively slowly, largely through diffusion. This results in a long, luminous flame, called diffusion flame, long flame, luminous flame, or yellow flame. Also see long flame burner.

Delta configuration - a configuration of three motor windings. Motors wired in a delta configuration have twelve wires protruding from the motor. These motors can be wired for operation at different voltages, depending on the available voltage.

Delta transformer - a three phase electrical transformer that has the start end of one winding connected to the finished end of the second. Power input lines come into the junctions of the windings. The circuit configuration resembles the Greek letter Delta Δ.

Delta T - as used in the heat loss and heat gain, it represents the difference between the inside and outside design temperatures. The difference in temperature across or through a heat exchange item.

Delta winding - a winding layout of some three phase motors, where the beginning of the windings is connected to the ending of the windings.

Demand - (1) the amount of power that a power company must supply at any given time or period. (2) The instantaneous consumption requirements of a device on an electrical power circuit. (3) The maximum amount of gas input required per unit of time, usually expressed in cubic feet per hour, or Btu/h.

Demand billing - the demand upon which billing to a customer is based, as specified in a rate schedule or contract. The billing demand need not coincide with the actual measured demand of the billing period.

Demand charge - that part of an electric bill based on the rate of electric power consumption in kilowatt-hours for a specified demand time (15 or 30 minutes) interval (peak) in addition to the cost for the total electricity consumed.

Demand defrost - the process by which a heat pump will go into defrost mode whenever there is sufficient frost or ice on the coil to warrant defrost. This type of defrost does not employ a predetermined time interval between defrost attempts.

Demand factor - average demand over a specific period divided by the maximum demand over the same period (e.g., monthly demand factor, annual demand factor).

Demand fired - a term describing a boiler that is fired when there is a demand for heat (e.g., the boiler does not maintain temperature at all times).

Demand interval - the period of time during which the electric energy flow or load is averaged in determining effective demand, such as 60 minutes, 30 minutes, 15 minutes, or instantaneous.

Demand load - the amount of power that would probably be needed at any given time. Generally, the minimum demand load is considered to be about 35% of the connected load.

Demand meter - a device that indicates or records the demand or maximum demand. Since demand involves both an electrical factor and a time factor, mechanisms responsive to each of these factors are required as well as an indicating or recording mechanism. These mechanisms may be either separate from or structurally combined with one another. Demand meters may be classified as follows:

Class 1 -- curve drawing meters,
Class 2 -- integrated-demand meters,
Class 3 -- lagged-demand meters.

De minimis - minimum. According to Section 608 of the Clean Air Act of 1990, the smallest quantity of refrigerant released in the course of making good faith attempts to recapture and recycle or safely dispose of refrigerant.

Dendrite - a slender threadlike spike of pure crystalline material, such as silicon.

Density - in the context of this book, density refers to the weight of a substance divided by its volume. Some readers will recognize this as being the same as the specific weight of a substance. Common English units for density, as defined herein, are lb/ft^3 (pounds per cubic foot).

Deodorizer - a device that absorbs various odors, usually by principle of absorption. Activated charcoal is a common substance used.

Depletion mode - the operation of a field effect transistor in response to an increase in the gate-to-source voltage in an effort to reduce the drain current.

Depletion region - an area on either side of a P-N junction that contains impurity ions that is depleted of carriers.

Depreciation - decrease in value of a fixed asset due to wear and tear or obsolescence.

Depression - the difference between wet and dry bulb temperatures.

Depressor fork - a device that when energized, moves to hold open the cylinder suction valves in certain compressor capacity modulation systems.

Derating factor - a factor used to reduce a current carrying capacity of a wire when there are more than three current carrying wires in a raceway or cable. This factor is used in other than the environment for which the value was established. These multipliers are found in Table 310 in the National Electric Code. Also referred to as an adjustment factor.

Derivative mode - the portion of the controller transfer function that changes its output signal in response to the rate of change in the process error. This mode is responsible for reducing the effects of fast process changes on the stability of the control loop by responding to the anticipated change in the process based upon its current rate of change. Also called rate mode.

Desiccant - a substance used to collect and hold moisture in refrigerating system. A drying agent. Common desiccants are activated alumina or silica gel.

Desiccator - a machine that applies vacuum, heat, or a chemical process to the process medium to remove unwanted liquid.

Design certificate - the process by which a product is evaluated and tested by an independent laboratory to affirm that the product design complies with specific requirements.

Design conditions - conditions that help to determine the worst case heating and cooling requirements for selecting HVAC equipment.

Design cooling conditions - summer outdoor design conditions listed for selected locations.

Design dry bulb temperature - the temperature the outside air is at, or above, 97.5% of the year. This is the outside temperature at which the design heating and cooling load is calculated.

Design heating conditions - winter outdoor design conditions listed for selected locations.

Design heat loss - the heat loss of a building or room at design indoor outdoor temperature difference.

Design load - the design heat loss plus all other heating requirements to be provided by the heating appliance.

Design model - model that includes all equipment, vessels, piping, valves, fittings, electrical, instrumentation, and HVAC. The main purpose of a design model is to develop the ultimate piping arrangement with a minimum number of drawings. A design model is constructed in two stages: the basic model and the final model.

Design pressure - highest or most severe pressure expected during operation. Sometimes used as the calculated operating pressure plus an allowance for safety.

Design temperature difference - the difference between the design indoor and outdoor temperatures.

Design temperature, indoor - the temperature to be maintained within the conditioned space.

Design temperature, outdoor - the outdoor temperature arbitrarily established as the maximum against which the system must be able to maintain the desired indoor conditions used when figuring heat loss and heat gain calculations.

Design value - is the amount of heating or cooling a system is designed to produce at maximum conditions.

Design water temperature - the average of the temperature of the water entering and leaving the boiler when the system is operating at design conditions.

Design water temperature difference - the difference between the temperature of the water leaving the unit and entering the unit when the system is operating at design conditions.

Design water temperature drop - the difference between the temperature of the water leaving the boiler and returning to the boiler when the system is operating at design conditions. In large systems employing sub-circuits the design temperature drop is usually taken as the difference in the temperature of the water entering and leaving each sub-circuit.

Design working pressure - the maximum allowable working pressure, in pounds per square inch gauge (psig), for which a specific part or component will withstand.

Desired value - the value of the controlled variable that is desired to maintain.

Desuperheater - an accessory that lowers the temperature of superheated steam to saturation conditions.

Desuperheating - cooling down a product down to its condensing saturated temperature. In the case of a refrigeration system, the superheated refrigerant in the condenser is cooled to the condenser saturation temperature.

Detail drawing - drawings that show enlarged views of different parts of the structure. They show parts that are unusual or have a special design, such as an arch or duct design.

Detector - a device that produces an electrical output that is a measure of the radiation incident on the device. Also see flame detector, smoke detector, and sensor.

Detector, leak - a device used to detect and locate refrigerant leaks.

Detent - as in a rotary switch, the means by which the switch is snapped into a given position and held there. Different types of detent actions are magnetic, bimetal, over-enter, and mercury. Also known as snap action.

Detergent - in lubrication, either an additive or a compounded lubricant having the properties of keeping insoluble matter in suspension thus preventing its deposition where it would be harmful. A detergent may also re-disperse deposits already formed.

Deterioration - becoming or being made worse; depreciation.

Development, pipe - view on pipe fabrication drawing that shows the unfolding of pipe intersections in one plane. Developments are used for the construction of template drawings for pipe intersections.

Deviation - the difference between the set point and the value of the controlled variable at any instant.

Device - a component that primarily carries, but does not utilize electrical energy. Switches, receptacles, fuses, and lamp holders are examples of devices.

Device port - the port on a service valve that is connected to the device, which is normally the compressor or the receiver.

Dew - condensed atmospheric moisture deposited in small droplets on cool surfaces.

Dew point - the temperature at which a vapor will start to condense into its liquid form.

Dew point, apparatus - that temperature that would result if the psychrometric process occurring in a dehumidifier, humidifier or surface cooler were carried to the saturation condition of the leaving air while maintaining the same ratio of sensible to total heat load in the process. Also known as effective coil temperature.

Dew point refrigerant - the saturated vapor pressure for zeotropic blends, as listed on a temperature pressure card. The dew point temperature indicates the pressure at which the first drop of liquid condenses from a vapor.

Dew point rise - increase in moisture content (specific humidity) of air expressed in terms of rise in dew point temperature.

Dew point temperature - the temperature at which vapor (at 100% humidity) begins to condense and deposit as free water. At the dew point temperature, the air is 100% saturated, the dry bulb and wet bulb temperature are the same, and the relative humidity is 100%.

Diac - a semiconductor often used as a voltage sensitive switching device.

Diagnosis - deciding the nature of trouble or difficulty by careful observation and examination.

Diagonal - a line running from opposite corners.

Diagonal bracing - where additional support is desired in sheet metal, it is placed diagonally in a brake and slightly crimped, then turned and similarly crimped between the other two opposite points, forming a criss-cross in the flat surface of the sheet metal.

Diagonal cutter - diagonal cutters are hand tools that are made for pulling cotter pins, spreading cotter pins, and cutting small gauge wire.

Diagram, electrical - drawing that indicates parts and layouts of electrical circuit.

Diagrams - non-scaled views showing arrangements of special system components and connections not possible to clearly show in scaled views. A schematic diagram shows circuit components and their electrical connections without regard to actual physical locations. A factual or pictorial wiring diagram shows circuit components and the actual electrical connections.

Dial stem thermometer - a type of thermometer that can be operated by a bimetallic strip or a bellows charged with a volatile fluid. Some dial stems have a sensing bulb connected to the bellows by a capillary tube. The temperature range is from -40° F to 250° F.

Diameter - the length of a straight line running through the middle of the circle. The distance across the widest part of a circle.

Diameter of a circle - a straight line drawn through the center of a circle from one extreme edge to the other. Equal to twice the radius.

Diametral pitch - the number of gear teeth per inch of pitch diameter.

Diaphragm - a flexible membrane usually made of thin metal, rubber, or plastic that separates two pressure differences.

Diaphragm type expansion tank - an expansion tank that contains a flexible diaphragm that separates the air and water in the tank.

Diaphragm valve - a solenoid operated, gas valve in which a membrane or partition separates the inlet and outlet. When the solenoid is de-energized, the main gas supply applies pressure to the top of the diaphragm to help hold the valve closed. When the solenoid is energized, a plunger opens a bleed port in the top side of the valve. As the gas bleeds off the top of the diaphragm, main gas pressure on the underside of the diaphragm opens the valve. These valves are generally used on atmospheric type gas burners for smooth light off. They feature variable, slow opening action and fast closing action. Also see solenoid.

Dictates - controls; regulates.

Dichlorofluoromethane - refrigerant commonly known as R-12. Chemical formula is CCl_2F_2. Cylinder color code is white. Boiling point at atmospheric pressure is -21.62° F.

Die - any of various tools used for cutting out, forming, or stamping material. Any of several component pieces that are fitted into a die stock to cut external threads on screws or bolts.

Die cast - a process of molding low temperature metals in accurately shaped metal molds.

Dielectric - the insulating material separating the conducting surfaces of a capacitor. Materials could be air, mica, plastic, wax impregnated paper or ceramic.

Dielectric constant - a number that represents the ability of a dielectric to develop an electrostatic field, compared to air, that have a value of 1.0. The higher the K value, the better the dielectric material.

Dielectric fluid - a fluid with a high electrical resistance.

Dielectric strength - a term used to describe the limit, without damage of an insulating material, to an applied voltage potential.

Dielectric stress - in a charged capacitor, the electron orbit of the atoms in the dielectric extend and are considered to be in tension, or dielectric stress.

Dielectric union - a union made of steel on one side and brass on the other. A dielectric (electrically insulating) material separates the two metals to prevent galvanic action from occurring.

Die stock - a tool used to hold dies with external threads.

Dies (thread) - a tool used to cut external threads.

Difference in potential - the voltage across two points of a circuit.

Differential - the smallest range through which the controller variable must pass in order to move the final control element from one to the other of its two possible positions, such as on to off or cut-in to cut-out.

Differential amplifier - this component helps overcome induced noise by detecting voltage difference between pairs of signal wires and amplifying the difference so noise is ignored.

Differential control - a control that monitors the difference between two factors. The controller then takes this information and converts in into an output signal to the controlled device that will vary in accordance with the difference of the two sensed values.

Differential gap - this term applies to two position controllers. It is the smallest range through which the temperature, pressure or relative humidity must past to make the controller move from one to the other of its fixed positions.

Differential pressure - the difference in pressure between two points.

Differential pressure control - a controller that measures and controls the difference between two separate pressures.

Differential thermostat - a device that provides the control functions for solar heating and domestic hot water systems by measuring the leaving water temperature at the collector outlet and comparing this with the temperature of the water in the storage tank.

Differential valve - a valve that controls the bypass line and opens when the pressure difference between the receiver and hot gas discharge exceeds a preset value.

Diffuse - the migration of molecules of one substance through the molecular structure of another material. Oxygen, for example, can slowly diffuse through the molecular structure of certain polymers.

Diffuse insolation - sunlight received indirectly as a result of scattering due to clouds, fog, haze, dust or other substances in the atmosphere.

Diffuse radiation - this is scattered sunlight from the entire "sky vault" that forms air molecules, dust, and water vapor.

Diffuse sound field - a space in which at every point the flow of sound energy in all directions is equally probable.

Diffuser, air - a circular, square or rectangular air outlet comprised of deflecting members generally located in the ceiling and designed to distribute air in varying directions and planes.

Diffuser, centrifugal - a small passage through which compressed vapor is discharged from the impeller to the volute in a centrifugal compressor.

Diffusion - distribution of air within a space by an outlet discharging supply air in a spreading pattern.

Diffusion brazing - is a brazing process that bonds metals by heating filler metal by either capillary attraction or diffusion of the filler metal.

Diffusion device - a device which mixes incoming supply air with air already in the conditioned space.

Diffusion flame burner - see long flame burner.

Diffusion length - the mean distance a charge carrier (free electron or hole) moves before recombining with another hole or electron. Distances are short, typically several micrometers to a few hundred micrometers. Cell efficiency improves with increasing minority carrier diffusion length.

Diffusity - the ability of earth material to take on or give off heat energy depending on its density and moisture content.

Diffusivity - thermal conductivity divided by product of density and specific heat. Formula is k/pc. Units are square feet per hour.

Digital - information in discreet quantities; usually referring to a display that uses discreet numbers.

Digital clamp-on meter - a numerical read-out instrument primarily designed to measure moderate levels of current. Many clamp-ons have some DMM capabilities, although these may be over restricted ranges.

Digital code - a system of symbols that represent data values and make up a special language that a computer or a digital circuit can understand and use.

Digital device - a device that has only two states of operation, on or off.

Digital input/output - input or output information translated into binary numbers (on-off signals) that can be processed by a digital computer.

Digital logic - circuit elements connected in such a manner as to solve problems using components that have only two states of operation.

Digital multimeter - a meter that uses electronic circuitry to provide a direct numerical readout of a measured value. Digital multi meters have differing levels of measurement capability.

Digital signal - discrete or discontinuous signals whose various states are discrete intervals apart.

Digital to analog converter - an electronic device used to interface digital computer based systems to the analog world. The DAC converts the magnitude of a binary word into an analog output signal. Usually abbreviated D-A, D/A or DAC.

Digital VOM - a volt-ohm-milliampere meter that uses direct current reading numerical display as opposed to a meter movement.

Dike - wall or enclosure around a tank or holding vessel to provide a backup system in case of ruptures. The dike must be able to hold the total contents of the enclosed vessels or tanks.

Dilemma - a situation requiring a choice between undesirable results.

Dilution air - air that enters a draft hood or draft regulator and mixes with the flue gases.

Dimension and extension lines - thin lines that show the extent and direction of dimensions on blueprints or plotted prints. These lines will tell you the length of an object and in which direction it is going. Direction lines usually end against extension lines with arrowheads or tick marks.

Dimensions, overall - the projected dimensions of a device, usually on horizontal and vertical planes, that can be used to determine whether the device will fit in an assigned space or can be moved through a designed passageway.

DIN rail - an industry standard mounting rail for relays and other modular controls.

Diode - a solid state device composed of both P-type and N-type material. When connected to a circuit one way, current will flow. When the diode is reversed, current will not flow. Also serves as a rectifier.

Diode test - the capability of some digital multi meters to determine whether a diode is functioning properly.

Dioxin - a family of chemicals involving chlorine, hydrogen and carbon atoms. It is formed during combustion when chlorine is present, but also as a by-product in some manufacturing processes, such as paper manufacturing. There are 75 forms with 2,3,7,8-TCDD, the most toxic. Diesel fueled vehicles and hospital waste incinerators are significant sources.

Dip soldering - a soldering process in which the heat comes from a molten metal bath.

Dip tube - tube that extends to about one half inch from the bottom of the receiver to assure that only liquid enters the liquid line at the receiver outlet.

Direct acting controller - a controller that increases the air pressure on a pneumatic control system with an increase in the temperature, pressure, etc., of the controlled variable.

Direct bleed leak port - an assembly within a pneumatic controller that incorporates a flat lever positioned in the exhaust stream of a flat tipped nozzle. This produces a non-linear output with negligible feedback.

Direct burner ignition - an ignition system in which the ignition source directly lights the main burner without the use of a pilot such as direct spark or hot surface ignition.

Direct connected - driver and driven, as motor and compressor, positively connected in line to operate at the same speed.

Direct current (DC) - an essentially constant value current that flows in only one direction in a circuit. The varying current in one direction is a pulsating direct current, which may be derived from rectified alternating current.

Direct current excitation current - adjusted to change the power factor of the synchronous motor to normal, over excited, or under excited depending on adjustment.

Direct current motor - requires direct current supply and primarily used for adjustable speed drive applications or motor vehicles.

Direct current resistance - the resistance offered by any circuit to the flow of direct current.

Direct digital control (DDC) - use of digital computer to perform required automatic control operations in a total energy management system.

Direct drive - a mechanical linkage where the shaft of the prime mover, or motor, is directly connected to the driven shaft or blower wheel, both operating at the same speed.

Direct expansion cooling - an air conditioning system that uses refrigerant in a coil to cool return air. The system also contains a refrigerant compressor and a heat rejection device (condenser).

Direct expansion evaporator - an evaporator coil using either an automatic expansion valve (AXV), a thermostatic expansion valve (TXV), capillary or fixed orifice plate type refrigerant control to supply liquid refrigerant at the correct pressure and boiling point to obtain the desired rate and heat absorption into the liquid refrigerant.

Direct field - the sound in a region in which all or most of the sound arrives directly from the source without reflection.

Direct-fired heater - a heater in which the products of combustion (flue gases) are intermixed with the medium being heated (e.g., air). Also see direct fired external and direct-fired internal heaters.

Direct-fired external heater - any oven heating system in which the products of combustion from the burners are discharged into the oven chamber by a circulating fan or blower, and in which the burners are in a combustion chamber effectively separated from the oven chamber. There are two types.

Non-recirculating - products of combustion are discharged into the oven chamber and are not returned to the combustion chamber.

Recirculating - oven chamber atmosphere is circulated to the combustion chamber and is in contact with the burner flame.

Direct-fired internal heater - any oven heating system in which the burners are within the oven chamber, and the products of combustion are in contact with the oven atmosphere.

Direct fired make up heater - a heater in which all the products of combustion generated by the fuel gas burning device are released into the outside air stream being heated.

Direct gain - a direct gain passive solar system utilizes south facing glazing to allow sunlight to enter a home or building. A large window area accepts direct sunlight while thermal mass serves as storage.

Direct injection mixing - a piping configuration in which hot water from a heat source is injected into a distribution system, and cool water from the return side of the distribution system flows back to be reheated by the heat source.

Direct insolation - the sunlight falling directly upon a collector.

Direct method - refers to a way of taking advantage of solar heating through sunlight that enters windows of a house or structure.

Direct radiation - the energy from the sun that reaches the earth directly.

Direct return system, hot water - a two-pipe hot water system in which the water, after it has passed through a heating unit, is returned to the boiler along a direct path so that the total distance traveled by the water from each radiator is the shortest feasible. There is, therefore, a considerable difference in the lengths of the several circuits composing the system.

Direct spark ignition (DSI) - use of an ignition spark, generated by a high voltage transformer, to light the main burner; applies to burners not using a pilot, usually small oil burners. Also, see proved spark ignition.

Direct vent appliances - appliances that are constructed and installed so that all air for combustion is derived directly from the outside atmosphere and all flue gases are discharged to the outside atmosphere. Often referred to as sealed combustion, these systems are usually installed horizontally through an outside wall. For furnace applications, these are considered to be Category IV.

Direct venting - the mechanical exhausting of the flue gases of a heating unit in a structure that does not have a suitable chimney.

Directivity factor - the ratio of the sound pressure squared at some fixed distance and on the axis of maximum response to the mean square sound pressure at the same distance averaged over all directions from the source.

Directly proportional - when one quantity increases or decreases, causing another quantity to do the same. As with a 120 volt to 24 volt rated step down transformer that has a voltage of 60 volts placed on the primary, the output would be proportional to the input and deliver an output voltage of 12 volts.

Direction of electron flow - the flow of electrons, in a circuit, from negative to positive potentials.

Dirt - refers to scale, sludge, flux, and metallic particles. These contaminants must be removed to protect the operation of the solenoid, compressor, expansion valves, or capillary tube and the refrigeration system as a whole.

Disc - the movable part of the valve that makes contact with the valve seat when the valve is closed and that varies the area controlling the flow. The three most common types of control valve discs are contoured, V-port, and quick opening. A contoured disc controls flow by a shaped end, sometimes called a valve plug, and is usually end-guided at the top or bottom (or both) of the valve body. A V-port disc has a cylinder (called a skirt or guide) that rides up and down in the seat ring; the skirt guides the disc, and by virtue of shaped openings in the skirt, varies the flow area. Quick opening discs are so machined that maximum flow is quickly achieved when the disc lifts from its seat; they can be either end-guided or guided by wings riding in the seat ring. Discs are sometimes built so that the part of the disc that comes in contact with the seat may be replaced. This type of disc is known as the renewable disc. A common example is an ordinary is an ordinary hot water faucet in the kitchen sink. The renewable discs are usually made of a composition material. Many valves having all metal or nonrenewable discs have to be "ground in" to restore the seating surfaces if it is damaged; those valves with the flat-type metal disc do not.

Disc, rupture - diaphragm placed between flanges on a vent line. The diaphragm has a specific breaking pressure and will fail before the safety limits of the system have been reached.

Discharge - output of a pump, fan or compressor. To remove energy from a capacitor.

Discharge bypass valve - a modulating valve that controls the hot gas bypass to automatically maintain a desired minimum evaporator pressure.

Discharge coefficient - the ratio of the actual flow rate of a gas from an orifice or port to the theoretical, calculated flow rate. Always less than one.

Discharge, hot gas line - see discharge line.

Discharge line - the refrigeration line that runs from the outlet of the compressor to the inlet of the condenser coil. Also called the hot gas line.

Discharge muffler - this device made up of a shell and with or without baffles smooth out and quiets the compressor discharge pulsations on reciprocating compressors. Location can either be internal or external of the compressor.

Discharge piping - piping attached to the outlet side of a safety valve that conveys steam to the atmosphere.

Discharge port - hole through the valve plate of the compressor where hot gas is discharged.

Discharge pressure - the pressure at the outlet or discharge of a fluid moving device. In an air conditioning condensing unit, the pressure of the refrigerant gas as it leaves the discharge side of the compressor. Discharge pressure depends on type of refrigerant, temperature and CFM of air moving across or through the condenser coil, suction pressure, amount of refrigerant charge, and how clean or dirty the condenser coil is. Also referred to as the high side pressure or head pressure.

Discharge shut-off valve - a manual valve installed on the compressor discharge line designed to isolate the compressor from the rest of the refrigeration system.

Discharge valve - a valve in the compressor valve plate that controls the flow of hot high pressure gas to the discharge line.

Disconnect switch - an electrical switch that is used to disconnect electrical power to the load device. Can be single pole, double pole and/or triple pole single throw arrangements. Also see master switch.

Disconnect means - a device, or group of devices, or other means by which the conductors of a circuit can be disconnected from their source of supply or power.

Discus compressor - a reciprocating compressor distinguished by its disc type valve system.

Discus valve - a reciprocating compressor valve design with a low clearance volume and larger bore.

Disengaging area - the surface area at the waterline of a steam boiler.

Disinfectants - one of three groups of antimicrobials registered by EPA for public health uses. EPA considers an antimicrobial to be a disinfectant when it destroys or irreversibly inactivates infectious or other undesirable organisms, but not necessarily their spores. EPA registers three types of disinfectant products based upon submitted efficacy data: limited, general or broad spectrum, and hospital disinfectant.

Disk - a circular plate with magnetic material on both sides, continuously rotated for reading or writing by means of one or more read/write heads mounted on movable or fixed arms; disks may be permanently mounted on a shaft, or as a package they may be removable and others placed on the shaft.

Dismantle - to take apart, to strip of covering.

Disperse - to break up and scatter in all directions.

Displacement, actual - the actual volume of gas or vapor at compressor inlet conditions, moved by a compressor per revolution or per unit of time.

Displacement, piston - the volume obtained by multiplying area of the cylinder bore by the length of the piston stroke.

Displacement theoretical - the total volume displaced by the working strokes of all the pistons of a compressor per revolution or per unit of time.

Display hold - when a digital meter continues to display the last reading even after the probes have been removed from the test circuit.

Disposable - usually a filter element intended to be discarded and replaced after one service cycle.

Disposable cylinder - a one-trip refrigerant cylinder that is not to be refilled.

Disposal - any process leading to and including the discharge, deposit, dumping or placing of any discarded appliance or component parts, into or on any land or water. The disassembly of any appliance for discharge, deposit, dumping, or placing of component parts into or on any land. The disassembly of any appliance for reuse of its component parts.

Dissimilar metals - metals of unlike composition tend to set up a galvanic action that results in rapid deterioration in one of the metals. Dissimilar metals should be insulated one from the other by membranes, caulking or other coatings.

Dissipation factor - the ratio of the conductance of a capacitor, in which the material is the dielectric, to its susceptence; or the ratio of its parallel reactance to its parallel resistance, or the ratio of the power loss to the circulating KVA. Used to indicate relative energy loss in a capacitor.

Dissociation - the breaking up of combustion products into combustibles and oxygen, accompanied by an absorption of heat. This usually occurs at high temperature, and it is one of the factors limiting the maximum temperature of the flame.

Dissolved air - molecules of oxygen, nitrogen, carbon dioxide, and other gases found in the air that are present within liquid water.

Dissolved gases - those gases that enter into solution with a fluid and are neither free nor entrained gases.

Distill - the process of separating constituents of a liquid by boiling it then condensing the vapor that's produced. Distillation can be used to purify water and other substances, or to remove one component from a mixture, as when gas is distilled from crude oil.

Distillate fuel oil - a general term applied to the lighter grades of fuel oil which are produced by distilling petroleum. Such oils are a condensed product of distillation, often referred to as overhead oils.

Distillation - separation process based on the relative volatility of the materials to be separated and on a change in phase of the original mixture, that may be boiled off and recollected by condensation in another vessel. There are many processes including simple distillation, multiple distillation, fractional distillation, and steam distillation.

Distilled water - water that is free of impurities by distillation --- first heated to a steam vapor and then cooled back to a liquid.

Distilling apparatus - the fluid reclaiming device used to reclaim used refrigerants. Reclaiming is usually done by vaporizing and then re-condensing the refrigerant.

Distortion - in welding, refers to the warping of a piece of metal.

Distributed control network - a group of independently functioning controllers under the supervision of a central controller (supervisory computer), yet having the capability at each controller of data analysis and control action.

Distributed power - generic term for any power supply located near the point where the power is used.

Distribution basin - shallow pan-type elevated basin used to distribute hot water over the tower fill by means of orifices in the basin floor. Application is normally limited to crossflow towers.

Distribution center - an electric panel used to distribute electric supply to several places in a large structure. It can be a fused or circuit breaker design.

Distribution controls - systems with help evenly and efficiently transfer the heating or cooling medium to the area where it is needed.

Distribution system - (1) those parts of a tower, beginning with the inlet connection, that distribute the hot circulating water within the tower to the points where it contacts the air for effective cooling. May include headers, laterals, branch arms, nozzles, distribution basins, and flow regulating devices. (2) ducts or pipes which carry air, water, or steam to a conditioned space.

Distributor - a fluid delivery manifold, particularly the one that distributes liquid refrigerant from the metering device to different parallel passes of an evaporator coil.

Distributor tester - also known as a field analyzer, this tester will test individual components, the control module, status indicator, and all four probes individually without being mounted on the cuber. This tester is designed as a bench tester for a shop, or a counter top tester for large dealers and distributors handling many spare parts.

District heating and cooling - use of a central utility system designed to provide heating and cooling to large residential and industrial areas.

Disturbed earth effect - denotes the heat content change that effects the earth temperature by the addition to or removal from the heat energy by the heat pump system.

Diversity - that quality or characteristic by which individual maximum demands occur at different times such as hourly, daily, weekly, monthly and yearly. Diversity depends on the application and placement of windows and other openings in the building envelope.

Diversity factor - ratio of the maximum probable demand to the maximum possible demand. This factor is used to determine a fuel gas demand that is more realistic than the maximum possible demand. Diversity recognizes the variable nature of appliance usage and is based on probability.

Divert - to turn in a different direction.

Diverter, draft - a fitting installed in the vent or flue of a heating appliance that will either (1) provide a draft if there is none in the flue (2) prevents a down draft from entering the appliance or (3) neutralizes and chimney effect in the flue.

Diverter tee - a special fitting resembling a tee with an internal baffle. This fitting is used to force a portion of the flow entering the tee through a branch circuit connected to its side port.

Diverting switch - a pneumatic switch for connecting a common air line to either of two branches. Multi position switches provide for several different air connections, usually including a vent connection for exhausting air from one air line.

Diverting valve - a three-way valve with one inlet and two outlets. A fluid entering the inlet port is diverted to either of the two outlet ports in any proportion desired by moving the valve stem. A diverting valve is double seated with a disc for each seat, so it can generally be used in mixing applications also.

Division wall - a wall used to divide the floor area of a building or structure into separate parts for fire protection, for different uses, or for restricted occupancy.

DOE heating capacity - the heat output rating of boilers up to 300,000 Btu/hr based on a federal government rating standard.

Dolly - a cart or rack with wheels, used to move heavy components or appliances.

Domain theory - a theory of magnetism that assumes that groups of atoms produced by movement of electrons align themselves in groups called "domains" in magnetic materials.

Dome hat - the sealed metal container for the motor compressor of a refrigerating unit.

Domestic equipment - air conditioning or refrigeration equipment normally operated by the consumer.

Domestic hot water - the heated water used for domestic or household purposes such as laundry, dishes, bathing, etc.

Domestic refrigeration - any cooling unit that is used with the home. It includes air conditioners, refrigerators, and freezers.

Domestic well - a well that is used for heat pump applications as well as for domestic water usage. The water used for the heat pump is not altered in any way with the exception of temperature.

Donor - a dopant, such as phosphorus, that supplies an additional electron to an otherwise balanced crystal structure.

Door - all operable opening areas, which are not glazing, in the building envelop including swinging and roll-up doors, fire doors, smoke vents and access hatches.

Door area - total area of door measured using the rough opening and including the door and frame.

Door schedule - a part of working drawings listing all doors required on a job, indicating sizes, types, locations and special requirements.

Door switch - normally open single pole switch used in appliances that operates by the opening and closing of the doors or panels.

Dopant - a chemical element added in small amounts to an otherwise pure crystal to modify its electrical properties. An n-dopant introduces more electrons than are required for the perfect structure of the crystal. A p-dopant creates electron vacancies in the crystal structure.

Doping - adding an impurity to a semiconductor to produce a desired charge.

Dormer - a space which extends from the roof line of the house, usually including one or more windows.

Dosimeter - a device that records the amount of something that a person is exposed to and then reads back at the end of a day; usually measures sound or radiation.

DOT identification numbers - four digit numbers, preceded by UN or NA, that are used to identify particular substances for regulation of their transportation.

DOT 39 - a federal government numbering system at identifies disposable refrigerant cyclinders. These cylinders should never be refilled or used as compressed air tanks.

Double acting - describes the circuitry in a thermostat permitting it to be used for both heating and cooling applications. Usually a SPST configuration.

Double-block and bleed - a valve arrangement in which a normally open vent valve just downstream from an automatic safety shut off valve (SSOV) allows any gas that leaks past the SSOV during shut down to escape to the outside atmosphere. A second SSOV just downstream from the vent valve blocks any gas that is not vented. When the SSOV's are open, the vent valve is closed. Also see bleeding.

Double break contact - a contact that breaks connections at two points. Consists of a bridge and a stationary contact points

Double cleat ladder - a ladder with a center rail to allow simultaneous two way traffic for employees ascending or descending.

Double cut file - a hand file that has two sets of cuts which cross one another on the face. Double cut files are used for taking off material quickly.

Double duct system - an HVAC system which distributes both hot and cold air streams to a conditioned space and mixes air at the zone as opposed to mixing multiple air streams at a central plant.

Double duty case - a commercial refrigerator that uses part of it for refrigerated storage and part of it is equipped with glass windows for display purposes.

Double extra strong - the schedule of wrought pipe weights in common use.

Double flare - a connection used on copper, aluminum, or steel tubing that folds tubing wall to a double thickness.

Double flow - a cross flow cooling tower where two opposed fill banks are served by a common plenum.

Double fold - a straight edged sheet that has been turned back 180 degrees and again turned another 180 degrees, then flattened.

Double glazing - two panes of glass, usually parallel, with an air space between to provide increased thermal and/or sound insulation.

Double heat transfer - the transfer of heat from the plant to the heated medium (usually liquid) and from the liquid to the air in the conditioned space.

Double hung window - a window having two vertically sliding sashes, each covering one half of the opening.

Double pole - a switch that opens and closes two sets of contacts when the switch is thrown.

Double pole breaker switch - a circuit breaker switch that is used to disconnect both hot wires with a single switch off function.

Double pole double throw switch (DPDT) - a switch that has two poles and two contacts for each pole, so two contacts are always closed and two contacts are always open. This switch makes it possible to control a variety of loads from one location.

Double pole single throw switch (DPST) - an electrical switch so arranged that both switches are simultaneously opened or closed.

Double reduction - a multiple reduction unit containing two stages of gear reduction housed in a single enclosure. The overall speed reduction (ratio) is the product of the gear ratios provided by the individual stages.

Double seated valve - a valve with two seats and discs arranged so that in the closed position there is very little fluid pressure forcing the stem toward the open or closed position. These valves require less power to operate than a single-seated valve of the same size. They often has a larger port area for a given pipe size and do not have a tight shut off.

Double shaft motor - a motor with a long shaft that extends out of both ends of the motor.

Double riser system - a system in which two vertical hot gas discharge risers are used when a single vertical hot gas discharge riser sized for 1500 FPM at a minimum load will have an excessive pressure drop at maximum load. The smaller riser is sized for 1500 FPM gas velocity at minimum load while the larger is sized so the velocity through both risers will not be less than 1500 FPM at maximum load conditions.

Double thickness flare - copper, aluminum or steel tubing that has been formed into a 2 wall thickness. This flare is usually 37° or 45°.

Double throw - a contact arrangement in which each contact form included is a break make. Where one contact opens its connection to another contact and then closes its connection to a third contact.

Double wye - a wye used where branches are to be made on both sides of the pipe at the same point.

Dowel pin - an accurately dimensioned pin pressed into one assembly part and slipped into another assembly part to insure accurate alignment.

Downdraft - excessive high air pressure existing at the outlet of chimney or stack that tends to make gases flow downward in the stack.

Down-feed one pipe riser (steam) - a pipe that carries steam downward to the heating units and into which the condensate from the heating unit drains.

Down-feed system, steam - a steam heating system in which the supply mains are above the level of the heating units that they serve.

Downflow furnace - a forced-air, central furnace with essentially vertical airflow, that discharges air at or near the bottom of the furnace; as distinguished from an up flow furnace.

Downstream - a point further from the source of fluid flow than at another point.

Downtime - the period of time when equipment is malfunctioning or off line because of maintenance.

Draft - the movement of air into and through a combustion chamber; breeching, stack, and chimney. Draft may be natural, resulting from the difference in density of the heated air rising through the stack and the cooler displacing air. Artificial draft may be provided by mechanical means. This could either be induced draft or forced draft. Draft is affected by chimney height, temperature difference, cross section, wind, make-up air opening in the furnace room and weather conditions. Also see forced, induced, mechanical, natural, and over fire draft.

Draft control - device used on gun type oil burners used to keep the flue pressure constant. Commonly referred to as a barometric damper.

Draft diverter - a fitting on the furnace between the heat exchanger and the venting system that is designed to (1) assure the ready escape of the products of combustion in the event of no draft, back draft, or stoppage beyond the draft diverter; (2) prevent a back draft from entering the appliance; and (3) neutralize the effect of stack action of a chimney or gas vent upon the operation of the appliance. Also see draft hood.

Draft, forced - air flow through a combustion chamber when pressure is applied to the supply.

Draft gauge - a gauge used to measure very small pressures (above and below atmospheric) and compare them to the atmosphere's pressure. Used to determine the flow of flue gas in a chimney or vent.

Draft hood - a device built into an appliance or made a part of the vent connector from an appliance, that is designed to (1) provide for the ready escape of the flue gases from the appliance in the event of no draft, back draft, or stoppage beyond the draft hood, (2) prevent a back draft from entering the appliance, and (3) neutralize the effect of stack action of the chimney or gas vent upon the operation of the appliance. Also see draft diverter.

Draft indicator - an instrument used to indicate or measure stack draft or combustion gas movement. Draft is measured in units of .1 inch of water column.

Draft, induced - air flow through produced in a combustion chamber when a vacuum is applied at the outlet of the flue.

Draft loss - the reduction of draft intensity caused by resistance to flow of flue gases through the boiler, breeching and chimney, plus any resistance to flow of air through the burner.

Draft proving switch - a device that measures the negative air pressure in an exhaust system and closes its electrical contacts whenever a safe negative pressure is maintained.

Draft regulator - a device that acts to maintain a desired draft by automatically controlling the chimney draft intensity. This device built into a fuel-burning appliance, or made part of a chimney connector, that functions to reduce excessive draft through an appliance to a desired value by admitting ambient air into the appliance chimney, chimney connector, vent, or vent connector.

Drain - to make or become gradually dry or empty.

Drain cock - a valve installed in the lowest point of a boiler or at low points of a heating system to provide for complete drainage of water from the system.

Drain, electronics - a collector of the majority carriers.

Drain pan - pan shaped panel or trough used to collect condensate from evaporator.

Drain pipe - a pipe for the removal of condensate, excess water or fluids.

Drain wire - in a cable an un-insulated wire laid over the component or components and used as a ground connection.

Draindown system - a type of active solar energy system in which freezing is avoided by having all fluid drain out of the collectors and exposed piping to waste at the end of each operating cycle.

Draw - the amount of current required to operate an electrical device.

Draw down - the difference between the static water level and the pumping water level in a water well.

Draw down, pressure tank - the amount of liquid the tank will deliver per minute within a specified pressure range before the pump starts.

Drawings - a graphic representation of the work to be done. They show the relationship of the material to each other, including sizes, shapes, locations, and connections. The drawings may include schematic diagrams showing such things as mechanical and electrical systems. They may also include schedules of structural elements, equipment, finishes, and other similar items.

Dress - the process of grinding or filing the head of a chisel or punch to eliminate the jagged edges caused by being pounded on by a hammer.

Drier - a device containing a drying agent or substance placed in the refrigerant circuit to keep the moisture level in a system at a safe level. Its purpose is to collect and hold excess water from the system.

Drift, controls - the tendency of a controller to have a deviation between the control point and the set point.

Drift, water towers - entrained unevaporated water carried from a cooling tower by air movement through it.

Drift eliminators - an assembly of baffles or labyrinth passages through which the air passes prior to its exit from the tower, for the purpose of removing entrained water droplets from the exhaust air.

Drift punch - a hand tool used to drive out keyways and to line up holes in matching material.

Drill - a tool for drilling and boring, and for install all types of screws. Drill sizes can be either fractional, number, or letter.

Drill bit - drill bits are used to do the actual drilling or boring of a hole. They are used for installation and repair work. Drill bits are available for working in metal, wood, plastic and masonry. Most have a straight shank.

Drill Manufacturer's Standard (DMS) - equivalent to standard twist drill or steel wire gauge numbers.

Drill rod - an accurately ground and polished steel rod.

Drilled supply well - small diameter holes bored into the earth by a screw-type shaft attached to a drilling rig, using a stream of water or drilling mud to flush out the material loosened by the drill. The hole is bored to find water up to 500 feet if necessary.

Drilling - cutting round holes in material by use of a cutting tool called a drill.

Drilled port burner - a burner in which the ports have been formed by drilled holes in a thick section in the burner head or by a manufacturing method that results in holes similar in size, shape and depth.

Drinking water cooler - an assembly that either employs or is used in conjunction with a mechanical condensing unit for the purpose of cooling drinking water.

Drinking water cooler capacity - the quantity of water it will cool in a given ambient temperature with a given incoming water temperature and a given outgoing water temperature, under steady state conditions.

Drip - a pipe, or a steam trap and a pipe considered as a unit, which conducts condensate from the steam side of a piping system to the water or return side of the system.

Drip leg - a dead-ended section of pipe (minimum 3 inches and capped) at the lowest point in a gas supply line that accumulates any foreign matter that may have gotten in the gas or piping system. It is this pocket that keeps this material from entering the gas controls.

Drip, masonry - projecting horizontal masonry course sloped outward to throw water away from the building.

Drip pan - a pan shaped to collect moisture condensing on an evaporator coil in an air conditioning or refrigeration system.

Drip-proof motor - a motor that has the ventilation openings in the end shields and shell placed so that drops of liquid falling within an angle of 15° from vertical will not affect performance. Usually used for indoor motors in fairly clean locations.

Drip, steam - a pipe or steam trap and a pipe considered as a unit, which conducts condensation from the steam side of a piping system to the water or return side of the system.

Drip valve - valve installed on drip legs for the removal of material from pipelines.

Drive - to turn, operate or cause a motor to do its intended work. Also the package of motor, gearbox, belting, sprockets, etc. necessary to turn, operate or cause a motor to do its work.

Drive method - refers to the method used to drive or power a reciprocal compressors, such as a motor.

Drive pulley - a pulley attached to the shaft of the prime mover and from which power is transmitted.

Driven - describes a component that does not have its own power, but derives it from a separate engine or motor.

Driven pulley - the pulley connected to the blower shaft in an air conditioning distribution system. This pulley can rotate either faster or slower than the motor, depending on the size relationship between the drive and driven pulleys.

Driver - a digital circuit element coupled to the output stage of a circuit to increase the power or current handling capability, or fan-out, of the stage.

Droop - lowering of the operating temperature of a thermostat, below the set point, due to artificial heating of the bimetal by the heat anticipator or other heating source.

Drop - the vertical distance that a projected air stream falls between the grille and the end of its throw.

Drop in replacement - an alternative refrigerant that can be installed directly into an existing refrigeration system and operated at similar conditions without component change.

Drop-out millivoltage - the minimum voltage that will hold a pilot safety device in the energized position and will allow gas to still flow to the main burner(s).

Drop out voltage - the voltage or amperage at which the magnetic field around the coil of a control is no longer strong enough to attract the armature or contacts in a potential relay.

Drop, service - the overhead service conductor's between the utility company's last pole and the customer's first point of attachment.

Drop, voltage - the electromotive force needed to cause the current to flow through a load.

Drum - a container with a capacity of 55 U.S. gallons.

Drum, boiler - an enclosed vessel in a water tube boiler which distributes water to the tubes.

Drum, refrigerant - usually considered to be a non-returnable refrigerant cylinder of the 30 pound variety.

Dry - to separate or remove a liquid or vapor from another substance. The liquid may be water, but the term is also used for the removal of liquid or vapor forms of other substances.

Dry air - air that contains no moisture vapor.

Dry air cooler - removes sensible heat from the dehydrated air whenever it leaves the dehydrator at an elevated temperature.

Dry back boiler - a scotch, fire-tube boiler with a ceramic baffle at the back of the furnace to direct the products of combustion from the combustion chamber to the second pass. The baffle is separate from the pressure vessel and is constructed of heat resistant material (generally refractory brick and insulation). Also see fire tube and scotch boilers.

Dry base boiler - a boiler design in which the boiler block rests on top of the combustion chamber.

Dry bulb - an instrument with a sensitive element that measures sensible ambient air temperature.

Dry bulb temperature - the air temperature as indicated by an ordinary thermometer. Usually expressed in degrees Fahrenheit.

Dry bulb thermometer - an instrument that measures air temperature, independent of humidity.

Dry capacitor condenser - an electrolytic electrical device (start capacitor) made of dry metal and dry insulation that is used to store electrons.

Dry cell battery - an electrical device that has no liquid in the cells and is used to provide a direct current potential difference by converting chemical energy into electrical energy. It uses different electrode materials than a wet cell and has a paste type electrolyte rather than a liquid one.

Dry chemical extinguisher - fire extinguisher that contains an extinguishing agent and a compressed, non-flammable gas, which is used as the propellant.

Dry compression - the compression of a vapor in a vapor-liquid vapor compression refrigeration system.

Dry contacts - contacts isolated from any source of power.

Dry expansion - a process of heat removal by a refrigerant in an evaporator fed by a flow control, responsive to temperature or pressure or both at some point in the evaporator, or to the difference between the high and low side pressures, and not to the liquid level in the evaporator; all entering refrigerant is evaporated before being recirculated.

Dry gas - a gas having a moisture and hydrocarbon dew point below any normal temperature to which the gas piping is exposed.

Dry ice - a refrigerating substance made of solid carbon dioxide that changes directly from a solid to a gas (sublimates). It sublimates at a temperature of -109° F.

Dry location - a location not normally subject to dampness or wetness. A location classified as dry may be temporarily subject to dampness or wetness, as in the case of a building under construction.

Dry media filter - a filter that traps and holds dust without being coated with a sticky material.

Dry pipe separator - a closed pipe perforated at the top, with drain holes on the bottom that remove moisture from the steam.

Dry return - a return pipe in a steam heating system that carries both air and water from the condensate.

Dry steam - superheated water vapor containing no more than .5% moisture.

Dry switch - a switch that is operated by a device, but is not electrically connected to that device, such as an auxiliary contact on a contactor.

Dry system - a refrigerating system that has the evaporator liquid refrigerant mainly in the atomized or droplet condition.

Dry type evaporator - an evaporator in which all the liquid refrigerant boils off while flowing through the coil, before reaching the suction line. Also referred to as a direct expansion evaporator.

Dry wall - interior wall construction consisting of plaster boards, wood paneling or plywood nailed directly to the studs without application of plaster.

Dry well - used in geothermal open loop heat pump systems, these wells are used to accept the return water and are constructed as large pits filled with sand and gravel.

Dual element fuse - a device that interrupts an electrical circuit after excessive current over an extended time period (time delay element) or after very excessive momentary current.

Dual-fuel burner - a burner using one prime fuel, but capable of using a standby fuel under peak load conditions. Changeover from one fuel to the other can be made automatically using an electric temperature controller or by use of a manual switch. Also see combination burner.

Dual path system - an air system that splits the mixed supply air into two streams. One stream is chilled and the other is heated.

Dual pressure control - a control that senses and is actuated by both high and low side refrigeration system pressures.

Dual purpose condenser - water cooled condenser that can be used either on city water or cooling tower water.

Dual run capacitor - an electrical storage device that has two run capacitors, one for the fan and one for the compressor, built into one case.

Dual temperature water system - a hydronic combination heating and cooling system which circulates hot and/or chilled water through common piping and terminal heat emitter units. These systems operate within the temperature and pressure limits of a low temperature water systems, with usual winter design supply water temperatures of about 100 to 150° F and summer water temperatures of 40 to 45° F.

Dual thermostat - a two temperature thermostat. Equivalent to two separate thermostats in one case that have two different set points.

Dual voltage motors - motors manufactured to operate on either of two specified supplied voltages.

Duct - (1) a tube, channel, or canal through which a fluid moves. (2) a protective pipe through which wires or cables are run.

Duct coil - a heat exchanging coil mounted inside of an air delivery supply duct.

Duct fan - see tubeaxial fan.

Duct furnace - a central furnace designed for installation in a duct of an air distribution system to supply warm air for heating. For air circulation, it depends on a blower not furnished as part of the furnace.

Duct heat gain - after leaving a unit, air is warmed on its way to the registers, so the air at the end of the air conditioning path is warmer than at the beginning.

Duct heater - an electric resistance heater mounted inside an air delivery duct that provides supplementary heat to the delivered air.

Duct loss - heat loss through the walls of ductwork to the space surrounding the ducts.

Duct section - a single piece or section of duct.

Duct system - a continuous passageway for the transmission of air, in addition to ducts, includes duct fittings, dampers, plenums, fans and accessory air handling equipment. Can be either a supply or return system.

Duct tape - this (initially sticky tape is unfortunately the most common material used to seal duct connections. Care must be taken when it's applied. Foe effective sealing, the surface it is applied to must be clean - free of dust, dirt, oil, or other substances. Duct tape has a tendency to lose adhesion with age, especially when used on ducts in unconditioned spaces.

Ductility - the property of elongation, above the elastic limit, but under the tensile strength. It is the ability of a metal to be formed or shaped. A measure of ductility is the percentage of elongation of the fractured piece over its original length.

Ductwork - round or rectangular passageways for conditioned and unconditioned air.

Dulong's formula - an equation for estimating the heat content of a pound of coal based on its content of carbon, hydrogen, oxygen and sulfur.

Dumb mixing valve - a mixing device such as a 3-way valve that does not sense temperature or flow rate, and thus is incapable of reacting to changes in operating conditions.

Dummy leg - piece of pipe or rolled steel sections that is welded or connected to the pipe in order to support the line.

Dummy terminals - terminals that have no effect on the operation of the relay or equipment, but provide connections points for wires. Dummy terminals could also be located on terminal strips.

Duplex receptacle - a double outlet receptacle used in house wiring. It provides outlets to connect lamps and appliances. The duplex receptacle is mounted in a electrical box in the wall section.

Dust - an air suspension (aerosol) of particles of any solid material, usually from crushing, grinding, abrading, or blasting with particle size less than 100 micrometers.

Dutch fold - two straight edged sheets are folded 180 degrees and butted. A drive seam is formed when a cap strip (drive) is driven over the Dutch fold.

Duty - a requirement of service that defines the degree of regularity of the load.

Duty cycle - the relationship between operating and rest time. If the duty cycle of a motor is 50%, then it means the motor runs in any combination of times up to 30 minutes, or 50% of an hour). When applied to motors, duty cycle is usually referred to as continuous or intermittent duty.

Dwelling - a building or portion thereof that contains not more than two dwelling units.

Dwelling units - a single unit providing complete, independent living facilities for one or more persons, including permanent provisions for living, sleeping, eating, cooking and sanitation.

DWV - drain, waste and vent type pipe. Recommended for above ground, no pressure applications only. Use sweat fittings only. Available in hard type only.

Dx coil - an evaporator coil where the expansion of refrigerant inside the coil happens directly. A dry type evaporator.

Dynamic ampli-check - a Honeywell® trademark; a feature in some infrared flame signal amplifiers; circuitry tests only the flame signal amplifier during burner operation and shuts down the burner if the amplifier fails (including a flame-simulating failure).

Dynamic balance - the condition under which moving parts of machinery will not vibrate excessively while in motion at any specific speed.

Dynamic braking - using a DC motor as a generator to produce counter torque and thereby produce a braking action. Applying direct current to the stator winding of an AC induction motor to cause a magnetic braking action.

Dynamic insertion loss - the reduction in sound power level, expressed in decibels, due solely to the placement of a silencer, duct lining or other attenuating device, due solely to the placement of these sound attenuating devices in the duct system.

Dynamic self check - a Honeywell® trademark. A feature in the rectification type flame signal amplifiers of some flame safeguard controls; circuitry tests all electronic components in the flame detection system (amplifier and detector) 60 to 240 times a minute during burner operation and shuts down the burner if the detection system fails (including a flame-simulating failure).

Dynamometer - a device that measures the power input and/or the power output of a mechanism.

Dynamotor - an electrical motor that combines both motor and generator action in one magnetic field, either with two armatures or with one armature having two separate windings.

Dytel - trade name for a refrigerant that has a red dye added to it to aid in the visual detection of leaks.

-E-

E - symbol for electromotive force.

Ec - efficiency, combustion.

EAROM - electrically alterable read only memory.

Eco - energy cut off.

ECM - electronically commutated motor.

ECOP - economic coefficient of performance.

EDP - electronic data processing.

EDR - equivalent direct radiation.

EER - energy efficiency ratio.

Effss - efficiency, steady state.

ELA - effective leakage area.

Elem - elementary.

Eli - a term used to help remember that voltage (E) leads current (I) in an inductive (L) circuit.

Emf - electromotive force.

Emi - electromagnetic interference.

EMT - electrical metallic tubing.

Encl - enclosure.

EP - electric pneumatic

EPDM - Ethylene Propylene Diene Monomer

EPR - evaporator pressure regulator.

EPROM - erasable programmable read only memory.

Equip - equipment.

EREC - Energy Efficiency and Renewable Energy Clearinghouse.

ETD - equivalent temperature design.

ETR - evaporator temperature regulator.

Early spark termination - a feature of some programming controls that allows ignition cutoff before the main fuel valve(s) opens, resulting in a period during which only the pilot is on.

Earth - terminology for zero reference ground.

Earth aquifer - the water quantity below the surface of the earth.

Earth coupled systems - a term referring to closed loop, geothermal heat pump systems.

Earth heat exchanger - piping buried in the soil for extracting or rejecting heat.

Earth temperature, maximum (Tmax) - the highest temperature the earth will reach in the summer season. Will depend on location and weather conditions.

Earth temperature, minimum (Tmin) - the lowest temperature the earth will become in the winter season. This temperature will vary with location and weather conditions.

Earth tester - used to measure the resistance of a ground rod, steel tower, or actual earth resistance.

Ebullator - a pointed or sharp edged solid substance inserted in a flooded type evaporator used to improve the evaporation of the refrigerant in the coil.

Eccentric - a circle or disk mounted off center. Eccentrics are used to adjust controls and connect compressor drive shafts to the connecting rods.

Eccentric reducer - a pipe fitting designed to change from one pipe size to another and to keep one edge of both pipes in line. These fittings should be installed so that the "in line" section of pipe is at the top.

Ecological - having to do with the study of human's environment.

Ecology - science of life balance on earth.

Economic balance point - the lowest outdoor temperature at which the heat pump is more economical to operate than the furnace.

Economic coefficient of performance (ECOP) - energy in desired output units converted in terms of economic costs divided by fuel input in units of energy purchased or produced, where energy is measured in equivalent, consistent units again converted in terms of economic costs of each energy stream.

Economizer - a system of controls and dampers on an air conditioning system that mixes varying volumes of outdoor air with air in the conditioned space by using the lower mixed air temperature to achieve "free cooling".

Economizer control - a system of ventilation control in which outdoor and return air dampers are controlled to maintain proper mixed air temperatures for the most economical operation.

Economizer cycle - a method of operating an air conditioning system to introduce extra outside air when the temperature and relative humidity are in the correct proportion.

Economizer, water – a system by which the supply air of a cooling system is cooled directly, indirectly or both, by evaporation of water or by other appropriate fluid in order to reduce or eliminate the need of mechanical refrigeration.

Eddy current - induced current flowing in a magnetic core that are erected by a varying magnetic field.

Eddy current tests - a test that uses induced currents flowing in a magnetic core that are used to find potential failure faults in different types of tubing.

Edison base fuses - these 15, 20, and 30 amp fuses have the same base configuration as a regular based incandescent bulb.

Effect, chimney - the tendency of air or gas in a duct or other vertical passage to rise when heated due to its lower density compared to that of the surrounding air or gas. In buildings, the tendency toward displacement (caused by the difference in temperature) of internal heated air by unheated outside air due to the difference in density of outside and inside air.

Effect, refrigerating - the rate of heat removed by a refrigerant in a refrigerating system; this is equal to the product of the mass rate of refrigerant flow in the system and the difference in specific enthalpies of the refrigerant at two designated points in the system, or two designated thermodynamic states of the refrigerant.

Effect, total cooling - the difference between the total enthalpy of the dry air and water vapor mixture entering a unit per hour and the total enthalpy of the dry air and water vapor (and water) mixture leaving and unit per hour, expressed in Btu per hour.

Effective area - the actual flow area of an air inlet or outlet. The gross area minus area of vanes or grille bars. Net free area. It is equal to the free area of the device times the coefficient of discharge.

Effective coil temperature - the temperature of the coil if all of the air is affected by the coil temperature. If the air coming off of the coil is higher than the temperature of the coil (assuming a cooling coil) then some of the air was not affected by the coil. This air is called by-pass air. Also known as the apparatus dew point temperature.

Effective Emittance of an air space - the combined effect of emittances from the boundary surfaces of an air space, where the boundaries are parallel and a dimension much larger than the distance between them.

Effective heat allowance - an allowance added to the test output of certain designs of radiation to compensate for a better distribution of heat within the heated space. Some agencies do not permit the use of effective heat allowance.

Effective temperature (ET) - the overall apparent temperature sensed by a person in the conditioned space. An index that is a combination of temperature, humidity and air movement.

Effective temperature difference - the difference the room temperature and the supply air temperature at the outlet to the room.

Effective total R value - the average thermal resistance of a wall, ceiling, floor, etc., after compensating for the presence of framing members.

Effective value - The effective value of an AC voltage or current is the value that will cause the same amount of heat to be produced in a circuit containing only resistance that would be caused by a DC voltage or current of the same value. Thus, an alternating current with an effective value of 1 ampere generates the same heat in a 10 ohm resistor as does a direct current of 1 ampere. The effective value is also called the root mean square, or rms, value because of the way it is derived. It is equal to the square root of the average value of the squares of all instantaneous values of current or voltage during one half cycle. In a pure sine wave, the effective valve is 0.707 (the sin of 45°) times the peak value. Effective value is used to rate AC voltage and currents. Household line voltage is the rms value.

Effective voltage - a value of an AC sine wave voltage that has the same heating effect as an equal value of DC voltage.

To convert from effective voltage to peak voltage use the formula: $Epk = 1.414 \cdot Eeff$.

To convert from effective voltage to average voltage use the formula: $Eav = 0.9 \cdot Eeff$.

Effectiveness of a heat exchanger - the ratio of the actual heat transfer across a heat exchanger to the maximum possible rate of heat transfer. Effectiveness is a unit less number between 0 and 1. Higher numbers are more desirable.

Efficacy - the total amount of light energy (lumens) emitted from a light source divided by the total lamp and ballast power (watts) input, expressed in lumens per watt.

Efficiency - the ratio of output power divided by input power usually expressed as a percentage. A measure of how well the electrical energy input to a motor is converted into mechanical energy at the output shaft. For example, an 85% efficient motor consuming 1000 watts of electricity will produce 850 watts of work (expressed in horsepower) and 15% heat. The higher the efficiency, the better the conversion process and the lower the operating costs.

Efficiency capacity curve - graph used to evaluate pump performance by plotting the pumps efficiency on the vertical axis and its flow rate on the horizontal axis.

Efficiency of a boiler - the ratio in percentage of heat absorbed by water or steam to the actual total heat available in the fuel.

Efficiency of a circulator - the ratio of the rate of mechanical energy imparted to the fluid divided by the rate of energy input to the circulator. When the rate of energy input is the wattage drawn by the circulator motor, the efficiency is often expressed as the "wire-to-wire" efficiency.

Efficiency, over-all - the ratio obtained by dividing the theoretical power required by the actual power required.

Efficiency, volumetric - the ratio obtained by dividing the actual volume of gas delivered by the displacement.

Effluent - the liquid part of sewage. The fluid leaving a component.

Egress - exit.

Egress window - a window with specific release hardware and minimum clear opening size to allow occupants to escape through the window in case of a fire.

Ejector - a device that uses high fluid velocity such as a venturi, to create low pressure or vacuum at its throat to draw in fluid from another source.

Elastic limit - the greatest stress that a material can withstand without a permanent deformation after release of the stress.

Elastohydrodynamic lubrication - in rolling element bearings, the elastic deformation of the bearing (flattening) as it rolls, under load, in the bearing race. This momentary flattening improves the hydrodynamic lubrication properties by converting point or line contact to surface-to-surface contact.

Elastomer - any elastic, rubber like substance such as natural or synthetic rubber.

Elbow - a fitting with two openings at a 90° angle to each other.

Elbow, long radius - elbow whose radius equals one and one half pipe diameters.

Elbow, short radius - elbow whose radius equals one pipe diameter.

Electric arc - used to ignite the gas or oil flame in gas or fuel oil appliances. This arc is provided by an alternating current step up transformer.

Electric boiler - boiler that has heat produced by electric resistance coils or electrodes.

Electric circuit - a path for electrons to follow. The circuit may be open or closed depending on the position of its switches.

Electric controller - device(s) used to govern in some predetermined manner the operation of a circuit or piece of electrical apparatus. Could be a relay, contactor or motor starter.

Electric current - the movement of electrical charges through a conductive material.

Electric defrosting - the use of electric resistance heating coils to melt ice and frost off evaporators during defrosting.

Electric energy - energy that is produced by a movement of electrons. The energy can be produced by chemical, light, thermal, or mechanical means.

Electric field - the magnetic region in space that has force and direction.

Electric filament lamp - a light source consisting of a glass bulb containing a filament electrically maintained at incandescence. A lighting unit, consisting of an electric filament lamp with shade, reflector, enclosing globe, housing, or other accessories, is also commonly called a lamp. In such cases, in order to distinguish between the assembled lighting unit and the light source within it, the latter is often called a bulb.

Electric forced air furnace - an appliance constructed of high resistance wire or other material that produces heat when a current is passed through it and a powered blower assembly. It uses a duct system to provide heat to more than one room.

Electric heating - in a house heating system, a system that uses heat from electrical resistance units to heat rooms.

Electric heating element - a unit assembly consisting of a resistor, insulated supports and terminals for connecting the resistor to the electric power.

Electric hydronic boiler - a boiler using electrical resistance heat, which often has a closed loop piping system to distribute heated water for space heating.

Electric ignition - ignition of a pilot or main burner by an electric spark generated by a transformer, that is either automatically or manually energized. Also see direct spark ignition.

Electric meter - a device used to measure some electrical characteristics of a circuit such as the voltage, amperage, resistance, or wattage.

Electric motor - a machine that transforms electrical energy into mechanical energy.

Electric-pneumatic relay - a small electrically operated diverting valve, specially designed or suitable for diverting air from one control line to another.

Electric power - the rate at which electricity is being used, measured in watts.

Electric potential - the potential for electron flow between a positive and a negative pole. Electrical potential is measured in volts.

Electric precipitator - a device for removing dust from the air by means of electrostatic charges induced on the dust particles.

Electric pressure - another term used to refer to electromotive force, potential difference, and voltage.

Electric resistance heating element – component typically made of nickel chromium (nichrome) used in an air conditioning system to provide heat in cold weather.

Electric switch - a device that opens or closes to control some load in an electric circuit. It can be opened or closed by temperature, pressure, humidity, flow, or manual means.

Electric thermal storage system - a system designed to use off peak electrical energy to heat water for subsequent use in heating a building. The energy is stored in the form of heated water.

Electric valve - a solenoid type (electrically operated) valve used to operate the final end use.

Electrical angle - the method specifying the exact instant in an alternating current cycle. Each cycle is 360°; therefore 180° would indicate one half cycle, 45° one eighth cycle.

Electrical charge - there are two types of electrical charges--negative and positive. Like charges repel and unlike charges attract.

Electrical circuit - an electrical path between two or more points; the interconnection of a number of devices in one or more closed paths to perform a desired electrical or electronic function. Also see dynamic ampli-check and dynamic self check; and alarm, bridge, conductivity, firing rate switching, flip-flop, high voltage, isolated limited secondary, line voltage, low voltage, open, short, solid state, and tuned circuits.

Electrical control circuit - a wiring diagram showing how the various pieces of the control circuit are to be connected for proper operation.

Electrical degree - one 360th of an alternating current or voltage cycle.

Electrical drawing - electrical drawings show the electrical layout. Plan views show the location of switches, receptacles, lights and other related electrical equipment.

Electrical horsepower - a measure of electrical power. One horsepower is equal to 746 watts.

Electrical impulse - the electrical current fed from one control instrument to another to indicate that a position change is required.

Electrical interlock - when the contacts of one device or circuit prevent the operation of some other device or circuit.

Electrical load - that part of the electrical system that actually uses the energy or does the work required.

Electrical plan - a schematic of the electrical location of the meter, distribution panel, switches, convenience outlets and special outlets such as door bells, telephones, and GFIC's.

Electrical power - power is measured in watts. One watt is equal to one ampere flowing with a potential of one volt. Watts = volts x amps. ($P = E \times I$)

Electrical pressure – the electromotive force (voltage) that pushes electrons through a wire.

Electrical rating - the electrical power (volts and amperage) carrying ability that a switch or relay can handle through its contact points.

Electrical resistance - the property of an electric circuit that determines the rate at which electrical energy is converted into heat or radiant energy.

Electrical shock - an electrical current that travels through a human body.

Electrical symbols - standardized simple drawings used for ease of communication in schematic diagrams.

Electrical timer - device used to open or close a set of electrical contacts at predetermined time intervals.

Electrically reversible motor - one that can be reversed by changing the external connections.

Electricity - energy that is capable of producing an electron flow. An unbalanced condition that results when an electron can be easily displaced from an atom.

Electrochemical - a gas sensor that uses a chemical reaction to produce a low voltage signal, providing a measurement of a gas concentration.

Electrode - either of two terminals of an electrical source. When ignition is energized, a spark crosses the space between the electrodes (the spark gap) and ignites the flammable gas/fuel oil air mixture. Examples are ignition electrodes, flame rods, or electron tube elements.

Electrode (welding) - is a thin metal rod coated with a special substance and used as a filler to join the metal to be welded.

Electrolysis - the production of chemical changes by passage of current through an electrolyte.

Electrolyte - a nonmetallic substance, used in dry and wet cells, that conducts an electrical current in solution by movement of ions rather than electrons.

Electrolytic capacitor - a plate or surface capable of storing small electrical charges. Common electrolytic capacitors are formed by rolling thin sheets of foil between insulating materials. Capacitor capacity is expressed in microfarads.

Electromagnet - it is made by winding coil of wire around a soft iron core. When an electric current is run through the wire, the coil becomes a magnet.

Electromagnetic energy - energy that has both electrical and magnetic characteristics. Solar energy is electromagnetic.

Electromagnetic device - a device that utilizes the magnetic field created when current flows through a conductor. Coils of wire are employed to strengthen this field to perform work such as opening or closing different sets of electrical contacts or mechanical valves. Contactors and solenoid valves are examples of electromagnetic devices.

Electromagnetic field - the space around an inductor that changes due to current flow through a coil. The field expands and collapses with applied AC.

Electromagnetic force - the force that is harnessed by an electromagnetic device. This force is used to perform work such as opening or closing mechanical valves or electrical contacts.

Electromagnetic interference - stray electrical or magnetic fields that may induce damaging current or interference in other electrical equipment.

Electromagnetic spectrum - an orderly arrangement of radiant energy by wavelength or frequency. In the visible spectrum, the eye is sensitive to radiant energy between 380 nanometers (violet) and 770 nanometers (red) by wavelength.

Electromagnetic wave - the radiant energy produced by oscillation of an electric charge. Included are radio waves; infrared, visible, and ultraviolet light waves; and X, gamma, and cosmic rays.

Electromagnetism - the phenomenon of a magnetic field generated around a wire by electrical current flowing through it.

Electromechanical - a term used to describe controls that contain both electrical and mechanical components.

Electromechanical relay - an electromagnetic switch in which the opening and closing of one or more sets of contacts are controlled by current through the relay coil. These actions control the operation of one or more circuits connected to the relay contacts. An electronic relay performs the switching by electronic means rather than as a mechanical device.

Electromotive force (EMF) - electrical force (voltage) that causes current to flow or move in an electrical circuit. Unit of measure is the volt. Can be either AC or DC.

Electron - the elementary particle or portion of an atom that carries a negative charge. Electrons surround the positively charged nucleus of the atom. Electrons are the means by which the transfer of electrical energy can take place.

Electron current flow - current flow that is assumed to be in the direction of electron movement from a negative (-) potential to a positive (+) potential.

Electron hole pair - the free electrons and its hole that are produced by thermal excitation of a valance electron.

Electroplate - the term used to indicate the application of a metallic coating on a surface by means of electrolytic action.

Electronic - of or pertaining to devices, circuits, or systems utilizing the motion or emission of currents of free electrons, especially in vacuum, gas, or photo tubes, and special conductors or semiconductors.

Electronic air cleaner - a device that produces a powerful electric field to ionize dirt and dust particles in the air. The particles are then collected on electrically charged plates.

Electronic charging scale - an electronically operated scale used to accurately charge refrigeration systems by weight.

Electronic combustion analyzer - an electronic combustion analyzer is used to measure various items related to the combustion process. They can measure % of CO_2, % of O_2, stack temperature, PPM of CO, and overall combustion efficiency.

Electronic control - a control or system using electronic amplifiers to build up control signals for system operations.

Electronic control diagnostics - trouble codes that may be referenced on an automatic climate control system to diagnose problems.

Electronic expansion valve (EXV) - a metering valve that uses a thermistor as a temperature sensing element that varies the voltage to a heat motor operated valve.

Electronic ignition - a system that saves fuel by eliminating the continuously burning pilot light. The pilot flame is ignited by an electric spark and the pilot burns only upon a call for heat.

Electronic leak detector - an electronic instrument that measures electronic flow across an gas gap. Electronic flow changes indicate presence of refrigerant gas molecules.

Electronic relay - electronic switch, such as a triac, that controls a power consuming device.

Electronic scales - weighing devices with displays that can be set to show, in ounces or pounds, the amount of refrigerant being withdrawn from the cylinder and put into the operating equipment.

Electronic sight glass - device that sends an audible signal when the refrigerant system is low in refrigerant.

Electronic sound tracer - an instrument used to detect leaks by locating the source of high frequency sound caused by the leak.

Electronic thermistor - electrical device that senses temperature changes to control an output source. Also see thermistor.

Electronic thermostat - thermostat that uses electronic components to accomplish various sensing, switching, timing, staging, and display functions.

Electronically commutated motor - a type of infinite variable speed motor which changes speeds by input signals such as static pressure, frequency or air flow requirements.

Electronics - the field of science that deals with electron devices and their uses.

Electro-pneumatic relays - small electrically operated air valves used in the operation of pneumatic controls.

Electrostatic - electricity at rest.

Electrostatic charge - the electrical energy located on the surface of insulating material or in a capacitor because of an excess or deficiency of electrons.

Electrostatic field - the space around a charged material in which the influence of the electrical charge is experienced.

Electrostatic filter - type of filter which gives a particle of dust an electrical charge. This causes particles to be attracted to plates so that they can be removed from the airstream or atmosphere.

Electrostatics - the science or study of electricity at rest.

Element - the basic materials that make up all other materials; they exist by themselves (such as copper, hydrogen, carbon) or in combination with other elements (water is a combination of the elements hydrogen and oxygen).

Element, bimetallic - one formed of two metals having different coefficients of thermal expansion such as are used in temperature indicating and controlling devices.

Element, control - usually refers to the measuring part of a controller that senses the change in temperature, pressure or humidity. It produces a corresponding effect of force, pressure or electrical impulse to cause the actuator to function.

Element, electric heating - a unit assembly consisting of a resistor, insulated supports, and terminals for connecting the resistor to electric power.

Element, hydronic - a name for the finned tube component in a baseboard convector unit. The name also applies to the device that creates heat in an electric boiler or water heater.

Elevation drawing - a drawing showing the vertical elements of a building, either exterior or interior, as a direct projection to a vertical plane; the vertical distance above or below some established reference point.

Elevation head - the head energy within a fluid due to its elevation. The higher the elevation, the greater the elevation head energy content. This is how you get the available water pressure for your house.

Eliminate - to rid of or to do away with.

Eliminator - stationary vanes or louvers in a duct or plenum that removes any liquid from the gas passing through it.

Ellipse - a closed curve in the form of a symmetrical oval.

Ell, street - elbow made with male threads on one end and female threads on the other.

Ells - prebent tubing, factory-designed as soldered or flared fittings that let refrigerant lines be routed between components. Also applies to sheet metal fittings.

Elongation - term used to describe the stretch factor of a filler metal.

Emergency heat - heat pump mode of operation that is utilized when the vapor compression heat pump system is not operating. Emergency heat is typically provided by electric, resistive type strip heaters but can also be natural gas or fuel oil type heat.

Emergency heat switch - control on a heat pump type sub-base used to energize auxiliary heat for a heat pump system.

Emission - the discharge or escape into the atmosphere of one or more air pollutants.

Emission rate - amount of a contaminant released into the air by a source in a specified amount of time.

Emissivity - the capacity of a material to emit radiant energy. Emittance is the ratio of the total radiant flux emitted by a body to that emitted by an ideal black body at the same temperature. A low emissivity implies a higher reflectance of infrared radiation.

Emitter base junction - the P-N junction formed by the emitter and the base regions.

Emitter, electrical - either the low potential electrode of a p-n-p transistor or the high potential electrode of a p-n-p transistor.

Emitter, hydronic - the heat rejecting element of a steam or hot water boiler system. A typical hot water radiator is a heat emitter.

Emitter N-region - that region of an NPN transistor that emits a relatively large number of electrons in response to a relatively small number of electrons coming from the P region.

EMT - electrical metallic tubing. Galvanized, light weight conduit.

Enclosure (electrical) - Type 1 --- General Purpose -- an enclosure intended for indoor use where there are normal atmospheres. The enclosure protects against accidental contact of personnel with the enclosed control. Type 2 --- Drip proof -- an enclosure intended for indoor use to protect the enclosed control against falling non-corrosive liquids and falling dirt. These enclosures must have provisions for drainage. Type 3 --- Dust-tight, Rain-tight, & Sleet (ice) resistant -- an enclosure intended for outdoor use to protect the enclosed control against wind blown dust and water. These enclosures have provisions for locking. Type 4 --- Watertight & Dust-tight -- an enclosure intended for indoor use to protect the enclosed control against splashing water, seepage of water, falling or hose directed water, and severe external condensation. These enclosures have provisions for watertight connectors and provisions for mounting external to the enclosure cavity. Type 4X --- Watertight, Dust-tight & Corrosion resistive -- same as type 4 with corrosion resistant construction. Type 7 Class I, Group A,B,C or D indoor Hazardous Locations Air -Break Equipment -- an enclosure intended for indoor use in the atmospheres and locations defined as Class I, Group A, B, C or D in the National Electric Code (NFPA 70). The letters A, B, C and D that indicates the gas or vapor atmospheres in the hazardous location must appear as a suffix to the designation. These enclosures meet UL requirements pertaining to Industrial Control Equipment for use in Hazardous Locations (UL698) and must be marked to show the class and the group letter designations. Type 7 enclosures are termed explosion-proof per NEMA and the National Electric Code.

Enclosure (motors) - open drip proof motors are for use in areas that are dry, clean and well ventilated. If installed outside, motors must be protected with a cover that does not restrict air flow. Damp or dirty conditions require totally enclosed construction. Hazardous conditions, no matter how slight, require an explosion proof (hazardous location) motor. A hazardous location (explosion proof) machine is a totally enclosed motor whose enclosure is designed and constructed to withstand an explosion of a specified gas or vapor that may occur within it and to prevent the ignition of the specified gas or vapor that may occur within the motor casing.

Enclosures of motors - induction type according to NEMA.

Open type (OT) - an open motor is self-ventilated, having no restriction to ventilation or to the enclosure other than necessitated by mechanical construction.

Drip proof (DP) - constructed with ventilation openings so that falling liquid at an angle of 15° F from vertical will not enter the motor.

Totally enclosed (TE) - constructed with no openings in brackets or frame. These include totally enclosed non-vented (TENV), totally enclosed fan-cooled (TEFC), totally enclosed air over (TEAO) explosion proof (XP).

Encrustation - the buildup of slimy orange-brown deposits on the water side of pipes caused by iron bacteria in the water.

End bell - the shield at the end of the motor that supports the bearings. Also called end plates or end shields.

End mount motor - an electric motor mounted with tabs or studs fastened to the motor housing end.

End play - refers to the looseness the rotor assembly has within the end plates. Excessive end play can result in problems such as fan blades hitting fan guards. Zero end play can result in a tight motor.

End shield - a shield secured to the frame of a motor to protect the windings and to support the bearings.

End suction pump - a pump with an inlet connection perpendicular to the plane of its impeller.

End switch - a switch built into a zone valve, damper motor or contactor that opens or closes its electrical contacts when the device reaches its fully open or closed position.

End wrench - the same as a pipe wrench except the movable jaw is offset to get into tighter or closer positions.

Endothermal - chemical reaction in which heat is absorbed.

Energize - to supply or put power to an electrical load.

Energized - connected to an energy source or containing residual or stored energy.

Energy - the ability or capacity to do work. Mechanical energy is expressed in foot-pounds or horsepower-hours; electric energy, in kilowatt hours; and heat energy, in British thermal units.

Energy audit - process of accurately determining the current energy consumption for a given area.

Energy bands - the grouping of energy levels that occur when atoms are bonded together in solid materials.

Energy charge - that part of an electric bill based on kWh consumption. Expressed in cents per kWh. Energy charge covers the cost of utility fuel, general operating costs, and part of the amortization of the utility's equipment.

Energy conservation - process, upon reviewing the calculations for determining heat loads, of instituting changes that will result in energy savings.

Energy cut-off - a protective device placed in series with the thermocouple in a control system that opens the thermocouple circuit thereby shutting down the burner operation in case of abnormally high temperatures.

Energy efficiency ratio (EER) - the ratio of the rated cooling capacity in Btu per hour divided by the amount of electrical power used in watts. Expressed in Btuh per watt.

Energy isolating device - a mechanical device that physically prevents the transmission or release of energy. Examples are manually operated electrical circuit breakers, disconnect switch, a line valve, a block, or any other device used to block or isolate equipment from energy sources. Push buttons, selector switches and other control circuit devices are NOT energy isolating devices.

Energy levels - the energy states of electrons in an atom.

Energy management control system - controllers used in a system that optimizes total energy usage in a building or residence.

Energy payback time - the time required for any energy-producing system or device to produce as much useful energy as was consumed in its manufacture and construction.

Energy recovery ventilation system - a device or combination of devices applied to provide the outdoor air for ventilation in which energy is transferred between the intake and exhaust airstreams.

Energy source - any source of electrical, mechanical, hydraulic, pneumatic, chemical, thermal, water, steam or other energy.

Energy utilization index - (EUI) a number that is used to compare energy usages for different areas. It is calculated by dividing the energy consumption (in Btu's) by the square footage of the conditioned space area.

Engine - prime mover; device for transforming fuel or heat energy into mechanical energy.

Engineering controls - systems that reduce potential hazards by isolating the worker from the hazard or by removing the hazard from the work environment. Methods include substitution, ventilation, isolation, and enclosure. This is preferred over administrative controls and personal protective equipment.

Enhancement mode - the operation of a field effect transistor that occurs with an increase in the gate-to-source voltage in an effort to increase the drain current.

Entering wet bulb temperature - the wet bulb temperature of the air actually entering the water tower, including any effects of recirculation. In testing, the average of multiple readings taken at the air inlets to establish a true entering wet bulb temperature.

Enthalpy - the total amount of heat (both latent and sensible) in one pound of a substance that is calculated from an accepted base of 32° F. For refrigeration the accepted base is -40° F and expressed in Btu/lb. It is the sum of the internal energy plus the quantity of the substance.

Enthalpy switch over - automatic switching or regulation of outside air and return air dampers. Total heat content of inside and outside air is compared before selecting either inside or outside air or a mixture of the two for ventilation that will require the least amount of refrigeration, humidification or dehumidification.

Entrained - describes a liquid that is absorbed and held, but not dissolved, as in air.

Entrained air - air bubbles that are carried along with the fluid flowing through a piping system.

Entrainment - the induced flow of room air by the primary air from an outlet, creating a mixed air path (commonly called secondary air motion).

Entrainment ratio - total air divided by the air discharged from the outlet.

Entrance cap - a weather proof, insulated cap for terminating the power line connections to a building; service head.

Entrance ell - a metal box to complete a 90° angle with conduit. The cover allows the electrician to pull wires through conduit in either direction.

Entrance, service - the part of the electrical installation from the service drop to the main service panel.

Entropy - the ratio of the heat added to a substance to the absolute temperature at which it is added. A mathematical factor used in engineering calculations. The amount of energy in a system.

Envelope - (1) the outer boundary of an air stream moving at a specific velocity (for example, a 50 fpm envelope). (2) The air barrier that separates the conditioned space from the outside and from unconditioned spaces like attics and garages.

Envelope component - major section of the entire envelope, such as the opaque walls above-grade, ceilings, slabs, floors, glazings, doors, or walls below-grade.

Environment - surrounding; conditions and influences surrounding and affecting development and behavior.

Environmental air - air that is supplied, returned, recirculated, or exhausted from spaces for the purpose of modifying the existing atmosphere within the building.

Environmental control system - the process of controlling the environment by heating, cooling, humidifying, dehumidifying, or cleaning the air.

Environmental factors - conditions other than indoor air contaminants that cause stress, comfort, and/or health problems (e.g., humidity extremes, drafts, lack of air circulation, noise, and overcrowding).

Environmental tobacco smoke (**ETS**) - mixture of smoke from the burning end of a cigarette, pipe, or cigar and smoke exhaled by the smoker (also secondhand smoke or passive smoking).

Enzyme - a complex organic substance originating from living cells that speeds up the chemical change in food. Enzyme action is slowed by cooling.

Epoxy - a synthetic plastic adhesive formed by mixing two resins together.

Equal friction method - a method of duct sizing where the selected duct friction loss value is used constantly throughout the design of a low pressure duct system.

Equal percentage characteristic valve - a valve characteristic in which the flow rate through the valve increases slowly as the stem starts to lift, and increases faster the further the stem lifts. This characteristic gives an equal percentage change over the previous flow for equal increments of stem movement. This characteristic is desirable when heat output is being controlled by varying the flow rate through the heat emitter.

Equalizer - a piping arrangement to maintain a common liquid level or pressure between two or more chambers.

Equalizer, crankcase double pipe - a system in which both the gas pressure and the oil level in all compressors connected in multiple are equalized.

Equalizer, crankcase single pipe - a pipe or tube connection between the crankcases of the compressors of a multiple compressor system, the function of the pipe or tube being to equalize the pressures between the crankcases.

Equalizer, external - in a thermostatic expansion valve, a tube connection from a selected control point in the low-side refrigeration circuit to the pressure sensing side of the control element such that the control-point pressure is transmitted to the actuating element (diaphragm or bellows). This connection provides a means for compensating for the pressure drop through accessories and the evaporator coil.

Equalizer, internal - in a thermostatic expansion valve, an integral internal part or passage that provides exposure of the actuating element (diaphragm or bellows) to pressure leaving the valve.

Equalizer line - a line that equalizes the gas or oil pressure in two or more pieces of equipment.

Equalizer tube - a device that is used to maintain equal pressure or equal liquid levels between two containers.

Equalizer valve - a device that regulates the flow of gases or liquids. It's used to balance pressures on either side of some recovery machines.

Equalizers - connections used with thermostatic expansion valves when the superheat setting of the expansion valve cannot control the amount of refrigerant that flows through the coil.

Equalizing - the balancing of refrigerant system pressures.

Equalizing line - line used to warm up the main steam line and equalize the pressure around the main steam stop valve.

Equation of state - a thermodynamic relation giving the functional dependence of the properties of the substance with each other.

Equilateral - a figure that has equal length sides.

Equilibrium - conditions that exists at saturation or "balanced" when the energy input equals the energy output. A state of rest or balance due to the equal action of opposing forces.

Equivalent direct radiation (EDR) - a unit of heat delivery of 240 Btu per hr.

Equivalent duct diameter - the equivalent duct diameter for a rectangular duct with side dimensions A and B is the square root of 4AB/3.1416 for the same cross sectional area.

Equivalent evaporation - the amount of water a boiler would evaporate, in pounds per hour, if it received feed water at 212° F and vaporized it at the same temperature and corresponding atmospheric pressure.

Equivalent length (EL) - pressure drop in valves and fittings expressed as the equivalent length of straight run pipe. It includes the actual length of pipe run, plus the equivalent length of tube, pipe or duct per fittings (ells, tees, couplings, take-offs, transitions and valves).

Equivalent resistance - a resistance value that would be the same value in a circuit as two or more parallel resistances of a circuit.

Equivalent temperature design - a heat gain value that takes into account the outside design temperature, the daily range of temperature and the type of material assembly.
Equipment (service) - all the necessary equipment that constitutes the main control and cut off of the electrical supply to the premises.

ER rating - energy rating number developed by the Canadian Standards Association to compare the thermal performance of windows. Measured in watts per square meter.

Ergonomics - applied science that investigates the impact of peoples physical environment on their health and comfort.

Erosion - the gradual destruction of metal or other materials by the abrasive action of liquids, gases, solids or mixtures thereof.

Erosion corrosion - a process where metal is removed from the inside surfaces of piping or piping components due to excessively high flow velocity.

Erratic - having no fixed course. Irregular or wandering.

Error - the difference between the correct value and the reading taken.

Escutcheon - a decorative, protective plate covering the body or wiring of a component.

Ester based synthetic oil - oil that is wax free, having a low floc point, and is used with hydrofluorocarbon (HFC) refrigerants. Any to a class of organic compounds corresponding to the inorganic salts formed from an acid by the replacement of hydrogen by an alkyl radical.

Ethane - refrigerant (R-170) sometimes added to other refrigerants to improve oil circulation. Ethane is a hydrocarbon made up of two carbon atoms and six hydrogen atoms.

Ethanes - organic compounds that have two carbon atoms in the molecule, all bounded by a single bond.

Ethylene glycol - a liquid chemical used as a water antifreeze.

Ethylene propylene diene monomer - a high quality rubber compound resistant to many chemicals.

Eutectic mixture, solution - that certain mixture of two substances providing lowest melting temperature of all the various mixes of the two substances.

Eutectic alloy - a metal with no melting range that goes directly from a solid to a liquid. Eutectic materials include Safety Silv 72 (eutectic at 1435°F), Stay Brite (eutectic at 430°F) and those alloys used in thermal overload relays.

Eutectic point - freezing temperature for eutectic solutions.

Eutectic salts - a special salt that melt at low temperatures by absorbing large quantities of heat. The most common eutectic salt is Glauber's salt. Are used as a heat storage medium for solar systems.

Eutectic solution - a mixture that melts and freezes at the lowest possible temperature for mixtures of the included substances.

Evacuate - to withdraw from, to make empty, remove the contents from. To remove, through the use of a vacuum pump, all noncondensables and moisture from a system.

Evacuation - removal of air (gas) and moisture from a refrigeration or air conditioning system. Since air is a non-condensable gas and contains moisture, it must be completely removed from a refrigeration system. Air and moisture are probably the worst contaminants for a refrigeration system.

Evaporate - the change of state from a liquid to a vaporous state.

Evaporating temperature - the temperature at which a liquid will vaporize at a given pressure.

Evaporation - the physical process of changing a liquid to a vapor through the addition of heat energy.

Evaporation loss, water tower - water evaporated from the circulating water into the air stream in the cooling process.

Evaporation test - a test that checks the operation of the low water fuel cut-off control.

Evaporative condenser - a device that uses open spray or spill water to cool a condenser. Evaporation of some of the water cools the condenser water and reduces water consumption.

Evaporative cooling - a system in which the absorption of latent heat by evaporating water cools the air it contacts. This system is used where the % of RH in the air is low. This system will add humidity to the air. Commonly known as a swamp cooler.

Evaporative cooling tower - a device that dissipates heat energy by the evaporation of water. The water is re-circulated to cool refrigeration condensers.

Evaporative equilibrium (of a wet bulb thermometer) - the condition attained when the wetted wick has reached a stable and constant temperature (when the instrument is exposed to air at velocities over 900 fpm, this temperature may be considered to approach the true wet bulb temperature.

Evaporative humidifier - a humidifier that provides moisture on a media surface through which air is forced. The air picks up the moisture from the media as a vapor.

Evaporator - that part of a refrigeration system in which the refrigerant absorbs heat and evaporates. The evaporator is considered to be on the low side of the refrigeration system. Design evaporating temperature must be lower than the air passing over or through the coil.

Evaporator coil - a device made of a coil of tubing which functions as a refrigerant evaporator.

Evaporator, direct expansion - an evaporator designed to cool a medium in direct contact with the evaporator.

Evaporator, dry type - an evaporator in which refrigerant is fed from a pressure reducing valve. Little or no liquid refrigerant collects in the evaporator.

Evaporator fan - a fan that moves air over heat exchange surfaces of evaporator coil so heat can be absorbed from the air stream.

Evaporator, flooded - an evaporator containing liquid refrigerant at all times.

Evaporator freeze up - situation in which an evaporator has become totally restricted by frost or ice buildup.

Evaporator pressure - the refrigerate pressure at which evaporation takes place. Also known as low side or back pressure.

Evaporator pressure regulator (EPR) - a valve located at the outlet of the evaporator that will close if the evaporator pressure falls below a given point. This control maintains a predetermined refrigerant pressure, and therefore temperature, in a cooling coil.

Evaporator pressure regulator valve - operates by bellows; refrigerant pressure on the diaphragm of the bellows compresses the spring and holds the valve open.

Evaporator superheat - the actual temperature of the refrigerant vapor at the evaporator outlet as compared to the saturated vapor temperature indicated by the suction pressure.

Evaporator temperature - the temperature at which the liquid refrigerant is changing to a vapor in the evaporator. If a high pressure liquid (R-22) enters the evaporator from the metering device at a pressure of 69 psig, it will change to vapor at the corresponding temperature of 40° F due to the decreased pressure and continue to evaporate due to the higher temperatures of air through the evaporator until all liquid has evaporated. The liquid should be completely changed to a vapor far enough ahead of the evaporator outlet to allow the vapor to be superheated.

Evaporator types - there are basically two types of evaporators - flooded and dry. The flooded type is an application where the liquid refrigerant level is maintained to the top of the heat exchange coil. A minimum amount of superheat is achieved in this type. The dry type evaporator coil achieves the heat exchange process with a minimum of refrigerant charge. This type of evaporator requires superheat to protect the compressor from liquid refrigerant entering it.

Even parallel system - parallel compressors of equal sizes mounted on a steel rack and controlled by a microprocessor.

Excess air - air that passes through the combustion chamber and flues in excess of the quantity that is theoretically required for complete combustion thereby preventing the formation of dangerous products of combustion. This term is usually stated as a percentage of the air required for complete combustion.

$$\text{Excess air} = \frac{\text{air supplied / theoretical air}}{\text{theoretical air}}$$

Exfiltration - unintentional or uncontrolled air leakage outward through openings in ceilings, walls, or floors of a building to unconditioned space or the outside.

Exhaust - a vent or opening for unwanted or used air.

Exhaust air - air being removed from any conditioned space and conveyed directly to the atmosphere by means of openings or ducts. The exhaust may serve specific areas, such as toilet rooms, or may be used in an economizer.

Exhaust fan - a fan that provides a suction to remove the air, gasses, and products of combustion from the furnace or oven. In some systems, it also brings in combustion air. Also called a ventilator.

Exhaust hood, full opening - an exhaust hood with an opening at least equal to the diameter of the connected vent.

Exhaust opening - any opening through which air is removed from a space that is being heated or cooled, humidified or dehumidified, or ventilated.

Exhaust port - that opening which carries the fluid to the downstream pressure of a fluid system.

Exhaust system - an assembly of connected ducts, plenums, fittings, registers, grills and hoods through which air is conducted from the space or spaces and exhausted to the outside atmosphere. Such exhausted air may or may not be treated before being discharged.

Exhaust valve - a movable port that provides an outlet for the cylinder gasses in a compressor or engine.

Exhaust ventilation - mechanical removal of air from a portion of a building.

Exhauster - a fan used to withdraw air under suction.

Exothermal - chemical reaction in which heat is released.

Expandable partitions - these are heavy weight vinyl folding parts located on the sides of a window unit installation. They expand so the same unit can fit windows that differ in width and also insulate the room from the outside.

Expanding pilot - a pilot that burns throughout the entire time the burner assembly is in service, whether the main burner is firing or not. Upon a call for heat, the pilot flame is automatically increased in size so as to reliably ignite the main burner. This pilot may be turned down automatically at the end of the main burner flame-establishing period.

Expansion - increase in volume due to increased temperature or decreased pressure.

Expansion bends - installed on boiler main steam lines to allow for expansion and contraction of the lines.

Expansion bolt - bolt with a casing arranged to wedge the bolt into a masonry wall to provide an anchorage.

Expansion, dry - where all of the refrigerant entering the evaporator is vaporized, as contrasted with a flooded evaporator.

Expansion joint - joint between two adjoining members arranged to permit expansion and contraction with changes in temperature.

Expansion loop - a coil or u-shaped bend in a length of tubing that provides flexibility for expansion or contraction of the tubing.

Expansion offset - an offset rectangular path within an otherwise straight run of tubing that can absorb the movement in the pipe due to thermal expansion or contraction.

Expansion point - in a mechanical refrigeration system, a restriction or orifice that regulates the flow of refrigerant into the evaporator coil. May be in the form of a thermal expansion valve or a capillary tube.

Expansion ratio - the ratio of the total volume at the end of the expansion process of a cycle to the volume in the cylinder at the beginning of the process.

Expansion tank - an overflow space in a hydronic system that provides for the increased volume of water as it is heated. Also see air cushion tank.

Expansion tank, bladder-type - a bladder-type expansion tank having a rubber bladder that separates the air from the liquid in the tank. No absorption of the air into the liquid is possible. This is the only expansion tank that should be used in closed loop or supply type heat source/sink systems.

Expansion, thermal - expansion of a material caused by heating.

Expansion valve - a device in a refrigeration system that maintains a pressure difference between the high side and the low side and is operated by pressure.

Expansion valve, automatic - an expansion valve designed to maintain a constant pressure in the evaporator regardless of superheat. Seldom used in air conditioning where loads generally fluctuate. Cannot be used on systems with multiple valve installations.

Expansion valve capacity - the refrigerating effect in Btu per hour, or tons, each to 12,000 Btu per hour, produced by the evaporation of refrigerant passed by the valve under specified conditions.

Expansion valve, float - a valve designed to maintain a constant liquid level in a flooded evaporator.

Expansion valve superheat - the difference between the temperature of the thermal sensing bulb and the temperature corresponding to the pressure at the outlet, or at the equalizer connection, when provided, of a thermostatic expansion valve.

Expansion valve superheat change - the change in superheat of a thermostatic expansion valve required to open the valve a predetermined amount.

Expansion valve, thermostatic - an expansion valve designed to meter the flow of liquid to a dry expansion evaporator at a rate sufficient to maintain a constant superheat in gas leaving the evaporator.

Expendable refrigerant system - a refrigerating system in the refrigerant is sprayed directly into the area to be cooled and must be replenished when consumed. No longer valid after July 1, 1992.

Exploded view - a diagram intended to depict the order of assembly and placement of parts in a mechanism.

Exploring tube - a small flexible tube attached to a halide torch in such a manner that air is continually drawn through the tube to the torch flame. When the free end of the searching tube is placed near a refrigerant leak, some of the refrigerant is carried to the flame where its presence is indicated by changing the color of the flame.

Explosion relief door - a door in a furnace or boiler setting designed to be opened by a predetermined combustion pressure.

Exposed area - the area of any wall, window, ceiling, floor or partition separating a heated room from the outdoors or from an unheated space.

Exposed junction - the junction in a thermocouple probe that extends beyond the sheath material. Provides complete contact with the medium being measured and a very fast response time.

Exposed surface - any building surface exposed to outside air temperature.

Exposed wall - wall with one side in the conditioned space and the other side outdoors or confining a non-conditioning space.

Explosimeter - an instrument that indicates the combustibility of air in an enclosed space.

Explosion - combination occurring within a confined environment at such a rapid rate as to cause detonation, called a fireside explosion. Also the rupturing of a vessel due to the build up of pressure of water or steam, called a waterside explosion.

Explosion heads - a protective device for relieving excessive pressure in a premix system by bursting of a rupturable disc.

Explosion proof - type of construction designed to contain an explosion and prevent its propagation to the outside atmosphere; usually implies the use of heavy metal enclosure.

Explosion proof motor - a totally enclosed motor whose enclosure is designed and constructed to withstand an explosion of a specified gas or vapor that may occur within the motor and to prevent the ignition of this gas or vapor surrounding the motor.

Explosive limits - see flammability limits.

Explosive mixture - a flammable mixture in a confined space.

Exposed area - the area of any wall, window, ceiling, floor, or partition separating a heated room from the outdoors or from an unheated space.

Exposure diversity - the exposure factor of a building's glass surfaces to sunlight. The average glass load is computed by averaging the hourly total glass load values from 8 a.m. to 7 p.m. An adequate exposure diversity is when the total glass load is relatively constant as the individual glass loads increase and decrease on an hourly basis.

Exposure limits - the concentration in workplace air of a chemical deemed the maximum acceptable. Meaning that most workers can be exposed at given levels or lower without harmful effects. Exposure limits in common use are threshold limit value - time weighted average (TLV-TWA), short term weighted average (STEL), and ceiling value (C).

Extended plenum - an air delivery system in which a main duct is run from the plenum, and branch ducts run from the main duct to the various zones.

Extended surface - the heat transfer surface, one or both sides of which are increased in area by the addition of fins, discs, or other means.

Extended surface filter - type of filter where the filtering media area is increased by pleating or baffling it.

Extended surface heating unit - a heating unit having a relatively large amount of extended surface that may be integral with the core containing the heating medium or assembled over such a core, the core or by both pressure and soldering. An extended surface heating unit is usually placed within an enclosure and therefore functions as a convector.

Extension rod - an extension rod is used to connect the socket to the drive handle to give the socket longer reach to gain access to recessed areas.

Exterior envelope - the elements of a building that enclose conditioned spaces through which thermal energy may be transferred to or from the exterior.

External drive - term used to indicate a compressor driven directly from the shaft or by a belt using an external motor. Compressor and motor are serviceable separately.

External equalizer tube - a tube connected to the low side of an expansion valve diaphragm and to the exit of the evaporator. It lets the evaporator outlet pressure help control the operation of the valve. Used on evaporator coils having an excessive pressure drop.

External heat gain - (1) the heat that both infiltrates and is conducted into a cooled space; (2) the heat that escapes, both by air leakage and conduction, from a heated space or building.

External heat loss - the amount of heat that flows from the inside of the structure to the outside. This is due to the difference between inside and outside temperatures, the material through which the heat flows, and entering air.

External heater - a heater in which the burners and combustion chamber are effectively separated from the oven chamber or the medium being heated. Also see direct fired external and indirect fired external heaters.

External static pressure (ESP) - the sum of the absolute values of the entering and leaving static pressures of moving air measured at an air handling piece of equipment.

External treatment - boiler water treated before it enters the boiler to remove scale-forming salts, oxygen, and noncondensable gases.

External VA rating - the volt amps that can safety be supplied to an external electrical load from a controller with a built in transformer.

Externally equalized valve - a valve that resembles the internally equalized type, but evaporator pressure on the underside of the diaphragm is supplied from the evaporator outlet, rather than the inlet.

Extinction pop - an undesirable flame characteristic in which burner flames strike back into a burner to burn there or to create a pop after the gas supply has been turned off.

Extinguishing media - the type of fire extinguisher or extinguishing method appropriate for use on a specific chemical. Some chemicals react violently in the presence of water, so other suggested methods, such as foam or CO_2, should be followed.

Extra hazard area - a location where Class A and Class B flammables exceed those expected in ordinary hazard areas.

Extra strong - pipe sizes corresponding to schedule 80.

Extraction steam - steam that is extracted from a steam turbine at a controlled pressure point for use at a different location for a specific application.

Extractors - a set of curved blades formed into a unit that can be adjusted to direct air flow into a branch duct or supply outlet to provide positive control of air volume and to smooth out air turbulence.

Extrinsic semiconductor - a semiconductor that has been doped.

Extruded - pushed out through a die. Bars of ice, metal rods, shapes and tubes are all made by this method.

Extruded polystyrene foam - a type of closed cell foam insulation board often used for insulation purposes.

Eye of an impeller - the center opening where fluid flows into the impeller.

-F-

F - abbreviation for farad, frequency, fluorine, or Fahrenheit.

FCD - final controlled device.

Fid - flame ionization detector.

Fig - figure.

FLA - full load amps.

FM - Factory Mutual or frequency meter.

FOB - free on board.

FPM - feet per minute.

FS - full scale.

Ft - foot or feet.

Ft/lb - foot-pound.

Ft/min. - feet per minute.

Fu - fuse.

Fabrication drawing - drawing used to represent configurations that must be made up in the shop or in the field.

Fabrication isometric - isometric drawing of pipe lines showing line designation, pipe size, elevations, pipe footage, elbow-to-elbow dimensions, weld numbers, piping configurations, valve types, pipe supports and symbols, and equipment locations and numbers.

Facade - the exterior face of a building; the architectural front.

Face - the finished contact surface of flanged-end piping and components (such as valves).

Face area - the total plane area of the portion of a grille, coil, or other items bounded by a line tangent to the outer edges of the openings through which air can pass.

Face and bypass - a heating system in which the mixed airflow is divided into two duct sections, one through the coil (face) and the other around the coil (bypass). Dampers work in opposition (face damper closes while bypass damper opens, and vice versa) to regulate the amount of air that is heated. Used with either a steam or hot water coil system. Demonstrates on-off control of coil and modulating control of airflow across coil.

Face-to-face - dimensions from the face of the outlet port to the face of the inlet port of a valve or fitting.

Face velocity - the velocity (or speed) of the air passing through an air outlet or air return grille. Measured in feet per minute (FPM).

Faced wall - a wall in which the masonry facing and backing are so bonded as to exert common action under load.

Facility management system - the integrated control of a building environmental system using computers or electronic devices. Also see building automation system.

Facing - part of valve body that connects to a companion flange.

Factor of evaporation - if 970.3 Btu is added to 1 pound of water that is at atmospheric pressure and 212° F, it will be converted into steam and the steam will be at atmospheric pressure and 212° F. This is termed "evaporation from and at 212° F", or, briefly, "from and at." The heat added to a pound of water by the heating vessel (from the time it enters until it leaves as steam) divided by 970.3 is the factor of evaporation.

Factor of safety, pressure vessels - the ratio of ultimate strength of material to maximum permissible stress when in use.

Factory built chimney - a chimney composed of listed and labeled factory built components assembled in accordance with the manufacturer's installation instructions to form the completed assembly.

Factory installed wiring - the wiring installed in a piece of equipment at the factory. This wiring is usually the connections between the components in the control panel and the system components in the unit itself.

Factual diagram - a wiring diagram that is a combination of the pictorial and schematic diagram.

Fahrenheit scale - a thermometric scale in which, under standard atmospheric pressure of 14.696 psi at sea level, boiling point of water is 212° F and freezing point is 32° F above zero. Abbreviation is F.

Fail position - the mode in which a heat pump will operate if the reversing valve is in the de-energized position.

Fail safe - the programmed or elapsed time after which a heat pump system will terminate its defrost cycle and return to the normal heating mode.

Fail safe control - a device that opens a circuit when sensing element fails to operate.

False defrost - the system goes into the defrost mode even if the outdoor has no frost and/or ice build up.

Fan - any device or machine used to set up artificial currents of air. An air moving device comprising a wheel or blade, and housing or orifice plate. Also see blower and burner motor; and exhaust, forced-draft, and induced-draft fans.

Fan and limit control - usually a combination control and safety device that turns a heater fan or blower on and off according to preset temperatures; and shuts off the main burner gas in case a malfunction causes combustion or chamber temperature to rise above a preset limit value.

Fan assisted combustion system - an appliance equipped with an integral mechanical means to either draw or force products of combustion through the combustion chamber and/or heat exchanger.

Fan, centrifugal - a fan rotor or wheel within a scroll type of housing and including driving mechanism supports for either belt or direct connection.

Fan coil - a complete unit located in the room being conditioned consisting of a coil through which hot or cold water is circulated, a fan that circulates room air through the coil, a filter to remove lint and dust, a cabinet, a grille, and a control system. The boiler or chiller supplying the water is located centrally within the building.

Fan control center - a transformer relay combination on a common mounting plate for mounting on a standard four inch square electrical box. These control centers can be used in remote control circuits for furnaces and air conditioners as well as numerous applications on other indicating and regulating equipment.

Fan cycle control - device used to control the condenser fan(s). When head pressure drops below acceptable limits, the fans are turned off.

Fan cycling - the use of a pressure controller to turn a condenser fan(s) on and off to maintain the correct pressure within the system.

Fan cylinder - cylindrical or venturi-shaped structure in which a propeller fan operates. Sometimes referred to as a fan stack on larger cooling towers.

Fan deck - surface enclosing the top structure of an induced draft cooling tower, exclusive of the distribution basins on a cross flow tower.

Fan economizer - a device that prevents the operation of the fan motor as a cold diffuser during the shutdown period after the coil has been defrosted.

Fan + fan - refers to the maximum combined input rating of two or more fan assisted category I appliances attached to a common vent.

Fan Laws of air moving equipment - the performance of all fans and blowers is governed by certain rules of physics known as fan laws. Cubic feet per minute (CFM), revolutions per minute (RPM), static pressure (SP) and horsepower (HP) are all related to each other in a known manner and when one changes, all others change. For example, when CFM is changed RPM, SP, and HP will also change. CFM is the variable most commonly changed in an air moving system, therefore the following example of fan law application is based on a change from an existing CFM to a new CFM.

To determine performance at a new CFM, first calculate the ratio of the new CFM to existing CFM (new CFM divided by the existing CFM):

$$\text{Ratio} = \frac{\text{CFM new}}{\text{CFM existing}}$$

To determine the new RPM, multiple ratio in step #1 times existing RPM:

$$\text{RPM new} = \frac{\text{CFM new}}{\text{CFM existing}} \times \text{RPM existing}$$

To determine new SP, multiply ratio of step #1 times itself, times existing SP:

$$\text{SP new} = \frac{\text{CFM new 2}}{\text{CFM existing}} \times \text{SP existing}$$

To determine new HP required, multiply ratio in step #1 times itself twice, and then times existing HP:

$$\text{HP new} = \frac{\text{CFM new 3}}{\text{CFM existing}} \times \text{HP existing}$$

Sample calculations: Existing conditions are 5000 CFM, 1000 RPM, 0.5" SP. 0.5 HP motor. New CFM desired is 6000.

$$\text{Ratio} = \frac{6000}{5000} = 1.2$$

Calculate new RPM: 6000/5000 = 1.2 X 1000 = 1200 RPM.

Calculate new SP: 6000/5000 = (1.2) 2 = 1.44 X 0.5 = 0.72" SP.

Calculate new HP: 6000/5000 = (1.2) 3 = 1.73 X 0.5 = .86 HP.

New conditions show that the CFM increased 20%, RPM increased 20%, SP increased 44%, and the HP increased 86%.

Fan limit control - a fan switch and a limit switch in a common enclosure that uses a common sensing element to sense bonnet or output temperature.

Fan max - refers to the maximum appliance input rating of a Category I appliance with a fan assisted combustion system that could be attached to the vent.

Fan min - refers to the maximum appliance input rating of a Category I appliance with a fan assisted combustion system that could be attached to the vent.

Fan mix burner - a mechanical draft burner that uses the energy of high pressure gas (10 to 50 psi) to completely premix the air and gas. Gas is permitted to escape through a row of small orifices on the edge of propeller like blades mounted on a free spinning shaft. Reaction of the escaping gas rotates the blades that are mechanically coupled to a fan that draws in the primary air. The speed of the fan varies with the gas pressure, thereby automatically adjusting the air volume to maintain optimum combustion. Also see mechanical draft and pre-mix burners.

Fan mixer - an air blower in which gas is admitted to the inlet to be mixed with air.

Fan + nat - refers to the maximum combined input rating of one or more fan-assisted category I appliances and one or more category I draft hood equipped appliances attached to a common vent.

Fan performance curves - charts that illustrate and relate the various factors affecting fan operation and efficiency. These graphs present static or total pressure and power input over a range of air volume flow rates at a stated inlet density and fan speed. Graphs show a range of air volume from zero air flow (shut-off) to maximum air flow (free delivery).

Fan pitch - the angle where the blades of a propeller fan make with the plane of rotation, measured at a prescribed point on each blade.

Fan, propeller - a propeller or disc type wheel within a mounting ring or plate and including driving mechanism supports for either belt drive or direct connection.

Fan relay - electro-mechanical control used to operate the furnace blower motor.

Fan relay coil - a magnetic coil that controls the starting and stopping of a fan through opening and closing of the fan relay contacts.

Fan shroud - convoluted housing in which a centrifugal (blower) fan operates. Also known as a scroll.

Fan switch - a plenum mounted, warm air furnace thermostat that energizes the fan on temperature rise and stops it when the temperature falls to the off setting of the control.

Fan, tube axial - a propeller or disc type wheel within a cylinder and including driving mechanism supports for either belt drive or direct connection.

Fan, vane axial - a disc type wheel within a cylinder, a set of air guide vanes located either before or after the wheel and including driving mechanism supports for either belt drive or direct connection.

Farad - a unit of electrical capacity. The capacity of a device (either a capacitor or a condenser) when charged with one coulomb of electricity, gives a difference of one volt of potential. A unit used to measure capacitance.

Farad experiment - silver chloride absorbs ammonia when cooled and releases ammonia when heated. This is the basis on which some absorption refrigeration systems operates.

Fast acting fuse - a fuse that opens on overloads and short circuits very quickly. This type of fuse is not designed to withstand temporary overload currents associated with inductive and capacitive loads.

Fast food freezing - method that uses liquid nitrogen or carbon dioxide to turn fresh food into long lasting frozen food. It is often referred to as cryogenic food freezing and/or freeze drying.

Fatigue resistance - the ability of a metal to bend over and over again without going beyond its elastic limit or breaking.

Fault directory - a look-up table or equipment label used to translate numeric or alpha-numeric fault codes into text messages.

Feathering - the point when a safety valve is about to lift.

Feed - the primary medium in a process. An example would be coal or gas.

Feed, boiler - water used to produce steam inside a boiler.

Feedback - in a control system, feedback is the signal (or signals) fed back from a controlled process to denote its response to the command signal. Feedback is derived by comparing actual response to desired response, and any variation is fed as an "error signal" into the original control signal to help enforce proper system operation. Systems employing feedback are termed "closed loop" systems, with feedback closing the loop.

Feedback control system - control system that is constantly correcting the condition. Also called a closed loop system.

Feedback potentiometer - the potentiometer in a modulating motor that forms part of a bridge circuit along with the controller potentiometer and balancing relay in the motor. When the wiper on the controller potentiometer moves due to a change in the controlled variable, the motor runs and drives the wiper on the feedback potentiometer in the proper direction to rebalance the bridge circuit. When the circuit is balanced, the motor stops.

Feeder screw - a threaded screw with a handle at one end to apply tension on an object being held in place, as in a yoke vise, or on other objects to be cut.

Feeders - a conductor or group of conductors between the service equipment and the final branch circuit over current device.

Feedwater - provides original and make-up water for the heating process.

Feedwater heater - used to heat feedwater before it enters the steam and water drum.

Feedwater lines - the lines leaving the open feedwater heater and going to the boiler.

Feedwater pump - this pump takes water from the open feedwater heater and delivers it to the boiler at the proper pressure.

Feedwater regulator - control used to maintain a normal operating water level that cuts down the danger of high or low water.

Feedwater treatment - a chemical mixture that is fed into the boiler to prevent scaling or corrosion on the shell and tubes. Can be internal using chemicals, or external, using water softeners.

Feedwater valve - a valve that automatically allows cold water to enter a closed hydronic system upon a drop in system pressure.

Feet of head - the term used to designate the flow resistance the pump must overcome to deliver the required amount of liquid. 2.307 feet of head = 1 psig. One foot of head is equivalent to 0.433 psig.

Feet per minute - a velocity measurement of a moving air stream. Usually expressed as FPM.

Female thread - the internal thread on fittings, valves, machine bodies, and the like.

Fenestration - all light-transmitting envelope component assemblies in a building wall or ceiling used for light transmittance, ventilation, entry, or exit where such component assemblies enclose conditioned space. From the Latin word, “fenestra,” meaning window.

FERC efficiency - electrical output plus one half the thermal heat utilized divided by the energy input based on the low heating value of the fuel (defined by the Federal Energy Regulatory commission).

Ferromagnetic - a highly magnetic material such as iron, nickel, steel, etc.

Ferrous metal - metals that contains iron as the major portion of their make-up.

Ferrule type cartridge fuse - a special fuse with round metal caps on a cylinder case.

Festoon lighting - outdoor string of lights suspended between points exceeding 15 ft.

F-factor - the perimeter heat loss factor expressed in Btu/h·ft·° F.

Fiber optics - long, thin, flexible fibers of glass or plastic through which data or images are transmitted using light.

Fiber stress - the unit stress, usually in pounds per square inch (psi), in a piece of material that is subjected to an external load.

Field - an area of magnetic force.

Field coil - a coil of insulated wire wrapped around an iron core. A magnetic field is produced by current flowing in the coil. Used in such devices as motors.

Field discharge resistor - used across a winding to prevent excessive voltage which helps to reduce the voltage induced into the rotor by the collapsing magnetic field when current is disconnected.

Field effect - the application of a perpendicular electric field to cause a varying effect on the conductivity of the current path.

Field effect transistor - a transistor controlled by voltage rather than current. The flow of current through the transistor is controlled by the effect of an electric field that is exerted by an electric charge in a region located close to the channel called the gate. Sometimes known as a unipolar transistor. May have either a P channel or an N channel in their construction.

Field, electric - the area or region around a current carrying conductor in which an electric force can act on a charge.

Field erected boiler - a boiler that must be erected in the field because of its size and complexity.

Field, magnetic - the area or region in which the north and south polarity (magnetic forces) that are created by a magnet can act on a charge.

Field, motor - the stator (stationary) part of a motor that furnishes the magnetic field surrounding the armature.

Field of force - the area around a magnet that is affected by the strength of the magnet.

Field pole - that part of the stator of a motor that concentrates magnetic field of field winding.

Field rectifier - a diode connected in series with the field of a shunt motor. It conducts on every other alternation of AC, to ensure that the field current is always in the same direction.

Field run - piping routes that are determined at the time of installation.

Field wiring - wiring that must be done at the installation site (in addition to factory wiring) in order to complete an installation.

Figure of merit - for a thermoelectric material used in a thermoelectric device whose operation is based on the Seebeck effect or the Peltier effect, the quotient of the square of the absolute Seebeck coefficient by the product of the electrical resistivity and the thermal conductivity.

Filament - the resistive element of a bulb that glows to produce light when current is forced through it.

File - a steel instrument, having cutting ridges or teeth, made by indentation with a chisel, used for abrading or smoothing other substances, as metals wood, etc. A hand tool used to scrape rough surfaces smooth.

File card - a tool used to clean material out of metal files.

File, cross-cut - a file having two sets of teeth crossing obliquely.

File, second-cut - a file having teeth of a grade next finer than bastard.

File, smooth - a file having teeth so fine as to make an almost smooth surface.

Fill - that portion of a cooling tower that constitutes its primary heat transfer surface. Sometimes referred to as packing.

Fill cube - (1) Counter flow: the amount of fill required in a volume one bay long by one bay wide by an air travel high. (2) Cross flow: the amount of fill required in a volume one bay long by an air travel wide by one story high.

Fill deck - one of a succession of horizontal layers of splash bars utilized in a splash filled cooling tower. The number of fill decks constituting overall fill height, as well as the number of splash bars incorporated within each fill deck, establishes the effective primary heat transfer surface.

Fill factor - the ratio of the maximum power a photovoltaic cell can produce to the theoretical limit if both voltage and current were simultaneously at their maxima. A key characteristic in evaluating cell performance.

Fill light - supplementary illumination to reduce shadow or contrast range.

Fill or wetted surface method - water in a cooling tower is spread out over a wetted surface while air is passed over it to enhance evaporation.

Fill sheet - one of a succession of vertically arranged, closely spaced panels over which flowing water spreads to offer maximum surface exposure to the air in a film-filled cooling tower. Sheets may be flat, requiring spacers for consistent separation; or they may be formed into corrugated, chevron, and other patterns whose protrusions provide proper spacing, and whose convolutions provide increased heat transfer capability.

Fill valve - a device that allows water to enter the machine, generally operated by a pressure switch.

Filler metal - alloys used in the process of joining pieces of metal together.

Fillet - (1) the curved line connecting two lines that form a angle. (2) A bead of solder at the point where the joined pieces meet.

Fillet weld - used to join metal with faces that form an angle.

Film factor - the relationship between the medium giving up heat and the heat exchange surface (evaporator). This relates to the velocity of the medium passing over the evaporator. When the velocity is too low, the film between the air and the evaporator becomes greater and becomes an insulator, that slows the heat exchange.

Film filled - descriptive of a cooling tower in which film-type film is utilized for the primary heat-transfer surface.

Filter, air - a device to remove solid material from a fluid (liquid or gas). In furnace filters, a porous material (fiberglass or foam plastic) that is installed in the air circulation system of a furnace or fan coil unit to remove dust particles and pollen. Some are disposable, whereas some may be cleaned and re-used.

Filter backed return grille - a return air grille located in the occupied zone that contains and supports an air filter.

Filter, electronics - in electronics, a selective network of resistors, inductors, or capacitors that offers comparatively little opposition to certain frequencies or to direct current, while blocking or attenuating other frequencies.

Filter, electrostatic - an electrical filtering device for the removal from air of smoke and other particles too small for the usual mechanical type throw-away filter. The very small particles are forced to adhere to collector plates because of the electric charges imparted by the powerpack.

Filter drier - device to absorb remaining moisture and catch any foreign particles circulating with the refrigerant after evacuation. Driers are filled with a desiccant which facilitates the moisture and acid removal.

Filtration - this is the method used to remove dust from air during the conditioning process. Air filtration is one of the items needed for a total comfort system.

Fin - an extended surface to increase the heat transfer area, as metal sheets attached to tubes.

Fin comb - a hand tool used to straighten out bent coil fins.

Fin tube convector - terminal unit which uses a finned tube radiation to exchange heat by convection.

Fins per inch - on heat exchange surfaces, the number used to determine the spacing between the metal fins attached to the refrigerant tubing. The larger the number of fins per inch, the closer the fins are to each other. High temperature refrigeration systems used for air conditioning applications typically have twelve to fifteen fins per inch on the indoor coil.

Final control element - last component in control chain. That portion of the controlled device that regulates the final control agent like a valve or damper.

Fine print notes - explanatory material is in the form of fine print notes. Fine print notes (FPN) are informational only, and are not enforceable as a requirement by the National Electric Code.

Fines - finely crushed or powdered material or fibers; especially those smaller than the average in a mix of various sizes.

Finish grade - the top surface of the lawns, walks or other improved surface after completion of grading operations.

Finish plaster - the final or white coat of plaster.

Finned tube - a heat exchange device consisting of a metal tube through which water, steam or refrigerant may be circulated. Metal plates or fins are attached to the outside of the tube to increase the heat transfer surface. Finned tube or fin tube may consist of one, two, or three tiers and are designed for installation bare, or with open type grilles, covers, or enclosures having top, front, or inclined outlets. Usually finned tube units are for use in other than residential buildings.

Finned tube baseboard convector - a type of convector that is usually composed of a copper tube with aluminum fins mounted in a steel enclosure located at the base of a walls.

Firebox - see combustion chamber.

Firebox boiler - a horizontal, fire-tube boiler with a small, internal combustion chamber of cubical design. The short, first-pass bank of tubes is connected between the rear of the combustion chamber and the rear of the boiler. The remaining banks of tubes extend the full length of the boiler, above the combustion chamber and the first-pass bank. The combustion chamber may be steel, refractory, or water-cooled refractory. Also see fire-tube, multi-pass, and water-leg boilers.

Fire brick - ceramic brick made to withstand high temperatures for ling chimneys, incinerators and similar structures used to line combustion chambers.

Fire efficiency finder/stack loss slide rule - a slide rule that uses the CO_2 percentage and temperature of a vapor to determine the efficiency of a fossil fuel heating system.

Fire damper - a temperature actuated damper in a duct that closes during a fire to prevent the spread of flames by the air system. Such dampers are fabricated to operate at predetermined temperatures and to have overlapping blade edges along the top and bottom of the frame and where contact is made with the other blades.

Fire diamond - the National Fire Protection Agency visual rating system that addresses the health, flammability, reactivity, and related hazards of a material that may exist due to the short term, acute exposure caused by a fire, spill, or similar emergency.

In position A (left side) is the health hazard in blue for the degree of hazard of the level of short term exposure.

0 = ordinary combustible hazards in a fire.
1 = slightly hazardous
2 = hazardous
3 = extreme danger
4 = deadly

In position B (top) is the flammability hazard in red for the degree of flammability of the susceptibility to burning.

0 = will not burn
1 = will ignite if preheated
2 = will ignite if moderately heated
3 = will ignite at most ambient conditions
4 = burns readily at ambient conditions

In position C (right side) is the reactivity, instability in yellow for the energy released if burned, decomposed, or mixed.

0 = stable and not reactive with water
1 = unstable if heated
2 = violent chemical change
3 = shock and heat may detonate
4 = may detonate

In position D (bottom) is special hazards in white.

Ox = oxidizer
W = use no water, reacts!

Fire door - a door and its assembly, so constructed and assembled in place as to give protection against the passage of fire.

Fire door assembly - the assembly of a fire door and its accessories, including all hardware and closing devices and their anchors; and the door frame, when required, and its anchors.

Fire extinguisher - typically a small hand held device that is used to put out small fires. There are five types of fire extinguishers --- A, B, C, D and K. Also see class of fire.

Fire point - the minimum temperature at which a liquid will produce sufficient vapors to flash near its surface and continue to burn. This temperature is usually 10 to 30° C higher than the flash point.

Fire proofing - material applied to structural elements or systems that provide increased fire resistance.

Fire rating - the classification indicating in time (hours) the ability of a structure or component to withstand fire conditions.

Fire rated doors - doors designed to resist standard fire tests, and labeled for identification.

Fire resistance rating - the time, in minutes or hours, that materials or assemblies have withstood a fire exposure as established in accordance with National Fire Protection Association Standard 251.

Fire safety switch - if the ignition bracket assembly overheats or if there is an external fire that effect the assembly, this switch will turn off the ignition sequence. Also referred to as a flame roll-out switch.

Fire side - in a boiler, the area exposed to flame and combustion gasses, as opposed to the water side.

Fire stop - incombustible filler material used to block interior draft spaces.

Fire stat - a thermostat that senses temperature in a return air duct and shuts down the air conditioning or ventilating fans when the air temperature goes above a preset value.

Fire tower - a vertical enclosure (containing a stairway) having a fire endurance rating sufficiently high to qualify as a fire escape.

Fire tube - the primary heat transfer surface area in a fire tube boiler. A tube having water on the outside and carrying the products of combustion on the inside.

Fire-tube boiler - a boiler in which the products of combustion pass through straight tubes surrounded by water and steam. The tubes may run horizontally or vertically.

The horizontal-return tube (HRT) type has a horizontal shell containing the tubes. The products of combustion pass from the external combustion chamber underneath the bottom of the shell to the end of the boiler, and return through the tubes. If the products of combustion are passed through the shell more than once, it is called a multi-pass boiler.

The vertical type has a vertical, cylindrical shell containing the tubes. The products of combustion pass from the internal combustion chamber below the shell, up through the tubes. In a submerged vertical type, the water level extends higher than the top ends of the tubes. Also see firebox, multi-pass, and scotch boilers.

Fire wall - a fire resistance rated wall, having protected openings, that restricts the spread of fire and extends continuously from the foundation to or through the roof.

Fireplace - a fire chamber and hearth constructed of noncombustible material for use with solid fuels and provided with a chimney. May be one of the following types:

Masonry fireplace - a hearth and fire chamber of solid masonry units such as bricks, stones, listed masonry units, or reinforced concrete, provided with a suitable chimney.

Factory built fireplace - a fireplace composed of listed factory built components assembled in accordance with the terms of listing to form the completed fireplace.

Fireside explosion - see explosion

Firing device - the burner capable of burning oil, gas, or coal.

Firing rate - combustion rate; the rate at which fuel, or an air fuel mixture is supplied to a burner or furnace, expressed in volume, weight, or heat units supplied per unit time. Also see high fire, low fire, and modulating fire.

Firing rate controller - a controller that positions the firing rate motor to automatically regulate the burner firing rate at a predetermined air fuel ratio in accordance with load demand. It may position the air and fuel supplies for low fire or for high fire, or it may be a proportioning (modulating) type that gradually varies the air and fuel supplies within limits to meet the load demand. Also see firing rate motor.

Firing rate motor - a modulating or a two-position (open-closed) motor that switches external devices at predetermined times to control the firing rate of the burner.

Firing rate switching circuit - a circuit in a programming control that switches external devices at predetermined times to control the firing rate of the burner.

Firing rate valve - an automatic control valve for regulating the input of fuel to a burner in response to load demand. Also called a burner-input control valve, combustion control valve, metering valve, or throttling valve. Also see butterfly, modulating, and motorized valves.

First hour rating - the volume of domestic hot water that a given water heater can deliver within the first hour of the draw (fill water). It is based on a specified hot water supply temperature and rate of heat input

First impression - the feeling someone forms about you when you are first introduced.

First in, first out (**FIFO**) - applies to inventory control where the first item in is the first to go out or be sold.

First out indication - a device (usually a lamp) for each burner in a multi-burner system. The device indicates the first burner having flame failure.

First stage oil valve - in oil burners with more than one firing rate level (stage), the automatic safety shutoff valve that opens first to admit the fuel required for the lowest rate of combustion. In direct spark ignition systems, the first stage may take the place of a gas pilot.

Fish paper - a thin, translucent, brownish vulcanized fiber paper, coated with oil, which has good electrical insulating properties used in transformers and motor windings. Resembles household wax paper.

Fish tape - a flexible wire (metal or fiberglass) that can be pushed through conduit and around bends. Used to pull wires through conduit or wall cavities.

Fishing - a means of pulling wires through an enclosed wall section or conduit by means of a single wire or rope.

Fit - degree of tightness or looseness between two mating parts.

Fitting - mechanical accessory, like a bushing or locknut, of a wiring system.

Fitting brushes - are used to clean out the insides of tubes that are about to be soldered or brazed. These brushes should only be turned clockwise.

Fittings - special pipe connectors used to join straight lengths of pipe and make connections to fixtures, piping turns and branch connections. Common fittings include elbows, tees, unions, couplings, reducer couplings, and bushings.

Five wire control circuit - in this circuit five wires are typically used to operate a heating-cooling-continuous fan application. This application uses two low voltage transformers -- one in the air conditioner and the other in the furnace. The 24 volt power from the furnace transformer goes to the Rh terminal of the thermostat sub-base. The white wire goes from the thermostats "W" terminal to the heating circuit terminal in the furnace. The 24 volt power from the air conditioners transformer goes to the Rc terminal of the thermostat sub-base. The yellow wire goes from the thermostats Y terminal to the 24 volt coil of the condenser contactor. The green wire goes from the thermostats G terminal to the 24 volt coil of the indoor fan relay. This type of application requires an isolated sub-base that separates the heating transformer power from the cooling transformer power.

Fixed - describes a control or part that cannot be adjusted.

Fixed bore orifice - an expansion device with a fixed diameter that does not adjust to varying load conditions.

Fixed differential - the difference between the cut-in and cut-out points of a control which cannot be readjusted.

Fixed ladder - a ladder that cannot be readily moved or carried because it is an integral part of a building or structure.

Fixed lead boiler - an operating sequence for a multiple boiler system in which a designated boiler is always the first boiler fired upon a demand for heat.

Fixed resistor - a nonadjustable resistor. The resistance value cannot be changed.

Fixture - a receptacle attached to a plumbing system in which water or other waste may be collected for ultimate discharge into the plumbing system.

Fixture branch - the drain from the fixture trap to the junction of the drain with a vent.

Fixture drain - a unit flow rate from a fixture. The unit flow rate is determined to be one cubic foot per minute (or 7.5 gpm). A design factor to determine the drain pipe size for a fixture.

Fixture stud - a special fitting used to connect a light fixture to an outlet box.

Flaker - device used in the production of chips, crystals, or flakes from a hot concentrated solution of melted crystals.

Flame - the visible or other physical evidence of the chemical process of rapidly converting fuel and air into products of combustion.

Flame arrestor - a wire screen, parallel plates or tubes placed in a gas air mixture line for temporarily stopping the progress of a flame front (flashback).

Flame blow off - the phenomenon that occurs when a flame moves away from a burner, often resulting in the flame being extinguished. A flame blows off when the air fuel mixture leaves the burner at a velocity greater than the velocity with which the flame front progresses through the mixtures.

Flame conductivity - the ability of the ionized gases of a flame to conduct an electric current.

Flame current - see flame signal

Flame detection system - the flame detector, flame signal amplifier, and flame relay (in a flame safeguard control) that together determine whether a sufficient flame is present to continue the operation of a burner.

Flame detector - the components of a flame detection system that detect the presence or absence of a flame. Also see detector and flame rod; and infrared, optical, rectification type, rectifying photocell, and ultra-violet flame detectors.

Flame detector relay - a control consisting of a flame detection electronic network and a switching relay that pulls in when flame is detected. This control is not a primary control because it does not sequence burner operation, nor provide a safe start check or safety shutdown. Do not confuse it with flame relay.

Flame envelope - the confines (not necessarily visible) of the combustion process (converting fuel and air into products of combustion).

Flame failure - a condition where the flame in a furnace or boiler goes out.

Flame failure device - a combustion safeguard to shut off the supply fuel, if ignition does not occur within a pre-determined time.

Flame failure response time - the time interval between the loss of flame and the dropping out of the flame relay in a flame safeguard control, which then de-energizes the automatic fuel valve(s).

Flame front - the plane along which combustion starts, or the base of the flame.

Flame propagation rate - the rate at which a flame front travels through a combustible mixture of fuel and air. It is a function of air fuel ratio, mixture temperature, and the ambient pressure. Also known as burning velocity, flame speed, and flame velocity.

Flame rectification - the phenomenon that causes a flame to conduct more current in one direction than in the other. It is due to the ionization of air in and around the flame, and to the difference in area of the electrodes in the flame envelope. (The ground area is at least four times the area of the flame rod). Because the flame electrode (the flame rod) is very small compared to the ground electrode (usually the main burner when it contacts the flame), larger current flows in one direction than the other and the flame current is rectified to DC. The system control allows the burner to operate only as long as it receives a DC signal from the flame rod.

Flame rectifier pilot - see pilot burner assembly.

Flame relay - the relay in a flame safe guard control that pulls in when a flame, or a condition simulating a flame, is detected. It drops out on loss of flame signal, causing the flame safeguard control to shut down the burner. Do not confuse it with flame detector relay.

Flame resistance - the property of materials or combinations of component materials that restricts the spread of flame as determined by the flame resistance tests specified by code.

Flame retention device - a device added to a burner that aids in holding the flame base close to the burner ports. This device concentrates the air flow pattern to create a better fuel/air mixture at the time of ignition.

Flame retention nozzle - a burner nozzle surrounded with small ports that act as pilots to re-light the main burner flame if it blows off. The velocity through the small ports is less, so the flame almost never blows off of them. Also called a stick-tight nozzle.

Flame rod - a metal or ceramic rod projected into the flame envelope to function as an electrode in a flame detection circuit. Also see conductivity flame rod system and rectifying flame rod.

Flame roll out - a condition where flame rolls out of a combustion chamber when the burner is turned on.

Flame safeguard control - a safety control that provides a means for starting the burner in the proper sequence, proving that the pilot or burner flame is established, and supervising the flame during burner operation. Safety shutdown occurs if the pilot or burner flame is not established, or if the flame goes out. Also see cutoff, ignition return, and modulating, on-off, primary, programming, recycling, and non-recycling controls.

Flame safeguard system - the set of equipment used to provide safe control of burner operation. This usually includes the flame safeguard control, the flame detection system, all controllers, all limits, and interlocks, all fuel valves, the ignition system, the firing rate control system, and any other auxiliary equipment.

Flame scanner - a safety device located on the front of the boiler that uses a flame sensor to sense whether the pilot light and/or main flame is lit.

Flame sensing element - see flame detector and sensor.

Flame signal - the current measured at the meter jack of a flame signal amplifier in a flame safeguard control. The current is produced by the flame detector when it senses a flame.

Flame signal amplifier - the part of the flame detection system that increases the flame signal to a magnitude sufficient to pull in the flame relay in the flame safeguard control. The amplifier may be a permanent part of the flame safeguard control, or it may be a replaceable, plug-in unit.

Flame simulator - a device used to substitute for the presence of flame in a flame detection circuit. It is used during troubleshooting, to check the performance of the flame detection system. Flame simulators are available for use with rectification or ultraviolet amplifiers. In infrared systems, the flame signal can be simulated with a wire.

Flame speed - see flame propagation rate.

Flame spread - the propagation of flame over a surface.

Flame spread rating - the measurement of flame spread on the surface of materials or their assemblies as determined by tests conducted in compliance with ASTM E84.

Flame test for leaks - a tool (halide leak detector) that is principally a propane torch and when an air/refrigerant mixture is fed into the flame, this flame will change from a blue color to a green color in the presence of a heated copper element.

Flame velocity - the speed at which a flame moves through a fuel air mixture. It varies with type of fuel used, fuel-air mixture ratio and the temperature of the mixture. Also known as flame propagation.

Flammability classification - refrigerants shall be assigned to one of the three classes --- 1, 2 or 3 in accordance with ASHRAE 34. For class 2 and 3, the heat of combustion shall be calculated assuming that combustion products are in the gas phase and in their most stable state.

Class 1 - refrigerants that do not show flame propagation when tested in air at 14.7 psia and 70° F.

Class 2 - refrigerants having a lower flammability limit of more than 0.00625 pound per cubic foot at 70° F and 14.7 psia and a heat of combustion of less than 8,174 Btu/lb.

Class 3 - refrigerants that are highly flammable, having a lower flammability limit of less than or equal to 0.00625 pound per cubic foot at 70° F and 14.7 psia and a heat of combustion greater than or equal to 8,174 Btu/lb.

Flammability limits - the maximum and minimum percentages of fuel in an air fuel mixture that will ignite easily and burn rapidly.

Flammable - capable of being ignited and burned; commonly used as a synonym for combustible.

Flammable, aerosol - an aerosol that yields a flame projection of eighteen inches at the full valve opening or a flashback at any degree of valve opening when tested.

Flammable, gas - a gas that at ambient temperature and pressure forms a flammable mixture with air at a concentration of 13% by volume or less; or a gas that at ambient temperature and pressure forms a range of flammable mixtures with air greater than 12% by volume, regardless of the lower limit.

Flammable, liquids - a liquid having a flash point below 100° F (38° C), except that this term does not include any liquid mixture having one or more components with a flash point at or above 100° F that make up 99% or more of the total volume of the mixture.

Flammable, solid - a solid that will ignite readily and continue to burn or is liable to cause fires under ordinary conditions or transportation through friction or retained heat from manufacturing or processing and that burns so vigorously and persistently as to create a serious transportation hazard.

Flange - rim on the end of a pipe, valve, or fitting for bolting to another piping element.

Flange union - a method for joining pipe wherein flanges are screwed onto the ends of the pipe and the flanges are then bolted together.

Flanges - the common means of connecting a circulator to piping. The pump's volute has integral cast flanges on its inlet and discharge ports. These bolt to matching flanges threaded onto the piping. An O-ring provides a pressure tight seal between the flanges.

Flanging - the welding of standard or fabricated flanges to the ends of pipe sections or vessels.

Flap type damper - a damper consisting of one or more blades, each pivoted about one edge, and all linked together for simultaneous operation. Also see opposed and parallel blade dampers.

Flapper valve - the type of valve used in refrigeration compressors that allows gaseous refrigerants to flow in only one direction. Also referred to as a reed valve.

Flare - copper tubing is often connected to parts of a refrigerating system by use of flared fittings. These fittings require that the end of the tube be expanded at a 45° angle. This flare is firmly gripped by fittings to make a strong leak proof seal.

Flare bonnet - copper insert used to convert an ordinary flare nut into a cap nut.

Flare cap nut - female nut that is used to seal off a male threaded fitting.

Flare elbow - a fitting used to connect two flare nuts of the same size, while providing an accurate bend of either 45° or 90°.

Flare fitting - a type of soft tubing connector that requires that the tube be flared to make a mechanical seal.
Flare nut - fitting used to a clamp tubing flare against another fitting.

Flare nut wrench - is a special box-end wrench used to help connect tubing using flare nut couplings. The flare nut wrench is like a box end wrench but the ends are cut out. The openings allow the box end to fit over the tubing and onto the flare nut.

Flare plug - fitting used to seal a flare nut or similar female threaded opening.

Flare stack - tower that allows for the burning of waste gases or any other burnable substances that have been discharged.

Flare tee - fitting that makes it possible to connect a branch onto an existing line.

Flare union - a fitting that is used to connect two flare nuts of the same size.

Flaring block - special tool used to hold tubing for the flaring process or for enlarging in the swaging process.

Flaring single thickness connection - a process of expanding or spreading tubing so that the tube end formed into either a 37° or 45° angle.

Flaring tool - usually, a two piece tool consisting of a block and a yoke, with which tubing is flared.

Flash back - the phenomenon that occurs when a flame front moves back through a burner nozzle (and possibly back to the air fuel mixing point). Flashback occurs because the flame propagation rate exceeds the velocity with which the air fuel mixture flows through the burner nozzle. This situation may create a popping sound after the gas supply has been shut off.

Flash back arrestor - a metal gauze, grid or any other portion of a burner assembly used to avert flashback.

Flash boiler - a boiler with very limited water capacity. Usually about one gallon of water per 1000 Btu/h net rating.

Flash chamber - the space between the expansion valve and evaporator from which the flash gas that normally occurs at the evaporator inlet is drawn off and bypassed around the evaporator.

Flash economizer - a heat recovery system used to reclaim the heat from the boiler blowdown water and used in conjunction with the continuous blowdown system.

Flash gas - this is the instantaneous evaporation of some of the liquid refrigerant in the evaporator which in turn cools the remaining liquid refrigerant to its desired evaporation temperature. Subcooling helps to prevent flash gas.

Flash point - the minimum temperature corrected to a pressure of 14.7 psia at which the application of a test flame causes the vapors of a portion of the sample to ignite under the conditions specified by the test procedures and apparatus. The flash point of a liquid shall be determined in accordance with ASTM D56 (Tag Closed Tester), ASTM D93 (Pensky-Martens Tester) or ASTM D 3278. For #2 fuel oil it is 100° F.

Flash, steam - the rapid passing into steam of water at a high temperature when the pressure it is under is reduced so that its temperature is above that of its boiling point for the reduced pressure.

Flash tank - a tank in which water continuously flashes into steam.

Flash tube - an ignition device, commonly used for ignition gas on range top burners. An air gas mixture from the burner body is injected into the end of a short tube. The mixture moves along the tube, is ignited by a standing pilot flame at the other open end of the tube and the flame travels back through the mixture in the flashtube to ignite the gas at the burner ports.

Flash weld - a resistance type weld in which the mating parts are brought together under considerable pressure and a heavy electrical current is passed through the joint to be welded.

Flashing - a thin imperious material placed in construction to prevent water penetration and/or provide water drainage.

Flat gauge glass - a type of gauge glass used for pressures over 250 psi.

Flat lock seam - see Acme lock.

Flat plate solar collector - devices that convert solar radiation to heat using a flat, fluid cooled plate in an insulated and glazed housing that are exposed directly to normal incident sunlight.

Flat plate heat exchanger - a modern design for heat exchangers in which the fluids flow through alternating spaces between specially shaped stainless steel plates to exchange heat from one medium to another.

Flat rate schedule - a listing of the amount of time allotted to do various repair jobs, for the purpose of estimating the labor charge on a job.

Flex duct - usually installed in a single, continuous piece between the register and plenum box. A flexible duct usually has an inner lining and an insulated coating on the outside.

Flex life - the ability of a conductor, wire or cable to withstand repeated bending.

Flexible connector - a fitting that permits some movement of the lines it connects. Examples would either be a duct or pipe connector or steam or water lines..

Flexible coupling - a mechanical device used to connect two rotating shafts. Its construction permits a small amount of misalignment between the two shafts.

Flexible duct - a duct that can be routed around obstacles by bending it gradually. Flexible duct will come in one of these styles: metallic, uninsulated (M-UN), metallic, insulated (M-I), nonmetallic, uninsulated (NM-UN), nonmetallic, insulated with porous inner core (NM-IP), and nonmetallic, insulated lined with nonporous inner core (NM-IL).

Flexible duct, pressure classes and velocity types - flexible ducts shall be rated according to these positive pressure standards: ½" wg and 1000 fpm, 1" wg and 2500 fpm, 2" wg and 4000 fpm, 3" wg and over 4000 fpm, 4" wg and over 4000 fpm, 6" wg and over 4000 fpm, and 10" wg and over 4000 fpm.

Flight - a run of stairs.

Flip-flop circuit - flip-flop circuitry is an essential part of electronic computers and is used wherever it is necessary to count pulses, such as in timing devices. One of the most common applications of these circuits are in computers for storing and manipulating data in binary code. A circuit that provides automatic alternation between two possible circuit paths.

Float and thermostatic trap - a float trap with a thermostatic element for permitting the escape of air into the return line.

Float glass - transparent glass with flat, parallel surfaces formed on the surface of a pool of molten tin.

Float switch - a switch that opens or closes in response to the liquid level in a tank.

Float trap - a steam trap operated by a float. When enough condensate has drained by gravity into the trap body, the float is lifted which in turn lifts the pin off its seat and permits the condensate to flow into the return until the float has been sufficiently lowered to close the port. Temperature does not affect the operation of the float trap.

Float valve - type of valve that is operated by sphere or pan that floats on liquid surface and controls level of liquid.

Floating action - movement of the controlled device either toward its open or its closed position until the controller is satisfied, or until the controlled device reaches the end of its travel or until a corrective movement in the opposite direction is required. Generally there is a neutral zone in which no motion of the controller device is required by the controller.

Floating control - a type of control in which two contacts are used, in the controller, with a neutral zone between. The actuator (valve or damper operator) may assume any position from minimum to maximum, to maintain the conditions within the neutral zone of the controller.

Floating flames - an undesirable burner operating condition, usually indicating incomplete combustion in which flames leave the burner ports or heat exchanger to "reach" for combustion air.

Floating head pressure - letting the head pressure (condensing pressure) fluctuate with the ambient temperature from season to season for lower compression ratios and better efficiencies.

Floating heat exchanger - heat exchange unit, similar to the shell and tube exchanger, having U-shaped tubes running parallel with the shell, that are allowed to float or move during operation.

Floating zone - a specified range in which the error between the measured value of a controlled variable and the target value is small enough that no open or close signal is generated by the controller. It is sometimes called the "neutral" zone.

Floc point - the temperature at which wax separates out (precipitates) from a mixture of 10% oil and 90% refrigerant (R-12). The floc point is a measure of an oil's relative tendency to separate wax when mixed with a oil-soluble refrigerant.

Flood - to fill with an abundance or an excess.

Flood back - liquid in the suction line, resulting from too much liquid being metered into the evaporator.

Flood rim – the surface of a fixture over which water would pour if the fixture's drain is plugged and the water supply was left on.

Flooded evaporator - one that is full of a liquid refrigerant at all times. Additional liquid is permitted to enter only to replace that which boils away.

Flooded suction - a situation where the liquid source is higher than the pump, and liquid flows to the pump by gravity. Preferable for centrifugal pump installations.

Flooded system - type of refrigerating system in which liquid refrigerant fills evaporator.

Flooded system, high-side float - refrigeration system that has a critical charge and operated by the float level of the high-side liquid refrigerant.

Flooded system, low side float - refrigerating system that has a low side float refrigerant control.

Flooding - term applied to a refrigeration system when the liquid refrigerant reaches the compressor.

Floor over unconditioned space - a floor which separates a conditioned space from an unconditioned space which is buffered from exterior ambient conditions including vented crawlspaces and unconditioned basements or other similar spaces, or exposed to exterior ambient conditions including open parking garages and enclosed garages which are mechanically ventilated.

Floor plan - this plan provides the largest amount of information about a structure. From this plan you can see the basic layout of the structure. Individual rooms, partition walls, cabinets, and mechanical and electrical layouts can be viewed using floor plans.

Floor furnace - a completely self contained, connected or vented furnace designed to be suspended from the floor of the space being heated taking air for combustion outside this heated space and means for observing the flame and lighting the appliance from the space being heated.

Flow - the measure of the fluid volume capacity of a pump. Given in gallons per hour (GPH) or gallons per minute (GPM). To convert GPH to GPM divide by 60.

Flow characteristics - see valve flow characteristic.

Flow chart - a diagram showing all the logical steps of a program or the sequence of operation of a component or piece of equipment. A program is made by writing down the successive instructions that will cause the piece of equipment to operate as it does.

Flow check piston - piston assembly, with an orifice in the center, that can operate as a metering device.

Flow check valve - a valve with a weighted plug that prevents backward flow and forward migration of hot water due to buoyancy.

Flow coefficient (Cv) - valve coefficient of flow representing the flow rate of water at 60° F in gallons per minute that will produce a 1 psi pressure drop through the valve. Cv values for valves are often listed in manufacturers' specifications.

Flow control valve, hot water - a specially designed check valve, usually installed in the supply pipe, to prevent gravity circulation of hot water within the heating system when the pump is not in operation.

Flow control valve, water tower - manually controlled valves that are used to balance flow of incoming water to all sections of the tower.

Flow diagram - (1) schematic diagram showing the elements and directions of a process. This diagram may indicate flow rate, temperatures, and pressures. (2) A block diagram that outlines the operation of a heating and air conditioning unit.

Flow displacement meter - instrument placed directly in the line to indicate the total amount of flow.

Flow element - portable device, consisting of an orifice set with plugged tapped valves and a plate, used for testing flow. Vertical pipe sections are considered part of the measuring device.

Flow hood - a diagnostic tool used to measure air flow through ducts, supply registers, and return grilles. Read outs can be in FPM or CFM.

Flow indicating controller - control valve and flow indicator used for controlling flow by connections to an orifice through a pneumatic transmitter.

Flow indicator - linear indicator or dial that shows the flow rate. The indicator can be of the differential or the direct hook-up type.

Flow, laminar - fluid flow in which each fluid particle moves in a smooth path substantially parallel to the paths followed by all other particles. The velocity of the flow is greatest at the center of the pipe.

Flow meter - instrument used to measure velocity or volume of fluid movement.

Flow point - the temperature at which capillary action will pull molten alloy into a joint. With some alloys this may occur below the liquid point of the alloy.

Flow rate - the volumetric flow such as gallons per minute (GPM) or cubic feet per minute (CFM) passing a given point in a given unit of time. Also, the fuel input to burner measured in suitable units; e.g., cubic centimeters per minute, gallons per hour, pounds per hour; referred to as the firing rate.

Flow regulation – the action of changing the flow rate through a device, often for the purpose of controlling heat output or adjusting a mixed fluid temperature.

Flow switch - a switch that is mounted inside a liquid carrying pipe. The switch will open or close its contacts when the flow is established in the piping circuit. When the flow stops, the contacts go back to their normally rested position. Also see sail switch.

Flow temperature - the temperature at which an alloy becomes liquidus.

Flow, turbulent - fluid flow in which the fluid moves transversely as well as in the direction of the tube or pipe axis, as opposed to streamline or viscous flow.

Flow velocity - the speed of an imaginary fluid particle at some point to a piping or ductwork system. Common English units for flow velocity are feet per second.

Flow work - the product of pressure and specific volume of a fluid in a given state in a flow process.

Flue - gas or air passage that usually depends on natural convection to cause the products of combustion (flue gases) to flow through it and exit to the outside atmosphere.

Flue, appliance - the passage(s) within an appliance through which combustion products pass from the combustion chamber of the appliance to the draft hood inlet opening on an appliance equipped with a draft hood or to the outlet of the appliance on an appliance not equipped with a draft hood.

Flue, chimney - the vertical passage(s) in a chimney for conveying combustion products delivered to it by a vent connector into the outside atmosphere.

Flue collar - the portion of an appliance designed for attachment of a drafthood, vent connector, or venting system. Also see flue pipe.

Flue connection - a passage for conducting the products of combustion from a fuel fired appliance to the vent or chimney.

Flue gas analyzer - instrument used to analyze the operation of fossil fuel burning equipment, such as oil and gas furnaces, by analyzing the flue gases for CO_2, stack temperature and O_2.

Flue gas loss - the sensible heat carried away by the dry flue gas, and the sensible and latent carried away by the water vapor in the flue gas. Also referred to as stack loss.

Flue gases - gaseous by-products of combustion, oxygen, nitrogen, water vapor and excess air in appliances flues or heat exchangers.

Flue outlet - the opening provided in an appliance for the escape of flue gases.

Flue lining - a system or material used to form the inside surface of a flue in a chimney or vent, for the purpose of protecting the surface surrounding surface from the effects of combustion products and conveying combustion products without leakage to the atmosphere.

Flue pipe - the duct connecting an appliance with the vertical flue of a chimney; also called breeching or vent connector. (A flue collar is the part of the appliance for attaching the flue pipe.)

Fluid - any substance that can flow; i.e., move and change shape without separating when under pressure; liquid or gas.

Fluid coupling - device that transmits drive energy-to-energy absorber through a fluid.

Fluid expansion device - using a bulb or sensor, tube, and diaphragm filled with a fluid, this device will produce movement at the diaphragm when the fluid is heated or cooled. A bellows may be added to produce more movement. These devices may contain vapor and liquid.

Fluid feeder - a device that uses a pump to automatically add fluid to a hydronic system whenever the pressure at some location in the system drops below a preset value.

Fluid flow - the movement of a fluid by a pressure difference created by mechanical means or difference in density created by the addition or removal of heat energy.

Fluid, heat transfer - any gas, vapor, or liquid used to absorb heat from a source at a high temperature and reject it to a lower temperature substance.

Fluid power - a Honeywell trademark for the actuators used with V5055 industrial gas valves to provide a family of automatic safety shut off valves; driven by a fluid (usually oil) pumped by an electric motor.

Fluid, primary - the refrigerant, to distinguish it from a secondary fluid or brine.

Fluid properties factor - a number that combines the fluid properties of density and viscosity of a given fluid into a single index for use in determining hydraulic resistance.

Fluid, refrigerating - fluid used to transfer heat between cold refrigerant and the substance or bodies to be cooled. Circulation of fluid without change of state, or by evaporation-condensation process at essentially equal pressures.

Fluid, secondary - fluid cooled by the refrigerant in indirect method of refrigeration.

Flume - a trough that may be either totally enclosed, or open at the top. Flumes are sometimes used in cooling towers for primary supply of water to various sections of the distribution system.

Fluorescent lamp - a lamp consisting of a tube filled with a vapor with electrodes at either end of the tube. When a given potential is imposed across the electrodes the release of free electrons from the electrodes causes a current flow to take place through the vapor. The vapor emits invisible ultra-violet light that in turn causes a fluorescent material coating (phosphor) on the inside of the tube to emit visible white light.

Fluorocarbon - a molecule that contains fluorine and carbon atoms.

Flush - (a) an operation to remove any material or fluids from refrigeration system parts by purging them to a waste collection tank using refrigerant or other fluids. (b) arranged with adjacent sides, surfaces or edges close together; even with.

Flute - groove, as on twist drills, reamers and taps that allow for the removal of material from the hole.

Flux, brazing and soldering - substance applied to surfaces to be joined by brazing or soldering to free them from metal surface oxides and to void the air space in the joint clearance area to protect from further oxidation during the heating process. This facilitates a quality sound joint.

Flux density - the number of lines of force per unit area in a magnetic material or circuit.

Flux, magnetic - the magnetic lines of force of a magnet that connect the north and south poles of the magnet or an energized electromagnet.

Flux meter - a device used to measure the strength of magnetic fields.

Fly ash - suspended ash particles carried in the flue gas.

Flywheel - a pulley (sometimes weighted) that is attached to the external crankshaft of a motor.

Foam - a type of fire fighting material (medium) consisting of small bubbles of air, water, and concentrating agents. Chemically, the air in the bubbles is suspended in the fluid. It clings to vertical and horizontal surfaces and flows freely over burning materials. Foam will put out a fire by "blanketing" it, excluding air and the escape of volatile vapors. Its flowing properties will resist mechanical interruption and reseal the burning material.

Foam leak detector - a system of soap bubbles or special foaming liquids brushed over joints and connections to locate leaks.

Foaming, boiler - the condition created in a steam boiler by the pressure of oil, dissolved solids or suspended sediment in the boiler water. The bubbles of steam formed under such conditions do not break at the water surface area at the water, but retain their shape and collect in a mass of foam which reduces boiler efficiency, causing wet steam and water carryover.

Foaming, refrigeration - the formation of foam or froth in an oil refrigerant mixture due to rapid evaporation of refrigerant dissolved in the oil when the pressure is suddenly reduced. This occurs when the compressor starts operating and, if large quantities of refrigerant were dissolved, large quantities of oil may boil out and be carried through the refrigerant lines.

Fog - suspended liquid droplets generated by condensation from the gaseous to the liquid state, or by breaking up a liquid into a dispersed state, such as by splashing, foaming, and atomizing.

Fogging - a reference to the visibility and path of the effluent air stream after having exited the cooling tower. If visible and close to the ground, it is referred to as fog. If elevated, it is normally called the a plume.

Foot candle - a unit of measurement of light; the light produced on a surface from a source of one candle at a distance to one foot.

Foot of water - the pressure created by a column of water one foot in height. Equivalent to 0.433 lb/per square inch.

Foot pound - a unit of work. A foot pound is the amount of work done in lifting one pound one foot.

Footing - that portion of the foundation of a structure that transmits loads directly to the soil.

Forbidden energy band - the band of energies that is located between the valence and conduction bands that contains non-permissible energy levels.

Force - is that which produces, or tends to produce motion. Is the accumulated pressure and is expressed in pounds. If the pressure is 10 psi on a plate of 10 sq. in. area, the force is 100 lbs.

Force feed oiling - a lubrication system that uses a pump to force oil to surfaces of the moving parts.

Force of adhesion - the result of attraction between the molecules of solder and the metal part being soldered.

Forced air furnace - a central furnace equipped with a fan or blower to provide the primary means for circulation of air. Types are (1) Attic type forced air type furnace designed for installation in an attic or in a space with low headroom that is normally occupied. (2) The downflow type is a forced air type furnace designed with airflow essentially in a vertical path, discharging air at or near the bottom of the furnace. (3) The horizontal type is designed with airflow through the furnace in a horizontal path. (4) The upflow type furnace is designed with airflow essentially in a vertical path, discharging air at or near the top of the furnace.

Forced convection - movement of a fluid by mechanical force such as fans or pumps.

Forced draft - mechanically produced airflow into and through the combustion chamber, blown in by a fan or blower located at the inlet air passage to the furnace. This air movement minimizes problems resulting from insufficient stack, wind, weather, and other conditions.

Forced draft burner - a gas burner in which combustion air is blown in by a motor driven fan. Commonly called a power burner. Examples are aspirating, gun type, mechanical pre-mix, and zero governor burners. Also see mechanical draft burner.

Forced draft cooling tower - cools water by mechanically forcing air through the tower.

Forced draft fan - a fan or blower that blows air into the combustion chamber.

Forced feed oiling - a lubrication system in a compressor that uses a pump to force the oil to the surfaces of the moving parts.

Forced hot water - hot water heating systems in which a pump is used to create the necessary flow of water.

Forced ventilation - mechanically produced air flow through a room or an area.

Forecooler - in an ice plant, a device for cooling the water for ice making before it enters the cans.

Foreign object - an object that does not belong where it is found.

Formaldehyde (HCHO) - a colorless, gaseous compound, used in the manufacture of resins and dyes, and as a preservative and disinfectant. Present in many synthetic materials.

Forming - method of pipe fabrication including bending, swaging, lapping, extruding, expanding, and belling.

Forward bias - voltage applied to a P-N junction diode so that the potential barrier is neutralized (a positive voltage on a P region or a negative voltage on a N region).

Forward break over voltage - the value of the anode-to-cathode voltage required for the silicon controlled rectifier to turn on.

Forward current - the current flow across a P-N junction when it is forward biased.

Forward switching voltage - that value of anode-to-cathode at which a four layered diode switches.

Forward voltage - as measured in a diode, it is the amount of excess voltage on the anode as compared to the voltage on the cathode.

Forward curved blower - a blower where the tips of the blades are inclined in the direction of rotation. This is the most common type of centrifugal blower. Normally used in residential heating and air conditioning systems and light duty exhaust systems where maximum air delivery and low noise levels are required. It is capable of pressures up to approximately 1.5" static pressure.

Fossil fuels - minerals or gases that are burned to produce heat. Typical types include fuel oil, natural gas, and coal.

Fouling - blocking by dirt, soot, scale, algae, or any other foreign matter.

Fouling factor - factor that determines loss of heat transfer due to deposits of foreign material in the water side of tubing in refrigeration condensers or chillers.

Foundation - any part of a structure that serves to transmit the load to the earth or rock below grade level.

Foundation plans - foundation and basement plans are found on the same site plan when a basement is part of the structure. Foundation plans are a key part of any working drawings and show a great deal of structural information. The plan includes the foundation walls, footings, piers, fireplaces, stairways and other built-ins.

Four pipe system - a piping system that provides either heating or cooling, consisting of one coil (for cooling) connected to the chiller and another coil (for heating) connected to the boiler. These two coils provide instant and individual changeover from heating to cooling, or cooling to heating, as necessary.

Four way switch - a switch used in conjunction with a three way switches when control is desired at three or more places.

Four way valve - a type of reversing valve used to automatically divert the refrigerant in a heat pump to the proper indoor or outdoor coils. Composed of a main slide valve and a pilot valve operating together.

Four wire control circuit - in this circuit four wires are typically used to operate a heating-cooling-continuous fan application. The red low voltage wire is the uninterrupted power from the transformer power side that goes to the thermostats R terminal. The white wire goes from the thermostats "W" terminal to the heating circuit terminal in the furnace. The yellow wire goes from the thermostats Y terminal to the 24 volt coil of the condenser contactor. The green wire goes from the thermostats G terminal to the 24 volt coil of the indoor fan relay.

Four wire three phase transformer - a system where the secondary of three transformers are star connected, and the fourth wire is connected from a neutral point. The voltage between any main wire and the neutral wire will be 57 percent of the voltage between any two main wires.

FPM - the velocity of air flow in feet per minute. Recommended velocities for a quiet system in a residential system are as follows for net free area:

Main ducts	700-900 FPM
Branch ducts	600 FPM
Branch risers	500 FPM
Return grilles	500 FPM
Supply registers	450 FPM

Fractional horsepower - horsepower value less than one.

Fractionation - when one or more refrigerants of the same blend leak at a faster rate than the other refrigerants in the blend, changing the composition of the blend. Fractionation is possible only when liquid and vapor exist at the same rate.

Fractions - in petrochemicals processes, the products are often broken down into components of various weights (gases, light liquids, heavy liquids) called fractions.

Frame, motor - usually refers to the NEMA system of standardized motor mounting dimensions. This is the supporting structure for the stator parts.

Frame size - usually refers to the NEMA system of standardized physical dimensions of motors. These dimensions consist of shaft length, shaft diameter, shaft height and location of base mounting holes.

Framing plans - these plans are required for complex structures but may be left out on smaller ones. This set of plans contains the framing for the roof, floors and various wall sections. They will also show the size and shape of the material used on the roof and walls.

Frangible disk - a circular (round and dished) device used on some refrigeration equipment to provide pressure release for safety purposes. The frangible disk suddenly breaks when a certain excessive pressure is reached.

Free air delivery - a situation where the conditions exist that there are no effective restrictions to air flow (no static pressure) at the inlet or outlet of an air moving device.

Free area - the amount of unobstructed air space through a grille, register or diffuser.

Free blowing drain - used to remove condensate from the main steam line.

Free cooling - a method of temperature control which uses the temperature of the outdoor air to reduce the temperature of the indoor air.

Free electrons - electrons located in the outer orbit of an atom that are easily removed and result in electrical current flow.

Free enthalpy - a thermodynamic property that serves as a measure of the available energy of a system with respect to surroundings at the same temperature and same pressure as that of the system. No process involving an increase in available energy can occur spontaneously.

Free wheeling - continued rotation of magnetic clutch on automotive compressor when clutch is disengaged.

Freeze - to form into ice. To be hardened or solidified by removing heat.

Freezestat - a control used to protect against freeze-up of heating coils, cooling coils, chiller bundles, or similar temperature applications.

Freeze up - the formation of ice in the refrigerant control device that may stop the flow of refrigerant into the evaporator. Frost on a coil may stop the airflow through the coil.

Freezer - any box used to store and maintain products at sub-freezing temperatures.

Freezer alarm - device used in many freezers that sounds an alarm (bell or buzzer) when freezer temperature rises above safe limits.

Freezer burn - a condition that applies to food that have not been properly wrapped and that has become hard, dry, and discolored.

Freezing - change of state from liquid to a solid.

Freezing point - the temperature at which a liquid will solidify upon removal of heat. The freezing temperature for water is 32° F at atmospheric pressure.

Freezing point depression - temperature at which ice will form in a solution of water and salt.

Freezing time - time for any complete freezing process to take place.

Freon - trade name for a family of synthetic chemical refrigerants manufactured by Dupont de Nemours Inc.

Frequency - the number of recurrences (vibrations, waves or cycles) of a periodic phenomenon in a unit of time. Expressed in Hertz (Hz). That number of Hertz that an electric current completes in one second. In the United States current is generated at the rate of 60 Hertz per second.

Fresh air intake - source of outside air feeding the return air plenum of an air distribution system.

Friction - resistance to relative motion set up between particles of two surfaces in contact with each other.

Friction head - in a hydronic system the friction head is the loss in pressure resulting from the flow of water in the piping system.

Friction loss - pressure loss caused by friction between the pipe wall and the flowing fluid.

Friction loss, bends - the static pressure loss in bends of ductwork, expressed as a dimensionless coefficient (Co) at a specified bend radius ratio. Co is a dimensionless coefficient which represents the ratio of the total pressure loss to the dynamic pressure in terms of velocity pressure.

Friction loss, straight - the static pressure loss in a straight run of duct material

Front header - device (line) connecting the steam and water drums by downcomer nipples.

Front seated - the position of a service valve when the compressor and gauge ports are open and the line port is closed.

Frost - frozen water vapor.

Frost back - a condition in which liquid refrigerant flows from a evaporator into the suction line as indicated by frost formation on the suction line.

Frost control, automatic - a control that automatically cycles refrigerating system based on frost formation on the evaporator.

Frost control, manual - a manual control used to change refrigerating system to produce defrosting conditions.

Frost control, semiautomatic - a control that starts defrost part of a cycle manually and then returns this system to normal operation automatically.

Frost free refrigerator - a refrigerated cabinet that operates with an automatic defrost during each cycle.

Frost line - the depth below the surface grade to which the soil can be expected to freeze in winter.

Frosting type evaporator - a refrigerating system that maintains the evaporator at frosting temperatures during all phases of the cycle.

Frozen - water in a solid form or machine parts seized due to lack of lubrication.

Frozen storage - the storage of an already frozen product at a constant temperature, usually 0° F (-18° C) or lower.

Frustrum - a figure formed by cutting off a portion of a cone or pyramid parallel to its base.

Fuel - defined as any substance, solid, liquid or gaseous, which may be relatively easily ignited and burned to produce heat. Practically all fuels contain carbon and hydrogen. Any material that is burned to supply heat or power.

Fuel air ratio - see air fuel ratio.

Fuel gas - any substance in a gaseous form when used for combustion.

Fuel oil - kerosene or any hydrocarbon oil as specified by U.S. Department of Commerce commercial standard CS12 or ASTM D396-01 or the Canadian government specification board, 3-GP-28, and having a flash point not less than 100° F.

Fuel oil burner, pressure atomizing or gun type - a burner designed to atomize the oil for combustion under an oil supply pressure of 100 psig.

Fuel oil burner, rotary type - a burner employing a thrower ring that mixes the oil and the air.

Fuel oil burner, vaporizing or pot type - these burners use the heat of combustion to vaporize the oil in a pool beneath the vaporizing ring, and this vapor rising through the ring ignites and maintains combustion in the burner.

Fuel oil heater - used to heat fuel so it can be pumped and is at the correct temperature for burning. This heater may either be electric or steam.

Fuel oil pump - a pump that takes fuel oil from the fuel oil tank and delivers it to the burner at the proper pressure.

Fuel piping system - all piping, tubing, valves and fittings used to connect fuel utilization equipment to the point of fuel delivery.

Fuel rich - the ratio of air to fuel supplied to a furnace that provides more fuel than the optimum air fuel ratio.

Full bore port - valve in which diameter of internal bore equals the inside diameter of pipe used on line.

Full floating - a mechanism construction in which a shaft is free to turn in all the parts in which it is inserted.

Full load - the nominal speed at which an induction motor operates under rated horsepower conditions. This will always be less than the synchronous speed and will vary depending on the rating and characteristics of the particular motor. For example, a four pole 60 hertz fractional horsepower motors have a synchronous speed of 1800 RPM, a nominal full load speed (as shown on the nameplate) of 1725 RPM, and an actual full load speed ranging from 1715 to 1745 RPM.

Full load amps (FLA) - line current (amperage) drawn by a motor when operating at rated load, voltage, and frequency. This is important for proper wire size and motor starter heater selection.

Full load torque - the amount of torque produced by a motor when it is running at full load speed at rated horsepower.

Full port ball valve - a ball valve in which the hole through the ball is approximately the same diameter as the pipe size of the valve.

Full reset - the ability of an outdoor reset control to reduce the water temperature supplied to the distribution system all the way from some maximum value at design load conditions down to room temperature at no load conditions.

Full scale - the upper limit of measurement or source value for which a given uncertainty specification applies, including any over ranging.

Full storage systems - a thermal storage system designed to store all the energy needed by the load, for a 24 hour period, during the preceding off peak period.

Full wave rectifier - a rectifier that uses two diodes to convert AC to DC. Both halves of the AC wave is used. A circuit that converts every alternating current alternation to direct current.

Fully halogenated CFC - a refrigerant having all hydrogen atoms in a hydrocarbon molecule replaced with chlorine or fluorine atoms.

Fully rated system - panel boards are marked with their short circuit rating in RMS symmetrical amperes. In a fully rating system, the panel board short circuit rating will be equal to the lowest interrupting rating of any branch circuit breaker or fuse installed. All devices installed shall have an interrupting rating greater than or equal to the specified available fault rating.

Fumes - smoke; aromatic smoke; odor emitted, as of flowers; a smoky or vaporous exhalation, usually odorous, as that from concentrated nitric acid. The word fume is so broad and inclusive that its usefulness as a technical term is very limited. Its principal definitive characteristic is that it implies an odor. The terms vapor, smoke, fog, etc., which can be more strictly defined, should be used whenever possible. Also defined as solid particles generated by condensation from the gaseous state, generally, after volatilization from molten metals, etc., and often accompanied by a chemical reaction such as oxidation. Fumes flocculate and sometimes coalesce.

Function switch - on a multi-function machine or instrument, the switch that selects the desired application.

Fungi - any of a group of parasitic lower plants that lack chlorophyll, including molds and mildews.

Fungicides - agents that inhibit growth, or destroy fungi.

Furnace - an enclosed chamber or structure, including a burner and combustion chamber, that is provided for the combustion of fuel. The heat produced is used for heating a building or for processing materials. Also see batch furnace, central, continuous, downflow, duct, forced air, horizontal, industrial, and up flow furnaces.

Furnace, central - self-contained appliance designed to supply heated air through ducts to spaces remote from or adjacent to the appliance location. May be one of the following types:

a. Forced air furnace - a furnace equipped with a fan or blower that provides the primary means for circulation of air.

(1) Downflow (counter flow) - a furnace designed with airflow discharge vertically downward at or near the bottom of the furnace.

(2) Horizontal furnace - a furnace designed for low headroom installation with airflow across the heating element or heat exchanger essentially in a horizontal path.

(3) Upflow furnace - a furnace designed with airflow discharge vertically upward at or near the top of the furnace. This classification includes "highboy" furnaces with the blower mounted below the heating element and "lowboy" furnaces with the blower mounted beside the heating element.

b. Gravity furnace - a furnace depending primarily on circulation of air by gravity.

Furnace, direct vent central - a system consisting of (1) a central furnace for indoor installation, (2) combustion air connections between the central furnace and the outdoor atmosphere, (3) flue gas connections between the central furnace and the vent cap, and (4) vent cap for installation outdoors, supplied by the manufacturer and constructed so that all air for combustion is obtained from the outdoor atmosphere and all flue gases are discharged to the outdoor atmosphere.

Furnace, duct - a central furnace designed for installation in a duct of an air distribution system to supply warm air for heating and that depends on a blower not furnished as part of the furnace for air circulation.

Furnace, enclosed - a specific heating, or heating and ventilating, furnace incorporating an integral total enclosure and using only outside air for combustion.

Furnace, floor - a self-contained indirect fired or electrically heated furnace suspended from the floor of the structure to be heated. This furnace is designed to take air for combustion from outside the space being heated and is provided with means for observing the flame and lighting the appliance from the space being heated. Types may be either a fan assisted or gravity.

Furnace pressure - the gauge pressure in a furnace combustion chamber. The furnace pressure is said to be positive if greater than atmospheric pressure, negative if less than atmospheric pressure, and neutral if equal to atmospheric pressure.

Furnace, warm air - a vented heating appliance designed or arranged to discharge heated air into a conditioned space.

Furring - wood or metal strips nailed to walls or ceilings to provide a base for an even plastered surface; in walls, provides an air space between lath and wall to prevent condensation.

Fuse - electrical safety device consisting of a strip of fusible metal placed in series in the circuit that melts when subjected to high temperatures. It is intended to protect against abnormal conditions of current. A current limiting device.

Fuse, enclosed cartridge type - a tubular fuse with terminals at each end with a casing that insulates the fusible element and contains the arc.

Fuse, plug type - a household type fuse having threaded Edison or S type base. Rated up to 30 amperes at 125 volts.

Fuse, puller - a specially insulated tool for removing cartridge type fuses from the junction box.

Fuse, replaceable element - a renewable cartridge fuse that has a replaceable link. When one of these fuses blows, the ends are unscrewed and a new link is inserted. Replacement links are available in current-limiting, slow, and fast-responding types.

Fuse, S type - dual element fuses that must be used with adapters. Each size of fuse has a corresponding adapter that cannot be removed from the fuse holder after it is installed. It is therefore not possible to replace a fuse with a higher rating. These type fuses have amperage ratings up to 30 at 125 volts.

Fuse, time delay type - an enclosed fuse having two elements that protects against a temporary over current condition.

Fusible - able to be united as if melted together.

Fusible disconnect - an electric switch designed to isolate part of the system when required. Provided with fuses for the protection of the equipment.

Fusible element - metal link in a fuse that overheats and melts when excess current flows through it. This opens up the circuit and stops current flow.

Fusible link - a metal strip that will melt at a specified temperature to protect a heating element against burnout if the limit switch fails. Such devices are to be replaced following each over temperature occurrence.

Fusible load center - an electric panel that supplies circuits with power and protects them with fuses.

Fusible plug - a plug or fitting made with a metal of a known low melting temperature, used as a safety device to release pressures at a predetermined temperature.

Fusion - the melting and joining together of separate objects. The use of heat to change a substance from the solid to the liquid or plastic state.

Fusion temperature - fusion is the actor process of melting by heat or the state of being fused or melted. For coal, the fusion temperature is reported as initial deformation (IDT), ash softening (AST), and ash fusion (AFT), the test being made under a reducing atmosphere.

Fuzzy logic - control logic that allows a controller to automatically modify the operating characteristics of the controller based on how the system is responding, and without human intervention.

-G-

G - gram.

Gal - gallon.

Gen - generator.

GFCI - ground fault circuit interrupter.

Gil - green indicator light.

Glb - globe.

Gov - governor.

GPH - gallons per hour.

GPM - gallons per minute.

Gr - gram.

Grd - ground.

Gagging - application of a clamp on a safety valve spindle to keep the valve in full closed position during the hydrostatic test.

Gain - the ratio of signal output to input, frequently referred to as span; a main factor in amplification.

Galling - a form of wear in which seizing or tearing of the gear or bearing surface occurs.

Gallium - a chemical element, atomic number 31, metallic in nature, used in making certain kinds of solar cells.

Galvanic action - the corrosion action between two metals of different electronic activity. This action is increased in the presence of moisture.

Galvanized - metal that has been zinc coated.

Galvanized sheet metal - sturdy material of various gauges used to fabricate duct systems. Galvanized sheet metal and fittings can be bought in many shapes such as round, oval, rectangular and flat sheet stock.

Galvanometer - instrument that measures small electric currents.

Ganged switches - refers to switches or controls mechanically linked together for identical operation.

Gap - space between objects.

Gas - the fluid form of a substance in which it can expand indefinitely and completely fill its container; form that is neither liquid nor solid; a vapor; any mixture of inflammable gases used for lighting or heating. Also see blast furnace, flue, liquefied petroleum, manufactured, mixed, natural, and zero gas.

Gas analyzer - instrument used to analyze the gases of combustion in relation to CO_2, O_2, excess air, CO and stack temperature to determine combustion efficiency.

Gas appliance furnace categories - classification of vented gas appliances for venting purposes. These categories are broken down into four areas:

Category I – an appliance that operates with a negative (non-positive) vent static pressure and with a vent gas temperature that avoids excessive condensate production in the venting system

Category II – an appliance that operates with a negative (non-positive) vent static pressure and with a vent gas temperature that may cause excessive condensate production in the venting system.

Category III – an appliance that operates with a positive vent static pressure and with a vent gas temperature that avoids excessive condensate production in the venting system.

Category IV – an appliance that operates with a positive vent static pressure and with a vent gas temperature that may cause excessive condensate production in the venting system.

Gas burner - a burner for use with gaseous fuel. Also see atmospheric, mechanical draft, raw gas, partial premix, premix, in-shot, and up-shot burners.

Gas chromatography - a method of analyzing and measuring components in a gas mixture, both qualitative and quantitative.

Gas calorimeter - instrument used to determine the Btu content of natural gas.

Gas cock - a manual quick closing shut off valve. Also known as an "A" or "B" cock.

Gas constant - a factor describing the weight of a specific volume of gas at a specific pressure. The coefficient R in the perfect gas equation PV = MRT.

Gas detector - test instrument that samples ambient condition for a particular type or group of gases. Does not normally include a sampling pump.

Gas, discharge - the hot superheated refrigerant gas leaving the compressor.

Gas, flash - the gas resulting from the instantaneous evaporation of refrigerant in a pressure reducing device to cool the refrigerant to the evaporation temperature obtaining at the reduced pressure.

Gas, flue - gasses that results from combustion.

Gas fuel - any substance that must be kept sealed in a container to prevent it from escaping into the air. There are two major types of fuel gases: natural and liquefied petroleum.

Gas, inert - a gas that neither experiences nor causes chemical reaction nor undergoes a change of state in a system or process; e.g., nitrogen or helium mixed with a volatile refrigerant.

Gas main - a pipe installed in a community to convey gas to individual services or other mains.

Gas mixer - a device for mixing gas and air, such as a fan, aspirator, or inspirator. In an injection type burner, it is the combination of mixer head, mixer throat, and mixer tube.

Gas mixing chamber - where air and gas mix before they enter the furnace in low pressure gas burners. This fuel gas mixture must then be between the upper and lower flammability limits so the mixture will ignite.

Gas monitor - similar to a gas detector but does not normally include a pump.

Gas, noncondensable - a gas in a refrigerating system that does not condense at the temperature and partial pressure at which it exists in a condenser, and therefore imposes a higher head pressure on the system.

Gas oil burner - a burner capable of burning either gas or oil-some types can burn both simultaneously. Also see combination burner.

Gas pack - typically a combination roof-top unit that heats in the winter by using gas as its fuel and cools in the summer by using electric power.

Gas pressure regulator - a device for controlling and maintaining a uniform outlet gas pressure.

Gas pressure switch - switch used to detect an adequate gas pressure before gas burners are allowed to ignite.

Gas, residual - the gas remaining in the clearance space of a compressor cylinder after the piston has reached the top of the stroke.

Gas sorption - devices used to reduce levels of airborne gaseous compounds by passing the air through materials that extract the gases. The performance of solid sorbents is dependent on the airflow rate, concentration of the pollutants, presence of other gases or vapors, and other factors.

Gas tracer - a gas having a powerful odor. Sometimes used in small quantities with odorless refrigerants to give warning of a leak.

Gas utilization equipment - any device that utilizes gas as a fuel or raw material or both.

Gas valve - a device for controlling the flow of gas to the burner in a gas fired furnace or heater.

Gas velocity - the speed of the gas in the piping or equipment. Usually stated in feet per minute or feet per second.

Gas vent - the piping and fittings used to convey flue gases to the outside atmosphere. May be one of the following types:

Type "B" - a vent for venting listed gas appliances with draft hoods and other category I gas appliances listed for use with type "B" gas vents. This type vent pipe would normally be used with natural and propane fired equipment.

Type "BW" - a vent for venting listed gas fired vented wall furnaces.

Type "L" - a vent for venting appliances listed for use with type "L" vents and appliances listed for use with type "B" gas vents. This type vent pipe would normally be used with oil fired equipment.

Special gas vent - gas vents for venting listed category II, III, and IV gas appliances.

Gas volume control - used to regulate the amount of gas needed to produce certain desired temperatures and conditions in domestic absorption automatic control refrigerators.

Gases - includes natural gas, manufactured gas, liquefied petroleum (LP) gas in the vapor phase only. Liquefied petroleum gas-air mixtures, and mixtures of these gases, plus gas-air mixtures within the flammable range, with the fuel gas or the flammable component of a mixture being a commercially distributed product.

Gasket - a resilient or flexible material used between mating surfaces of refrigerating unit parts or of refrigerator doors to provide a leak proof seal.

Gasket, foam - a joint sealing device made of rubber or plastic foam strips.

Gasohol - a blend of 10% anhydrous ethanol (ethyl alcohol) and 90% gasoline, by volume.

Gate - a device that has multiple inputs and a single output. There are five basic types of gates, the and, or, nand, nor, and inverter. A gate can also be one terminal of some electronic devices, such as SCR's, triacs, and field effect transistors.

Gate valve - this valve is intended for use as a stop valve. It gives the best service when used in the fully open or closed position. There is less obstruction and turbulence within the valve and, therefore a correspondingly lower pressure drop than other valves. With the valve wide open , the wedge or disc is lifted entirely out of the fluid stream, thus providing a straight flow area through the valve. The wedge or disc in the gate valve is available in several forms: solid wedge, split wedge, flexible wedge, and double disc parallel seat. Gate valves should not be used as throttling valves except in an emergency.

Gauge - (1) term used to denote the physical size of a wire and sheet metal thickness. (2) instrument for measuring pressure or liquid level.

Gauge, compound - an instrument used for measuring pressures both above and below atmospheric pressures.

Gauge, draft - a vacuum gauge used to determine the negative pressure of a heating unit vent.

Gauge, gas manifold pressure - a direct reading gauge for measuring gas manifold pressure of a vapor burning heating unit. A more convenient substitute for a "U" tube manometer. Measures in increments of inches of water.

Gauge, glass - often referred to as the water glass; so arranged that the water in the boiler stands at roughly the same level as water in the glass. This glass is not found on hot water boilers. Actually, water in a boiler under pressure is about 4 inches higher because of expansion.

Gauge, high pressure - an instrument used for measuring pressures in the range of 0 to 500 psig.

Gauge, liquid level - a gauge mounted on or in a vessel to indicate the liquid level within the vessel.

Gauge, low pressure - an instrument used for measuring pressures in the range of 0 to 50 psig.

Gauge lugs - attached to the gauge ring on the receding die threader. Each lug has a size printed on its face. The lug size and the pipe size to be threaded must correspond. The positioning of the lug also positions the work holder jaws.

Gauge manifold - a field service tool equipped with a compound gauge (typically blue low side) and a pressure (typically red high pressure) gauge. It has a series of hoses and valves to facilitate the flow of fluids through it for the purpose of servicing refrigeration and air conditioning units.

Gauge port - opening or connection provided for a service technician to install a gauge.

Gauge pressure - pressure expressed in pounds per square inch gauge (psig). Readings in psig measure only pressures above atmospheric. Gauges are calibrated to read zero psig at atmospheric pressure.

Gauge, standard - an instrument designed to measure pressures above atmospheric.

Gauge glass - a glass tube used to determine the liquid level in a component or piece of equipment.

Gauge glass blowdown valve - valve used to remove any sludge and sediment from gauge glass lines.

Gauge rings - movable rings located at the rear of the work holder. The gauge ring has four lugs, one of each pipe size for which the threader can be used. Rotating the gauge ring will move the position of the work holder jaws.

Gauge vacuum - an instrument used to measure the pressures below atmospheric pressure.

Gauss - a unit of measurement of magnetic flux density or an electromagnetic unit of magnetic induction. One gauss represents one line of flux (one Maxwell) per square centimeter.

Gear - toothed wheels that transmit rotary motion from one shaft to another.

General lighting - lighting designed to provide a substantially uniform level of illumination throughout an area, exclusive of any provision for special lighting.

General purpose area - all areas where explosion proof or intrinsically safe construction is not required. Most areas where gas detection equipment is installed.

General purpose motor - it is designed with standard operating characteristics and mechanical construction for use under usual service conditions. Has a service factor rating.

General purpose relay - these relays are typically applied in both power and pilot duty circuits where dependable, low cost, quiet operation is desired. Equipment applications range from HVACR units to vending machines and appliances. A universal mounting base allows interchange with a wide variety of relays used by other HVAC manufacturers. They are generally rated for up to 15 amperes.

Generating efficiency - electrical energy output from the engine or turbine generator divided by the energy input to the prime mover, in consistent units.

Generator - a rotating electric machine to provide a large source of electrical energy. A generator converts mechanical energy to electric energy.

Generator, absorption - the basic part of an absorption system; a heat exchanger provided with means of heating used to drive refrigerant out of a solution.

Generator shell - the upper shell of a lithium-bromide absorption chiller, containing the generator and the condenser, where water vapor is separated from the lithium-bromide solution in the absorption cycle.

Generic name - a designation or identification such as code name, code number, or brand name used to identify a chemical by other than its original name.

Geothermal heat pump - a heat pump system that uses the earth as a source of heat during the heating mode of operation and as a heat sink during the cooling mode of operation.

Geothermal well - an underground well that is used in geothermal heat pump systems. Water is drawn from the top of the well and returned to the bottom of the same well after flowing through the systems heat exchanger.

Germanium - a gray-white, hard and brittle chemical element used in the manufacture of semiconductor materials such as transistors and diodes.

Giga - a numerical prefix denoting one billion. (As a gigacycle is one billion cycles).

Gigawatt - one billion watts. One million kilowatts. One thousand megawatts.

Gilbert - a unit of measurement of magnetomotive force (MMF).

Gland - an adjustable part that compresses packing into a stuffing box.

Glass, gauge - device foe showing liquid level.

Glass, sight - glass tube used to indicate the liquid level in tanks, bearings, and similar equipment.

Glass stem thermometer - a popular type of thermometer, this type of thermometer has a glass stem mounted in a metal case and can easily fit into a pocket. Glass stem thermometers will most often read from -30° F to 120° F. The tubes are filled with either mercury or red fluid.

Glaubers salt - the most common type of eutectic salt. Used to store energy.

Glazing, buildings - all areas, including the frames, in the shell of a conditioned space that let in natural light including windows, clerestories, skylights, sliding or swinging glass doors and glass block walls.

Glazing, glass - the act or process of fitting glass or panels into the sash or frame of a window. Glass in a window or door.

Glazing, solar- the outer transparent sheet or specialized coating applied to the surface of the absorber plates. This is what allows it to collect heat efficiency.

Glide - the difference in temperature between the dew point (condensation) and the bubble point (evaporation) for a zeotrope or near azeotrope refrigerant. Also referred to as temperature glide.

Glitch - an unwanted pulse or logic state, usually caused by poor design and/or propagation delays.

Global warming - often called the greenhouse effect. In global warming, tropospheric pollutants like CFC's, HCFC's, HFC's, carbon dioxide, and carbon monoxide absorb and reflect the earth's infrared radiation. This causes re-radiation back to the earth, and a gradual increase in the earth's average temperature.

Global warming potential (GWP) - a GWP number is assigned to each refrigerant indicating the risk of a refrigerant contributing to the global warming effect.

Globar - a conductive ceramic from which high temperature flame rods are made. It has a maximum temperature rating of 2600° F (1425° C).

Globe valve - a valve with a movable disc-type element and a stationary ring seat that changes the direction of the fluid flowing through it. These valves can be opened and closed substantially faster than a gate valve because of the shorter lift of the valve. When valves are to be operated frequently or continuously, the globe valve provides the more convenient operation. The seating surfaces of the globe valve are less subject to wear and the discs and seats are more easily replaced than on a gate valve. The different types of disc and seating arrangements for globe valves are the plug disc, narrow seat (or conventional disc), needle valves, and the composition disc.

Glow bar - a bar of high temperature material, usually spirally machined to provide a long electrical path, heated with an electrical current to incandescence to ignite gas burner flames.

Glow coil - a device made of a platinum alloy that automatically reignites a pilot light if it goes out.

Glycol - antifreeze solution used in the water loop of geothermal heat pumps.

Glycol water solution/forced air furnace - furnace with 50% glycol and 50% distilled water solution that passes through a tube and fin heat exchanger to distribute heat through the furnace duct system.

Goggles - safety glasses with side shield to prevent objects from getting into your eyes.

Government anchor - a steel anchor, inserted through a hole in the web of a steel beam, used to anchor a wall bearing beam to masonry construction.

Government lock - a transverse joint formed by placing a right angle flange with an S type slip arrangement over the end of one duct section and fastening by button punching. A second duct section with a short 90 degree right angle flange is inserted into the first flange arrangement, then the higher flange is turned over the inserted flange to form a tight lock. Also known as a pocket or box lock.

Governor, zero - a regulating device that is normally adjusted to deliver gas at atmospheric pressure within its flow rating.

Grace period - the time required to build up an explosive charge in a burner; the maximum time the un-ignited combustible reactants can safely be put into and retained in a combustion chamber.

Grade - (1) a reference plane representing the average of finished ground level adjoining the building at all exterior walls. (2) refers to the slope of a pipe line. (3) grade of fuel oil as in #2 grade. (4) refers to the size, heating value, and the ash content of coal.

Grade beam - horizontal, reinforced concrete beam between two supporting piers at or below ground supporting a wall or structure.

Grades of fuel oil - voluntary commercial standards recommended by the U.S. Bureau of Standards for different classification of fuel oils, based on characteristics such as specific gravity and viscosity.

Gradual switches - manual switches used in a pneumatic control system to vary the air pressure in a line to any value from zero to full supply pressure.

Graduate relay - a pneumatic relay capable of maintaining in one air line, usually the branch line, an air pressure proportional to the pressure in another line, usually the pilot line. It may be direct acting, increasing branch line pressure on an increase in pilot line pressure, or it may be reverse acting, reducing branch line pressure on an increase in pilot line pressure.

Graduate switch - a pneumatic manual switch for adjusting the air pressure in a line to any value from zero to full supply pressure.

Graduated acting - term applied to a control instrument or device that functions to give throttling control --- that is, operates between full on and full off position.

Graduated cylinder - device used for measuring an exact refrigerant charge. Such cylinders are very accurate. Commonly referred to as a Dial-a-charge.

Grain - a unit of weight and equal to 1/7000th of a pound and equal to .06480 grams. It is used to indicate the amount of moisture in the air.

Gram - a metric unit of weight equal to .035 ounces.

Graphic display - a master control panel that pictorially and usually colorfully traces the relationship of control equipment and the process operation. It permits an operator to check at a glance on the operation of a distance control system by noting dials, valves, scales, and lights.

Graphite - a soft, black, shiny form of carbon having a laminar structure. Used for lead in pencils, lubricants and electrodes. It may be of natural or synthetic origin.

Grate area - grate surface area measured in square feet, used in estimating the fuel burning rate of coal.

Grates - where the combustion process starts in a coal fired furnace.

Gravity - referring to the acceleration due to gravity which is 32.16 feet per second per second.

Gravity convection - is the natural movements of heat through a fluid that occurs when a warm fluid rises and cool fluid sinks under the influence of gravity. Also known as thermosiphoning.

Gravity drain back system - a type of active solar energy system in which freezing is avoided by having all water drain out of the collectors and exposed piping into a holding tank at the end of each operating cycle.

Gravity feed - supplying oil to a burner from a gravity tank, using the force of gravity.

Gravity, fuel oil - an indication of the heat content of a fuel. ASTM standards for fuel oils density to be 7.1 pounds and 139,4000 Btu's per gallon.

Gravity furnace - a furnace that delivers heat by the natural tendency of warm air to rise and cold air to sink. A gravity type furnace that contains no fans or filters.

Gravity hot water - hot water heating system where circulation is promoted by the difference in density between hot and cold water.

Gravity purging - filling a hydronic system from the bottom and allowing the air in the piping to rise up to, and out of high vent ports.

Gravity, specific - the specific gravity of a solid or liquid is the ratio of the mass of the body to the mass of an equal volume of water at some standard temperature. At the present time a temperature of 4° C (39° F) is commonly used by physicists, but the engineer uses 60° F. The specific gravity of a gas is usually expressed in terms of dry air at the same temperature and pressure as the gas.

Grease - a lubricant composed of an oil or oils thickened with a soap, soaps or other thickener to a semi-solid or solid consistency.

Grease fittings - located on individual bearings or bearings of larger motors that provide a port by which grease is added to these units during lubrication. These fittings have a ball check valve built into them to prevent the grease from escaping once filled.

Greenfield - a flexible metallic conduit used in connection of electrical power to machinery and applications that require bends of the conduit at various angles.

Greenhouse effect - the retention of heat from sunlight at the earth's surface caused by atmospheric carbon dioxide that admits shortwave radiation emitted by the earth.

Grid - network of transmission lines, substations, distribution lines, and transformers used in central power systems.

Grille - an ornamental or louvered opening placed at the end of an airway. It permits air flow, but prevents large objects from entering the duct. These may be located in the sidewall, ceiling or floor.

Grind - to remove metal by means of an abrasive wheel, often made of carborundum. Used chiefly used where accuracy is required.

Grommet - a plastic metal or rubber doughnut shaped protector for wires or tubing as they pass through hole in object.

Gross capacity, heat pump - the gross capacity of a heat pump in the heating mode. It is the total amount of heat energy transferred to the inside air or material. This is made up of the heat energy picked up from a heat source (air or water) plus the heat energy equivalent of the electrical energy required to operate the heat pump.

Gross conditioned floor area - the horizontal projection of that portion of interior space which is contained with exterior walls and which is conditioned directly or indirectly by an energy using system, and which has an average height of five feet or greater, measured from the exterior faces.

Gross exterior wall area - the normal projection of the building envelope wall area bounding interior space which is conditioned by an energy using system and which separates conditioned space from unconditioned space, semi-heated space, exterior ambient conditions or earth. The gross wall area of walls consists of all opaque wall areas, including foundation walls, between floor spandrels, peripheral edges of floors, vertical glazing areas and door areas, where such surfaces are exposed to exterior ambient conditions and enclose a conditioned space including interstitial areas between two such spaces.

Gross floor area - the sum of the areas of the several floors of the building, including basements, cellars, mezzanine and intermediate floored tiers and penthouses of headroom height, measured from the exterior faces of exterior walls or from the center line of walls separating buildings.

Gross output - a rating applied to boilers. It is the total quantity of heat that the boiler will deliver and at the same time meet all limitations of applicable testing and rating codes.

Gross roof/ceiling area - a roof/ceiling assembly shall be considered as all components of the roof/ceiling envelope through which heat flows, thus creating a building transmission heat loss or gain, where such assembly is exposed to exterior ambient conditions and encloses a conditioned space. The assembly does not include those components that are separated from a heated and/or cooled space by a vented air space. The gross area of a roof/ceiling assembly consists of the total interior surface of such assembly, including overhead glazing.

Ground - (1) an electrical term meaning to connect to the earth or other large conducting body to serve as earth thus making a complete circuit. A common return to a point of zero potential, such as a furnace or air conditioners metal chassis. (2) a nailing strip fixed in a masonry or concrete wall as a means of attaching wood trim or furring strips.

Ground clamp - a mechanical clamping device to connect a ground wire or conduit to ground.

Ground coil, earth coil - a heat exchanger buried in the ground that may be used either as an evaporator or as a condenser.

Ground circuit - the wiring or body metal that makes up the ground of an electrical circuit.

Ground fault - conducting connection, intentional or accidental, between any conductor of an electrical system and the conducting material (metal conduit, metal cabinet, etc) that encloses the conductors as well as any conductors that are grounded or may become grounded.

Ground fault circuit interrupter (GFCI) - safety device designed to protect a circuit from over current and to protect people from potentially hazardous ground faults arising from the use of defective appliances or portable tools. It uses a fast acting circuit breaker that senses small imbalances in the circuit caused by current leakage to ground and, in a fraction of a second, shuts off the electricity. The GFCI continually matches the amount of current going to an electrical device against the amount of current returning from the device along the electrical path. Whenever the amount "going" differs from the amount "returning" by approximately 5 milliamps, the GFCI interrupts the electric power within as little as 1/40 of a second.

Ground fault circuit path - an electrically conductive path from the point of a ground fault on a wiring system through normally non-current carrying conductors, equipment, or the earth to the electrical supply source.

Ground fault current path, effective - an intentionally constructed, permanent, low impedance electrically conductive path designed and intended to carry current under ground fault conditions from the point of a ground fault on a wiring system to the electrical supply source.

Ground joint - the interconnecting pieces of a union.

Ground loop, double "U" bend, vertical - a multiple pipe system in vertical holes, connected to divide the liquid flow among the loops.

Ground loop, four layer, horizontal - a two two-layer pipe loop located in the same trench, one above the other.

Ground loop, single layer, horizontal - one continuous length of pipe laid horizontally in the earth.

Ground loop, single "U" bend, vertical - a single run of pipe in one or more vertical holes, connected in series where all the liquid travels through all portions of the loop.

Ground loop, two layer, horizontal - a single loop that doubles back on itself to reduce the earth area needed.

Ground potential - the potential of the earth. A circuit, terminal, or chassis is considered to be at ground potential when it is used as a reference point for other potentials in the circuit.

Ground, short circuit - a fault in an electrical circuit allowing electricity to flow into the metal parts of the structure.

Ground wire - an electrical wire that will safely conduct electricity from a structure into the ground.

Grounded - refers to the condition of having a proper earth ground or some conducting body that serves in place of the earth.

Grounded circuit - a circuit that is intentionally or unintentionally shorted to ground.

Grounded conductor - a system or circuit conductor that is intentionally grounded.

Grounded effectively - intentionally connected to earth through a ground connection or connections of sufficiently low impedance and having sufficient current carrying capacity to prevent the buildup of voltages that may result in undue hazards to connected equipment or to persons.

Grounded junction - a junction is grounded if the measuring junction of a thermocouple probe is in contact with the sheath material.

Grounding - to remove the possibility of electrical shock. Protection against static charges that sometimes build up on operating equipment.

Grounding conductor - a conductor used to connect equipment or the grounded circuit of a wiring system to a grounding electrode or electrodes; usually colored green.

Grounding electrode conductor - the conductor used to connect the grounding electrode to the equipment grounding conductor, to the grounded conductor, or to both, of the circuit at the service equipment or at the source of a separately derived system.

Group 1 refrigerant - a nontoxic, nonflammable refrigerant.

Group 2 refrigerant - a toxic and moderately flammable refrigerant.

Group 3 refrigerant - a very flammable refrigerant.

Grout - a cement like material injected between the well casing and the drilled hole for a well. Grout adds rigidity and prevents contamination from entering the well.

Guide, pipe - type of hanger support that allows a pipe to move along its length but not sideways.

Guide, valve - that part of the valve disc that keeps the disc aligned with the valve seat. Top or bottom guides on a valve disc are usually so located that they do not influence flow but merely accomplish the centering function. Valve guides often have the added function of determining the valve flow characteristic. These are known as skirted guides and usually have notches or V's cut into them to characterize flow.

Guide vanes - vanes used to produce capacity control in a centrifugal compressor. Also known as pre-rotational guide vanes.

Gun type burner - a power burner in which a fan, or blower, blows air at high velocity past turbulators. The gas, at regulated pressure, is released into the turbulent airstream-typically from a ring at the mouth of the burner (in which case it is also called a ring burner). The gas escapes through orifices around the periphery of the ring. Gas and air are adjusted simultaneously to maintain the proper air fuel ratio as the firing rate changes. In high pressure gun burners, the gas is ejected at pressures up to 15 psi from the spuds located in the center of the turbulent airstream instead of at the outside. Also see forced-draft and raw gas burners.

Gusset plate - a steel or wood plate used to connect two or more members or to add strength to a framework.

Gutter - a bent channel used to carry away the water from the eave of a sloping roof.

Gypsum based underlayments - a pourable self leveling mixture containing gypsum based cement and washed sand. It is specifically intended for use as a thin slab material.

Gypsum thin slab - a type of construction for radiant floors heating in which a slab of poured gypsum underlayment ranging from 1.25 to 2 inch thick is poured over tubing that has been fastened to a subfloor.

-H-

H - abbreviation for magnetic intensity or henry.

HC - hydrocarbon

HCFC - hydrochlorofluorocarbons.

HDD65 - heating degree-days base 65° F.

HEPA - high efficiency particulate arrestance.

HFC - hydrofluorocarbons.

HFE - hydrofluoroethers

Hg - mercury.

HLF - heating load factor.

Hp - horsepower or high pressure.

HSF - high side float.

HSI - hot surface ignition.

HSPF - heating seasonal performance factor.

Ht - height.

HTM - heat transfer multiplier.

HVACR - heating, ventilating, air conditioning and refrigeration.

HVU - heating ventilating unit.

HWR - hot water return.

HWS - hot water supply.

Hz -hertz.

H_20 - water.

Habitable space - space in a building intended for regular human occupancy, which for indoor air quality discussions is assumed to be other than an industrial or factory environment.

Hack saw - a fine-toothed saw used to cut nonferrous, ferrous and plastic material.

HACR breaker - a circuit breaker that has been tested and found suitable for use on heating, air conditioning, and refrigeration equipment. The circuit breaker is marked with the letters HACR. Equipment that is suitable for use with HACR circuit breakers will be marked for use with HACR type circuit breakers.

Hair - human hair is used in some humidisats to sense humidity, since it contracts and expands with changes in the amount of humidity in the air.

Half bridge - two active elements or strain gauges.

Half wave rectifier - the simplest type of rectifier circuit that contains a diode and converts or changes alternating voltage into direct voltage every other alternating current alternation to direct current.

Halide leak detector - a device used to detect refrigerant leaks in a system. A burner is equipped with a source of fuel, a mixing chamber, a reactor plate, and an exploring tube. The reactor plate surrounds the flame. When the open end of the exploring tube is held near a refrigerant leak, some of the refrigerant is drawn into the mixing chamber when its presence changes the color of the flame. Also see flame test for leaks.

Halide refrigerants - a family of synthetic refrigerants that contains halogen chemicals. These refrigerants include R-12, R-22, R-500, and R-502.

Halide torch - a device used to detect refrigerant leaks in a system. A burner equipped with a source of fuel, a mixing chamber, a reactor plate, and an exploring tube. The reactor plate surrounds the flame. When the open end of the exploring tube is held near a refrigerant leak, some of the refrigerant is drawn into the mixing chamber where its presence changes the color of the flame.

Hall effect - the change in conduction of a special semiconductor device when the device is in the influence of a magnetic field. It is used in clamp-on meters to read DC current.

Halogenated - usually applied to hydrocarbons that are combined with fluorine, bromine, chlorine, iodine; vinyl chloride, methyl chlorine, halon, and chloroform.

Halogens - any of the five chemically related nonmetallic elements that include fluorine, chlorine, bromine, iodine, and astatine.

Halon (Freon 1301) - a fire suppressant as well as a non-combustible gas frequently used in fire extinguishing systems.

Hammer - a hand tool used for striking another object.

Hand hole - small hole in a vessel used for hand access for maintenance or adjustment. Hand holes are found on both firetube and watertube boilers that are removed when cleaning the water side of the boiler.

Hand notcher - a tool used for cutting notches along the edge of a piece of sheet metal.

Hand seam - a tool used to form the seams on a piece of sheet metal.

Hand shut-off valves - valves that make it possible to quickly isolate sections of a system for servicing.

Hand truck - a two-wheeled piece of equipment that can be used for moving heavy objects.

Hanger - device attached to walls or other structure for support of system piping.

Hardcopy - any printed or plotter paper output.

Hard drawn tubing - copper tubing which has been specially hardened to keep it stiff.

Hard drawn copper wire - copper wire that has been drawn to size and not annealed.

Hard flame - a flame with a hot, tight, well defined inner cone.

Hard soldering - see brazing.

Hard solders - silver bearing brazing alloys that have a melting temperature that ranges from 1100° F to 1500° F (593° C to 816° C).

Hard start kit - a kit consisting of a start relay (potential relay) and a start capacitor used to provide high starting torque. When added to a PSC motor circuit it changes it into a CSR motor circuit. Required when a refrigeration system utilizes a TXV as the metering device. This is because the TXV doesn't have good equalization characteristics on shut down between the high and low sides of the system..

Hard starting - refers to the difficulty of a motor or compressor in starting that may be due to low voltage or excessive pressure differential.

Hard wired logic - specific wiring connections between components that force a control system to operate in a predetermined manner.

Hard wiring - permanent, line voltage wiring.

Hardening - the heating and cooling of certain iron based alloys for the purpose of producing a hardness that is superior to that of the untreated metal.

Hardness - (1) the total mineral content of the water and affects the scaling of the inside of pipes surfaces. (2) the resistance of a substance to surface abrasion.

Hardware - the mechanical, magnetic, electronic, and electrical devices from which a computer is fabricated. The assembly of material forming a computer.

Harmonics - wave forms whose frequencies are multiples of the fundamental (60 Hz or 50 Hz) wave. The combination of harmonics and the fundamental wave causes a nonsinusoidal, periodic wave. Harmonics in power systems are the result of nonlinear effects. Typically, harmonics are associated with rectifiers and inverters, arc furnaces, arc welders, and transformer magnetizing current. Both voltage and current harmonics occur.

Harmonic filter - a combination of inductive, capacitive, and resistive components designed to suppress harmonic frequencies. A filter can include any two, or all three, of these components.

Harness - a term used to describe a group of conductors laid parallel or twisted, usually with many breakouts, laced together or pulled into a rubber or plastic sheath, used to interconnect electrical circuits.

Hartford loop - the loop in the return main of the boiler piping that maintains the water level in the boiler.

Harvest cycle - the part of an ice cuber operation where the cubes are melted out of the mold and placed in the harvest bin.

Hastelloy - trade name for a hard, non-corroding metal alloy.

Hazard communication rule - requires chemical manufacturers and importers to assess the hazards associated with the chemicals in their workplace. Material handling data sheets, labeling, and training are all the result of this law.

Hazardous area - an area where combustible gases are usually present or likely to be. Generally such areas are classified as hazardous and are the opposite of general purpose areas. Reference National Electric Code (NFPA 70) Article 500 for hazardous areas.

Hazardous location - any location considered to be a fire hazard for flammable vapors, dust, combustible fibers or other highly combustible substances. The location is not necessarily categorized in the building code as a high hazard use group classification.

Hazardous material - is any substance or mixture of substances having properties capable of producing adverse effects on the health and safety of a human being. A hazardous material is one that has one or more of the following characteristics: (1) has a flash point below 140° F, closed cup, or is subject to spontaneous heating; (2) has a threshold limit value below 500 ppm for gases and vapors, below 500mg/m for fumes, and below 25 mppcf (million particles per cubic foot) for dusts; (3) has a single dose oral LD50 below 50 mg/kg; (4) is subject to polymerization with the release of large amounts of energy; (5) is a strong oxidating or reducing agent; (6) causes first degree burns to skin in short time exposure, or is systemically toxic by skin contact; or (7) in the course of normal operations, may produce dusts, gases, fumes, vapors, mists, or smokes which have one or more of the above characteristics.

Head - a measure of pressure representing the height of water. Usually applies to centrifugal pumps. Indicates the height of a column of water being lifted by the pump, neglecting friction losses in piping. For water divide head in feet by 2.31 to get pressure in pounds per square inch (psi). Usually expressed in feet of water, inches of mercury, or millimeters of mercury.

Head, available - the pressure available to overcome friction in the system and to circulate the fluid.

Head capacity curve - graph used to evaluate a pumps performance by plotting the pumps head value on the vertical axis and the flow rate on the horizontal axis.

Head, dynamic - in flowing fluids, the sum of the static and velocity heads at the point of measurement. Also known as total head.

Head, friction - head required to overcome friction of the interior surface of a conductor and between fluid particles in motion.

Head loss - the loss of pressure in a flow system measured using a length parameter such as inches of water or inches of mercury.

Head pressure - pressure that exists on the condensing side of the refrigerating system. Also called discharge pressure or high side pressure. It is that pressure between the discharge side of the compressor and the inlet of the metering device.

Head pressure control - the pressure operated control that opens the electrical circuit if the high side pressure becomes excessive.

Head, static - the pressure of fluid expressed in terms of height of column of the fluid, such as water or mercury.

Head, static discharge - static head from the centerline of the pump to the free discharge surface.

Head, static suction - the head from the surface of the supply source to the centerline of the pump.

Head, thermal - difference in pressure between the weight of the column of warmed water on the supply side and the equal column of cooler water on the return side of a hot water heating system. This is the only force in a gravity hot water heating system.

Head, tool - the working end of a tool or implement. An attachment to or part of a machine that holds or contains the operative device.

Head, total static - in flowing fluid, the sum of the static and velocity heads at the point of measurement. Static head from the surface of the supply source to the free discharge surface.

Head, velocity - in a flowing fluid, the height of fluid equivalent to its velocity pressure.

Head pressure - high side pressure in a refrigeration system; pressure from the compressor discharge to the metering device.

Head pressure controller - either a pressure operated control that opens the electrical circuit when the head pressure exceeds certain preset values or a pressure controller than cycles the condenser fan on and off to maintain correct head pressure.

Head velocity - height of a flowing fluid that is equivalent to its velocity pressure.

Header - a manifold or supply pipe to which a number of branch pipes are connected.

Header, chimney - a beam set at right angles to floor or roof joists to provide support and framing around the opening.

Header, close, geothermal - a piping system where the supply and return headers for multiple loop vertical systems are spaced close together with the multiple loops arranged in a circle around the headers.

Header, construction - a horizontal framing member placed over the rough opening of a window to prevent the weight of the wall or roof from resting on the window frame. A beam across the top of the opening for a window or door. Headers rest on jack studs and support the weight of the wall above the window unit. They are often constructed out of doubled 2x6's, 2x10's or 2x12's.

Header, reverse return, geothermal - a piping system where the vertical loops are arranged in a straight line, and the headers between are connected to supply and return from opposite ends. Supply is to the closest loop, return is from the distant vertical loop.

Headroom - physical space above a valve, pipe, platform, etc.

Heat - a form of energy which causes substances to rise in temperature. Heat always flows from a body of higher temperature to a body of lower temperature. Energy associated with random motion of molecules. Also see latent heat, sensible heat, specific heat, total heat and heat of the liquid.

Heat affected zone - that portion of the base metal which has not been melted but which has been altered by heating, welding, brazing, soldering, or cutting.

Heat alarm - a single or multiple station alarm responsive to heat.

Heat anticipator - a heater, either fixed or adjustable, in the thermostat that prematurely terminates the furnace on cycle to assure that room temperature does not exceed the thermostat set point due to thermal inertia. Must be set to the amp draw of the heating low voltage circuit amperage. Designed to minimize system overshoot.

Heat block - insulation material installed between the cell mount of an infrared flame detector and the sighting pipe to reduce heat conduction to the cell.

Heat capacity - the amount of heat necessary to raise the temperature of a given mass one degree. Numerically, the mass multiplied by the specific heat.

Heat coil - a device, made of tubing or piping, designed to transfer heat to a cooler substance by using transfer fluids.

Heat conductor - a material capable of readily conducing heat. The opposite of an insulator or insulation.

Heat content - the amount of heat, usually stated in Btu's per pound (calories per gram), absorbed by a refrigerant in raising its temperature from a predetermined level to a final condition and temperature. Where change of state is encountered, the latent heat necessary for the change is included.

Heat dissipation - conduction of heat away from a solid state switch by use of a heat sink. This is necessary to prevent damage with currents of 3 or more amperes.

Heat effect factor - an unscientific allowance given to the test output of some types of terminals units (heat emitters) when establishing catalog ratings. This allowance is intended to give credit for improved heat distribution obtained from the terminal unit.

Heat emitter - a generic term for a device that releases heat from a circulating stream of heated water into the space to be heated. Examples include convectors, radiators, and fan coils.

Heat energy - total energy of all the molecules in a given substance.

Heat exchanger - any device for transferring heat from one medium to another. A direct heat exchanger is a self-contained device that transfers heat generated in the combustion chamber directly through the walls of the heat exchanger to the heating medium. An indirect heat exchanger contains a heating medium and transfer heat to another medium separately contained in close contact with or directed through the heat exchanger.

Heat exchanger, double pipe - one in which two pipes are arranged concentrically, one within the other, and in which one fluid flows through the inner pipe and the other through the annulus between the other.

Heat flow - heat flows from a warmer to a cooler substance. The rate depends upon the temperature difference, the area exposed and the type of material.

Heat flux - the rate of heat flow across a unit of area. The common English units for heat flux are Btu/hr/ft2.

Heat fusion - a process that will permanently join sections of plastic pipe together.

Heat gain - the rate at which a structure gains heat by infiltration, solar radiation, occupant respiration and lighting (in Btu's) for a space to be conditioned, at the local summer outdoor design temperature and a specified indoor design condition.

Heat, humid - ratio of increase of enthalpy per lb of dry air with its associated moisture to raise the temperature under conditions of constant pressure and constant specific humidity.

Heat intensity - heat concentration in a substance as indicated by the temperature of the substance through the use of a thermometer.

Heat interchanger - a device to transfer heat from the liquid refrigerant to the suction gas; also known as a heat exchanger.

Heat lag - when a substance is heated on one side, it takes time for the heat to travel through the substance. This time is called heat lag.

Heat, latent - heat characterized by a change of state of the substance concerned, for a given pressure and always at a constant temperature for a pure substance, i.e., heat of vaporization or fusion.

Heat leakage - the flow of heat through a substance is called heat leakage.

Heat load - amount of heat, measured in Btu, which is added or removed during a period of 24 hours.

Heat load, water tower - total heat to be removed from the circulating water by the cooling tower per unit time.

Heat loss - the rate at which a structure loses heat (in Btu's) from a space to be conditioned, at the local winter outdoor design temperature and a specified indoor design condition.

Heat loss coefficient - a value that indicates the rate of heat loss (usually for an entire building) per degree of temperature difference between the inside and the outside air temperature.

Heat loss factor - the rate of heat transfer through a specific material for every one degree in temperature difference.

Heat, mechanical equivalent of - an energy conversion ratio of 778.177 ft/lb = 1 Btu.

Heat motor actuator - a device that moves the stem of a valve when supplied with a suitable voltage. The necessary movement is produced by the expansion and/or contraction of a thermal element that is heated by an internal resistor.

Heat of combustion – the heat evolved when the substance combines rapidly with oxygen.

Heat of compression - the mechanical energy of pressure transformed into energy of heat.

Heat of condensation - the latent heat energy involved in the transition from the solid to the liquid state.

Heat of fusion - the heat released in changing a substance from a liquid state to a solid state. The heat of fusion of ice is 144 Btu per pound.

Heat of reaction - heat per unit mass or per mol of one of the reagents or products of reaction in a chemical reaction; exothermal if given off, endothermal if absorbed.

Heat of respiration - the process by which oxygen and carbohydrates are assimilated by a substance. Also when carbon dioxide and water are given off by a substance.

Heat of the liquid - the heat (Btu) contained in a liquid due to its temperature. The heat of the liquid for water is 0 at 32° F and increases 1 Btu approximately for every degree rise in temperature.

Heat of vaporization - the latent heat energy required to change a substance from a liquid to a gas.

Heat output ratings - one of several ratings describing the rate of heat output of a heat source such as a boiler.

Heat pipe - a heat exchanger capable of transferring heat over a relatively long distance at a low temperature difference.

Heat pump - a compression cycle system used to supply heat or cooling to a temperature controlled space. This system may also remove heat from the same space.

Heat purging - removing additional heat from a boiler by maintaining circulation through it after the burner has stopped firing.

Heat rate - measure of generating station thermal efficiency, generally expressed in Btu per net kilowatt-hour, or lb steam/kWh.

Heat reclaim - the process of reusing discharged heat from such sources as exhaust fans, condenser coils, and hot water drains to do useful work.

Heat recovery system - produces and stores hot water by transferring heat from condenser to cooler water.

Heat recovery ventilators - mechanical ventilation systems that exchange air between the inside and outside of a structure while recovering heat from the exhaust air.

Heat rejection effect, condensing - that portion of the total refrigerant heat rejecting effect of a condenser which is used for condensing the entering refrigerant vapor to a saturated liquid at the entering refrigerant pressure.

Heat rejection effect, subcooling - the total refrigerant heat rejection effect less the condensing heat rejection effect.

Heat rejection effect, total refrigerant - total useful capacity of a refrigerant condenser for removing heat from the refrigerant circulating through it.

Heat relay - devices used to stage the elements of an electric heating system on and off.

Heat, sensible - a term used in heating and cooling to indicate any portion of heat that changes only the temperature of the substance involved.

Heat sequencer - a time delay relay used to sequence the on and off operation of several electrical strip heaters along with the indoor fan operation.

Heat sink - a device, usually a metal plate or fin-shaped object, to transfer a rise in temperature away from a switch into the surrounding air, into a liquid or into a larger mass.

Heat source - the source of the heat that is supplied to the occupied space. In the case of a geothermal heat pump system operating in the heating mode, the heat source is the earth.

Heat, specific - the ratio of the quantity of heat required to raise the temperature of a given mass of any substance one degree to the quantity required to raise the temperature of an equal mass of a standard substance (usually water at 59° F) one degree.

Heat sink - relatively cold surface capable of absorbing heat.

Heat source - the thermal reservoir from which energy is withdrawn through heat transfer.

Heat storage capacity - the physical property of materials (mass) located inside the building envelope to absorb, store and release heat.

Heat storage medium - usually liquid or rocks. These are used to store heat from the collectors so that it is available for future use.

Heat stress index - a numerical index of the relative degree of heat stress imposed by various combinations of metabolic heat production with environmental radiant heat, air temperature, air velocity, and relative humidity of the air. This index is an indication of the relationship between environmental conditions and physiological strain or the feeling of being uncomfortable.

Heat tape - electric resistance wires embedded into a flexible housing usually wrapped around a pipe to keep it from freezing.

Heat, total - the sum of sensible and latent heat. Usually referred to as enthalpy.

Heat transfer - the movement of heat from one body or substance to another. Heat may be transferred by radiation, conduction, convection or a combination of the three.

Heat transfer coefficient - the amount of heat passing through a given thickness and area of a material, per unit of time and temperature difference.

Heat transfer film coefficient - the heat transferred between a surface and a fluid per unit time per unit area driven by a unit temperature difference between the surface and the fluid in contact with it in Btuh·ft2·°F.

Heat transfer fluid - medium used to transfer heat from the collector to the heat storage tank.

Heat transfer liquid - the operating or thermal storage liquid in a mechanical system, including water or other liquid base, and additives at the concentration present under operating conditions used to move heat from one location to another. Refrigerants are not included as heat transfer liquids.

Heat transfer medium - usually air or liquid. These items are heated by the sun as it passes through the collectors. The heat gathered is used to heat the building or it may be transferred and stored in a heat storage medium.

Heat transfer multiplier (HTM) - a heat gain value that is equal to U times ETD. It takes into account the heat flow characteristics of the material assembly, outside design temperature, and daily range of temperatures. It is expressed in Btu/h per square feet.

Heat transfer plate - a metal plate that fits over heating tubing and provides lateral heat conduction away from the tubing.

Heat transmission - any time rate of heat flow; usually refers to conduction, convection, and radiation combined.

Heat transmission coefficient - any one of a number of coefficients used in calculating heat transmission through different materials and structures, by conduction, convection, and radiation.

Heat treatment - the application of a series of heating and cooling cycles to a metal or alloy to bring about a desired condition.

Heat unit - a given amount of heat, such as Btu, calorie, or therm.

Heat, vital - heat generated by fruits and vegetables in storage due to ripening.

Heat wheel - a heat recovery device which uses loosely packed wire mesh or ceramic material in a rotating wheel to absorb heat from one air stream and reject it to another air stream.

Heated bimetal valve - a heat motor valve in which the heat generated by a small resistance heater causes a bimetal element to actuate the valve.

Heated slab - concrete slab-on-grade floor containing wires, cables, pipes, or ducts that transfers heat to the conditioned space.

Heated space - enclosed space within a building that is heated by a heating system whose output capacity exceeds 10 Btu/h x ft2 or is capable of maintaining a space dry-bulb temperature of
50° F or more at design heating conditions.

Heater - an apparatus for giving heat or warmth; a furnace, oven, stove, radiator, or similar device. Also see air, booster, central, direct fired, external, indirect fired, internal, make up air, radiant tube type, space, and unit heaters.

Heater, infrared - heater that converts flame energy into infrared radiant energy.

Heater, make-up air - a heat exchanger whose only function is to increase the temperature of outside make-up air to that of the air re-circulating in the system.

Heater, room - a heater designed to warm only the room in which it is installed, without use of ducts.

Heater section - section of a solder melting type overload relay that is connected in series with the motor.

Heater, unit - a self contained heater, designed to heat only the non- residential space in which it is installed.

Heaters - an internal component part of a motor starter device that senses the heat generated by current flow. If this heater exceeds its must hold value, it opens a normally closed set of contacts. These contacts then disconnect the motor from the electrical circuit preventing further heat generation. Most motor starters are manually reset.

Heating capacity - the rate of heat output of a heat source, usually expressed in Btu/hr.

Heating coil - a heat transfer device that releases heat.

Heating control - a device that controls temperature of heat transfer unit which releases heat.

Heating curve - another name for reset line.

Heating degree days - each degree that the average temperature is below 65° F produces one heating degree day.

Heating effect factor - an arbitrary allowance added to the test output of some types of terminal units when establishing the catalog ratings. This allowance is intended to give credit for improved heat distribution obtained from the terminal unit.

Heating element, electric - a unit assembly consisting of a resistor, insulated supports, and terminals for connecting the resistor to the electric power.

Heating element, gas - all parts of an appliance that are in contact with the flames and/or hot flue gases and with the medium that is being heated and which transmit heat from the flames and hot flue gases to the medium. Generally, the heating element is on the exterior surface of a combustion chamber.

Heating load - a heating load is the rate at which heat energy must be supplied to a space in order to maintain a certain condition designed for it.

Heating medium - a substance such as air, steam, water, or "heat transfer fluids" used to convey or transfer heat from the boiler, furnace, or other source of heat to the heating units from which the heat is dissipated.

Heating mode - the operating phase of a system that is adding heat energy to an occupied area.

Heating, regenerative - a heat transfer process for utilizing heat which must be rejected or absorbed in one part of a cycle, to perform a useful function in another part of the cycle.

Heating seasonal performance factor (HSPF) - total heating output of a heat pump during its normal annual usage period for heating (in Btu's) divided by the total electric energy input during the same period (in watt-hours).

Heating surfaces - all surfaces which transmit heat directly from the flame or flue gases to the medium being heated.

Heating system, direct fired - a heating system in which combustion takes place in the air being blown into the building. The outdoor air temperature is increased by direct contact with the flame of the heater. Efficiency of this system is 100%.

Heating system, duct heater - a heating system in which the heater is installed directly in the distribution duct of the central air conditioning or heating system. May be either electric, gas-fired, or oil-fired.

Heating system, electric heater - a heating system that consists of one or more stages of resistive heating elements installed in a duct or central furnace.

Heating system, high-pressure steam - a steam heating system employing steam at pressures above 15 psig.

Heating system, high temperature water - a heating system in which water having supply temperatures above 350° F is used as a medium to convey heat from a central boiler, through a piping system, to suitable heat distributing means.

Heating system hot water - a heating system in which water having supply temperatures less than 250° F is used as a medium to convey heat from a central boiler, through a piping system, to suitable heat distributing means.

Heating system, hot water coil - a heating system in which hot water is supplied by a central hot water boiler. Hot water coils are used to heat air.

Heating system, indirect fired - a heating system in which combustion takes place in a boiler or furnace. The fuel is burned in a combustion chamber and the flue gases do not mix with the incoming air.

Heating system, radiant - a heating system in which only the heat radiated from panels is effective in providing the heating requirements. The term radiant heating is frequently used to include both panel and radiant heating.

Heating system, low-pressure steam - a steam heating system employing steam at pressures between 0 and 15 psig.

Heating system, medium temperature water - a heating system in which water having supply temperatures between 250° F and 350° F is used as a medium to convey the heat from a central boiler, through a piping system, to suitable heat distributing means.

Heating system, panel - a heating system in which heat is transmitted by both radiation and convection from panel surfaces to both air and surrounding surfaces.

Heating system, perimeter warm air - a warm air heating system of the combination panel and convection type. Warm air ducts embedded in the concrete slab of a house or structure that does not have a basement, around the perimeter, receive heated air from a furnace and deliver it to the heated space through registers placed in or near the floor. Air is returned to the furnace from registers near the ceiling.

Heating system, radiant - a heating system in which only the heat radiated from panels is effective in providing the heating requirements. The term radiant heating is frequently used to include both panel heating and radiant heating.

Heating system, split - a system in which the heating is accomplished by means of radiators or convectors supplemented by mechanical circulation of air (heated or unheated) from a central point. Ventilation may be provided by the same system.

Heating system, steam - a heating system in which heat is transferred from the boiler or other source of heat to the heating unit by means of steam at, above, or below atmospheric pressure.

Heating system, vacuum - a two pipe steam heating system equipped with the necessary accessory apparatus that will permit operating the system below atmospheric pressure.

Heating system, vapor - a steam heating system that operates under pressures at or near atmospheric and which returns the condensate to the boiler or receiver by gravity.

Heating system, warm air - a warm air heating plant consisting of a heating unit (electric or fuel burning furnace) enclosed in a casing, from which the heated air is distributed to various rooms in a building through ducts.

Heating unit - radiators, convectors, baseboards, finned tubing, coils embedded in floor, wall, or ceiling, or any device that transmits the heat from the heating system to the room and its occupants.

Heating value - the number of Btu's released by the combustion, at constant pressure, of one cubic foot of gas when the products of combustion are cooled to the initial temperature of the gas and air, when the water vapor formed during combustion is condensed, and when all necessary corrections have been applied.

Heavy ends, hydrocarbon oils - the heavy molecules or larger molecules of hydrocarbon oils.

Heavy fuel oil - Grades 5 and 6, used in commercial and industrial burners that generally requires preheating. Heating values range from 146,800 to 155,900 Btu's per gallon.

Heel jaw - the bottom stationary serrated jaw of a pipe wrench.

Height - linear measurement of one dimension as from the base to the top of an object.

Height, water tower - On cooling towers erected over a concrete basin, height is measured from the elevation of the basin curb. Nominal heights are usually measured to the fan deck elevation, not including the height of the fan cylinder. Heights for towers on which a wood, steel, or plastic basin is included within the manufacturer's scope of supply are generally measured from the lowermost point of the basin, and are usually overall of the tower.

Heli-arc welding - joining of metal by using high frequency electric current. An inert gas (usually helium) surrounds the welding to prevent oxidation.

Helical stripe - a continuous, colored, spiral stripe applied over the outer perimeter of an insulated conductor for circuit identification purposes.

Helix coil - a piece of bimetal formed into a helix-shaped coil that provides for more surface area and a longer travel when heated.

Hem lock - a straight edged sheet that has been turned 180 degrees and the fold flattened.

Henry - an electrical unit denoting the inductance of a circuit in which a current varying at the rate of one ampere per second produces an electromotive force of one volt. Symbol is "L".

HEPA filter - an abbreviation for high efficiency particulate arrestance. These filters have an efficiency of 99.90 or higher and are capable of removing particles as small as 0.01 microns. HEPA filters are used when a high degree of filtration is desired or required. Also known as an absolute filter.

Hermetic - totally sealed, especially against the escape or entry of air. In HVAC/R applications, it means sealed by gaskets or welds, as in refrigeration compressors.

Hermetic analyzer - this is an instrument that can be used for rocking frozen, wedged compressors free by reversing motor rotations. Has a bank of starting and run capacitors, and has the ability to check motor windings for shorts, opens, continuity and grounds. Some analyzers can be used as an ammeter, voltmeter, ohmmeter, and as a capacitor or relay analyzer.

Hermetic compressor - a compressor and electrically driven motor sealed within the same casing.

Hermetic motor - compressor drive motor sealed within the same housing as the compressor and cooled by the refrigerant gas.

Hermetic system - a refrigeration system which has a compressor driven by a motor contained in a compressor dome or housing.

Hermetically sealed unit, refrigeration - a sealed hermetic type condensing unit is a mechanical condensing unit in which the compressor and compressor motor are enclosed in the same housing with no external shaft or shaft seal, the compressor motor operating in the refrigerant atmosphere. The compressor and compressor motor housing may be either the fully welded or brazed type, or of the service sealed type, the housing is permanently sealed and is not provided with means of access for servicing internal parts in the field. In the service sealed type, the housing is provided with some means of access for servicing internal parts in the field.

Hermetically sealed unit, window - an insulating glass unit that is sealed against moisture. The unit is made up of two light panels, separated by a roll formed metal spacer tube which contains a moisture and/or solvent absorbing material. The unit is then completely sealed, creating a moisture free air space.

Hertz (Hz) - a unit of frequency equal to one cycle per second.

Heterojunction - zone of electrical contact between two dissimilar materials.

Hex head - type of bolt or other fastener with a 6 sided head for use with a wrench.

Hexagon - a six-sided figure with each of its sides forming a 60° angle.

Hexagonal shoulder - a 6-sided edge.

Hg - chemical symbol for mercury. A heavy silver white metallic element. It is the only metal that is liquid at ordinary room temperature. Symbol is Hg for the Latin term hydrargyrum.

Hickey - a device used to bend conduit, also, a special fitting for attaching light fixtures to outlet boxes.

Hickey punched - when two or more pieces of metal are to be joined by slipping together, the union can be made tight by the use of a tool that forces a deep indentation into all pieces of metal at one time. These indentations are generally spaced on about 18 inch centers.

Hidden lines - are light, dashed lines used on blueprints and plot plans to show edges that are not seen from a particular view. You have to look at another plan in the set to find these edges. Hidden lines are used to help the worker understand the drawing.

High boy furnace - a furnace in which air is drawn in through the sides or bottom and discharged out the top.

High capacity unit - a factory made assembly consisting of dual duct modulating dampers and integral volume control station together with a temperature mixing station to supply air volumes above 2000 cfm from an external source and static pressures greater than .40" H2O. These units are designed for furred-in application and distribution ducts to provide sound attenuation and air delivery to the occupied room.

High efficiency particulate arrestance (HEPA) - a filter type that can achieve efficiencies of 99.97% by removing particulates of 0.3 microns or larger in diameter from an airstream. Glass asbestos is used as the filtering media. Typical applications include hospital operating rooms, research laboratories, pharmaceutical plants, computer rooms, and any other applications where atmospheric contamination must be prevented. May also be referred to as absolute filters.

High efficiency gas furnace - forced air gas fired furnace that uses recycling of combustion flue gases through a secondary heat exchanger to achieve efficiencies above 85%.

High fire - the maximum firing rate of a burner or furnace.

High fire interlock - an interlock that proves the damper is open for pre-purge.

High fire switch - see high fire interlock.

High leg - a configuration in a delta wiring connection where one transformer is larger and center tapped. Caution should be used with the connection as it could give you approximately 170 volts instead of the assumed 120 volts. Also called the wild leg, stinger leg or bastard leg.

High limit - a controller that features normally closed contacts and a sensing element inserted in a furnace (or boiler) that shuts down the burners or electric heater regardless of room thermostat demand when circulating air or water exceeds preset limits.

High-low fire - provision in a burner for either of two firing rates, high or low (no rate in between), selected according to load demand by the firing rate controller. A system is normally wired for guaranteed low fire start.

High mica - condition that exists when the mica between the copper segments of a commutator extends beyond the outer level of the copper.

High point vent - devices located at the high point(s) of a piping system that can release air that may accumulate at these high points.

High-pressure appliance - an appliance that uses a refrigerant with a boiling point between -50° C and 10° C at atmospheric pressure (29.921 inches of mercury). This includes but is not limited to appliances using refrigerants R-12, R-22, R-114, R-500 or R-502.

High-pressure boiler, hot water or steam - a boiler furnishing hot water at pressures in excess of 160 pounds per square inch (psig) or at temperatures in excess of 250° F (121° C); a boiler furnishing steam at pressures in excess of 15 psig.

High pressure cut out - a normally closed electrical control switch operated by the high side refrigerant pressure that automatically opens an electrical circuit if excessive head pressure or condensing pressure is reached.

High-pressure gauge - an instrument that measures pressures on the high side of the refrigeration system.

High-pressure gun burner - see gun type burner.

High-pressure switch - see pressure switch

High pressure unit - a factory made assembly consisting of either a dual duct modulating damper or a single duct pressure reducing damper and an integral volume control station together with a temperature mixing and sound attenuation system. These units are designed to deliver air when supplied from an external source with static pressure greater than 0.4"water column. Sometimes referred to as a mixing box, dual duct box, or a constant volume box.

High probability systems - any system in which the basic design or the location of components is such that a leakage of refrigerant from a failed connection, seal or component will enter an occupancy classified area, other than the machinery room.

High side - that part of a refrigerating system that are under condensing or high side pressure. Also refers to the outdoor unit, which contains the condenser fan motor, compressor, outdoor coil, and possibly a receiver, mounted on a base unit.

High side charging - the process of introducing liquid refrigerant into the high side of a refrigeration system. The acceptable manner for placing refrigerant into a non-operating system.

High-side float - a refrigerant control mechanism that controls the level of the liquid refrigerant in the high pressure of the mechanism.

High side pressure - the pressure contained in the high side of the refrigeration system. This pressure would be from the outlet of the compressor (hot gas discharge), through the condenser coil, out the liquid line to the inlet of the metering device. Also referred to as head pressure.

High speed fuses - fuses with no intentional time-delay in the overload range and designed to open as quickly as possible in the short-circuit range. These fuses are often used to protect solid state devices.

High temperature refrigeration - a refrigeration temperature range starting with evaporator temperatures no lower than 35° F, a range usually used in air conditioning cooling systems.

High temperature water system - a hydronic heating system that operates at temperatures over 350° F and usual pressures of about 300 psi. The maximum design supply water temperature is usually about 400° F, with a pressure rating for boilers and equipment of about 300 psi. the pressure temperature rating of each component must be checked against the system's design characteristics.

High vacuum - a vacuum greater than 29 inches of mercury.

High vacuum pump - a machine that can create a vacuum in the 1000 to 1 micron range.

High voltage circuit - a circuit involving a potential of more than 600 volts; not defined by Underwriters Laboratories Inc.

High voltage probe - used to extend the range of a multi meter or oscilloscope to read up to 40Kv AC or DC.

Hip roof - a roof that slopes upward from all four sides of a building.

Hi-Re-Li system - a patented heat pump cycle developed by Westinghouse Corporation.

Hold back thermostat - outdoor ambient thermostat used to determine the number of supplementary electric strip heaters that will be energized during the defrost mode.

Holding charge - the refrigerant charge that is supplied in a new condensing (high side component) unit.

Holding charge - the small amount of refrigerant or nitrogen that is shipped within the sections of a split system air conditioning unit. This ensures that the system is leak free upon delivery. A loss of the holding charge is an indication that there may be a leak in the system.

Holding coil - that part of the magnetic starter or relay that causes the device to operate when energized.

Holding contacts - a set of relay contacts used for the purpose of maintaining current flow to the coil of a relay. Generally used when the relay is initially energized by a momentary switch closure.

Holding current - the amount of current needed to keep an SCR or triac turned on.

Holding relay - used with a short cycling timers, this relay energizes and changes holding contacts during short cycling recovery time.

Holdover - in an evaporator, the ability to stay cold after heat removal from the evaporator stops. A material used to store heat in latent or sensible form.

Hole - a vacancy that has been caused by a valence electron breaking its covalent bond and becoming a conduction electron. It has a positive charge.

Hollow cone nozzle - a nozzle used in fuel oil burners where the spray pattern is concentrated on the outer edge of the spray.

Hollow tube gasket - a sealing device made of rubber or plastic with tubular cross section.

Hollow set screw (Allen) wrenches - used to remove and install set screws that are recessed.

Hollow wall - a wall built of masonry units so arranged as to provide an air space within the wall, and in which the facing and backing of the wall are bonded together with masonry units.

Home run distribution system - a distribution system in which each heat emitter (as in the case of a hydronic system) has its own supply and return tube routed from a common manifold. This same situation applies to the natural gas supply line that is run from a common manifold bar to each appliance.

Homojunction - the zone of contact between the n-layer and the p-layer in a single material, the two layers having been created by doping the basic crystal with other substances.

Hone - a fine grit stone used for precision sharpening.

Hook jaw - the top movable serrated jaw of a pipe wrench.

Hood - an air-intake device used to capture by entrapment, impingement, adhesion or similar means, grease and similar contaminates before they enter a duct system. There are two classifications of hoods:

Type I -- a kitchen hood for collecting and removing grease and smoke.

Type II -- a general kitchen hood for collecting and removing steam, vapor, heat or odors.

Hook seam - a sheet metal seam where the edges of adjoining sheets are turned 180 degrees in opposite directions, interlocked and dressed with a mallet to tighten. The joint is formed with top sheet lapped in the direction of flow. The joint is usually filled with mastic material to permit a small amount of movement due to temperature changes. Also known as a clinch lock seam.

Hoop stress - the stress in a pipe wall (usually in pounds per square inch), acting circumferentially in a plane perpendicular to the longitudinal axis of the pipe and produced by the pressure of the fluid in the pipe.

Hopper windows - a window sash opening inwards, either hinged at the top, bottom or sides.

Hopscotching - a troubleshooting procedure for electric circuits that is accomplished by jumping from one component to another to determine which components are open or closed.

Horizontal boiler - see water tube boiler.

Horizontal flow - a system with return air from one side flowing horizontally through a heat exchanger and out the other side as supply air.

Horizontal furnace - a forced air, central furnace designed with essentially horizontal airflow.

Horizontal ground loop - on geothermal heat pump systems, horizontally installed ground loops. Used in locations where there is ample land surface or where digging large trenches is neither practical nor possible.

Horizontal joint - type of brazing or solder connection in which the tubing and the fitting are on the same plane and are heated equally.

Horizontal panel radiator - a panel radiator that is wider than it is tall, and has tubes oriented in the horizontal direction.

Horizontal return tube (HRT) boiler - see fire tube boiler.

Horizontal rotary oil burner - see rotary oil burner.

Horizontal stripe - a colored stripe running horizontally with the axis of a conductor, sometimes called a longitudinal strip, used as a means of circuit identification.

Horsepower (HP) - as applied to an electrical motor, the horsepower is an index of the amount of work the motor can produce in a period of time. For example, one horsepower equals 33,000 foot pounds of work per minute, with suitable gearing and neglecting all losses, can lift 33,000 feet in one minute, or one pound 33,000 feet in a minute. Assuming 100% efficiency, 746 watts of electrical power will produce one horsepower. Horsepower maybe determined by using one of the following formulas.

$$HP = \frac{\text{Watts output}}{746}$$

$$HP = \frac{\text{Torque (in/oz.)} \times \text{RPM}}{1{,}000{,}000}$$

$$HP = \frac{\text{Foot pounds torque} \times \text{RPM}}{5250}$$

Horsepower, boiler - a measure of boiler capacity, equal to the production of 33,475 Btu's per hour.

Horsepower capacity curve – a graph used to evaluate a pumps performance by plotting the pumps horsepower requirements on the vertical axis and its flow rate on the horizontal axis.

Hortensphere - spherical pressure storage tank.

Hose bib - a name for a valve that allow a common garden hose to be attached.

Hot - carrying a current or danger of shock.

Hot bending - the bending of pipe to a predetermined radius after heating to a suitable temperature.

Hot circuit - that portion of a circuit not a ground potential, and which is electrically insulated, and is usually at a potential (voltage).

Hot contacts - the set of contacts in a stack relay that are in the closed position when the flue pipe is hot.

Hot deck - the heating section of a multizone system.

Hot gas - the refrigerant vapor leaving the compressor.

Hot gas bypass - a piping system in a refrigerating unit that moves hot refrigerant gas from condenser into the low side of the system.

Hot gas defrost - a defrosting system in which hot refrigerant gas from the high side is directed through the evaporator for a short period of time and at predetermined intervals in order to remove frost from the evaporator.

Hot gas discharge line - a tubing, pipe or line in a refrigeration system that is connected from the outlet of the compressor to the inlet of the condenser coil.

Hot junction - that part of a thermocouple where the two wires that make up the thermocouple are joined together and heated to produce electrical energy by the thermoelectric effect.

Hot lead - a wire or conductor connected to high side, not to ground or neutral terminal.

Hot line - incoming, live (not grounded) wire(s) used to power a device, usually designed L1.

Hot load - a function of the amount of compression when a pipe is at its hottest operational temperature. Hot load is greater than cold load if pipe moves down when the system goes into operation. Sometimes the temperature change from cold to hot causes other supports to take up the load and the pipe rises locally so that its hot load is less than cold load.

Hot pull down - the process of lowering the refrigerated space to the design temperature after it has been allowed to warm up considerably over the temperature.

Hot refractory hold in test - a test, for a flame safe guard control using a rectifying photocell, a cadmium sulfide cell, or an infrared flame detector, to make certain that hot refractory will not cause the flame relay to stay pulled in after the burner flame goes out.

Hot refractory override test - see hot refractory saturation test.

Hot refractory saturation test - a test, for a flame safe guard control using an infrared flame detector, to make certain that radiation from hot refractory does not mask the flickering radiation of the flame itself. Also called hot refractory override test.

Hot surface igniter - a gas burner ignition system in which a silicon carbide or silicon nitride strip resistive element is used as a heat source to ignite the gas.

Hot water heating system - a heating system in which water is used as the medium by which heat is carried through pipes from the boiler to the heating units.

Hot wire - an electrical lead that has a voltage difference between it and another wire either another hot wire or a neutral wire.

Hot wire anemometer - an instrument that measures air speed by changes in electrical resistance of a wire exposed to the air current.

Hot wire relay - a relay that is opened or closed by a thermal element that senses the starting current of the motor. The relay allows a starting capacitor or starting winding to drop out or drop in the starting circuit. This type of relay also has a built in means of overload protection.

Hot wire valve - a heat actuated valve in which the valve is opened and closed by expansion and contraction of a taut wire heated by an electric current passing through it.

Housing - any protective shell or covering.

Hub end - end connection, caulked or leaded, used on valves, fittings, and pipe (mainly for water supply and sewage lines).

Huddling chamber - that part on a safety valve that increases the area of the safety valve disc, thus increasing the total upward force, causing the valve to pop open.

Humid - damp or moist. Containing much moisture.

Humid climate - climate in which the following conditions occur: (a) 67° F or higher wet bulb outdoor ambient temperature for 3,500 or more hours during the warmest six consecutive months of the year. (b) 73° F or higher wet-bulb outdoor ambient temperature for 1,750 or more hours during the warmest six consecutive months of the year.

Humidification - the process of increasing the water vapor content of the conditioned air.

Humidifier - a device used to add to and control the humidity in a confined space.

Humidifier, air washer - a device which adds moisture to air adiabatically using a heated water spray chamber.

Humidifier, atomizing - a device which adds moisture to air by spraying small water particles directly into the air stream.

Humidifier, central - a device that humidifies air to be circulated through ducts in an air conditioning system.

Humidifier, evaporative pan - a device which adds moisture to air by evaporating water from an open container.

Humidifier fever - a respiratory illness that may be caused by exposure to toxins from microorganisms found in wet or moist areas in humidifiers and air conditioners. Also called air conditioner or ventilation fever.

Humidifier, room spray type - an air humidifier that sprays water directly into the room.

Humidifier, steam grid - a device which adds moisture to air by injecting steam through an orifice in a dispersion manifold enclosed in a heating duct.

Humidifier, wetted media - a device which adds moisture to air by evaporating water from a wetted medium using a fan or built in air supply.

Humidify - to add water vapor to the atmosphere. To add water vapor or moisture to any material.

Humidifying effect - the latent heat of evaporation of water at the average evaporating temperature times the weight of the water evaporated per unit of time.

Humidistat - a normally open electrical primary control device that makes (or closes) on humidity decrease used to control humidity. It uses a moisture sensitive element (usually nylon, human hair, biwood or an electronic element) to control a mechanical linkage that opens and closes an electrical switch.

Humidity - moisture or dampness. Relative humidity is the ratio of vapor present in air to the greatest amount possible at a given temperature.

Humidity, absolute - weight of water vapor per cubic foot of dry air, expressed as grains of moisture per pound.

Humidity, percentage - the ratio of the weight of water vapor associated with a pound of dry air to the weight of water vapor associated with a pound of dry air saturated at the same temperature.

Humidity, specific - weight of water vapor in one pound of dry air, expressed as grains of moisture per pound.

Humidity, ratio - the number of pounds of water contained in air on a per pound dry air basis. Means the same thing as moisture content or specific humidity.

Humidity, relative - the amount of moisture in the air expressed as a percentage of the maximum amount that the air is capable of holding at a specific temperature.

Hunting - a continuous oscillation of the temperature, pressure, or relative humidity above and below the control point. Very rapid hunting is usually termed "oscillation".

HVAC - heating, ventilating, and air conditioning.

HVAC drawing - drawing that shows ducting location and supports, openings, elevations to bottom of ducts, configuration, and duct size.

HVACR - heating, ventilating, air conditioning and refrigeration.

HVACR circuit breaker - a circuit breaker with a built in trip delay commonly used in air conditioning installations due to the power surge on start-up.

Hydraryrum - mercury. From the Latin term for liquid silver.

Hydraulic coupling - coupling between the drive element and fan or pump.

Hydraulic equilibrium - the condition where the head added by the systems circulator exactly equals the head dissipated by viscous friction of the fluid flowing through the system.

Hydraulic fluid - fluid serving as the power transmission medium in a hydraulic system. The most commonly used fluids are petroleum oils, synthetic lubricants, oil-water emulsions, and water-glycol mixtures. The principal requirements of a premium hydraulic fluid are proper viscosity, high viscosity index, anti-wear protection (if needed), good oxidation stability, adequate pour point, rust inhibition, resistance to foaming, and compatibility with seal materials. Anti-wear oils are frequently used in compact, high pressure, and capacity pumps that require extra lubrication protection.

Hydraulic motor - a device that converts fluid power into mechanical force and motion. It usually provides rotary motion. Motors are either fixed displacement, in which the quantity of fluid per unit of output motion is constant or variable displacement, in which the quantity of fluid per unit of output motion can be adjusted.

Hydraulic oil - an oil specially suited for use as either the specific gravity or the API gravity of a liquid.

Hydraulic pump - a device that converts mechanical force and motion into fluid power. Pumps are either fixed displacement or variable displacement types.

Hydraulic resistance - a means of representing the ability of piping components to remove head energy from a flowing fluid. The greater the hydraulic resistance of a component, the more head loss it creates for a given flow rate.

Hydraulic resistance diagrams - a drawing composed of hydraulic resistors that represent an assembly of piping components.

Hydraulics - branch of physics having to do with the mechanical properties of water and other fluids or liquids in motion.

Hydrocarbons - any compound whose molecular structure is made up exclusively of molecules of hydrogen and carbon. Hydrocarbons compounds can be gas, liquid, or solid, depending on the temperature and structure. All petroleum products are derived from combinations of hydrocarbons.

Hydrochloric acid - acid and acid fumes that are given off when a fluorocarbon refrigerant is being burned.

Hydrochlorofluorocarbons (HCFC) - molecules created when some of the hydrogen atoms in a hydrocarbon molecule are replaced with chlorine or fluorine atoms. Because they have a shorter life than CFC's, HCFC's are less harmful than CFC's to stratospheric ozone.

Hydrodynamic lubrication - a system of lubrication in which the shape and relative motion of the sliding surfaces causes the formation of a fluid film having sufficient pressure to separate the surfaces.

Hydrofluorocarbons (HFC) - molecules created when some of the hydrogen atoms in a hydrocarbon are replaced with fluorine. Because HFC's contain no chlorine, they don't destroy ozone but they may contribute to global warming.

Hydrofluoroether (HFE) - a nontoxic, secondary fluid used in refrigeration systems. HFE's have a zero ozone depletion potential.

Hydrogen - a basic element present in gas, coal, and fuel oil.

Hydrolyzing action - the corrosive effect within a refrigeration system caused by the chemical reaction between moisture and the refrigerant.

Hydrolysis - the splitting of chemical compounds by the reaction with water.

Hydrometer - a floating instrument used to measure specific gravity of a liquid. Specific gravity is the ratio of the weight of any volume of a substance to weight of equal volume of substance used as a standard.

Hydronic - a type of heating system that circulates a heated fluid, usually water, through baseboard coils. The circulating pump is usually controlled by a thermostat.

Hydronics - the science of heating and cooling with liquids.

Hydrostatic design stress - the estimated maximum tensile stress in the wall of the pipe in the circumferential orientation due to internal hydrostatic pressure that ca be applied continually with a high degree of certainty that failure of the pipe will not occur.

Hydrostatic pressure - extremely high and dangerous pressure created by the expansion of a liquid in an enclosed space, such as a cylinder.

Hydrostatic test - a water test, not to exceed 1.5 times the maximum design working pressure, applied to a pressure vessel. The recommended water temperature for this test is 70° F.

Hygrometer - (1) an instrument used to measure the degree of moisture in the atmosphere or (2) an instrument used to measure the specific gravity or "charge" of the electrolyte of a storage battery.

Hygroscopic - the ability of a substance to absorb and retain moisture and change physical dimensions as its moisture content changes.

Hygrostat - a control that actuates equipment when a specified humidity is attained.

Hysteresis - a lag in the production or the collapse of a magnetic field, due to friction between the molecules of the magnetized substance.

Hysteresis circuit - a special circuit of the differential thermostat that prevents the pump from being unnecessarily turned on and off immediately following an initial pump turn-on or turn-off event.

Hysteresis loop - a graphic curve that shows the value of magnetizing force for a particular type of material.

Hysteresis test - an electrical test used to check the integrity of centrifugal chiller tube bundles.

-I-

I - symbol used to designate the "intensity" of current.

IAQ - indoor air quality.

Ic - integrated circuit.

Ice - a term used to help remember that current (I) leads or is in front of voltage (E) in a capacitive circuit.

ICM - integrated controlled motor.

Icr - inductance, capacitance, resistance.

Id - inside diameter.

IFM - inducer fan motor or indoor fan motor.

Ime - ice melting equivalent.

Imp - impedance.

In - inch or inches.

In/lb - inch pounds.

I/O - input/output.

IPLV – integrated part load valve.

IPS - iron pipe size.

IPT - iron pipe thread.

IR - infrared radiation.

Inv - inverse.

I2r - formula for power in watts. I refers to the "intensity" of current measured in amps and the r refers to the resistance measured in ohms.

IBR gross output - an industry standard rating for boiler heat output that includes the heat transferred to water flowing through the boiler but does not include heat loss from the boiler jacket. This output is expressed in Btu/hr.

IBR net output - the IBR gross output rating minus 15 percent for the assumed heat loss of distribution piping, and delay in heat delivery due to the thermal mass of the boiler. This output is expressed in Btu/hr.

Ice - frozen water; the state of H_2O below 32° F or 0° C.

Ice bank - a thermal accumulator in which during off-peak periods of refrigeration demand, ice is formed, and in which, during peak periods of refrigeration demand, compressor capacity is supplemented by melting ice.

Ice box - a refrigerator or cooler in which the heat absorber is ice instead of an evaporator.

Ice cans - a mold in which large blocks of ice are formed by suspension in a chilled brine tank.

Ice cream cabinet - a commercial refrigerator that operates at approximately 0° F and is used for storage of ice cream.

Ice harvest switch - located at the end of the water plate, it resets the timer on the status indicator card after each ice harvest. It is used only with a status indicator.

Ice maker - a cyclic type automatic has separate and sequential water fill, freezing and harvesting phases of the ice making process.

Ice melting effect - see ice melting equivalent.

Ice melting equivalent (IME) - the amount of heat absorbed by melting one ton of ice at 32° F. This would absorb 288,000 Btu's (144 latent heat of fusion for ice x 2000 pounds) in a 24 hour period.

Ice point - the freezing temperature of water. The freezing temperature of water is 0° C, 32° F, or 273.16° K.

Ice ring - an accumulation of ice at the bottom of the outdoor coil due to incomplete defrost operation of a heat pump system.

Ice storage bin rating - the capacity of the ice bin of an automatic ice maker storage bin. The average weight of ice contained in the bin when the bin fill device terminates the ice making process.

Ice water thickness - a commercial classification applied to molded cork covering and similar insulation for refrigerant and chilled water lines. Somewhat thinner than "brine thickness".

Identification plate - a metal plate or sticker that provides information such as manufacturer, part number, and specifications. Frequently mounted on the outside housing of compressors and motors.

Identified - refers to equipment that is suitable for a specific purpose, function, use, environment, or application.

Identified conductor - the identified conductor is the insulated grounded conductor. For sizes #6 or smaller, the insulated grounded conductor shall be identified by a continuous white or gray outer finish or by three continuous white stripes on other than green insulation along its entire length. For sizes larger than #6, the insulated grounded conductor shall be identified either by a continuous white or gray outer finish or by three continuous white stripes on other than green insulation along its entire length, or at the time of installation by a distinctive white marking at its termination.

Identified terminal (capacitors) - the terminal on a capacitor that is marked to show that it is the terminal that is connected to the plate nearer to the outside casing. This terminal usually gets power from a supply side wire.

Identified terminal (electrical) - the identification of terminals to which a grounded conductor is to be connected shall be substantially white in color. The identification of other terminals shall be a readily distinguishable different color.

Idler - a pulley used on some belt drives to provide the proper belt tension and to eliminate belt vibration.

Ignite - to set fire to, to heat to a great degree.

Igniter - any electrical or mechanical device and related equipment used to ignite a fuel air mixture. Also, a burner (like a pilot) using gas or oil to ignite very large gas, oil, or pulverized coal burners in very large water tube boilers (normally field erected).

Ignition - the act of starting combustion. Also see automatic, continuous, direct spark, electric, inherent, intermittent, interrupted, manual, proved, proved spark, torch, and unproved ignition.

Ignition arch - this is made of refractory material that absorbs the heat from the fire and radiates it back to the green coal.

Ignition, automatic - a heating fuel ignition system that ignites fuel at the beginning of the cycle only, not constantly throughout the cycle.

Ignition bracket assembly - this assembly is made up of a bracket, an ignition transformer, a solenoid gas valve, an electric igniter, a fire safety switch, a twist lock receptacle, a terminal block and a burner latch.

Ignition electrode - a metal conductor used to generate a spark across a narrow gap to ground in order to ignite a fuel air mixture.

Ignition interference - electrical noise, caused by ignition current feeding through the flame to the flame rod, which is superimposed on the flame signal (flame rod detectors only). The interference may be additive or subtractive. (It may increase or decrease the flame signal.) Subtractive interference may cause the flame relay to drop out; severe additive interference may damage the flame relay unless an arc gap protector is in the circuit.

Ignition interference test - a test to make certain that ignition current is not interfering with the proper operation of a flame rod. See ignition interference.

Ignition point - the temperature at which a fuel/air mixture will light and continue to burn.

Ignition return - immediate restoration of ignition, without pre-ignition purging or interlock checks, after loss of the main burner flame. If the main burner flame is not re-established, safety shutdown occurs. Also called relight.

Ignition spark response test - a test for a flame safe guard control using an ultraviolet flame detector, to make certain that the ignition spark is not actuating the flame relay.

Ignition temperature - a term generally used to mean the lowest temperature that will cause a gas to ignite and burn. This is not a precise term or value and one subject to several variables. Used only as an indication. The ignition temperature and flammability limits of a fuel-air mixture, together, are a measure of the potential for ignition.

Ignition timing - a period of time during which ignition means are on. This provides a means of establishing and stabilizing the flame to ensure to ensure proper burning of the fuel when the main burner is firing.

Ignition transformer - a device that provides a high voltage spark (usually 6,000 to 10,000 volts) that will cause ignition.

Ignition trials - a period of time during which the pilot and main burner must be ignited (main burner only if using direct spark ignition). Also see pilot flame establishing period and main burner flame establishing period.

Ignition velocity - see flame propagation rate.

Igniter - any device used to light gas. A spark igniter uses an electric spark generated across an air gap for this purpose.

Illumination - the density of the luminous flux on a surface; it is quotient of the flux by the area of the surface when the latter is uniformly illuminated.

IME - ice melting equivalent. The amount of heat absorbed in melting ice at 32° F. It is equal to 144 Btu's per pound (latent heat of fusion) or 288,000 (144x2000) Btu's per day.

Immerse - to dip, stick or plunge into, as in a liquid.

Immersion - the act of putting an object into a liquid so that it is completely covered by the liquid.

Immersion type controller - a controller constructed with a flexible tube and a remote measuring element, immersed in a well in a pipe or tank.

Impact insulation class (**IIC**) - a single figure rating that provides an estimate of the impact sound insulating performance of a floor ceiling assembly.

Impact pressure - the total pressure of a flowing gas, including velocity and pressure above atmospheric.

Impact tool - a hand tool that converts a hammer blow to useful leverage.

Impact tube - the center tube of a pitot tube that measures both pressure and velocity.

Impedance - the total resistance to the flow of an AC current as a result of resistance, reactance or capacitance in the circuit. It is expressed in ohms and is designated by the symbol "Z".

Impedance protected motor - a motor designed to not burn out in less than 15 days under locked rotor (stalled) conditions, in accordance with UL Standard 519.

Impedance relay - a single pole single throw normally closed relay that has a high resistance coil that is wired to prevent compressor start-up after a safety control opens. Equipment operation will not start until a manual reset is performed. Also called a lockout relay or control.

Impeller - a series of rotating blades or vanes similar to the old-fashioned paddle wheel. In an open impeller, the impeller blades rotate between the stationary walls of the blower housing. The walls tend to channel the air so that most of it flows out through the tips of the blades, but some air slips out sideways from between the blades and short circuits back to the impeller inlet. A closed impeller has cover plate discs attached to the sides of the blades, and thus short-circuiting is minimized.

Impingement - a condition in a gas or oil furnace when the flame strikes the sides of the combustion chamber, resulting in poor combustion efficiency.

Impingement steam - steam that strikes the boiler heating surface, causing erosion of the boiler metal.

Impingement target burner - a burner consisting simply of a gas orifice and a target, with the gas jet from the orifice entraining combustion air in the open and the mixture striking and burning on the target surface. No usual burner body is used.

Implosion - a term describing the rapid and violent collapse of vapor pockets as the pressure around them rises above the vapor pressure of the fluid.

Impurities - dirt or other foreign matter mixed in a substance so as to make it unclean.

In contacts - those relay contacts that complete circuits when the relay coil is energized. Also referred to as normally open contacts.

In line pump - a centrifugal pump which has its inlet and its discharge in a straight line.

In phase - two waveforms of the same frequency that pass through their minimum and maximum values at the same time and polarity.

In sight - the NEC specifies that equipment shall be "in sight from," "within sight from," or "within sight" the equipment shall be visible and not more than 50 feet away.

Incandescence - the state of a material with such a high temperature that it gives off light.
Incandescent lamp - a lamp which gives light due to a filament glowing at a white heat.

Inch - an English standard measure. There are 12 inches in one foot. One inch (1 in.) is equal to 25.4 millimeters (25.4 mm) in the metric measuring system.

Inches of mercury - a unit used in measuring pressures in which atmospheric pressure is equal to 29.921 inches of mercury column.. One inch of mercury column equals a pressure of 0.491 lb/in.

Inches of water - a unit of pressure equal to the pressure exerted by a column of liquid water one inch high (25.4mm) at sea level and at a temperature of 39.2° F. One inch of water column equals a pressure of 0.578 ounce per square inch (.036 psi). One inch mercury column equals about 13.6 inches water column. Approximately 1/28 of one psi.

Inch ounce - a measure of the torque for a small motor. One ounce of force applied at 1" from the centerline of the shaft.

Inch pound - the amount of power needed to raise one pound one inch.

Incinerate - to burn to ashes, burn up.

Inclined - making an angle with another line or plane.

Inclined gauge - see inclined manometer.

Inclined manometer - an instrument that measures low gas pressure by means of a diagonal column of water or a special specific gravity fluid.

Incombustible material - material which will not ignite or actively support combustion in a surrounding temperature of 1200° F during an exposure of 5 minutes, and will not melt when temperature of the material is maintained at 900° F for a period of at least 5 minutes.

Incompatible - not suited to be used together. Not in harmony or agreement.

Incomplete combustion - combustion in which the fuel is only partially burned, and is capable of being burned further under proper conditions. It results from an inadequate amount of air for the quantity of fuel provided. It can also occur, even in excess air conditions, if mixing is poor and/or the flame is chilled below the ignition temperature before the full combustion reaction can take place. An example is the burning of carbon, producing carbon monoxide (CO); with more air, carbon dioxide (CO_2) would result.

Incompressible - the ability of liquids to transmit large pressures without themselves being compressed into smaller volumes.

Increaser - coupling that has a larger opening at one end (used to increase size of pipe opening).

Increment - amount of increase or addition.

Independent light off - separate ignition of each burner in a multi- burner system. Each burner can be started or stopped by itself.

Indicating controller - a controller equipped with a scale and an indicating pointer. It often shows the set point as well as the actual temperature of the medium being controlled.

Indicator, electrical - that part of the electrical system that shows if it is "on" or "off" (a light bulb) or indicates a specific quantity (like an electrical meter).

Indicator compounds - chemical compounds, such as carbon dioxide, whose presence at certain concentrations may be used to estimate certain building conditions.

Indirect conditioned space - an enclosed space within a building that is not a heated or cooled space, whose area weighted heat transfer coefficient to heated or cooled spaces exceeds that to the outdoors or to unconditioned spaces; or through which air from heated or cooled spaces are transferred at a rate exceeding three air changes per hour. Enclosed corridors between conditioned spaces shall be considered as indirectly conditioned space.

Indirect fired heater - a heater in which the products of combustion (flue gases) are not mixed with the medium being heated. Also see indirect fired external, indirect fired internal, and radiant tube type heaters.

Indirect fired external heater - any oven heating system in which the oven atmosphere is kept separate from the combustion gases, and in which the burners and combustion chamber are outside of the oven chamber. There are three types:

Recirculating - oven atmosphere is returned to the heater to be reheated.

Non-recirculating - oven atmosphere is not returned to the heater.

Internal radiator - products of combustion from the heater are circulated through radiator tubes located within the oven chamber.

Indirect fired internal heater - any oven heating system in which the oven atmosphere is kept separate from the combustion gases, and in which the burners are contained in gas tight radiators within the oven chamber. There are two types:

Explosion resisting - constructed to withstand explosion pressure from ignition of a gas air mixture in the radiators.

Non-explosion resisting - with gas tight radiators that are not designed to withstand an internal explosion.

Indirect system - a solar heating or cooling system in which the solar heat is collected outside of the building and moved inside through ducts or piping, with fans or pumps through ducts or piping.

Indirect water heater - a coil or bundle of tubes, usually copper, surrounded by hot boiler water. The domestic water is within the tube and is heated by transfer of heat from the hot boiler water surrounding the tube.

Indirectly conditioned space - enclosed space within a building that is not heated or cooled space, whose area-weighted heat transfer coefficient to heated or cooled space exceeds that to the outdoors or to unconditioned space, or through which air from heated or cooled space is transferred at a rate exceeding three air changes per hour.

Indoor air - that air that occupies the space within the interior of a house or other structured buildings.

Indoor air quality (IAQ) - the state of the environment in a space occupied, or intended to be occupied, by human beings specifically concentrating on the constituents in the air that are not considered to be part of normal, fresh outside air.

Indoor coil - a heat pump component that is either an evaporator or a condenser, depending upon whether the system is operating in the heating or cooling mode.

Indoor design temperature - the indoor air temperature used when calculating the design heat loss. The indoor design temperature is usually assumed to be 70° F.

Indoor fan relay - an electric relay that starts and stops an indoor fan on cooling, electric heating, and heat pump systems.

Indoor outdoor temperature difference - the temperature of the indoor air minus the temperature of the outdoor air.

Indoor unit - the component located inside the house which contains the indoor heat exchanging coil, fan, motor, filtering and possibly a source of heat.

Induced air - air that flows into a furnace because the furnace pressure is less than atmospheric pressure. This includes air brought into a furnace by entrainment in a high velocity stream.

Induced current - the current that flows as the result of an induced electromotive force.

Induced draft - air movement into and through the combustion chamber and breeching, produced by a partial vacuum within the combustion chamber created by a fan at the outlet of the chamber.

Induced draft burner - a gas burner that uses an induced draft to bring in combustion air. Also see induced draft and mechanical draft burner.

Induced draft cooling tower - a cooling tower in which the flow of air is created by one or more fans drawing the saturated air out of the tower.

Induced draft fan - a fan or blower that produces a negative pressure in the combustion chamber, causing a suction that draws in air.

Induced magnetism - the ability of a magnetic field to produce magnetism in a metal.

Induced voltage - the potential that causes induced current to flow through a conductor that passes through a conductor that passes through a magnetic field.

Inductance - the inherent property of an electrical circuit that opposes a change in current. The property of a circuit whereby energy may be stored in a magnetic field. Inductance is present only when the current is changing.

Induction - the phenomenon of a voltage, magnetic field, or electrostatic charge being produced in an object by lines of force from the source of such fields.

Induction air - the process of drawing room air into the projected air stream due to the velocity of the projected air stream. Also called aspiration.

Induction coil - a device for obtaining high voltage. It contains a primary coil with only a few turns of heavy wire and a secondary coil with many turns of fine wire.

Induction extraction turbine - a steam turbine with the capability of either exhausting or admitting a supplemental flow of steam through a intermediate port in the casing, thereby maintaining a process heat balance. Turbines of the extraction and induction/extraction type may have several case openings, each passing steam at a different pressure.

Induction motor - an alternating current motor that operates on the principle of rotating magnetic field. The rotor has no electrical connection, but receives electrical energy by transformer action from the field windings.

Induction start induction run motor - a motor that starts and runs solely on the torque generated by the difference in magnetic field between the start and run windings. Also known as a split phase motor.

Induction turbine - a steam turbine with separate inlets for steam at two pressures, with an automatic device for controlling the pressure of the secondary steam induced into the turbine and means for directly regulating the flow of steam to the turbine stages below the induction opening.

Induction unit - a factory made assembly consisting of a sound attenuator, a manual control damper and one or more nozzles used to develop a high jet velocity of the externally supplied primary air. The high velocity air induces secondary or room air to mix with the primary air providing a high volume of total air to be delivered from the unit. In some types the secondary air is drawn through a coil and either heated or cooled.

Inductive circuit - any circuit that contains at least one magnetic load.

Inductive heating - a method of heating conductors in which the material to be heated is placed inside a coil of wire and high frequency alternating current voltage is applied.

Inductive load - a lagging load; a load that is predominantly inductive, so that the alternating current lags behind the alternating voltage, i.e., the current does not change direction until after the voltage does. Power consumption for inductive loads is calculated by volts times amperes times power factor of the load.

Inductive pick up - the generation of a voltage in one conductor by induction from another nearby conductor.

Inductive reactance - electromagnetic induction in a circuit creates a counter or reverse (counter) EMF (voltage) as the original current changes. It opposes the flow of alternating current.

Inductor - a coil of wire that has the property of inductance and is used in a circuit for that purpose. Also called coils or chokes.

Industrial - a term used to describe manufacturing and processing industries.

Industrial air conditioning - air conditioning for other than human comfort conditions.

Industrial furnace - a process type heater that changes the characteristics of the material being processed. It is usually operated at temperatures above 1200° F.

Industrial oven - a process type heater normally used for drying. It is generally operated at temperatures from 400° F to 800° F.

Industrial process refrigeration - complex, custom appliances used in the chemical, pharmaceutical, petrochemical, and manufacturing industries designed for a specific purpose. Temperature range is usually from -70° F to +60° F. These systems are usually field fabricated and installed which introduces other problems and supervision by the customer. The majority of industrial refrigeration systems use NH_3 (ammonia) as the refrigerant.

Inert gas - a gas, usually nitrogen, that is chemically inactive in a refrigeration system. Used as a substitute for refrigerant in drying a system or while shipping components.

Inertia - that property of a body (mass) at rest to resist movement and of a body in motion to continue in motion, so that some external force must be applied to move a body at rest and to slow down or stop a body in motion.

Inerting - introducing an inert medium into a furnace, pulverizer, or other enclosure to reduce the oxygen content of the resulting mixture below the limits of flammability.

Inerts - noncombustible substances in a fuel.

Infiltration - air leakage inward through openings in walls, floors, or ceilings from unconditioned space or from outside of a building. It results from the forces of wind, temperature differences, and HVAC operation due to pressure differences.

Infinite resistance - term used to describe a circuit with an energized power supply but with no current flow. Also referred to as an open circuit.

Infinity - without end, limits or bounds. Numberless and immeasurable. Symbol is ∞. This symbol is shown on electrical meters when trying to ohm out a motor or relay coil that has an open in it.

Inflammable - capable of being ignited and burned; synonym for flammable and combustible.

Infrared - an area in the electromagnetic spectrum extending beyond red light from 760 nanometers to 1000 microns.

Infrared burner - a type of burner in which the main burner flame heats a ceramic or metal surface which then emits infrared radiation. Infrared does not heat the air that passes through, but rather heats objects which it hits.

Infrared flame detector - a detector that responds to the infrared radiation (wavelength between 0.75 and 1000 microns) emitted by a flame. Also see lead sulfide cell.

Infrared heater - a heater that directs a substantial amount of its energy output in the form of infrared energy into the area to be heated. Such heaters may be of the vented or unvented type.

Infrared humidifier - a humidifier that has infrared lamps with reflectors to reflect the infrared energy onto the water. The water evaporates rapidly into the duct air steam and is carried throughout the conditioned space.

Infrared lamp - an electrical device that emits infrared rays. These are invisible rays just beyond red in the visible spectrum.

Infrared radiation (IR) - invisible, electromagnetic waves of low frequency (long wavelength-between 0.75 and 1000 microns). They are generated by thermal agitation and radiated by everything with a temperature above absolute zero (minus 273° C). The hotter the object, the greater the infrared radiation. Infrared radiation comprises about 90% of the light spectrum, ranging from the low end up to the white visible light range. Also see hot refractory hold in test and hot refractory saturation test.

Infrared rays - the rays that transfer heat by radiation.

Infrared spectroscopy - an analytical method using infrared absorption for assessing the properties of used oil and certain contaminants suspended therein.

Inherent ignition - feedback of part of the heat generated by the flame to ignite more of the combustible mixture; it must keep the fire going after ignition is terminated.

Inherent protector - a temperature and current sensitive device placed in a hermetic compressor for motor overload protection.

In. Hg. - the abbreviation used to indicate inches of mercury in vacuum measurements. Hg is the chemical symbol for mercury. To pull a complete vacuum in a refrigeration system, the pressure inside the system must be reduced to 29.921 inches Hg. vacuum.

Inhibitor - a substance added to oil, water, gas, etc. An item used to prevent unwanted action such as rusting, foaming, etc.

Initial resistance - a new, clean filter's resistance to flow.

Initiate - to begin an action or sequence of events such as a defrost cycle.

Injection burner - see inspirating burner.

Injection control device - a device such as a variable speed pump or modulating valve that regulates the rate at which hot water is injected into a circulating distribution system.

Injection mixing - a method of controlling the temperature of a hydronic distribution system by adding (injecting) hot water and simultaneously removing cooler water. The faster the injection rate, the warmer the distribution system becomes and the greater the heat output.

Injector - jet used for pumping.

Inlet - that part of a valve through which a fluid enters.

In-line ammeter - a current measuring device that is connected in series with the circuit being tested.

Inline circulator - a circulator constructed such that its inlet and outlet ports lie on a common centerline.

Inline fan (tubular centrifugal fan) - a fan where air flow is developed as in a centrifugal blower, but after leaving the impeller the air is contained in a tubular housing and, by means of turning vanes, is discharged in an axial direction. Employs single inlet centrifugal wheels, usually with backward inclined blades. The tubular centrifugal fan has performance characteristics similar to a centrifugal blower and the compact physical configuration of the tubeaxial fan. Can be vertically or horizontally mounted, thus providing a simpler installation by minimizing need for duct turns and transitions.

Inline fuse - a low amperage fuse placed in the low voltage control circuit that protects the control transformer from excessive current.

Inner cone - the smaller flame that does most of the heating in oxy/acetylene brazing.

Inner core/liner - that portion of insulated flexible duct which determines the internal dimensions and inner physical form.

In phase - two waveforms of the same frequency which pass through their minimum and maximum values at the same time and polarity.

Inorganic compounds - those compounds that contain no carbon atoms. Refrigerant examples would be ammonia and water.

Input combustion control - see firing rate controller.

Input output board - a solid state electronic board that receives a signal from a microprocessor and initiates a response to cycle compressors or initiate a defrost cycle.

Input impedance - resistance or opposition to current flow in an AC or DC circuit.

Input rate - (1) the power that is applied to a piece of equipment or to the terminals on the piece of equipment to which the power is being applied. (2) The fuel rate going into a heating appliance.

Input rating - the gas burning capacity of an appliance in Btu/hour as specified by the manufacturer. Appliance input ratings are based on sea level operation up to 2,000 feet elevation. For operation at elevations above 2,000 feet, input ratings should be reduced at the rate of 4% for each 1,000 feet above 2,000 feet.

Input sensor - a device that can detect the presence, or absence, of a particular type of stimulus. It may also be a transducer, converting the stimulus to some other form of energy.

Input signal - the variations of the input voltage or current. The alternating current portion of the signal.

Input transducer - a transducer that changes some other form of energy into electrical energy.

Input voltage - the amount of voltage connected to a device or circuit.

Inrush current - in a solenoid or coil, the steady-state current drawn from the line when the armature is in its maximum open position. Allowable inrush can be up to 10 times the pilot duty rating (running current).

Inscribe - to draw one figure within another figure.

Insertion element - a tube-shaped thermostat sensing element designed for use in liquids or in air.

Insertion thermostat - controllers with extended elements that can be inserted into a duct or other enclosure in which temperature is to be maintained.

Insertion loss - the positive or negative change in acoustic power transmission that results when the element is introduced.

Insertion type controller - a controller mounted on a duct or pipe, with its rigid measuring element inside the duct or pipe.

Inshot burner - a gas burner in which the gas is delivered from the side(s) of the combustion chamber, resulting in a horizontal flame.

Inside calipers - a hand tool used to measure material on the inside. Hole diameter can be measured by placing a caliper inside of it.

Inside diameter - measurement across the widest part of a circular object, from inside surface to inside surface.

Inside micrometer - a hand tool used to measure the inside diameter of objects such as rings and cylinders.

Inside stop - the inside edge of a double hung window. The expandable partitions of a window air conditioner expand to touch each side of the inside stop.

Inside wrench - a wrench that grips a pipe or fitting from the inside to prevent scoring of the outer surface, especially threads and chrome.

In sight - where the National Electric Code specifies that one equipment shall be "in sight from," "within sight from," or "within sight," etc., of another equipment, the specified equipment is to be visible and not more than 50 feet from the other.

Insolation - the solar energy received by a given surface.

Inspection - the measuring and checking of equipment or installations to determine whether they have been serviced or installed per specifications.

Inspection plates - hand hole covers which, when properly removed, make possible the inspection of the internal parts of equipment.

Inspirating burner - a gas burner that uses venturi action to bring in combustion air. Also called an injection burner or venturi mixing burner. Also see atmospheric burner and venturi.

Inspirator mixer - a mixing device in which a jet of gas is used to entrain primary air at a rate proportional to the gas flow rate. Also see inspirating burner.

Instability - the characteristic of a control system with just enough variation in maintaining the set point to distinguish it from sluggishness or cycling.

Installation diagram - a diagram that shows little internal wiring but gives specific information as to terminals, wire, sizes, color coding, and breaker or fuse sizes.

Installation instructions - a manufacturers printed instructions in accordance with the terms of their listing which advises the method and materials to install the piece of equipment or part.

Instantaneous hot water heater - one than contains no more than one gallon of water per 4,000 Btu/h of input.

Instantaneous peak - the maximum demand at the instant of greatest load, usually determined from the readings of indicating or graphic meters.

Instantaneous voltage - a value at any instant (time) along a waveform.

Instrument - used broadly to denote a device that has measuring, recording, indicating, and/or controlling abilities.

Instrument air - separate air system that operates instrumentation in a plant. Instrumentation air and utility air must be kept separate.

Instrument channel - channel shaped tray that supports instrument lines. Similar to an electrical tray.

Instrument piping - the tubing, piping and electrical wiring used to measure pressure, heat sensing and electrical instruments.

Instrumentation - the application of industrial instruments to a process or manufacturing operation.

Instrumentation drawing - drawing which establishes points of connection, size, location, type, and other necessary information for the installation and operation of instruments.

Insulate - to bar the flow of electrical current by the use of a non-conducting material.

Insulated junction - the measuring junction in a thermocouple that is insulated from the sheath material.

Insulating glass unit - a combination of two or more panes of glass with sealed air space(s) between panes --- dual panes, triple panes, and quad panes. Also referred to as thermopanes and sealed units.

Insulation - a material having a resistance to heat flow. Materials having high R-values or low U values, the better the heat retarding ability of the material.

Insulation baffle - a rigid material, resistant to wind driven moisture, allows air to flow freely into the attic or crawl space and to prevent insulation from blocking the ventilation of these spaces, or the loss of insulation. These baffles could be made of sheet metal or wax impregnated cardboard.

Insulation classes - electric motor insulation is rated by its temperature capability for reasonable motor life. The classes are Class A (105° C), Class B (130° C), Class F (155° C) and Class H (180° C). These are total temperatures, not temperatures over ambient.

Insulation, brine thickness - a commercial classification applied to molded core covering and similar insulation for refrigerant and brine lines.

Insulation, electric - materials which are poor conductors of electricity and which are used to cover wires and components to prevent short circuits and accidental shock hazards.

Insulation, fill - granulated, shredded, or powdered material, prepared from vegetable, animal, or mineral origin. It can come in bulb or batt form.

Insulation, ice water thickness - a commercial classification applied to molded core covering and similar insulation for refrigerant and chilled water lines.

Insulation pins - a spindle-shaped metal or other material with serrations near the pointed end for attaching washer type insulation retainers.

Insulation resistance - that property of an insulating material which resists electrical current flow through the insulating material when a potential difference is applied.

Insulation, sound - acoustical treatment of fan housings, supply ducts and other parts of the system and equipment for isolation of vibration, or to reduce transmission of noise.

Insulation tape - a plastic like tape used to insulate bare or exposed wire.

Insulation, thermal - a substance that is used to retard or slow down the flow of heat transfer through a wall or partition.

Insulator - substances containing very few free electrons and requiring large amounts of energy to break electrons loose from the influence of the nucleus.

Intake (valve) - opening through which a fluid is drawn into a component.

Integral - whole number as integral horsepower motors are motors of one horse power and larger.

Integral horsepower motor - any motor greater than one horsepower.

Integral tank - any fuel tank furnished by the manufacturer as a component part of the complete assembly; either attached to the appliance, or intended for separate mounting adjacent to the appliance.

Integrated circuit - a circuit that incorporates multiple transistors and other semiconductors to a single circuit, sometimes called a "chip".

Integrated circuit board - electronic circuit made from transistors, resistors, etc., all placed into a package called a "chip," since all circuits are on one base of semiconductor material.

Integrated demand - the average kilowatt demand determined by an integrating demand meter or by the averaging of a load curve over any interval of stated length. It is the average of the continuously varying instantaneous demands during a specified demand interval.

Integrated part load value (IPLV) - single number figure of merit based on part load EER or COP expressing part load efficiency for air conditioning and heat pump equipment on the basis of weighted operation at various load capacities for the equipment.

Integrated system - a solar heating or cooling system in which the solar heat is absorbed in the walls or roof of a dwelling and flows to the rooms without the aid of complex piping or ducts.

Integrator - a calculating device used on differential pressure flow meters to determine hourly or daily flow rates.

Intelligent mixing device - a mixing device that senses fluid temperature and is able to adjust one or more flow rates based on how the sensed temperature compares to a set target value. Such devices may be electronic or use non-electric thermostatic actuators.

Interchangeable - a part that has been made to specific dimensions and tolerates and is capable of being fitted into equipment in place of a similarly made part.

Interconnecting wire - a type of wire for external use in electronic equipment where exposed to physical abuse. Interconnecting wire encompasses both control and power circuits.

Intercooler - the cooling coil between stages in a multistage compression system.

Interface - the interrelationship or means of communication between any two (or more) functions, modules, components, or communication devices.

Interference - the thermal contamination of a tower's inlet air by an external heat source. Example would be the discharge plume of another water tower.

Interior circuits - a radiant floor tubing circuit that is not run adjacent to exterior (or otherwise exposed) walls.

Interior design drawing - drawing which includes information of wall, floor and ceiling finishes, and indicates material, colors, etc. Also shown on these drawings are furniture or movable objects.

Interlock - an automatic controller or switch which monitors the physical state of a required condition and furnishes proof to the flame safe guard control that the condition is proper for burner operation to start or to continue. Also see high fire, lock out, low fire, pre-ignition, re-cycle, running, start, and valve closed interlocks.

Interlocking systems - an arrangement of interlocks to insure that all inter-related parts of an air conditioning or refrigeration system are operating properly together.

Intermittent cycle - a cycle that repeats itself at different intervals.

Intermittent ignition - ignition by an energy source which is automatically energized each time there is a call for heat, and which is maintained during the entire run period. This ensures that a means of igniting the incoming fuel will always be present.

Intermittent pilot - a pilot automatically ignited each time there is a call for heat and maintained during the entire run period. It is shut off with the main burner at the end of heat demand.

Internal combustion engine - an engine driven by expansion resulting from the burning of fuels confined within the engine.

Internal compressor overload - an overload that is embedded in the windings of a motor. Some internal overloads break the power to the motor directly (direct style) while others merely open a set of contacts that is wired into an electric control circuit (pilot duty style).

Internal control - the methods and procedures adopted within a business to safeguard assets and control its operation.

Internal energy - the sum of all the kinetic and potential energy contained in a substance due to the states of motion and separation of its several molecules, atoms and electrons. It includes sensible heat (vibration energy) and that part of the latent heat that is represented by the increase in energy during evaporation.

Internal feedwater line - perforated line located at the normal operating water line (NOWL) in the boiler that distributes the relatively cool feed water over a large area to prevent thermal shock to the boiler metal.

Internal heat exchanger - a heat exchanger contained within a storage tank and relying on natural convection for heat exchange with the water in the storage tank.

Internal heat gain - heat produced within the conditioned space by people, electric appliances, lights, cooking and/or other heat and moisture producing factors.

Internal heater - a heater in which the burner(s) and combustion chamber are within the oven chamber and in contact with the medium being heated. Also see direct fired internal and indirect fired internal heaters.

Internal mix oil burner - see atomizing oil burner.

Internal overflow - a pipeline located in an open feedwater heater that prevents the water level from exceeding a fixed level and flooding the system.

Internal overload - a heat sensitive control imbedded in motor windings; opens in case of excessive winding temperatures, or current draw.

Internal process heat - heat from industrial activities and sources such as wastewater, boiler flue gas, coolants, exhaust air, and some waste materials. This heat is normally wasted unless equipment is included to extract it for further use.

Internal threads - female threads or threads that are not exposed.

Internal treatment - the addition of chemicals directly into the boiler water to control pitting, scale, and caustic embrittlement.

Internally equalized - describes a thermostatic expansion valve that senses evaporator inlet pressure under its diaphragm.

Internally equalized valve - in a thermostatic expansion valve, an integral part or passage that provides exposure of the actuating element (diaphragm or bellows) to pressure leaving the valve.

Internally fused - this is the fusible link built inside of a run capacitor that helps stop motor winding damage if a capacitor shorts out.

Interpolation - a mathematical method of estimating a numerical value that is know to be between two other stated values. It is based on proportioning.

Interrupt - to stop a process or cycle in such a way that it can be resumed at a later time.

Interrupted ignition - ignition by an energy source which is automatically energized each time there is a call for heat, and cut off automatically after flame is proved or at the end of a predetermined time.

Interrupted pilot - a pilot automatically ignited each time there is a call for heat and cut off automatically at the end of the main burner flame establishing period.

Interrupted rating - the highest current at rated voltage that a device is intended to interrupt under standard test conditions.

Interrupter - a switching device that opens and closes a circuit.

Interruptible supply - refers to a fuel supply agreement whereby the customer's supply may be cut off in case of shortage; usually available at a reduced rate. Could be either the electrical or natural gas supply.

Interrupting capacity - the maximum short-circuit current that a circuit breaker can safely interrupt.

Interrupting rating - the highest current that a fuse or circuit breaker will permit under ideal or test conditions.

Interstage differential - in a multistage HVAC system, the change in temperature at the thermostat needed to turn additional heating or cooling equipment on.

Interzone heat transfer - heat movement across interior partitions, floors, and ceilings separating building areas that are kept at different temperatures.

Intrinsic semiconductor - a semiconductor that has no impurities in its crystal structure.

Intrinsically safe - an instrument designed and constructed to practically preclude the possibility of igniting a combustible mixture.

Invar - a metal that is a alloy of iron and nickel. Used as a bimetal for thermostat and temperature sensing devices.

Inverse - the value of one divided by some quantity, such as 1/Rt for finding parallel resistances.

Inversely proportional - when one quantity increases or decreases, causing another quantity to do the opposite.

Inverted trap - a trap in refrigeration piping to prevent slugging of refrigerant or oil back to the compressor.

Inverter - (1) as used in relation to power supplies, a device that changes DC to AC or AC to DC. (2) a device that alters the frequency of an electronically altered sine wave, which will affect the speed of an alternating current motor.

Inverting buffer - a one input gate circuit, the output of which is the opposite of the input. Such a stage can drive a number of outputs, also different types of outputs, with minimum loading effects.

In wall evaporator - this type of evaporator is used mainly on domestic freezers where the tubing is attached to the outside walls of the inner liner.

I/O device - input/output device. A keyboard, magnetic disk, printer, CRT terminal or similar devices that transmit data to or receives data from a computer or secondary storage device. In a more general sense, any digital device, including a single integrated circuit chip, that transmits data to or receives data or strobe pulses from a computer.

Ion - an atom that has lost or gained some electrons. It may be positive or negative depending on the net charge.

Ion exchange - the occurrence when one ion or group of ions replaces another ion or group of ions in a solution.

Ionic bonding - a type of atomic bonding that occurs when one atom donates valence electrons to another atom.

Ionization - a process by which electrons are detached from atoms or molecules, thus originating ions and free electrons. This is achieved by collisions or by high energy radiation.

Ionization voltage - the potential at which a material ionizes. The electrical potential at which an atom gives up an electron.

IP units - inch pound units are based on the English measurement system.

IR drop - an electrical term indicating the loss in a circuit expressed in amperes times resistance (IxR) or voltage drop.

Ired - an infrared light emitting diode, the output of which is in the infrared portion of the frequency spectrum. These are usually made of gallium arsenide or gallium arsenide plus silicon.

Iron pipe size (IPS) - the size of the pipe measured on the inside diameter.

Iris slide - a thin plate with a small aperture.

Is, iso - prefix meaning constant; as isothermal, constant temperature; isentropic, constant entropy; isobaric, constant pressure; etc. In chemicals, one having different characteristics but with the same number and kind of atoms.

Isentropic - a process carried out reversibly without energy interchange as heat. Also a process carried out with no entropy change.

Isolate - to set apart from others, place alone or separate. To find the problem in an electrical circuit.

Isolated - not easily accessible. Requiring special ways of clearance to obtain entry.

Isolated contacts - a term usually applying to relays (or switches) which are supplied with no instrument power to them and available to the user for switching power circuits of the person's choice.

Isolated ground receptacle - a grounding type device in which the equipment ground contact and terminal are electrically isolated from the receptacle mounting means.

Isolated limited secondary circuit - a circuit of limited energy derived from an isolated secondary winding of a transformer having a maximum capacity of 100 volt amperes (VA) and an open circuit secondary voltage rating not exceeding 1000 volts.

Isolation flanges - special flanges that contain a built-in shut-off valve.

Isolation relays - components used to prevent stray unwanted electrical feedback that can cause erratic operation.

Isolation transformer - a transformer with a 1:1 turn ratio used to isolate an AC power line from equipment with a chassis ground. A transformer where the secondary winding is electrically isolated from its primary winding.

Isomers - molecules that have the same numbers of the same atoms, but the atoms are arranged differently in their structure. Even though isomers of the same compound have equal numbers of atoms of the same element, they have very different physical properties. ASHRAE appends a lower case letter to the number designation to distinguish different isomers as in R-134 and R-134a. One letter is used for ethanes (R-134a) while two is used for propanes (R-236fa).

Isometric - (1) a process carried out at a constant volume. (2) architectural or mechanical drawing showing both horizontally and vertically views.

Isometric drawing - this is a pictorial in which the sides of the object are shown 120E apart.

Isopiestic process - a process carried out at constant pressure.

Isosceles triangle - any triangle that has two equal length sides, hence two equal angles.

Isothermal - changes of volume or pressure under conditions of constant temperature.

Isothermal block - a barrier that keeps the temperature at the thermocouple junction the same as the junction between the thermometer and the thermocouple. No additional voltage is created by the second junction and measurements are not effected.

Isothermal process, expansion and contraction - any constant-temperature process, such as expansion or compression of gas, that is accompanied by heat addition or removal at a rate just adequate to maintain the constant temperature.

Isotherms - lines connecting points having the same temperature. Isotherms are frequently used to display the results of finite element analysis of heat transfer through solid materials.

Isotron - trademark name of refrigerant manufactured by Pennsalt chemicals.

Iteration - a design process that entails making successively refined estimates until a mathematically stable operating condition is determined.

I-V curve - a graphical presentation of the current versus the voltage from a photovoltaic cell as the load is increased from the short circuit (no load) condition to the open circuit (maximum voltage) condition. The shape of the curve characterizes cell performance.

-J-

Jack stud - vertical framing members, generally 2x4's, which form the inside of the window or door rough opening. They support the header and run down to the sole plate.

Jacket - the outer non-metallic coating or covering applied over insulated wire or cable.

Jacketing - surrounding by a control bath or stream of fluid for temperature control or heat absorption.

Jalousie - a shutter or blind with fixer or adjustable slats that exclude rain and provide ventilation, shade, and visual privacy.

Jamb - the main vertical members forming the sides of a window or door frame. Side jambs are the vertical member at either side of a door or window frame. The head jamb is the horizontal piece across the top.

Jaw-type pipe wrench - wrench with serrated jaws to grip pipe.

J box - a metal or plastic box where electrical connections are made.

Jellif alloy "K" - a chromium aluminum alloy iron wire that is resistant to oxidation up to 2200° F (1200° C); used for flame rods in flame detectors.

Jet burner - a burner in which streams of gas or air-gas mixtures collide in air at some point above the burner ports and burn there.

Jet cooling system - jet pump is used to produce a vacuum so water or refrigerant may evaporate at relatively low temperatures. These systems usually require a large condenser and have a low efficiency to remove heat.

Jet pump - device that uses the energy of another pressure for pumping a second fluid.

Jet velocity - this is the maximum velocity of air coming out of an outlet.

Job made ladder - a ladder that is fabricated by employees, typically at the construction site. Not commercially manufactured.

Jog mode - a system of motor operation in which the motor turns only as long as the start button is pressed. It stops when the start button is released.

Joint - the space between adjacent surfaces or the place where two members or components are held together by nails, fasteners, cement, mortar, etc.

Joint, adhesive - a joint made in plastic pipe by the use of an adhesive substance that forms a continuous bond between the mating surfaces without dissolving either one of them.

Joint, bellows - an item of equipment used to compensate for the expansion and contraction of a run of pipe. The device is built with a flexible bellows that stretches or is compressed as necessary to accept the movement of the piping.

Joint, brazed, high temperature - a gas tight joint obtained by the joining of metal parts with metallic mixtures or alloys which melt at temperatures below 1500° F, but above 800° F.

Joint clearance - the space between two parallel surfaces of a joint.

Joint compound - a substance applied to threaded fittings to ensure a leak proof joint.

Joint, flanged - a joint made by bolting together two flanged ends.

Joint, flared - a metal-to-metal compression joint in which a conical spread is made on the end of a tube that is compressed by a flare nut against a mating flare.

Joint, insulating - a fitting made of material that will not conduct electricity, thereby preventing the flow of current through a piping system.

Joint, leaded - a joint that is sealed with poured hot lead.

Joint, mechanical - a gas tight joint obtained by the joining of metal parts through a positive holding mechanical construction (such as flanged joint, screwed joint, or flared joint).

Joint, pipe - the point of connection between two piping elements.

Joint, plastic adhesive - a joint made in thermoset plastic piping by the use of an adhesive substance which forms a continuous bond between the mating surfaces without dissolving either one of them.

Joint, plastic heat fusion - a joint made in thermoplastic piping by heating the parts sufficiently to permit fusion of the materials when the parts are pressed together.

Joint, plastic solvent cement - a joint made in thermoplastic piping by the use of solvent or solvent cement that forms a continuous bond between the mating surfaces.

Joint, slip - a joint in which the provision for expansion and contraction consists of a cylinder that moves in and out of the main body of the device.

Joint, soldered - a gas tight joint obtained by the joining of metal parts with metallic mixtures or alloys that melt at temperatures below 800° F.

Joint, solvent cement - a joint made in thermoplastic piping by the use of a solvent or solvent cement that forms a continuous bond between the mating surfaces.

Joint, welded - a gas tight joint obtained by the joining of metal parts in the plastic or molten state.

Joist - one of the series of parallel beams of timber, reinforced concrete, or steel used to support floor and ceiling loads, and supported in turn by larger beams, girders or bearing walls.

Jogging - the quickly repeated closure and opening of the circuit to start a motor from the rested position for the purpose of accomplishing small movements of the driven machine.

Joule - a scientific unit of heat. The amount of heat needed to raise the temperature of 1 Kilogram of water 1° C is 4.187 Kilojoules. The quantity of heat equal to one watt-second. The basic unit of thermal energy.

Joule-Thomson effect - a change in temperature of a gas on expansion through a porous plug from a high pressure to a low pressure.

Joule's Law - the law which states the relationship between the heat produced by current flow and the amperage, resistance and time span involved.

Journal - that part of a rotor that is in contact with or supported by a bearing in which it revolves.

Journal bearing - a sliding type of bearing having either rotating or oscillating motion and in conjunction with which a journal operates. In a full or sleeve type journal bearing, the bearing surface is 360° in extent. In a partial bearing surface is less than 360° in extent, i.e., 150°, 120°, etc.

Journal, crankshaft - that part of the shaft or axle that rotates or angularly oscillates in or against a bearing or about which a bearing rotates or angularly oscillates.

Journeyman - a tradesperson who has served his or her apprenticeship training or learning of 3-5 years and is capable of completing an installation without supervision. These duties include the installation and servicing of residential or commercial applications.

Jumper - a short length of wire used to complete a circuit temporarily or to bypass part of a circuit. Also the act of using a jumper.

Jumper bars - flat copper pieces used to connect different motor terminals for easy change from one voltage to the other.

Jumper out - to temporarily remove an electrical component from an electrical circuit by connecting a wire across the component's contacts. For test purposes only.

Junction - a point in a circuit where the current branches out into other sections.

Junction block - an accessible block for the connection of two or more wires for the convenience of checking electrical circuits.

Junction box - a box used to connect sections of electrical conduits or wires, at the same time providing access to the enclosed conductors through a removable cover on the junction box for connecting and branching the enclosed conductors.

Junction capacitance - the effective capacitance that a P-N junction produces.

Junction diode - a diode that is made by joining together two pieces of semiconductor material.

Junction field effect transistor - a field effect device that uses a reversed biased P-N junction gate.

-K-

K - Kelvin, kilo.

KB - 1000 British thermal units (ASME)

Kg - kilogram.

Khz - kilohertz.

Ko - knockout or Kilohms.

KPa - kilo-pascals

Kv - kilovolt.

Kva - kilovolt amps.

Kvah - kilovolt ampere-hour.

Kvar - kilovar.

Kw - kilowatt.

kWh - kilowatt hour.

K factor - the ability of a material to conduct heat. Expressed in Btu's per hour per square foot, per inch of thickness of the material.

Kanthal A-1 - a stainless steel alloy used in the manufacture of flame rods. It has a maximum continuous operating temperature rating of 2462° F (1350° C).

Kata thermometer - a large bulb alcohol thermometer used to measure air velocities or atmospheric conditions by means of the cooling effect.

Kelvin scale - the thermometer scale on which the unit of measurement equals the Centigrade degree and according to which absolute zero is 0° degree, the equivalent of -273.16° C. Water freezes at 273.16 K and boils at 373.16 K. The degree symbol is not used with the Kelvin temperature scale.

Kerf - a groove or cut made by a saw.

Kerosene - an oil or liquid product of petroleum that does not emit a flammable vapor below a temperature of 115° F when tested in a Tag closed cup tester (ASTM D56).

Key - a small piece of metal partially in the shaft and partially in the pulley to prevent rotation of either the shaft or the pulley from each other.

Keyway - (1) the recess or slot in a shaft that holds the key. (2) a slot used to interlock slabs of masonry or concrete walls built at different times.

Kick-back diode - a diode that is made by joining together two pieces of semi-conductor material.

Kick space heater - a small horizontal fan coil convector that mounts in the recessed space at the base of a kitchen or bathroom cabinet.

Kiel - a crayon used for making marks on pipe. Kiel, unlike ordinary chalk, will not crumble.

Kilo - prefix meaning one thousand. Twenty kilovolts means twenty thousand volts.

Kilocycle - one thousand cycles.

Kilogram - one thousand grams.

Kilohertz - one thousand hertz.

Kilojoule - in refrigeration work, the unit used to measure quantities of heat.

Kilometer - a metric unit of linear measurement at equal 1000 meters.

Kilopascal - a metric unit of measurement for pressure used in the HVAC industry. There are 6.89 kilopascals in 1 psi. a pressure equal to 1000 Newton's of force distributed over one square meter of area.

Kilovolt - (**KV**) one thousand volts.

Kilovolt-amperes (**KVA**) - the product of voltage and amperes (power), multiplied by 1000.

Kilowatt - a unit of electrical power equal to 1000 watts. Equivalent to 3,414 Btu.

Kilowatt hour - 1000 watts per hour. Common unit of measurement of electrical energy for home and industrial use. For example, a 100 watt lamp operated for 1000 hours would consume 100 kilowatt hours, (100 watts x 1000 hours = 100 kWh). At a billing rate of $0.10/kwh, this lamp would cost $10.00 (100 kWh x $0.10/kwh) to operate. Power is priced by the kWh.

Kinematic viscosity - the time required for a fixed amount of oil to flow through a capillary tube under the force of gravity. The unit of kinematic viscosity is the stoke or centistokes (1/100 of a stoke). Kinematic viscosity may be defined as the quotient of the absolute viscosity in centipoises divided by the specific gravity of a fluid, both at the same temperature.

Kinetic energy - energy that exists because of movement.

Kinetic trap - a steam trap that operates on the difference in the flow characteristics of steam and condensate.

King valve - a manual liquid line service valve located at the outlet of the receiver.

Kirchhoff's current Law - the sum of the current flowing into any point or junction of conductors of a circuit is equal to the sum of the currents flowing away from that point.

Kirchhoff's voltage Law - in any current loop of a circuit the sum of the voltage drops is equal to the voltage supplied to that loop; or taken with proper signs (- or +), the algebraic sum of the voltage sources and voltage drops in a circuit is equal to zero.

Klixon™ - generic term used for a line voltage thermal overload.

Knife switch - a manual open electrical circuit control, with lever mounted poles. Used mainly in experimentation.

Knocked down - prefabricated, but not assembled; said of items, delivered to the job site for assembly there.

Knockout - a removable portion in the side of a box or cabinet. During installation, it can be readily taken out with a hammer, screwdriver, or pliers so wires, cables, or fittings can be attached.

Knockout drum - vessel used to separate a mixture of gas and liquid by means of a series of baffles. The impact of the material with the baffles frees the gas, thereby allowing it to rise while the heavier substances flow to the bottom of the vessel. Piping then removes the gases from the top of the unit and the liquid from the bottom.

Knurl - a pattern of dents in a turned surface with a knurling tool to produce a better hand grip.

Krypton - an inert, colorless gas used in place of air in between panes of glass in insulating glass units to increase the insulation value. Krypton provides better insulation value than argon.

KW demand - the maximum rate of electrical power usage for a 15 or 30 minute interval in a commercial building for each billing period. A utility meter records this maximum rate, and customers are billed for this peak rate usually once per month.

KWh consumption - the amount of electrical energy used over a period of time. The number of kWh used per month. Often called consumption.

-L-

L - symbol for inductance.

LAL - lowest achievable level.

Lbs - pounds.

LCD - liquid crystal display.

Led - light emitting diode.

Lel - lower explosive limit.

Lgth - length.

LNG - liquefied natural gas.

LOEL - lowest observed effect level.

LPC - low pressure control.

LPG - liquefied petroleum gas.

LR - locked rotor.

LRA - locked rotor amps.

LRSV - liquid receiver service valve.

LSF - low side float.

LWCO - low water cut off.

LTW - low temperature water.

L1 - incoming power line. Usually 120 volts.

L1-L2 - incoming power supply. Usually 240 volts.

L1, L2, L3 - incoming three phase power supply lines.

Label - any written, printed, or graphic material displayed on or affixed to containers of hazardous chemicals. A label should contain identity of the hazardous chemical; appropriate hazard warnings; and name and address of the chemical manufacturer, importer, or other responsible party.

Label diagram - a type of circuit diagram that contains the component arrangement diagram, a wiring diagram, and a legend. Label diagrams are generally placed in a convenient location inside the unit, such as on the inside of the control panel.

Labeled - devices, equipment or materials to which have been affixed a label, seal, symbol or other identifying mark of a nationally recognized testing laboratory, inspection agency or other organization concerned with product evaluation that maintains periodic inspection of the production of the above labeled items and by whose label the manufacturer attests to compliance with applicable nationally recognized standards.

Lacquer - a protective coating or finish which dries to form a film by evaporation of a volatile constituent.

Ladder diagram - a schematic circuit diagram drawn in the form of a vertical ladder. The outer vertical lines represent the electrical supply conductors. The horizontal steps represent individual circuits with all component devices.

Lag - a delay in the effect of a changed condition at one point in the system, on some other condition to which it is related. Also, the delay in action of the sensing element of a control, due to the time required for the sensing element to reach equilibrium with the property being controlled; i.e., temperature lag, flow lag, etc.

Lagging - the thermal insulation applied directly to pipes, ducts and boiler jackets.

Lagging phase angle - the angle by which current lags voltage (or voltage leads current) in an inductive circuit.

Laminar flow - a classification of fluid flow where streamlines remain parallel as the fluid moves along the pipe. Laminar flow is more typical at very low flow velocities.

Laminated core - thin plates of laminated steel that are layered together to form the center, or core, of a low voltage control transformer.

Laminated glass - a type of safety glass where two or more pieces of glass are bonded together with a clear vinyl interlay.

Laminated wood - wood built up of plies or laminations that have been joined either with glue or mechanical fasteners. Usually, the plies are too thick to be classified, as veneer, and the grain of all plies are parallel.

Laminations - thin sheets of soft silicon-alloy iron or steel used in the construction of electrical machines and components to reduce heat losses due to eddy currents.

Lamp - an electrical load device that converts electrical energy into light energy.

Lamps, steri - a lamp that gives of a high intensity ultraviolet ray and is used to kill bacteria. It is often used in food storage cabinets.

Lantern gland (packing) - a packing ring inside a stuffing box that has perforations for the introduction or removal of oil.

Lap joint - joint between two wood members in which the same width and depth of the members is retained.

Lapped seam - a sheet metal seam where the two edges of a sheet metal panel are overlapped with the top edge pitched in the direction of water drainage. The lapped seam may be left loose to permit free movement or it may be soldered, riveted or secured with adhesives.

Lapping - smoothing a metal surface to a high degree of refinement or accuracy using a fine abrasive. Usually done on compressor valves and crankshaft seals.

Large port burner - see spreader burner.

Laser - laser stands for light amplification by stimulated electromagnetic radiation. A device that emits a very intense red light beam.

Laser beam - a small diameter high intensity beam of light. This light could be used for leveling for creating 90° angles.

Last in, first out (LIFO) - a method of inventory control in which the last item placed into inventory is the first one selected for shipment or purchase.

Latching - a characteristic of a P-N switch that allows it to remain in the on position after the triggering signal has been removed.

Latching contacts - contacts that are energized by the latching relay coil so that the contacts hold power to the coil. Typical applications would include impedance relays or start - stop stations.

Latching current - the amount of current to an SCR at the turn-on point so that the silicon controlled rectifier will latch on.

Latching relay - a relay that must be reset either manually or electrically after being energized. Another common term is an impedance relay.

Latent heat - a term used to express the heat energy required to change the form of a substance without changing its temperature. The heat energy used to change water to steam, ice to water, water to ice and remove moisture from the air is latent heat. It requires 970 Btu to change one pound of water to steam. It also takes 970 Btu's to remove one pound of moisture from the air. This heat cannot be measured by a thermometer. Latent is the Latin word for "hidden". Expressed as Btu per pound.

a. Fusion - change of state of element or compound from solid to liquid, or reverse.
b. Vaporization - change of state from a liquid to a gas.
c. Sublimation - change of state from a solid to a gas.

Latent heat of condensation - amount of heat released (lost) by a pound of a substance to change its state from a vapor (gas) to a liquid.

Latent heat of fusion - the amount of heat energy, in Btu's required to change the state of one pound of a liquid to one pound of solid at the same temperature.

Latent heat of melting - the amount of heat energy, Btu's, that must be removed to change the state of one pound of solid to one pound of liquid at the same temperature.

Latent heat of sublimation - the process in which a substance changes directly from a solid to a vapor, without passing through the liquid phase.

Latent heat of vaporization - amount of heat required, per pound of substance, to change its state from a liquid to a vapor (gas). In the case of water, it requires 970 Btu's to change 1 pound (approximately 1 pint) to a vapor at atmospheric pressure and at sea level.

Lathe - a machine used to shape metal or other materials by rotating against a tool.

Law - a rule established by authority. A body of rules the practice of which is authorized by a community or state.

Law of conservation of energy - the amount of energy in existence is constant. The machines that we build and operate do not produce energy; they merely change it from one form to another.

Law of corresponding states - all gases having the same reduced properties are at the same state.

Law of electric charge - see law of magnetism.

Law of magnetism - that like magnetic poles repel and that unlike magnetic poles attract.

Layout tools - sheet metal tools that are used to mark, measure and square sheet metal.

LC50 - the lethal concentration for 50% of tested animals, and sometimes referred to as the median lethal concentration. It is a primary measure of acute toxicity by the inhalation of gases. It most commonly is measured with rats for exposures of four hours. A number of federal regulations and most building, fire, and mechanical codes deem substances to be toxic for one hour LC50 concentrations of 200-2,000 ppm and highly toxic for less than 200 ppm. Typical LC50 concentrations for one hour are double those for four hours.

Leach bed - a private waste removal system consisting of porous pipe that allows the liquid waste to escape into the ground.

Lead - the phase of one alternating quantity that is ahead of the other measured in angular degrees. Also a wire connection.

Lead acid cell - a secondary cell that has positive and negative plates made of lead peroxide and lead and has a liquid electrolyte of sulfuric acid mixed with water.

Lead end - end of motor where the electrical connections are located.

Lead/lag control - a control technique that automatically rotates the order of operation for boilers, pumps or air conditioning units in an attempt to keep the operating hours of each piece of equipment approximately equal.

Lead sulfide cell - an infrared sensor using lead sulfide (PBS) as the sensing material. The resistance of the lead sulfide decreases when exposed to infrared radiation, resulting in increased current flow.

Leaders - (1) the vertical pipes that connect gutters and roof drains to the ground or horizontal storm drain. (2) the portion of a radiant panel tubing circuit that is not within the radiant panel, but instead connects the tubing within the radiant panel to a manifold station.

Leading phase angle - the angle by which current voltage (or voltage lags current) in a capacitive circuit.

Leak check - an operation performed on a complete gas piping system and connected equipment prior to placing it into operation following initial installation and pressure testing or interruption of gas supply to verify that the system does not leak.

Leak detector - a device or instrument such as a halide torch, an electronic leak detector, or soap solution used to detect leaks.

Leakage - the time rate of air volume loss expressed in cubic feet per minute.

Leakage current - the small amount of current leaked when a very sudden increase of resistance opens a set of contacts. This leakage maintains the temperature of the thermistor and prevents it from returning to a low resistance.

Leakage path - a path, normally of high resistances, through which an undesirable stray current (leakage current) flows to ground; caused by buildup of moisture, soot, or accumulated dirt.

Leakage rate - the maximum rate at which a fluid is permitted or determined to leak through a seal. The type of fluid and the pressure differential across the seal determine leakage rate.

Lean mixture - an air-fuel mixture in which an excess of air is supplied in proportion to the amount of fuel. Also see air rich.

Led - a light emitting diode that will produce light when current flows through it.

Left hand rule - (1) to determine the direction of the magnetic field around a single conductor, point the thumb of the left hand in the direction of current flow (- to +) and the fingers will extend around the conductor in the direction of the magnetic field; (2) to determine the polarity of an electromagnetic coil, extend the fingers of the left hand around the coil in the direction of current and the thumb will point to the north polarity; (3) to determine the direction of the induced current flow in a generator conductor, hold the thumb, forefinger, and middle finger of the left hand at right angles to one another, point the thumb in the direction of motion of the conductor, the forefinger in the direction of the magnetic field (N to S), and the middle finger will point in the direction of induced voltage.

Left hand threads - left handed threads are turned counterclockwise or backwards to tighten and clockwise to loosen.

Legend - that section of an electrical wiring diagram that explains any abbreviations, color codes, and other symbols that appear in the diagram.

Length, water tower - for cross flow towers, length is always perpendicular to the direction of air flow through the fill (air travel), or from casing to casing. For counter flow water towers, length is always parallel to the long dimension of a multi-cell tower, and parallel to the intended direction of cellular extension on single cell towers.

Lenz's Law - the induced counter electromotive voltage in any circuit is always in such a direction that it will oppose the force that produces it.

Level - a tool used to establish a true horizontal or vertical line.

Level glass - reading device directly connected to a vessel from the low to high points of a level variation. The liquid in the vessel finds its own level and may be observed through a glass in the instrument.

Level indicating controller - instrument that indicates a vessel's liquid level and regulates it by a signal to a control valve.

Level indicator - dial or linear indicator that shows the level of liquid in the vessel.

Level switch - see float switch.

Lever truck - a long-handled, two-wheeled device that can be used to lift and assist in moving heavy objects.

Lever type tubing bender - tool calibrated to allow the making of accurate short radius bends up to 180°.

Life cycle - a test performed on a material or configuration to determine the length of time before failure in a controlled, usually accelerated, environment.

Life cycle cost - the cost of the equipment over its entire life including operating and maintenance costs.

Lift - (1) valve stem travel or the height of a vertical riser or length of tubing. (2) To elevate a fluid from one level to another.

Lift, pressure drop due to - the difference in pressure between the top and the bottom of a column of fluid due to the weight of the fluid.

Lift slab - a method of concrete building construction in which floor (and roof) slabs are cast, usually at ground level, and then raised into position by jacking.

Lifting flames - flames which lift from or blow off the pilot or main burner ports. Caused by an overrating of the burner, too hard a primary air adjustment, or both.

Ligament - the portion of a boiler tube sheet between the tube holes.

Light emitters - those devices that convert electrical energy into light energy.

Light emitting diode - a semi-conductor device that emits incoherent light formed by the P-N junction. Light intensity is roughly proportional to the electrical current flow.

Light fuel oil - Grades 1 and 2, which are distillate fuel oils, and used predominantly in domestic heating equipment. Heating values range from 132,900 to 141,800 Btu per gallon. Most common value is 140,000 Btu's per gallon for #2.

Light off - the procedure of igniting a burner or system of burner. Also see independent, sequential, and unison light-off.

Lightweight concrete - concrete that uses a lightweight aggregate in place of a more dense stone aggregate. Lightweight aggregates include vermiculite, expanded shale, and polystyrene beads.

Limit action - the contacts of a limit controller open to shut down a burner (and in some instances make an alarm circuit) because of an abnormal or unsafe condition.

Limit control - a controller that continuously monitors a condition (such as temperature, pressure, or liquid level) in a controlled medium and responds immediately to shut down the system if a dangerous, predetermined conditions occurs. It is normally set beyond the operating range of the controlled equipment.

Limit of flammability - the minimum and maximum limits of concentration by volume of a flammable substance in air below and above which flame propagation does not occur upon contact with a source of ignition.

Limit shutdown - a condition in which the system has been stopped because the value of the temperature or pressure has exceeded a pre-established limit.

Limit switch - an electrical safety switch, usually on a heater, that will open to turn off a heat source if the sensed temperature rises to the set point.

Line - a colloquial term for power line. The conductors carrying power from the generating equipment. Any wire, pipe, etc., or system of these, carrying fluid, electricity, etc.

Line break overload - an overload that breaks the power going to the motor and is most commonly used on small motors.

Line charge - the amount of refrigerant that must be added to a system's normal charge to compensate for the added length of its lines.

Line, condenser drain - that part of the refrigerant piping between the outlet of the condenser and the inlet to the liquid receiver.

Line cord - the power supply cable from the appliance to a wall outlet.

Line, discharge - that part of the refrigerant piping between the outlet of the compressor to the inlet to the condenser.

Line drop - the voltage drop due to resistance, reactance, or leakage in an electrical conductor.

Line duty device - a protective device that opens the motor winding circuit under conditions of excess current or temperature.

Line port - the port on a service valve that connects to either the liquid line or the suction line. The other two ports are the device and service ports.

Line pressure - pressure in a line measured in pounds per square inch and rated in terms of static pressure or for use with water, oil, or gas.

Line set - a term used for the insulated suction and liquid line pre- made by the manufacturer.

Line starter - a contactor that has built in overload protection. It is used to operate and protect single and three phase motors.

Line, suction - that part of the refrigerant piping between the outlet of the evaporator and the inlet to the compressor.

Line tap - a valve that can be attached to a straight run of pipe to tap into that line for pressure testing, charging and evacuation procedures.

Line valve - a 3 way stem type service valve that may be placed in a liquid line to service the line and used to pump down the system. Line valves may also be placed in the suction line. See service valves.

Line voltage - the voltage of the power supply to the building, as distinguished from control voltages. In homes this is usually 120 or 240 volts.

Line voltage circuit - a circuit involving a potential of more than 30 volts and less than 600 volts (generally 120 or 240 volts AC). Underwriters Laboratories Inc. also calls this a high voltage circuit.

Line voltage thermostat - using building line voltage to control loads such as blower fans and heating elements without an intervening relay.

Line wiring diagram - sometimes called a ladder diagram, this type of diagram show the power consuming devices between the lines. Usually, the right side of the diagram consists of a common line.

Linear characteristics - a valve flow characteristic, approximating a straight line relationship, which results in equal volume changes for equal movements of the valve stem, regardless of the percentage of valve opening. Output that varies in direct proportion to input.

Linear, meter - generally used to mean a read-out of a meter in which all divisions are equal.

Linearity - the deviation from a straight line. The change in one quantity that is directly proportional to the change in another quantity.

Lined ductwork - duct sections that have been fabricated with inside acoustical insulation to reduce noise transmission within the duct system, or as thermal lining where ductwork is exposed to weather.

Liner - see radiation shield.

Liner, refrigeration - the enclosure forming the interior of a general refrigerated compartment and/or some freezer compartment(s). The complete liner comprises the compartment liner in the cabinet, the exposed breaker strip surfaces and the door liner(s).

Lines of force - same as magnetic flux. See flux.

Lining - those interior surfaces of a combustion chamber that are exposed to combustion during use of the device.

Linkage - a mechanical arrangement for transferring motion in a desired manner, consisting of solid pieces with movable joints. The connections through which an actuator opens, closes, or positions a valve or damper; usually consists of a push rod, two crank arms, and two ball joints.

Lintel - a beam placed over an opening or recess in a wall that supports the wall construction above. When referring to masonry fireplaces, that horizontal, noncombustible member, usually of masonry or steel that span the opening of a masonry fireplace to support the load above.

Liquation - the tendency of the lower melting point contents of a filler metal to separate and flow away from the higher melting contents during heating. When this happens an unmelted “skull” of alloy is often visible. Liquation occurs when brazing with alloys that have a wide melting range, especially if they are slowly heated through the range.

Liquefaction - change of state to liquid, generally used instead of condensation in case of substances ordinarily gaseous.

Liquefied natural gas (LNG) - natural gas that has been cooled until it becomes a liquid.

Liquefied petroleum (LP) gas - fuel gases, including commercial propane (predominantly propane or propylene) or commercial butane (predominantly butane, isobutane, and/or butylene), which are sold in bottles under pressure as a liquid and called bottled gas. These gases are higher hydrocarbon gases normally obtained as a by-product of oil refineries or by stripping natural gas. Liquefied petroleum gases are usually obtained as a byproduct of oil refinery operations or by stripping natural gas, and are commercially available as butane, propane or a mixture of the two. Propane has a boiling point of -40° F (40° C) at atmospheric pressure and a heating value of approximately 21,560 Btu per pound or 2,500 Btu per cubic foot. Butane has a boiling point of 32° F at atmospheric pressure and a heating value of approximately 21,180 Btu per pound or 3,200 Btu per cubic foot. LP gas is odorized so that gas leakage can be detected by one’s sense of smell.

Liquid - the fluid form of a substance in which it can assume the form of its container while retaining its independent volume. Form that is either a gas or solid.

Liquid absorbent - a chemical in a liquid form that has the property to take on or absorb moisture.

Liquid charge bulb - usually refers to the power element of temperature controls and thermostatic expansion valves. The power element and remote bulb are sometimes charged with a liquid rather than a gas.

Liquid charging - the process of charging a refrigeration system with liquid. Usually done when the system is not operating.

Liquid collector design - a solar collector design in which either water or an antifreeze solution is passed through the collector and heated by the sun.

Liquid cooled condenser - condensers that use liquid to absorb the heat that the system is rejecting.

Liquid de-superheater - valve that permits small flow of refrigerant to enter low side of systems to cool suction gas.

Liquid eliminator - a device for the separation of liquid droplets entrained in a flowing gas stream.

Liquid fill - a temperature sensing bulb, either on a thermostatic expansion valve or temperature controller, that is completely filled with liquid. Any change in temperature causes the liquid to expand or contract, thus causing valves to open or close or contact opens to change position.

Liquid filter - a very fine strainer placed in the liquid line to remove foreign matter from the refrigerant.

Liquid floodback - the process by which liquid refrigerant enters the compressor via the suction line. This can result in major damage to reciprocating compressors.

Liquid indicator - a device located in the liquid line that provides a glass window through which liquid may be observed.

Liquid line - the tube or pipe that carries liquid refrigerant from the condenser outlet or liquid receiver outlet to the inlet of the refrigerant metering control device.

Liquid line service valve - valve installed in the liquid line. It usually has a gauge port connection for connecting to high side hose of service manifold.

Liquid nitrogen - nitrogen in liquid form that is used as a low temperature refrigerant in chemical (or expendable) refrigerating systems.

Liquid source heat pump - heat pump systems that utilize liquid as the source of heat while operating in the heating mode.

Liquid receiver - cylinder connected to the condenser outlet for the storage of liquid refrigerant in the system.

Liquid receiver service valve - a two or three way manually operated valve located at the receiver outlet used for installation and service operations. Also called a king valve.

Liquid refrigerant charging - the process of allowing liquid refrigerant to enter the refrigeration system through the liquid line to the condenser and evaporator.

Liquid return line - the line that carries the liquid refrigerant from the outlet of the condenser to the inlet of the receiver. Also see condensate line.

Liquid sight glass - a glass bull's eye installed in the liquid line to permit visual inspection of the liquid refrigerant. Used primarily to detect bubbles in the liquid, indicating a shortage of refrigerant in the system. Also called a liquid indicator.

Liquid slugging - a large amount of refrigerant liquid in the compressor cylinder clearance space, usually causing immediate damage.

Liquid source heat pump - heat pumps that use liquid as the source of heat while operating in the heating mode.

Liquid stop valve - a magnetically operated valve generally used to control the flow of liquid to an evaporator, but may also be used whenever on-off control is permissible. An electrical solenoid winding controls the action of the valve. Also called a pump down valve.

Liquid to air heat pump - liquid source heat pumps that are used to ultimately treat air in either the heating or cooling mode.

Liquid-to-gas ratio - a ratio of the total mass flows of water and dry air in a cooling tower.

Liquid to liquid heat pump - liquid source heat pumps that are used to ultimately treat another liquid.

Liquid to suction heat exchanger - a heat exchanger where warm liquid is cooled by cold gas to increase subcooling.

Liquid vapor valve - a dual hand valve on refrigerant cylinders that is used to release either gas or liquid refrigerant from the cylinder.

Liquid, volatile - one that evaporates readily at atmospheric pressure and room temperatures.

Liquidity - the ability of a firm to readily convert its assets into cash.

Liquidus - the lowest temperature or point where melted solder or brazing filler material starts to flow and is completely liquid. Below the liquidus point the material is part liquid and part solid.

Liquor - solution used in absorption refrigeration. Solvent can have a relatively high or low concentration of solute.

Lissajous figure - the pattern seen on the screen of an oscilloscope when sine wave voltages are applied to both the vertical and horizontal inputs of the oscilloscope at the same time. The sine waves can differ in amplitude, frequency, and/or phase.

Listed - equipment or materials included in a list published by an organization acceptable to the authority having jurisdiction and concerned with product evaluation that maintains periodic inspection of production of listed equipment or materials and whose listing states either that the equipment or material meets appropriate standards or has been tested and found suitable for use in a specified manner.

Listing mark/label - a printing, tag or other marking device of the listing agency permanently and prominently affixed to the listed product that certifies compliance to the applicable standard and contains recommendations relating to installation and maximum physical conditions of use.

Liter - metric unit of volume that equals 61.03 cubic inches.

Litharge - lead powder mixed with glycerin to seal pipe thread joints.

Lithium bromide - chemical commonly used as the absorbent in absorption cooling systems. Water would then be used as the refrigerant.

Live load - all loads on structures other than dead loads; includes the weight of persons occupying the building and free standing material.

Live steam - superheated high-energy steam direct from the boiler.

Living unit - one or more rooms designed or used as living quarters providing complete, independent living facilities for one or more persons, including permanent provisions for living, sleeping, eating, cooking, and sanitation.

Load - (1) the power consumed by a machine or circuit in performing its function. (2) A device that converts electrical energy into another form of energy such as heat, light, sound or magnetism. (3) The impedance to which energy is being supplied. The required rate of heat removal. Heat per unit of time that is imposed on the system by a particular job. Also see capacitive, full, inductive, and net load.

Load bearing wall - a wall capable of supporting an imposed load in addition to its own weight.

Load calculation - a series of studies performed to determine the heating and cooling requirements of your home. An energy load analysis uses information such as the square footage of your home, window or door areas, insulation quality and local climate to determine the heating and cooling capacity needed for your furnace, heat pump or air conditioner. When referring to heating, this is often known as a heat loss analysis, since a home's heating requirements are determined by the amount of heat lost through the roof, entry ways and walls.

Load carrying capacity - property of a lubricant to form a film on the lubricated surface, which resists rupture under given load conditions. Expressed as maximum load the lubricated system can support without failure or excessive wear.

Load center - generally the service entrance. A point from which branch circuits originate.

Load change - the change in the usually uncontrollable heat gain or loss caused by lights, machinery, people, outside air temperature variations or solar effect.

Load curve - a curve on a chart showing power (kilowatts) supplied plotted against time of occurrence, and the varying magnitude of the load during the period covered.

Load diversity - the difference between the sum of the peaks of two or more individual loads and the coincident or combined peak load, usually measured in kilowatts.

Load factor - the ratio of the average load in kilowatts supplied during a designated period to the peak or maximum load occurring in that period. Typical formula is expressed as the kilowatt hours consumed over the specified period divided by the product of the kilowatt peak demand registered times the number of hours in the period. The kilowatt-hours and peak generally are on a net output basis. The peak generally is for a sixty minute demand interval.

Load management - the control of electrical loads to reduce kW demand and kWh consumption.

Load management system - the set of devices that, when installed, effectively reduces kW and kWh consumption.

Load matching - trying to always match the capacity of the refrigeration or air conditioning system with that of the heat load put on the evaporator.

Load programmer - any device that turns loads on and off on a real time, time interval, or kW demand basis.

Load, refrigeration - the amount of heat per hour that the refrigeration system is required to remove at design conditions.

Load relay - a relay in a flame safeguard control circuit that starts the burner operating sequence.

Load shedding - shutting down non-critical electrical equipment to a pre-selected level, when a peak electric use period approaches. This is done to prevent paying an excessive electric rate that is based on the highest electric usage during a preset billing period.

Load usage - the sum of the air change, product and miscellaneous loads on a refrigerator; the sum of the loads exclusive of the wall heat gains.

Loading - the effect when a meter is connected into a circuit that causes the meter to draw current from the circuit.

Location plan - reference drawing which appears on the pipe support detail giving coordinate information and location dimensions in two perpendicular directions from the building columns, pipe centerline elevations, building area, and area number.

Localized controllers - independent energy control device located near the system it is controlling.

Lock fold - a straight edged piece of sheet metal that has been turned 180 degrees.

Lock out - a condition that occurs when an electronic ignition system fails to sense a flame, and the control box will not energize anything until it is reset.

Locked rotor - in a motor, the steady-state current taken from the line while the rotor is locked and the rated voltage (and frequency in alternating current motors) is applied to the motor. Allowable locked rotor current can be up to 6 times the running load amperes.

Locked rotor amperage - the amperage flowing in the circuit to a motor driven apparatus when the motor is locked to prevent its movement. This amperage is with rated voltage and frequency applied and is typically as much as six times the full load current of the motor and four to five times the full load current in hermetic compressors.

Locked rotor current - the amount of current produced when voltage is applied to a motor and the rotor is not turning.

Locked rotor torque - the amount of torque produced by a motor at time of initial start.

Locking cam - secures the workholder on the pipe when pushed away from the threader and releases the workholder when pushed toward the threader.

Lockout (control) - a control feature, when the control reaches its cut-out point it stays cut-out until manually reset. Also see safety shutdown.

Lockout (safety) - the process of locking an energy isolating device, in accordance with an established procedure, to ensure that the energy isolating device and the equipment being controlled cannot be operated until the lockout device is removed. Usually the main electrical switch is placed in the open position to safely perform equipment maintenance.

Lockout device - a device that uses a lock, either key or combination, to hold an energy isolating device into a safe position to prevent the energizing of a machine or equipment.

Lockout interlock - a running interlock used with a primary or programming control which will cause safety shutdown (will not recycle) if an improper condition occurs. The interlock may be a combustion air controller (airflow switch) or a fuel pressure switch. See running interlock.

Lockout relay - a manually reset relay that prevents the system compressor from operating if an unsafe condition, usually pressure or temperature, has been established. To reset, the power to the system (line or low voltage) must be interrupted and then restored.

Lockout switch - a time delay device in a flame safeguard control which trips and locks out the control (safety shutdown) if improper conditions occur. It must be manually reset to restart the system.

Lockout switch timing - the length of time (usually 15 or 30 seconds nominal) that the heater in a lockout switch must be energized before it causes the switch to trip (pop out).

Lock/out tag/out - see lockout safety.

Lockout time - the time a flame safeguard control takes to be electrically inoperative. See lockout switch timing.

Lockshield valve - a valve that mounts on the outlet of a heat emitter and serves as a combination isolation valve, flow balancing valve, and drain valve.

Lodestone - a natural magnetic iron ore that attracts and holds pieces of iron and steel. Name used for natural magnets used in early times.

Logarithm - a mathematical expression dealing with exponents showing the power to which a value called the base must be raised in order to produce a given number.

Long flame burner - a burner that purposely has poor mixing of fuel and air (using secondary air only) so that the fuel takes more time to burn, thus producing a long, luminous flame. Also called a diffusion flame burner, luminous flame burner, or yellow flame burner. Also see delayed mixing.

Long nipple - pieces of pipe in graduations of up to 12 inches in length.

Long term burst test - the internal pressure at which a pipe or fitting will break due to a constant internal pressure held for 100,000 hours (11.43 years).

Long term hydrostatic strength - the estimated tensile stress in the wall of the pipe in the circumferential orientation (hoop stress) that when applied continuously will cause failure of the pipe at 100,000 hours (11.43 years). These strengths are usually obtained by extrapolation of log-log regression equations or plots.

Longitudinal-drum boiler - see water tube boiler.

Longitudinal joint - a seam running the length of a fabricated item.

Longitudinal section - longitudinal sections are through the wider or longer part of the building.

Loop - a closed electric circuit. Also see open and closed loop.

Loop vent - used in single story houses or on the top floor of a multi-story building, in which the vent branch goes back into the stack vent at a point above all drain inlets.

Loose house - construction type that has no insulation, storm doors, storm windows, weather stripping or vapor barriers. In regards to infiltration and construction standards, is considered to be two air changes per hour.

Loss factor - the loss factor of an insulating material is equal to the product of its dissipation and dielectric constant.

Loudness - loudness is the subjective human definition of the intensity of a sound. Human reaction to sound is highly dependent on the sound pressure and frequency.

Louver - a series of vanes that permit directional adjustment of air flow. They are usually installed on outside grilles or intake openings to impede water entry into ducts or elbows, and to minimize turbulence. They may also include indoor grilles to eliminate light penetration.

Louver type damper - a damper consisting of several blades, each pivoted about its center, and all linked together for simultaneous operation. Also see opposed and parallel blade dampers.

Low ambient control - a device for maintaining adequate head pressure when the temperature of an air cooled condenser falls below normal due to low ambient temperatures.

Low boy furnace - a short upflow furnace. This furnace is approximately 4 foot high, and the air intake and discharge are both at the top.

Low emissivity glass - low “E” glass has a transparent coating applied to a glass surface to separate long wave (heat) energy and short wave (light) energy thus acting as a thermal mirror. The long wave is reflected back to the heat source. The short wave is allowed to pass through the coating.

Low fire - the minimum firing rate of a burner or furnace.

Low fire interlock - an interlock that proves the damper is closed prior to ignition trials.

Low fire start - the firing of a burner with the fuel controls in a low fire position to provide a minimum of incoming fuel from the main fuel valve(s) to facilitate a smooth start and to provide a safe operating condition during light-off. In a system with guaranteed low fire start, interlocks are used to prevent start-up if the burner is not in the low fire start position.

Low fire start/high fire run - operation sequence in which the burner starts in the low fire position. After the flame is proven, the burner goes to the high fire rate for the duration of the run period.

Low fire switch - see low fire interlock.

Low head boiler - see water tube boiler.

Low high low pre-purge - a pre-purge period during which the air damper is driven open for the purge period and then back to low fire position for the burner light-off. This is called low/high low proven pre-purge if suitable interlocks indicate that the damper actually reached the open position during pre-purge, and then closed again before ignition trials.

Low limit - a controller that shuts down the system if a condition drops below its minimum value for safe operation. When used where the boiler also heats the domestic hot water supply, this control maintains adequate boiler water temperature at times when the thermostat is not calling for heat. Also see limit.

Low limit, chiller - a protective device to prevent a system, or any part of it, from operating below the setting of that device. Used to prevent freeze-up in a water chiller or excessively low temperature in ductwork.

Low link control - a switch operated by the temperature of the water in a boiler and used to start the burner at any time the water temperature drops to some prescribed minimum. This control is used if the boiler is supplying domestic hot water as well as heat for the building.

Low loss refrigerant fittings - any device that connects hoses, appliances, or recovery or recycling machines, and that is designed to close automatically or to be closed manually when it is disconnected.

Low oil temperature switch - a temperature actuated device arranged to effect the safety shut down of a heavy oil burner or to prevent it from starting when the fuel oil temperature falls below the limits required to maintain the viscosity range recommended by the burner manufacturer.

Low pressure appliance - any appliance that uses a refrigerant with a boiling point above 50° F at atmospheric pressure. This includes, but is not limited to equipment utilizing R-11, R-113, and R-123.

Low pressure boiler, hot water or steam - a boiler constructed for maximum working pressures of 15 psig steam and up to 160 psig for hot water. Hot water boilers are limited to 250° F (121° C) operating temperature. Operating and safety controls and relief valves, which limit temperatures and pressures, are ancillary devices required to protect the boiler and prevent operation beyond design limits.

Low pressure control - motor protection device that senses low side pressure. Control is wired in series with the motor and will shut off during periods of excessively low suction pressure.

Low pressure switch - see low pressure control.

Low probability systems - any system which the basic design or the location of components is such that a leakage of refrigerant from a failed connection, seal or component will not enter an occupancy-classified area, other than the machinery room.

Low side - that portion of a refrigerating system that is under the lowest evaporating pressure. This is from the metering device outlet to the compressor suction valve.

Low side charging - the process of introducing refrigerant vapor or gas into the low side of the system. Usually reserved for the addition of a small amount of refrigerant after service repairs.

Low side float valve - a refrigerant control valve operated by level of liquid refrigerant in the low pressure side of the system.

Low side pressure - pressure in the cooling side (evaporator) of the refrigerating system.

Low side pressure control - a device used to keep the low side evaporating pressure from dropping below certain preset pressure.

Low temperature refrigeration - a refrigeration temperature range starting with evaporator temperatures no higher than 0° F for storing frozen foods.

Low temperature water system (LTW) - this hydronic heating system operates within the pressure and temperature limits of the ASME Boiler and Pressure Vessel Code for low pressure boilers. The maximum allowable working pressure for low pressure boilers is 160 psig, with a maximum temperature limitation of 250° F. The usual maximum working pressure for boilers for LTW systems is 30 psi, although boilers specially designed, tested, and stamped for higher pressures are frequently used. Steam to water or water to water heat exchangers are also used for heating low temperature water. Low temperature water systems are used in buildings ranging from small, single dwellings to very large and complex structures.

Low voltage - control voltage usually used for domestic room thermostats. It is usually considered to be between 18 to 30 volts, but the standard is 24volts.

Low voltage circuit - a circuit involving a potential of not more than 30 volts AC (42.4 volts peak) or 30 volts DC in which the current is supplied by a primary battery or by a standard Class 2 transformer or other suitable transforming device, or by a suitable combination of transformer and fixed impedance having the output characteristics of a Class 2 transformer. A circuit derived from a line voltage or a high voltage circuit, by connecting series resistance to limit the voltage and current, is not considered to be a low voltage circuit.

Low voltage control - a control used to open or close electrical circuits as temperature or pressure limits are reached.

Low voltage protection - a magnetic relay circuit so connected that a drop in voltage causes the motor starter to disconnect the motor from the line.

Low voltage thermostat - a thermostat whose switches are designed to operate a 24 volt load.

Low water - a situation when ever the water level in the gauge glass is below the normal operating water level (NOWL).

Low water fuel cutoff - a device to shut down the burner when the water level in a boiler falls to a pre-determined, dangerously low level.

Lower explosive limit - denotes the minimum concentration of combustible gas or vapor (% by volume in air) that will ignite and burn if an ignition source is present.

Lower sash bottom rail - this part of a double hung window closes down on the top of an installed window air conditioner unit.

Lower sash top rail - this is part of a double hung window where at the inside stop a bracket is placed when a window air conditioner unit is installed. This bracket pushes against the lower sash top rail. It keeps the window from being raised.

Lower zone - the room volume that is below the stratification level created by displacement ventilation.

Lowest achievable level - the lowest amount of emissions possible of CFC and HCFC refrigerants.

Lowest observed effect level - the lowest concentration at which sensitization occurs in tests, normally to test animals treated with epinephrine to simulate the effects of stress.

LP gas - liquefied petroleum, a substance (most commonly propane and butane) used as a gas for fuel. It is transported and stored in the liquid state.

LP gas-air mixtures - liquefied petroleum gases distributed at relatively low pressures and normal atmospheric temperatures which have been diluted with air to produce desired heating value and utilization characteristics.

L type copper - medium thickness tubing suitable for low pressure applications.

Lubricant - any substance interposed between two surfaces in relative motion for the purpose of reducing the friction and/or the wear between them.

Lubrication - the act of making slippery or smooth in order to reduce friction. This is usually done with an oil or grease.

Lubrication, force-feed - a lubricating system in which the lubricant is forced to the various bearing surfaces by an oil pick-up or pump, opposed to splash lubrication.

Lubricated plug valve - a manual shutoff valve with a tapered plug, which uses a lubricant to hydraulically lift the plug from its seat and to seal the valve against leakage in the closed position.

Lubricity - ability of an oil or grease to lubricate. Also called film strength.

Lug - mounting or brackets (usually three) extending from a motor shell.

Lugs - terminals or terminal places on the end of a wire or equipment to facilitate rapid connection.

Lumen - a unit of measurement for the flow of light.

Lumens per watt (lpw) - a measure of the efficiency of a light source in terms of the light produced for the power consumed. For example, a 100 watt lamp producing 1750 lumens gives 17.5 lumens per watt. Examples are:

Edison's first lamp	1.4 lpw
Incandescent lamps	10-40 lpw
Fluorescent lamps	35-100 lpw
Halogen lamps	20-45 lpw
Mercury lamps	50-60 lpw
Metal halide lamps	80-125 lpw
High pressure sodium lamps	90-140 lpw

Luminaire - a complete lighting unit consisting of a lamp or lamps together with the parts designed to distribute the light, to position and protect the lamps and to connect the lamps to the electric power supply.

Luminous flame burner - a burner that uses only secondary air for the combustion of the gas and therefore produces a yellow flame.

-M-

M - meter, mega, milli or mutual inductance.

Ma - milliampere (one-thousandth of an ampere).

Man - manual.

MAWP - maximum allowable working pressure.

Max - maximum.

Mbh - thousands of British thermal units per hour. (45 Mbh = 45,000 Btu).

Mbtu - 1000 British Thermal Units.

Mbuth - 1000 British Thermal Units per hour.

MCA - maximum circuit ampacity.

MCC - maximum continuous current.

MCS - multiple chemical sensitivity.

Mache - mechanical.

Meg - prefix meaning one million (1,000,000).

Mega - prefix meaning one million (1,000,000).

MEP - mean effective pressure.

MFD - microfarad.

Mho - an electrical unit of conductivity, of a body with the resistance of one ohm. The reciprocal of ohm.

Min - minimum.

Misc - miscellaneous.

MMF - magneto motive force.

MOP - maximum operating pressure or maximum overcurrent protection.

MOPD - maximum operating pressure differential.

MOS - metal oxide semiconductors.

Mot - motor.

MRHL - manual reset high limit.

MRT - mean radiant temperature.

MSD - most significant digit. The leftmost digit in a display.

MSDS - material safety data sheet.

MSS - minimal stable signal.

MSST - maximum safe storage temperature.

Mtg - mounting.

MTW - machine tool wire.

Mv - millivolt.

MVAC - motor vehicle air conditioner.

Mach number - ratio of the velocity at a point in a fluid to the velocity of sound at that point at a given instant of time.

Machine room - area where commercial and industrial refrigeration and heating machinery, except evaporators, are located.

Machine tool - is a power driven tool that is used to cut and shape metal such as in making ductwork.

Machined seat - a surface that has been ground to a perfectly level plane.

Machinists file - are used for filing and finish work.

Magnehelic - a dry type diaphragm operated differential pressure gauge which employs a calibrated spring loaded horseshoe magnet lever operated from the differential pressure on the diaphragm, causes rotation of a highly magnetic permeable helix which positions a pointer on the pressure scale. The magnehelic pressure gauge operated by magnetic field linkage only, hence it is extremely sensitive and accurate, its construction design makes it resistant to shock and vibration. The helix rotates on anti-shock mounted sapphire bearings and the zero calibration screw is located on the cover. There are approximately 30 available pressure ranges for this instrument. The instrument minimum accuracy is "2% of full dial range.

Magnesium (Mg) - a metallic element that is light silver-white, easy to work with, burns with a hot, white light -- used in photographic flash bulbs.

Magnet - a body that has the property of attaching or repelling magnetic materials.

Magnetic across-the-line starter - a motor starter or switch that allows full line voltage to the motor windings when engaged.

Magnetic circuit - a complete path for magnetic lines of force from a north to a south polarity.

Magnetic clutch - a device operated by magnetism to connect or disconnect a power drive.

Magnetic contactor - a multiple switch assembly that includes an electromagnetic which, when energized, closes all switches in unison. Magnetic contactors are used with single and three phase motors that have overload devices built into their windings.

Magnetic field - magnetic lines of force that extend from a north polarity and enter a south polarity to form a closed loop around the outside of a magnetic material.

Magnetic field strength - amount of magneto motive force per unit length. Other terms for magnetic field strength are magnetic field intensity and magnetizing strength.

Magnetic flux - the rate of flow of magnetic energy across or through a surface (real or imaginary).

Magnetic gasket - a sealing material that adheres due to small magnets inserted in the gasket.

Magnetic hysteresis - internal friction resulting from subjecting a ferromagnetic substance to a varying magnetic field.

Magnetic materials - metallic materials such as iron, nickel, and cobalt that exhibit magnetic properties.

Magnetic overload - an overload device that senses the current draw of a load by the magnetic field produced, which is proportional to the current draw. The device will open a set of contacts on the high current draws and allow them to close when the ampere draw returns to normal.

Magnetic poles - area of concentrated lines of force on a magnet that produce north and south polarities.

Magnetic saturation - a condition that exists in a magnetic material when an increase in magnetizing force does not produce an increase in magnetic flux density around the material.

Magnetic starter - a multiple switch assembly that includes overload devices. An electromagnet is included to activate the switches in unison. If the overload devices sense an overload condition at the motor, they will open the circuit to the coil, opening all switches to stop the motor. Magnetic starters include a reset button to reinstate motor operation after it has been shut down by the overloads.

Magnetic starters - an on-off switch that electrically opens and closes a contact much the same as a relay. Magnetic starters have provisions for overload protection.

Magnetic valve - see solenoid valve.

Magnetism - an electrical force that causes a magnetic field to attract ferrous metals, or where like poles of a magnet repel and unlike poles attract each other.

Magnetize - molecular rearrangement in a material, which converts it into a magnet.

Magneto - an AC generator that uses a permanent magnet for its field magnet.

Magneto motive force (mmf) - a force that produces magnetic force around a magnetic device.

Main - the primary circuit that supplies all the others. In relation to air ducts, electrical, plumbing, and piping.

Main air - see combustion air.

Main burner - the burner (as distinguished from the pilot burner) that delivers fuel to the combustion chamber (and mixes it with the air) where it is burned to provide heat to the desired space.

Main burner flame establishing period - the period of time in ignition trials during which the flame safeguard control permits the main fuel valve(s) to be open before the flame detector is required to detect the main burner flame.

Main circulating loop - the section of an oil handling system which delivers the oil from storage to the branch circuits and returns the unused oil to the storage tank.

Main distribution cabinet - the enclosure in a facility that houses utility metering equipment, the main electrical disconnect switch, and the branch circuit over current protection. Electrical service is distributed throughout the building from this cabinet.

Main flame trial - see main burner flame-establishing period.

Main fuel valve(s) - an automatic, or manual reset, safety shutoff valve(s) which controls the fuel input to the main burner and is energized by the flame safeguard control. Also see manual reset, motorized, solenoid, and safety shutoff valves.

Main header - that part of the system that connects boilers in battery and then distributes the steam to where ever it is needed.

Main limit - one of two high limit contacts connected in series with the automatic gas valve of a heating unit. May be located or be part of a fan-limit control.

Main line - a large pipe that distributes fluid to smaller branch pipes.

Main object line - these are the heavy lines used to outline an object or structure. It is a heavy continuous line that shows all edges and surfaces.

Main truck line - pipe or duct for distributing fluids such as air, water, or steam to various branch ducts and collecting fluids from various branches.

Main vent - the principle pipe of a plumbing venting system.

Main windings - the run winding of an electric motor.

Mains - when referring to pneumatic controls, this term usually refers to the air lines which supply primary air pressure at 15 to 20 psig.

Maintenance power - electric energy supplied by an electric utility during scheduled outages of the co-generator.

Maintenance time reminders - a computer program which monitors the time a machine has been operating since its last maintenance check, and will produce an appropriate report to the operator.

Maintaining contact - also known as a holding or sealing contact. It is used to maintain the coil circuit in a relay control circuit. The contact is connected in parallel with the start push button.

Major diameter - the widest measurement from the outside edges of the threads.

Majority carrier diffusion current - the current flow carried by the diffusing majority carriers and crosses a P-N junction.

Majority carriers - a carrier, either an electron or a hole, having the greatest number in a semiconductor material.

Make - a term used to indicate that an electrical control device, such as a switch or a set of points, is in the on or closed position. To complete an electrical circuit or establish contact.

Make up - to apply joint compound to the male threads and tighten the fitting onto the pipe.

Make-up air - air drawn into a conditioned space and that has not been previously circulated through the system and is used to replace that which has been removed by an exhaust system. For combustion appliances, air supplied to support the combustion process.

Make-up heater - a heater that raises the temperature of outdoor air that is brought into a building to compensate for air removed by exhaust fans or by other methods.

Make-up water - water that is added to a boiler, tank, or some other container to replace water which has been lost by evaporation, drift, windage, blowdown and leakage, to maintain the proper water level.

Make-up water line - the water connection to the boiler or system for filling or adding water when necessary.

Male thread - external thread on pipe, fittings, and valves for making connections that screw together.

Malfunction - failure to operate as it should.

Malleable - term used to describe pipe fittings that are made of annealed cast iron. Annealing relieves brittleness and improves its tensile strength.

Mandatory - required, necessary, commanded by authority. The terms “shall” and “shall not” are used when a code statement is mandatory.

Mandrel - a nail like pin designed to break off a bind rivet when the rivet head is formed to fasten two pieces together.

Manhole - opening found on the steam and water side of a boiler that is used for cleaning and inspecting of the boiler.

Manifold - a pipe with one inlet and several outlets or with one outlet and several inlets, for connecting with other pipes. In a multi-burner system, the tube or pipe that distributes fuel to the individual burners. Also see header.

Manifold, discharge - a device used to collect compressed refrigerant from the various cylinders of a compressor.

Manifold, gas - the conduit of an appliance that supplies gas to the individual burners.

Manifold gauges - see manifold, service.

Manifold pressure - the gas pressure in an appliance manifold that is upstream of burner orifices.

Manifold station - a combination of a supply manifold and return manifold that services two or more floor heating circuits.

Manifold, service - a device equipped with gauges and manual valves, used by service technicians to charge, evacuate, recover refrigerant, measure system pressure and analyze system operation in refrigeration systems.

Manifold type distributor - device used to distribute refrigerant evenly to parallel circuits in an evaporator coil.

Manifolding - in direct expansion or dry evaporators, the method of circulating the refrigerant through separate rows of tubes.

Manipulated variable - that quantity or condition which is regulated by the automatic control system in such a matter that it causes the desired change in the controlled variable. The manipulated variable is a characteristic of the control agent.

Manometer - an instrument for measuring the pressure of gases or vapors. A column of incompressible liquid, such as mercury, is displaced an amount proportional to the magnitude of the gas pressure. Thus, the height of the column indicates the pressure.

Manometer, inclined - a manometer on which the liquid tube is inclined from horizontal to produce wider, more accurate readings over a smaller range of readings.

Manometer, "U" tube - an instrument used for measuring pressures only slightly above or below atmospheric. The manometer consists of a U shaped tube partially filled with liquid, usually water, mercury, or light oil, so constructed that the amount of displacement of the liquid levels indicates the pressure being exerted on the instrument.

Manometer, vertical/inclined - the combination inclined/vertical manometer is constructed of solid transparent plastic block. The upper portion of the block has two tube connection chambers, each with its own adjustable valve and vent tubing connector. The inclined fluid column has scales of 0 to 1.0 or 2.0" while the vertical fluid column has scales to 5 or 10". The scales behind the inclined and vertical legs of the manometer read in inches of water. The gauge oil recommended by the manufacturer shall be used. A bubble level is located at the top of the manometer block for leveling the instrument, and the bottom of the block has an adjustable foot fulcrum base and an adjustable leveling screw. Also at the bottom is a horizontal fluid chamber with an adjustable screw for making zero reading adjustments. Models may have magnets included for attachment to ductwork in lieu of base support.

Manual - (1) done by hand; not automatic. (2) Any reference book for installation or service work.

Manual air vents - devices that, when manually opened, allow air to be released (vented) from a piping system.

Manual changeover - the process by which a system is switched over from the heating to the cooling mode. The switch must be repositioned manually.

Manual controller - a controller operated by hand at the location of the controller.

Manual ignition - ignition by an energy source that is manually energized, and when the fuel to the main burner is turned on only by hand and ignited under supervision of the operator.

Manual ranging - used where the operator must select the proper range on a measurement instrument.

Manual reset - (1) the manual operation required after safety shut down before the system can be restarted. This as preferred where unexpected restarting would be hazardous, as on saws, conveyors, or compressors. (2) the adjustment on a proportioning controller which shifts the proportioning band in relationship to the set point to eliminate droop or offset errors.

Manual reset high limit - a safety control that opens a normally closed set of contacts when a set temperature is detected at its probe. A button on the device must be pushed to close the contacts once they have opened.

Manual reset valve - a safety shut off valve that is opened manually and held in the open position electrically or pneumatically. It is closed automatically by the control system and cannot be opened manually until power is restored by the control system.

Manual shutoff valve - a manually operated valve in a fuel line (main or pilot) for the purpose of completely turning on or shutting off the fuel supply. Also see lubricated plug valve.

Manual start - the burner is started by operator action, such as pushing the start button on a controller; sometimes required by local codes to ensure that an operator is actually present on start up.

Manual starter - a switch that is operated manually and which is used to start motors. The switch includes an overload device that will disengage the electrical circuit to the motor in the event the motor becomes overloaded.

Manual switch - a switch that is manually operated as with a typical toggle light switch.

Manual system - a system which requires all manual operation; a system in which the burner is purged, started, ignited, sequenced, modulated, and stopped manually.

Manually ignited burner - a burner in which fuel to the main burner is turned on only by hand and ignited under supervision.

Manufactured gas - a gas made artificially for use as fuel; made by distillation or cracking of oil or coal, by the steam-carbon reaction, or by a combination of these three processes. Examples are coke oven gas and carbureted water gas. Also known as town gas.

Manufacturer - a business that purchases raw materials and transforms them into finished products.

Man way - large nozzle and hinged flange on a vessel used to provide access to vessel internals.

Mapp gas - methyl acetylene-propadiene stabilized fuel gas. Used in the soldering and brazing process.

Mar - to bruise or damage.

Marine water box - a water box on a chiller or condenser with a removable cover.

Masking - the presence of a background noise designed to cover up other unwanted noises.

Masonry anchors - expanding devices that accept a fastener for mounting objects on a masonry surface.

Masonry opening - the space in a masonry wall left open for windows or doors. The same as a rough opening in a frame wall. The header (or lintel) in a masonry opening is usually a steel beam.

Mass - a quantity of matter cohering together to make one body which is usually of indefinite shape.

Mass spectrum analysis - an absorption machine factory leak test performed using helium.

Master - a trade person possessing the qualifications of a journeyman and also the knowledge of the physical laws affecting his/her work and installations. An expert in his/her field.

Master controller - an instrument whose variable output is used to change the set point of a sub-master controller. The master may be a humidistat, pressure controller, manual switch, transmitter, or thermostat.

Master fuel trip - a device for rapid automatic shutoff of all fuel, including igniters; may have manual or automatic initiation.

Master switch - the manual switch through which power is applied to a flame safeguard control; the principal disconnect means. Also called a disconnect switch.

Mastic cement - An adhesive paste used in the fabrication and sealing of thermal insulation on piping fittings, equipment, and duct work. It spreads easily and dries permanently. Its applications include new ducts and old, sheet metal seams, and ductboard. Not all mastics are created equal. Some are toxic, some are water-based, and some take longer to cure, making them more suitable for new construction than retrofit. A non-hardening, adhesive waterproof material (usually with a synthetic rubber, bituminous or similar material used as a base) for sealing joints between sheet metal panels when required.

Mat foundation - continuous reinforced concrete foundation constructed under entire building as a unit. Also known as a raft foundation or floating foundation.

Matched system - a heating and cooling system comprised of products that have been certified to perform at promised comfort and efficiency levels when used together, and used according to design and engineering specifications.

Material safety data sheet - an OSHA form that must be in English and include material identity (chemical and common names); ingredients in mixtures, if they are hazardous; physical and chemical characteristics; physical hazards (fire, reactivity); health hazards (signs and symptoms of exposure); medical conditions aggravated by exposure; primary route(s) of entry into the body; OSHA, ACGIH, and venting exposure levels and TLV's; carcinogenicity statement per NTP, IARC, or OSHA; general precautions for safe handling and use; emergency response procedures and first aid; and appropriate work practices.

Material takeoff - a listing of materials including number of items, descriptions, quantities, and prices for various aspects of a project. These are usually made from the details, fabrication drawings or models.

MATT - for efficient combustion, it is necessary to have the proper mixture, atomization, temperature, and time to complete combustion.

Matter - any material that makes up the world. Anything that occupies space and has weight. Can either be a solid, a liquid, or a gas.

Maximum allowable charge - the amount of liquid refrigerant that the base of the compressor can safely contain without slugging on start-up.

Maximum allowable working pressure (MAWP) - this is determine by the design and construction of the boiler in conformance with the ASTM Code.

Maximum allowed upward heat flux - the maximum rate of heat transfer from a radiant panel, (often measured in Btu/hr/ft2), based on limiting the average surface temperature of the panel to a predetermined value.

Maximum continuous current (MCC) - a value that is determined by the manufacturer of refrigerant motor compressors under high load (high refrigerant pressures) conditions. This value is established in the National Electric Code as being no greater than 156% of the marked rated amperes or the branch circuit selection current.

Maximum display counts - the highest value a meter can display in any range. For example, a 4000 count meter can display from 0000 to 3999.

Maximum operating pressure (MOP) - limits the pressure to the compressor to protect it from overloading or overamping the compressor after defrost or pull down.

Maximum overcurrent protection - a term used with equipment that have a hermetic motor compressor(s). Maximum over current protection is the maximum ampere rating for the equipment's branch circuit over current protective device. The ampere rating is determined by the manufacturer, and is marked on the nameplate of the equipment. The nameplate will also indicate if the over current device can be a HVAC/R type circuit breaker, a fuse, or either.

Maximum power rating - the maximum power that a device can handle safety. Measured in watts.

Maximum surge pressure - the safe pressure for the switch housing but which may damage the mechanism by continuous or repetitive application of pressure.

Maximum working pressure - the maximum pressure at that a piping system may be operated in accordance with the provisions of the piping code. It is this pressure that is used in determining the setting of pressure relieving or pressure limiting devices installed to protect the system from accidental excessive pressure.

Mean - the mid point in any ranked series of numbers. The mid-point between the maximum and the minimum values of any event.

Mean annual earth temperature - the mean average temperature of the earth as it changes throughout the year. The largest factor for earth temperature change is sunshine.

Mean effective pressure (MEP) - the average pressure on a surface when a changing pressure condition exists.

Mean radiant temperature (MRT) - the mathematically average of the temperatures of all wall, ceiling, and floor surfaces in a room. In some definitions of MRT, the temperature of each surface has a weighting factor based on the location in the room at which the MRT is being determined.

Mean temperature - the average temperature for a given day.

Means of egress - a continuous and unobstructed path of travel from any point in a building or structure to a public way and consists of three separate and distinct parts: (a) the exit way access, (b) the exit way and 8 the exit way discharge. A means of egress comprises the vertical and horizontal means of travel and shall include intervening room spaces, doors, hallways, corridors, passageway, balconies, ramps, stairs, enclosures, lobbies, escalators, horizontal exits, courts, and yards.

Measurable resistance - the actual resistance of a circuit or component measured with an ohmmeter.

Measured gas - gas that has passed through and the volume of which has been measured by a meter, or gas that has been otherwise measured such as by weight or volume.

Measured variable - the variable such as temperature, relative humidity or pressure measured by a sensing element.

Measurement junction - the hot junction on the thermocouple used to measure the unknown temperature.

Mechanical atomizer - a device that breaks fuel up into tiny droplets without using an atomizing medium. Also see atomizer.

Mechanical-atomizing oil burner - see atomizing oil burner.

Mechanical bender - a tool that provides extra leverage for bending tubing, while preventing crimping.

Mechanical cycle - a cycle that has a repetitive series of mechanical events.

Mechanical draft - the pressure difference, created by machinery such as a fan or blower, sufficient to supply all, or part of, the required combustion air into the combustion chamber. See forced and induced drafts.

Mechanical draft burner - a gas burner relying on machinery to bring in combustion air. Several methods are used to get air into the combustion chamber.

Forced draft burner (commonly called a power burner) - combustion air is blown in by a motor driven fan. Examples are aspirating, gun type, mechanical premix, and zero governor burner.

Induced draft burner - a fan at the outlet of the combustion chamber creates a slight partial vacuum within the chamber; the higher atmosphere pressure outside will force air into the combustion chamber.

Fan-mix burner - air is drawn in by a fan, which is driven by high-pressure fuel gas escaping from the small orifices in driver arms attached to the fan blades.

Mechanical draft tower - a cooling water tower that employs a fan or series of fans to move large volumes of air over its wetted surfaces. These towers can be forced or induced draft.

Mechanical drawings - these plans show the heating, ventilation, cooling, and plumbing systems. Sometimes these plans will be shown on separate drawings sheets.

Mechanical duty (motors) - cooling by means of a fan inside the motor housing.

Mechanical efficiency - the ratio of work done by a machine to the energy used to do it.

Mechanical energy - the total energy possessed by an object based on its height above the earth, its speed, and in the case of a fluid -- its pressure. The total mechanical energy content of a fluid is called head.

Mechanical equivalent of heat - the ratio of heat energy to mechanical energy. 1 Btu = 778 foot- pounds. This is sometimes called Joule's equivalent.

Mechanical exhaust system - equipment installed in and made a part of the vent, which will provide a positive induced draft.

Mechanical interlock - a scheme where a device is energized, and the operation of that device mechanically operates a switch that energizes a second device.

Mechanical joint - a connection between pipes, fittings, or pipes and fittings, which is neither screwed, caulked, threaded, soldered, solvent cemented, brazed nor welded. Also, a joint in that compression is applied along the centerline of the pieces being joined. Some joints are part of a coupling, fitting or adapter.

Mechanical linkage - a combination of mechanical devices for transferring motion within a mechanical system. Shafts, gears, drive wheels, cams, and other similar devices are used in mechanical linkage systems.

Mechanical premix burner - a power burner where the air and gas are mixed, under pressure, in a motor driven blower. Simultaneous modulation of air and gas allows adjustment of the air fuel ratio over a wide range, so an extremely short hot flame can be produced. Also see forced draft and premix burners.

Mechanical refrigeration - a term usually used to distinguish a compression system from an absorption system.

Mechanical seal - a device mounted in the pump housing and/or on the pump shaft, to prevent leakage from the pump. It has a rotating part and a stationary part with highly polished touching surfaces. Has excellent sealing capability and life, but can be damaged by dirt or grit.

Mechanical stoker - a device consisting of a mechanically operated fuel feeding mechanism and a grate, and is used for the purpose of feeding solid fuel into a furnace, distributing it over the grate, admitting air to the fuel for the purpose of combustion, and providing a means for removal or discharge of refuse. The stoker is used with firebox and water tube boilers only.

Mechanical steam trap - a trap that operates on the differences of the densities between condensate water and steam.

Mechanical ventilation - the mechanical process of supplying air to, or removing air from, any space.

Media - the material in a filter that traps and holds impurities.

Medium - the porous material that performs the actual process of filtration. The plural of this word is “media”.

Medium, heating - a solid or fluid, such as water, steam, air or flue gas, used to convey heat from a boiler, furnace or other heat source, and to deliver it, directly or through a suitable heating device, to a substance or space being heated.

Medium temperature refrigeration - refrigeration where evaporator temperatures are 32° F or below and normally used for preserving fresh food.

Medium temperature water system - this hydronic heating system operates at temperatures between 250 and 350° F, with pressures not exceeding 160 psi the usual design supply temperature is approximately 250 to 325° F, with a usual pressure rating of 150 psi for boilers and equipment.

Mega - prefix for one million.

Megacycle - one million cycles.

Megahertz - one million cycles per second.

Megavolt - one million volts.

Megawatt - one million watts.

Megger - an instrument (megohm meter) that can detect very high resistances, in million of ohms. Megger relates to megohm or 1,000,000 ohms.

Megohm - a measure of electrical resistance equal to a million ohms.

Megohm meter - an instrument for measuring extremely high resistances (in the millions of ohms range).

Melt - to change state from a solid to a liquid. This change of state requires latent heat.

Melting point - the temperature at atmospheric pressure, at which a substance will melt.

Melting range - the temperature range between the liquidus and solidus points. The greater the difference between these two points, the greater the melting range.

Melting time - the amount of time required to melt the fuse link during a specified over current.

Membrane waterproofing - system of waterproofing masonry walls with layers of felt, canvas or burlap and pitch.

Mercaptan - a group of organic compounds resembling alcohols, but with sulfur replacing the oxygen of the hydroxyl group. This is the chemical that is put into natural gas for scent detection.

Mercoid bulb - see mercury bulb.

Mercury - electrically conductive, silver colored metal that is liquid at room temperature. Liquid commonly used in a liquid thermometer. Chemical symbol is Hg.

Mercury barometer - a measuring device for atmospheric pressure. It consists of a hollow glass tube, sealed on one end and open at the other that is filled with mercury and turned upside down, with the open end placed in a dish half filled with mercury.

Mercury bulb - an electrical circuit switch that uses a small quantity of mercury in a sealed glass tube to make or break an electrical contact with the terminals inside the tube.

Mercury switch - an electric switch made by placing a large globule of mercury in a glass tube having electrodes arranged in such a way that tilting the tube will cause the mercury to make or break the circuit.

MERV rating - the MERV (minimum efficiency reporting valve) rating is the "overall" reporting valve of an ASHRAE standard 52.2 evaluated air filter. The MERV is a single number that is used, along with the air velocity at which the test was performed. This was to simplify the extensive data generated by the method of testing. MERV is expressed on a 20 point scale and is derived from the particulate size efficiency for each of the groups. MERV values of 1 to 4 are less than 20% efficient, 5 to 8 are <20 to 35%, 9 to 12 are 40 to 75%, 13 to 16 are 80 to 95+%, and 17 to 20 are for HEPA and ULPA filters that have a 99.97 to 99.999% efficient rating. The MERV is based on the worst case performance of a filter through all six stages of dust loading and all particles 0.3-10 microns. Because the rating represents the worst case performance, specifiers can use it to assure performance in applications where a maximum particle count must be maintained over the filter's entire life.

Met - term applied to the heat released from humans at rest. It equals 18.4 Btu/ft2/hr.

Metal fatigue - a weakening of metal due to repeated bending or constant vibration.

Metal oxides - oxidized ferrous particles which are very old or have been recently produced by conditions of inadequate lubrication. Doing trend analysis on oils are important for equipment life and maintenance.

Metallic bonding - the method by which loosely held atoms are bound together in metals.

Metallurgy - is the term for the science and technology of metals and their characteristics.

Meter - a metric unit of linear measurement equal to 39.371 inches or an instrument for measuring rates or integrating rates over a period of time. 100 cm.

Meter, electrical - an instrument used to measure various characteristics of an electric circuit such as ohms and volts.

Meter, flow - an instrument used to indicate the rate of a liquid flow in the system. The instrument is calibrated for the specific application.

Meter, multiplier - a resistor connected to a meter (in series) for purposes of providing greater range capacity.

Metered balancing valves - specialized valves that allow the flow rate in a branch piping circuit to be precisely adjusted to a desired value.

Metering device - a valve or small diameter tube that restricts fluid flow. The metering device regulates refrigerant flow to the evaporator to match the evaporator's capacity. The metering device separates the high and low side of the refrigeration system. The coldest point in the refrigeration system.

Metering valve - see firing rate valve.

Methane - a hydrocarbon gas with the formula CH_4, the principal component of natural gas. Molecular weight 16.043, specific volume is 23.80 cubic feet/lb., and a boiling point of -259.0° F.

Methanol drier - an alcohol type chemical used to change water in refrigerating system into a non-freezing solution.

Methyl Chloride (R-40) - a chloromethane chemical once commonly used as a refrigerant. The chemical formula is CH_3CL. Cylinder color code is orange. The boiling point at atmospheric pressure is -10.4° F.

Methylacetylene-propadiene - a stabilized gas often referred to as MAPP gas.

Metric system - a decimal system of measures and weights, based on the meter and gram. Length of one meter is 39.371 inches.

Mezzanine - an intermediate level between the floor and the ceiling of any story, and covering not more than thirty-three (33) per cent of the floor area of the room in which it is located.

Mg - milligram (1/1000, 10-3) of a gram.

MHO - the reciprocal of resistance.

Mica - an insulating material that can be cut into thin sheets. It is used as the dielectric in some capacitors and for insulation in vacuum tubes. In motors, mica is used to separate the copper segments of the commutator from each other. In boiler applications it is used to protect the flat gauge glass from the etching action of steam and water.

Micro - one millionth part of unit specified.

Microamp - one millionth of an ampere.

Microammeter - a meter that has the ability to measure and display microampere currents.

Microampere - one one-millionth (0.000001) of an ampere.

Microbar - a unit of pressure equal to one one-millionth (0.000001) of the pressure exerted by the atmosphere.

Microbial contaminants - fungal, bacterial, or viral organisms, and the toxins they produce, or particles bearing such organisms or toxins that are airborne or deposited on indoor surfaces and that can cause disease, irritation, or allergic reaction. Also referred to as biological contaminants or biocontaminants, and when carried by the airstream may become bioaerosols.

Microbubble resorber - a type of deaerator that separates and collects microbubbles from the system's fluid. It lowers the dissolved air content of the water, enabling it to absorb any residential air in the system.

Microbubbles - very small bubbles that give water its cloudy appearance.

Microfarad - unit of capacitor electrical capacity equal to one millionth of a farad. A common rating of both start and run capacitors. The "electrical size" of the capacitor rating. Refers to the size of the plate area of the capacitor. Symbol is μ or mfd.

Microgram - one millionth of a gram.

Microhenry - one millionth of a Henry.

Microinch - one millionth of an inch.

Micromanometer (hook gauge) - an instrument designed to read small differences in air pressure accurately and usually have a wide scale range. Most scales read from 0 to 4.0 inches water column in hundredths of an inch on the vertical scale and in thousandths of an inch on a vernier scale and are readable to 0.001 inches. There is more than one variation of this instrument. The most common type contains two glass vials about 2 to 3 inches in diameter. A pointed needle or hook is positioned by a micrometer adjustment until the point dimples the water surface but does not break the surface tension. The difference in level is determined in micrometer readings. Another variation of this instrument has a single vial or well and an inclined scale. The well is positioned by a micrometer or vernier adjustment. It is very important that all micrometers, including the electronic units, be accurately leveled before and after each reading. The solid state electronic hook gauge will measure positive, negative, or differential pressures to ".00025 inch water column over a 0 to 2 inch water column range. It can also be used with a pitot tube for precision measurement of low air velocities.

Micrometer - a micrometer is used to measure the outside of an object. An object is placed inside the micrometer structure, then the thimble is tightened down to place the object against the anvil. The micrometer is read by reading the highest figure visible in the thimble: 1 = 0.100, 2 = 0.200, and so on. Add to this number the visible vertical number: 1 = 0.025, 2 = 0.050, and so on. Now add these two numbers together and arrive at the total thickness of the material being measured.

Micron (μ) - a metric unit. A term used in measuring extremely small particles. For length it is one millionth (1/1,000,000) of a meter or 39 millionth of an inch (.000039"). For pressure it is one 25,400 (1/25,400) of an inch of mercury. There are 25,400 microns in one inch. Relatively speaking, a grain of salt is about 60 microns and the naked eye can see particles to about 20-40 microns. A micron is also known as a micrometer.

Micron gauge - instrument for measuring vacuums levels very close to a perfect vacuum. These instruments measure the last part of an inch of vacuum in microns.

Microorganism - a microscopic organism, especially bacterium, fungus, or a protozoan.

Microsecond - one millionth of a second.

Microswitch - an electrical circuit control that opens and closes on a very slight motion.

Microvolt - one one-millionth (0.000001) of a volt.

Mid seated (cracked) - a position on a valve that allows refrigerant flow in all directions.

Migrate - to move from one area to another.

Mil - one one thousandth of an inch.

Mil foot - standard measurement of resistance of wire. A mil foot of wire is one circular mil in diameter and one foot long.

Mildew - fungi growth to cover an area with superficial fungal stains.

Mile - an English measure of distance. One mile equals 5,280 feet (1.609 meters) or 1.609 km.

Mill files - mill files are always single cut. They are used primarily for sharpening.

Milli - one-thousandth (0.001) 1000 millivolts equals 1 volt.

Milliammeter - a meter that measures current in thousands of an ampere.

Milliampere - a unit of electrical current equal to one thousandth of an ampere.

Milligram - 1/1000 of a gram.

Millimeter - 1/1000 of a meter.

Millimeter of mercury - a unit of pressure equal to the pressure exerted by a column of mercury one millimeter high at sea level and at temperature of 0° C (32° F).

Millisecond - one thousandth of a second.

Millivolt - a unit of electrical voltage equal to one thousandth (1/1000) of a volt.

Milliwatt - one one-thousandth (0.001)of a watt.

Min/max/average - A feature of some digital multimeters or clamp-on ammeters that will record the minimum, maximum, and average of all measurements made during a time period.

Mineral oil - a traditional refrigerant lubricant used in CFC and HCFC systems. Oil derived from a mineral source, such as petroleum, as opposed to oils derived from plants and animals.

Minimum circuit ampacity (MCA) - a term used with equipment that has a hermetic motor compressor(s). Minimum circuit ampacity is the minimum ampere rating value used to determine the proper size conductors, disconnecting means, and controllers. Minimum circuit ampacity is determined by the manufacturer, and is marked on the nameplate of the equipment.

Minimum firing rate - the lowest input rate for a burner or a process.

Minimum fresh air requirements - the amount of fresh, outdoor air needed to replenish interior space air to satisfy local legal codes.

Minimum on-time - the shortest period of time that a load can be energized when it is being duty cycled.

Minimum stable signal (MSS) - the best superheat setting which will provide constant or little temperature change at the thermostatic expansion valve temperature sensing element while the system is running.

Minimum supply temperature - the lowest temperature allowed as the supply temperature from a boiler being operated by an outdoor reset control. The minimum is typically set to prevent the boiler from operating with sustained flue gas condensation.

Minipeeper ® - a Honeywell trademark. A compact, ultra-violet flame detector which consists mainly of a high power sensing tube called a ultraviolet power tube.

Minor diameter - the smallest diameter measuring from the inside of the threads.

Minority carrier drift current - the current flow across a P-N junction carried by the minority carriers and is accelerated by the potential barrier.

Minority carriers - current carriers (either electrons or holes) that are in the minority in a specific layer of semiconductor material. It is the diffusion of minority carriers through the cell barrier that creates a voltage, and constitutes a current, in a photovoltaic device. The process becomes more efficient with increasing minority carrier diffusion length.

Misalignment - not properly aligned, not straight.

Miscible - the extent to which liquids and gases can be mixed or blended. Capable of being mixed in all proportions.

Miscibility - the ability of different materials to mix together.

Mist - suspended liquid droplets in the air generated by condensation from the gaseous to the liquid state or by mechanically breaking up a liquid by splashing or atomizing.

Mitigation - procedures and processes used to reduce the effects of any IAQ problem situation.

Mixed air - a combination of return air and outdoor air. The outdoor air may be introduced to meet outdoor ventilation requirements, or to reduce mechanical cooling when the outdoor air conditions are suitable. That air in a ventilation system that is composed of return air and outdoor air before it has been conditioned.

Mixed gas - a mixture of natural and manufactured gases. Mixed gases result from the increased distribution of natural gas, through transcontinental pipe lines, into areas having existing manufactured gas facilities.

Mixer - a device that combines things so that the resulting substance is uniform in composition, whether or not the separate elements can be distinguished. See gas, fan, aspirator, and inspirator mixers.

Mixer face - the air inlet end of the mixer head.

Mixer head - that portion of an injection type burner, usually enlarged, into which primary air flows to mix with the gas stream.

Mixer throat - that portion of the mixer which has the smallest cross-sectional area and which lies between the mixer head and the mixer tube.

Mixer tube - that portion of the mixer that lies between the mixer throat and the burner head.

Mixing box - a container, located at the room being conditioner, in which hot and cold air is mixed as required to maintain the desired room temperature.

Mixing dampers - two interlinked sections of a damper. As one section opens, the other section closes.

Mixing device - any device in which two or more streams of fluid are intentionally mixed together to create a fluid temperature between the higher and lower entering fluid temperatures.

Mixing reset control - the process of automatically adjusting the supply temperature to a hydronic distribution system based on outdoor reset control logic, and accomplished by mixing hot water from a heat source with cooler return water from the distribution system.

Mixing valve - a 3 way valve with two inlets and one outlet, designed specifically for mixing applications. The proportions of the fluid leaving the outlet can be varied by moving the valve stem. It cannot be used for diverting applications because it has only one disc and two seats.

Mixture - a blend of two or more components that do not have a fixed proportion to each other, and that, however well blended, keep their individual chemical characteristics. Unlike compounds, mixtures can be separated by physical methods like distillation. One example is a near-azeotropic blend of refrigerants.

Ml - milliliter. One thousandth of a liter. A metric unit of capacity, for all practical purposes equal to 1 cubic centimeter. One cubic inch is about 16 ml.

MMbtu - an abbreviation for one million (1,000,000) Btu's.

mm Hg - millimeters of mercury. A measure of pressure in millimeters of a mercury column above a reservoir, or difference of level in a U-tube.

Mode - a particular functioning arrangement or condition.

Mode door - a baffle that directs incoming air through either the heating or cooling coil depending upon the function desired.

Model number - a number that identifies a set of identical products or components.

Modulate - to adjust by increments and decrements; to modify by varying a second condition; the firing rate position during which the firing rate motor is under control of an external series 90 controller.

Modulating - a type of device or control that tends to adjust by increments (minute changes) rather than by either full on or full off operation.

Modulating boilers - a boiler that adjusts its heat output from some maximum value down to a significantly lower level based on the instantaneous demand for heat by the load.

Modulating control - a method of control in which the controlled device (in this case a valve or damper) may assume any position between fully open and fully closed. Also see proportioning control.

Modulating controller - a proportioning controller that automatically positions the device it controls anywhere between its extreme positions in response to changes in the controlled medium. Also see firing rate and proportioning controllers.

Modulating fire - the process of varying the firing rate as the load varies. This results in longer burner operation because the firing rate is lowered as temperature or pressure builds up toward the cutoff point. It decreases on/off operation and increases the efficiency of the system.

Modulating motor - an electric motor, used to drive a valve or damper, which can position the valve or damper anywhere between fully open or fully closed in proportion to changes in the controlled medium. Also see firing rate motor.

Modulating pressure control - located at the highest part of the steam side of the boiler. Sends a signal to the modulating motor that controls the firing rate.

Modulating range - it is the change in the controlled condition necessary for the controller to position the modulating actuator from one extreme to the other.

Modulating refrigeration cycle - refrigerating system of variable capacity.

Modulating valve - an automatic control valve designed so that the valve opening is varied within narrow limits throughout the entire range from high fire to low fire position by means of a proportional actuator. Also see firing rate valve.

Module - a combination of components in a circuit that are combined in a replacement package. Replacement is by the module in its entirety.

Module, water tower - a pre-assembled portion or section of a cooling tower cell. On larger factory assembled towers, two or more shipped modules may require joining to make a cell.

Modutrol motor - Honeywell® trademark. See modulating motor.

Moisture barrier - a substance that will prevent the flow of water, but not water vapor.

Moisture content - amount of water vapor in a given amount of air, usually expressed in grains of moisture per pound of dry air. There are 7000 grains of moisture per pound of moisture.

Moisture determination - an action using instruments and calculations to measure the relative or absolute moisture in an air conditioned space.

Moisture indicator - instrument containing a chemical disk that is sensitive to moisture and gradually changing color to reflect these changing conditions. Used to measure and indicate moisture content of a refrigerant.

Moisture vapor seal - a tight barrier located on the interior of a wall section placed on the outside of the insulation to prevent pushing of moisture through the insulation by vapor pressure.

Mol - the unit of weight for mass.

Mold - infestation of fungal growth to the point of destruction or disintegration of the base on which the mold is growing.

Molecular - pertaining to or consisting of molecules.

Molecular motion - the movement of molecules within a substance.

Molecular sieve - a refrigerant drying method employing an adsorbent having pores of uniform molecular dimensions to aid in the collecting of unwanted water.

Molecular weight - the average weight of a molecule of a substance. It is measured against a n oxygen atom.

Molecule - the smallest particle that a compound can be reduced to before being broken down into its basic elements.

Molex connector - connectors that provide an easy means of disconnecting and connecting wires. Can be used over and over.

Mollier's diagram - graph showing the relationship between the thermal properties of fluids. These would include refrigerant pressure, heat and temperature properties. Also known as an enthalpy chart.

Molly bolts - expanding metal fasteners used to anchor light loads to hollow wall sections.

Molten alloy relay - an overload device that opens a set of contacts by thermal energy. This type of device allows the temperature produced by the starting current of a load to be transferred to a molten alloy device. When it reaches a certain temperature, it will melt the solder around the device, causing it to slip and open the contacts; when it cools it will harden again and the relay must be manually reset.

Moly - molybdenum disulfide, a solid lubricant and friction reducer, colloidally dispersed in some oils and greases.

Momentary loss of power - a short interruption of power to a circuit or system.

Momentary switch - a spring loaded electrical control that will make or break a circuit only when it is held in place.

Monel - a trademark name for metal alloy consisting chiefly of copper and nickel.

Monitor - a device that senses a condition requiring attention and gives an alarm without initiating corrective action.

Monitor top - unit built by General Electric that which has a cylindrical condenser surrounding the motor compressor mounted on the top of the cabinet.

Monochlorodifluoromethane - a HCFC refrigerant known as R-22. Chemical formula is $ChClF_2$. Cylinder color code is green. Boiling point is -41.4° F at atmospheric pressure. See R-22.

Monoflo tee® fitting - a fitting used in one pipe heating systems to divert hot water from the supply line to the individual heaters or radiators. Manufactured by the Bell and Gossett Corporation.

Monoxor® - an instrument used to measure the concentration in parts per million (ppm) of carbon monoxide (CO) in a vapor sample. Manufactured by Bacharach.

Montreal Protocol - an agreement signed in 1987 by the United States and 22 other countries, and updated several times since then, to control releases of ozone-depleting substances like CFC's and HCFC's, and determine the eventual phase out of their use.

Motive power - term used to express the source from which a device such as a diaphragm valve, relay, or motor obtains its power (electric or fluid).

Motor - a device used to create a rotating motion and drive components that require rotating motion. Electric energy is changed to mechanical energy by magnetism, which causes the motor to turn.

Motor 2 pole - a 3600 rpm electric motor (synchronous speed). Full load speed of 3450 rpm.

Motor 4 pole - a 1800 RPM electric motor (synchronous speed). Full load speed of 1725 RPM.

Motor 6 pole - a 1200 RPM electric motor (synchronous speed). Full load speed of 1150 RPM.

Motor burn out - a condition in which the insulation of an electric motor has deteriorated by overheating.

Motor, capacitor - a single phase induction motor with an auxiliary starting winding connected in series with a capacitor for better starting characteristics.

Motor, capacitor start/induction run - a motor having high starting and breakdown torque, and medium starting current. Used on hard starting applications such as compressors, positive displacement pumps, farm equipment, etc.

Motor, capacitor start/capacitor run - a motor similar to the capacitor start motor except that the run capacitor and run winding are designed to remain in the circuit at all times.

Motor control - device to start and/or stop a motor at certain temperature or pressure conditions.

Motor, induction – this is the most common motor type used in the HVAC/R industry. Speed remains relatively constant as load changes. Several kinds of induction motors include the shaded pole, permanent split capacitor, split phase start-induction run, capacitor start-induction run, capacitor start-capacitor run and three phase.

Motor mounting - manner in which motor is attached to equipment. Can be one of the following types:

1. Bolted - motor is attached to frame with removable bolts.
2. Rigid - motor solidly fastened to equipment through a metal base that is bolted, cast or welded to the motor shell or frame.
3. Resilient or cradle - motor shell is isolated from the base by rubber rings on the ends shields to reduce transmission of vibration to the driven equipment.
4. Face or flange - drive end has a flat mounting surface, machined to standard dimensions, with holes to allow easy, secure mounting to driven equipment. Commonly used on pumps, oil burners and gear reducers.
5. Stud - extended through bolts: motor has bolts extending from the front or rear, by which it is mounted. Often used on small, direct drive fans and blowers.
6. Yoke - tabs or ears are welded to motor shell, to allow bolting motor to a fan column or bracket.

Motor, permanent split capacitor (PSC) - performance and applications similar to shaded pole but more efficient, with lower line current and higher horsepower capabilities.

Motor service factor - a factor used as a multiplier over and above an electric motor's normal operating design parameters, as indicated on the nameplate, under which it can operate.

Motor, shaded pole - a small induction motor with a shading pole used for starting. This type of motor has a very low starting torque.

Motor, split phase start/induction run - moderate starting torque, high breakdown torque. Used on easy starting equipment, such as belt driven fans and blowers, grinders, centrifugal pumps, etc.

Motor start up current - a condition where motors draw more current in the first few seconds after they are turned on than they do after they are up to speed. Clamp-on meters are used to measure this start up current.

Motor starter - high capacity electrical switching device that contain a coil, contacts and overload protective devices.

Motor, three phase - operates on three phase power only. Provides high starting and breakdown torque, high efficiency, medium starting current, simple, rugged design, and has a long life. Designed for commercial applications.

Motor vehicle air conditioner – mechanical vapor compression, open drive compressor appliances used to cool the drivers or passengers compartment of a motor vehicle, including agricultural and construction vehicles. This excludes appliances using HCFC-22 (R-22) refrigerant, and devices using hermetic or semi-hermetic compressors.

Motorized damper - a damper that has an actuator linked to it in such a way as to make it modulating, proportional, normally open or normally closed.

Motorized valve - an automatic control valve that is completely opened by the rotation of an electric motor, and is generally automatically closed by a spring or other mechanical means when the electrical circuit is broken. Also see firing rate motor, firing rate valve, and safety shutoff valve.

Mount - a support leg.

Mouse trap - another name for the balancing relay in a modulator motor. See balancing relay.

Movable contact - the member of a contact pair that is moved directly by the actuating system.

MSDS - material safety data sheet. A fact sheet summarizing information about material identification, hazardous ingredients, health, physical and fire hazards, first aid, chemical reactivities and incompatibilities, spill, leak, and disposal procedures, and protective measures required for safe handling and storage. OSHA has established guidelines for descriptive data that should be concisely provided on a data sheet to serve as the basis for written hazard communication programs. The thrust of the law is to have those who make, distribute, and use hazardous materials responsible for effective communications.

MSST - maximum safe storage temperature.

Mud drum - vessel in the lower part of a boiler where sludge and other impurities collect after settling out of water.

Muffler - sound absorbing chamber in a mechanical system used to reduce the sound of gas pulsations.

Mullion - the horizontal or vertical stationary member forming a division between adjoining doors or windows.

Mullion heater - an electrical heating element mounted in the mullion and used to keep the moisture around refrigerated doors from condensing, sweating or freezing.

Multi-burner system - a heating system with more than one main burner.

Multi-conductor - more than one conductor within a single cable complex.

Multi-drop - a form of digital communication in which all digital devices are connected in parallel to a common bus.

Multi-family structure - building of three stories or fewer above grade containing three or more living units other than townhouses, including a manufactured building (modular).

Multi-flame tip - brazing tip that produces several small flames, which tend to wrap around the object or tubing being heated.

Multi-fuel burner - see combination burner.

Multifunction counters - instruments that measure alternating electrical signals. Provide frequency, period, frequency ratio, time interval, and totalize functions. Key parameters include frequency range, least significant digit displayed, resolution, accuracy, and stability.

Multimeter - a meter that measures two or more electrical quantities, such as volt, ohm and milliamperes. There are two types: analog and digital. Either type is available in bench or handheld models. Key parameters of multimeters are voltage, current, and resistance ranges, number of digits displayed, resolution, and accuracy. Also called a VOM or DMM.

Multi-pass boiler - a horizontal, fire tube boiler in which the flue gases are passed back and forth through the boiler shell two or more times by means of a tube arrangement or baffles. Each change in direction is called a pass. In order to maintain the flue gas velocity, each succeeding pass is reduced in total cross section to compensate for the reduced flue gas volume as the gases cool. Also see firebox, fire tube, and scotch boilers.

Multiple evacuation - a procedure for removing the refrigerant from the system. A vacuum is pulled, a small amount of refrigerant is allowed into the system, and the procedure is repeated. This is also called a "triple evacuation" because the procedure is usually done three times.

Multiple evaporator system - a refrigerating system with two or more evaporators connected in parallel. Also referred to as multiplexing.

Multiple junction cell - a photovoltaic cell containing two or more cell barriers, each of which is optimized for a particular portion of the solar spectrum to achieve greater overall efficiency in converting sunlight into electricity.

Multiple loop control - one controller used to control several independent HVAC systems.

Multiple pass – the process that applies to recovery/recycle units that removes the refrigerant from the system being repaired and circulates through the recovery/recycle system several times or for a time to remove contaminants other than noncondensables and other types of refrigerants. The refrigerant is then pumped into the recovery cylinder.

Multiplexed evaporators - the connection of several evaporators to a single compressor. Evaporator pressure regulator valves make it possible to operate each evaporator at different temperatures.

Multiplexing system - a refrigerating mechanism in which several evaporators are connected to one condensing unit.

Multiplier (voltmeter) - a resistance placed in series with the movement of a voltmeter to increase the measuring range of the instrument.

Multiplier (ohmmeter) - the multiplying factor used to determine the actual resistance value of the item being tested. For example, the actual ohmic value measured by the meter is 35 ohms and the meter is being used on the R times 100 scale. The actual resistance would be 3500 ohms. (35 x 100 = 3500)

Multi-port burner - a burner having a number of nozzles from which fuel and/or air are discharged into the combustion chamber; as distinguished from a single port burner.

Multi-position - a control arrangement that provides two or more points of control between full off and full on.

Multi-range meter - a meter that has two or more ranges to measure an electrical quantity.

Multi-speed motor - one which can be operated at any of two or more definite speeds, each being practically independent of the load. In the case of multi-speed permanent split capacitor (PSC) and shaded pole motor, the speeds are dependent on the load. Speed changes are used to vary the output of the conditioning equipment. Speed changes could be achieved by regulation of internal or external resistance or by frequency changes.

Multi-stage thermostat - a temperature control that sequences two or more switches in response to the amount of heating or cooling demand.

Multi-tapped transformer - a transformer that uses different turn ratios between the primary and secondary windings to achieve different voltage and amperage values.

Multi-vibrator - a term that describes a group of relaxation oscillators that generally use two active devices, and which produce non-sinusoidal waveforms. The two active devices are RC coupled to each other; one conducts while the other is cut off. Frequency is determined by the RC times of the coupling circuits.

Multi-zone - a heating and cooling unit that is equipped to condition more than one zone in a structure.

Muriatic acid - a cleaning solution known as hydrochloric acid.

Mushroomed - term used to describe the head of a chisel or other tool that is flattened and spread out from being struck.

Muntins - these are the parts of a casement window that separates panes of glass. They may be made of metal or wood.

Mutual inductance (M) - when two coils are located close together so that the magnetic flux of the coil affect one another in terms of their inductance properties.

-N-

NA - not applicable.

Nc - national course thread.

N/c - normally closed.

Neg - negative.

Nf - national fine thread.

N/o - normally open.

No - number.

NO - the chemical symbol for nitric oxide.

NO_2 - the chemical symbol for nitrogen dioxide.

Nom - nominal.

NOWL - normal operating water level.

NPSHA - net positive suction head available.

NPSHR - net positive suction head required.

Nr - not recommended.

NRE - net refrigeration effect.

Ntc - negative temperature coefficient thermistor.

N type material - semiconductor material with an excess of electrons that move from negative to positive when a voltage is applied.

Nand gate - a "not" and logic gate. An "and" gate, followed by an inverter, which inverts an "and" gate output to the opposite value.

Nameplate - plates on motors, furnaces, air conditioning and other appliances that provide important information describing the characteristics of that particular piece of equipment. Items could include, but not limited to model number and serial number, voltage, amperage, RLA, FLA, LRA, wire size and fuse size etc.

Nano - prefix meaning one one-billionth (0.000000001).

Naphthenic - a type of petroleum fluid derived from naphthenic crude oil, containing a high proportion of closed ring Methylene groups. A term describing mineral oils used in refrigeration systems. Napthenic oils are not compatible with most of the newer HFC refrigerants developed to lessen ozone depletion.

NARMS's - nearly azeotropic refrigerant mixtures. Zeotropic blends with a small temperature glide.

Nat + nat - refers to the maximum combined input rating of two or more drafthood equipped appliances attached to the common vent.

National Electric Code (NEC®) - a national code written for the purpose of safeguarding persons and property from the hazards arising from the improper use of electricity and published by the National Fire Protection Association. The NFPA 70 is used by insurance inspectors and many government bodies regulating building codes. It does not become law until adopted by federal, state, and local laws and regulations. It is not intended as a design specification nor an instruction manual for untrained personnel.

National fine threads - threads that conform in shape and dimension to standards established by the industry.

National Fuel Gas Code - a model code developed by the National Fire Protection Association (NFPA 54) that is widely recognized in the U.S. It specifically covers the fuel supply, ventilation, and exhaust requirements of combustion-type heat sources.

National pipe thread - the standard tapered pipe threads used on piping, fittings, and valves in the U.S.

Natural convection - heat transfer between a solid surface and a moving fluid when the fluid's motion is created only by natural processes such as the tendency of a warm fluid to rise.

Natural draft - the pressure difference resulting from the tendency of hot gases to rise up a vertical flue or chimney, thus creating a partial vacuum in the furnace. Natural draft is primarily a function of the vertical flue height and the temperature difference between the inlet and outlet points of the flue.

Natural draft burner - a gas burner that depends on a natural draft to bring in combustion air. Also see atmospheric burner and natural draft.

Natural draft cooling tower - a cooling tower in which the flow of air depends on natural air currents or a breeze. Generally applied where the spray water is relatively hot and will cause some convection currents.

Natural gas - a mixture of several combustible gases and, usually, a small percentage of inert gasses, obtained from geologic formations in nature. These gasses consist mostly of a mixture of organic compounds (normally methane, butane, propane, and ethane). The heating value of natural gases varies between 900 to 1400 Btu per cubic foot; the usual range for use is 1000 to 1050 Btu per cubic foot. Properties include an ignition temperature of 1170° F, maximum burning speed is 25 inches per second, upper explosive limit (UEL) is 14% by volume, lower explosive limit (LEL) is 4% by volume, and a specific gravity of between 0.4 and 0.8. Chemical formula is $CH4$.

Natural magnet - metallic materials that have magnetic properties in their natural state.

Natural return - process by which air is returned to the air handler without a physical connection between the return grille and the air handler.

Natural ventilation - the unaided movement of air into and out of an enclosed space through intentionally provided openings, such as open windows and doors.

Near-azeotropic blend - a blend that acts like an azeotrope, but has a small volumetric composition change and temperature glide as it evaporates and condenses.

NEC Class 1 wiring - wiring methods and materials required by the National Electric Code for voltages not exceeding 600 (line voltage) except where specifically permitted. This does not involve remote control, low voltage relay switching, low energy power and signal systems, or communications systems as defined by the code.

Needle bearing - a rolling type of bearing containing rolling elements that relatively long compared to their diameter.

Needle nose pliers - pliers with long, tapering jaws used for gripping small objects, especially wires.

Needle point valve - type of valve having a needle point plug and a small seat orifice for low flow metering.

Negative - a non-positive, below zero value. An electric polarity sign indicating an abundance of electrons.

Negative electrical charge - the result of electrons joining atoms. The electrical charge of a body that has an excess of electrons.

Negative pressure - a condition that exists where less air is supplied to a room than is exhausted. Less than atmospheric pressure. Also see furnace pressure.

Negative resistance - a resistance that seems to operate opposite from normal resistance, in which an increase of voltage causes an increase of current. The condition in a generator in which output voltage increases with a decrease of load resistance.

Negative temperature coefficient thermistor (NTC) - electronic thermistor that decreases in resistance as temperature increases. These are composed of metal oxide ceramics, are low cost, and the most sensitive temperature sensors. They are also the most nonlinear.

Negative terminal - the terminal on a battery or power supply that has an excess of electrons, and is said to be negatively charged.

Negotiable instrument - a written promise, such as a promissory note, check or draft, to pay a desired amount of money at a specified time for a particular person or his order.

NEMA ratings - electrical control device rating of voltage, current, horsepower, and interrupting capability given by NEMA.

Neon test light - a lamp consisting of a glass bulb filled with neon and two or more insulated electrodes. When the neon is ionized by a potential difference, current flows through the neon, creating a red glow. Used to check for the presence of voltage in a circuit.

Neoprene - a synthetic rubber that is resistant to hydrocarbon oil and gas.

Net effective volume - that portion of the total structure volume within which the circulating water is in intimate contact with the flowing air.

Net heating capacity - the useful heat exchanged between the refrigerant vapor and the eater being heated. This value is the product of the mass flow rate of water, the specific heat of water, and the temperature difference of the water.

Net load - heat requirements exclusive of piping losses and pickup.

Net oil pump pressure - the difference between crankcase pressure and oil pump pressure.

Net output - for a boiler, the usable gross output per hour from a boiler after subtracting piping and pickup losses.

Net positive suction head available (NPSHA) - the total head available to push water into a circulator at a given point in a piping system. This indicator is totally determined by the piping system and the fluid it contains, and does not depend on the circulator. It is expressed in feet of head. To avoid cavitation, the NPSHA must be equal to or greater than the NPSHR of the selected circulator.

Net positive suction head required (NPSHR) - the minimum total head required at the inlet of the circulator to prevent cavitation. This value is specified by the pump manufacturers.

Net rating - the rating of a boiler which indicates the net load that may be connected satisfactory to it, provided the piping losses and pickup load do not exceed the standard allowances.

Net refrigeration effect - the refrigeration ability of a refrigerant. It is found by subtracting the enthalpy of the liquid entering the metering device from the total heat of the vapor as it leaves the evaporator. This refrigeration effect is measured in Btu's per pound of refrigerant flow.

Network - a series-parallel circuit.

Network control unit - device which monitors or controls designated equipment in a distributed control system and shares data with other control units.

Network expansion unit - device which performs a local control function in a distributed control system, such as starting or stopping a fan.

Neutral - neither positive nor negative. That part of a 115v circuit that carries current, but is at zero electrical potential.

Neutral bus bar - the bare bus bar that all of the neutral wires are attached to. The neutral is at the same potential difference as ground.

Neutral conductor - the circuit completion wire in a 2 wire system. A neutral wire has zero voltage.

Neutral flame - the flame that results from supplying equal amounts of oxygen and fuel.

Neutral point - a predetermined point in the range of a control instrument in which no action takes place.

Neutral pressure - equal to atmospheric pressure. Also see furnace pressure.

Neutral wire - the balance wire in a 3 wire electrical distribution system. The grounded conductor which is usually colored white.

Neutralization number - a measure of the total acidity or basicity of an oil. This includes organic or inorganic acids or bases or a combination of the two.

Neutralizer - a substance used to counteract the effects of corrosive acids. To render chemically harmless.

Neutron - a particle in the nucleus (center) of an atom that has no electrical charge or is neutral.

Newton - force exerted on an object that has a mass of 1 kilogram and gravitational acceleration of 1 m/sec2.

Newton/meter2 - metric unit of measurement for pressure. Also called a Pascal.

Newtonian fluid - a fluid with a constant viscosity at a given temperature regardless of the rate of shear. Single grade oils are Newtonian fluids. Multigrade oils are NON-Newtonian fluids because viscosity varies with shear rate.

Nichrome - nickel chromium alloy used to make resistance heating elements.

Nickel cadmium cell - an alkaline cell that has a longer life than carbon zinc dry cells.

Night set back - adjustment of a temperature control to a lower or higher temperature to save operating costs.

Nipple - a short pipe that is threaded (male) on the outside on both ends. Used for joining piping elements.

Nipple, close - nipple that shows no unthreaded section of pipe between the threaded ends.

Nitration - nitration products are formed during the fuel combustion process. Most nitration products are formed when an excess of oxygen is present. These products are highly acidic, form deposits in combustion areas and rapidly accelerate oxidation.

Nitric oxide (NO) - a colorless gas, formed during combustion that is irritating to skin, eyes, and the respiratory tract. An unstable gas that will quickly change to NO_2 as it cools in a flue.

Nitrogen - a colorless, odorless gas used in pressure testing and purging of refrigerant piping. Symbol is N_2. It is used in the brazing process because it replaces the oxygen in the air in the tube or pipe and reduces the formation of scale (copper oxide). Molecular weight is 28.01, specific volume is 13.8 cubic feet/lb., and a boiling point of -320.4° F.

Nitrogen dioxide (NO_2) - a reddish-brown, mildly poisonous gas formed through combustion that is highly irritating and damaging to the respiratory tract. Often found in smog or transportation exhaust fumes. When combined with rain can produce acid rain. Molecular weight is 46.005, specific volume of 4.7 cubic feet/lb., and a boiling temperature of 70.1° F.

Nitrogen oxides (NO_x) - the class of compounds including nitrogen dioxide and nitric oxide that are formed through combustion of fuels.

Nocturnal radiation - the loss of energy by radiation to the night sky.

Node - in schematic drawings, the dot in the center of the cross created by connecting wires.

No frost freezer - a low temperature refrigerator cabinet in which no frost or ice collects on produce stored in the cabinet.

Noise - (1) electrical or electronic interference with the signal of interest. (2) any undesired sounds, usually of different frequencies resulting in an objectionable or irritating sensation.

Noise criteria - this is the outlet sound rating. It is based on the sound absorption value of a specific room and other set standards.

Noise criteria curve - a single number method, derived from subjective witness testing, used to predict the acceptability of random sounds. It is the most commonly used method used in the air conditioning industry.

Noise dosimeter - instrument used to measure sound in DbA.

Noise reduction - a reduction or loss in sound transmission from one space to another space, through one or more parallel paths of either ceiling furred, plenum ceiling, through partitions, or along ventilation ducts.

Noise reduction ratio - ratio of the ability to diminish stray electrical noise or interference in circuits.

Nomenclature - a system of special terms or symbols, like those used in science.

Nominal - the average rating of voltage, current, power, etc., during normal operation of an electric circuit.

Nominal body rating - see valve body rating.

Nominal R-value - the thermal resistance of insulation as specified by the manufacturer according to recognized trade and engineering standards.

Nominal size - an approximate size that may be greater or less than the actual size.

Nominal size tubing - tubing measurement which has an inside diameter the same size as iron pipe of the same stated size.

Nominal voltage - value assigned to a circuit or electrical system to designate its voltage (for example 120, 208, 240, 277, etc)

Non-adhering sludge - residue formed when feedwater chemicals are added to boiler water containing scale forming salts.

Non-adjustable - type of pressure switch that must be matched to the refrigerant system, unlike adjustable switches which can be used to switch different systems.

Non-automatic extraction turbine - a steam turbine that has opening(s) in the turbine casing for the extraction of steam without a means for controlling its pressure.

Non-bearing wall - a wall that does not support vertical loads other than its own weight.

Non-code installation - a functional refrigerating system installed where there are no local, state, or national codes in force.

Non-condensable gas - any gas, usually in a refrigeration system, that cannot be condensed at the temperature and pressure at which the refrigerant will condense, and therefore requires a higher head pressure.

Non-condensables - foreign gases mixed with a refrigerant that cannot be condensed into the liquid form at the temperatures and pressures at which the refrigerant condenses. These non-condensables create higher head pressure on the refrigeration system along with higher amp draw on the compressor.

Non-condensing turbine - a steam turbine designed to operate with an exhaust steam pressure at any level that may be required by a downstream process, where all condensing takes place.

Non-conductor - an insulator or material that does not conduct electricity.

Non-ferrous - group of metals and metal alloys that contain no iron. These usually refer to copper and copper alloys, aluminum, titanium, etc.

Non-flammable - will not support combustion.

Non-frosting evaporator - an evaporator that never collects frost or ice on its surface.

Non-inductive load - a load that has only resistive qualities with no inductive qualities. An electric heater and incandescent lighting are two common types of non-inductive loads.

Non-miscible - when two substances, such as water and oil, are incapable of mixing.

Non-polar - a system or substance without opposite extremes, as of magnetism or electric charge.

Non-polycyclic organic matter - air pollutants from wood burning stoves considered to be health hazards.

Non-recycling control - a flame safeguard control that does not recycle after shutdown. See cutoff.

Non-renewable energy sources - all energy sources that are not renewable energy sources including natural gas, oil, coal, wood, liquefied petroleum gas, steam and any utility supplied electricity.

Non-reversing - a device that can be operated in only one direction.

Non-sinusoidal wave - any wave that is not a sine wave, such as a triangular, square or pulse wave.

Normal care - the periodic tasks usually performed to operate and maintain an appliance, such as air, fuel, pressure, and temperature regulation; and cleaning, lubrication, and resetting of controls.

Normal charge - the prescribed volume of refrigerant in a refrigeration system.

Normal fuel supply pressure - the pressure at the fuel service connection for which the fuel burning system is designed.

Normal operating procedure - the use of a machine or equipment to perform its intended production or function.

Normal operating water level (NOWL) - water level carried in the boiler gauge during normal operation (approximately one-third to one half glass).

Normalizing - process in which an alloy or carbon steel is heated to a suitable temperature above the transformation range and is subsequently cooled in still air to room temperature.

Normally closed - a controlled device that automatically assumes a closed position when no operating force is applied. It is the opposite of normally open. For example, a control valve which automatically closes when the air pressure or electrical current is removed.

Normally open - a controlled device which automatically assumes an open position when no operating force is applied. For example, a control relay that requires electrical current to close it.

North pole, magnetic - that end of a magnet from which magnetic lines of force flow.

NO_X - oxides of nitrogen (NO, NO_2, and N_2O_4). They react with the moisture in the respiratory tract to produce acids that corrode and irritate tissue, causing congestion and pulmonary edema. Symptoms of acute exposure can develop over six to twenty-four fours. Chronic exposures to low levels can cause irritation, cough, headache, and tooth corrosion.

Nozzle - any piece of pipe (stub-in) that is welded to a vessel or piece of equipment and has a flanged end onto which the pipeline with a similar flanged connection can be bolted. It is also possible to use a long neck welded flange as a nozzle. Nozzles provide for the attachment of a piping system to a vessel, column, or tower. Also, any device, fixed or adjustable, that controls a flow rate or discharge pattern from a pipe or line by means of special contour or size of an orifice.

Nozzle adapter - a pipe fitting that connects the nozzle to the nozzle line. This piece is usually made of brass with both ends female threads.

Nozzle assembly - a complete unit made up of the nozzle line, nozzle adapter, nozzle, electrodes and static pressure plate. Also known as a drawer assembly as it slides in and out of the blast tube.

Nozzle line - section of steel tubing that connects the outlet of the pump with the inlet of the nozzle adapter fitting.

Nozzle mixing burner - a burner in which the fuel and air are not mixed until just as they leave the burner port, after which mixing is usually very rapid. The flame cannot flash back in this type of burner.

Nozzle, oil - the device on the end of the oil fuel pipe used to form the oil into fine droplets by forcing the oil through a small hole to cause the oil to breakup. The oil is also forced into definite swirl patterns to mix with air for complete combustion. Oil nozzles are rated by the gallons per hours, spray angle and spray pattern.

Nozzle, water tower - a device used for controlled distribution of water in a cooling tower. Nozzles are designed to deliver water in a spray pattern either by pressure or by gravity flow.

N-silicon - silicon containing a minute quantity of impurity, or dopant, such as phosphorus, which causes the crystalline structure to contain more electrons than required to exactly complete the crystal structure. There is no electrical imbalance, however.

Nucleus - the core or center part of an atom, that contains protons having a positive charge and neutrons having no electrical charge.

Null - zero

Number 2 fuel oil - the most common type of fuel oil used for residential and light commercial applications. It has a heating capacity of approximately 140,000 Btu's per gallon.

Nut driver - hand tools that have a socket head used primarily to drive hex head screws on equipment.

Nuts - internally threaded metal pieces, usually square or hexagonal in shape, that are used with a bolt.

N type impurities - impurities that have more than four valence electrons that will donate free electrons to a semiconductor.

Nylon - synthetic material used as a sensing element in some humidistats.

Nylon strap fastener - this tool is used to install nylon strap clamps around flexible ducts. It automatically cuts the strap off when a pre-set tension is reached.

-O-

OA - outside air.

OC - on center.

OCV - open circuit voltage.

OD - outside diameter.

ODP - ozone depletion potential or open drip proof.

ODS - oxygen depletion safety.

ODW – ozone depletion weight.

OEM - original equipment manufacturer.

OCN - optimal climate normals.

OPAO - open, air over.

OPC - oil pressure control.

OS&Y - outside stem and yoke.

Oz - ounce

Oakum - a material used for sealing cast iron pipe joints.

Obscure glass - mainly used for decoration, diffusion, or privacy. The pattern is rolled into the hot glass during glass manufacturing.

Obsolete - out of date. No longer in use or practice.

Obtuse angle - an angle with more than 90°.

Occupant density - the number of persons in a space per unit of net occupied (or intended to be occupied) area.

Occupants - people or animals in a space.

Occupied space - the floor area of a space intended for human occupants and mathematically defined by the floor area encompassed by the inside surfaces of its walls.

Occupied zone - the region within an occupied space between planes 3 and 72 inches above the floor and more than 2 feet from the walls or fixed air conditioning equipment. An area controlled by one thermostat that is being occupied.

Octagon - an 8 sided figure with each side forming a 45° angle.

Octagon base - cement foundation associated with the support and footing area of columns (towers) or other vertical vessels.

Octal plug - an eight pin male connector with location key for proper orientation.

Octave - frequency difference between harmonic vibrations.

Octyl alcohol-ethyl hexanol - additive in absorption machines that reduces surface tension in the absorber.

Odor - that property of air contaminants that affect the sense of smell.

Odor threshold - the lowest concentration of a materials vapor (or a gas) in air that can be detected by smell.

Odorant - a substance added to an otherwise odorless, colorless and tasteless gas to give warning of gas leakage and to aid in leak detection.

Off cycle - that part of a refrigeration cycle when the system is not operating.

Off cycle losses - heat losses from a heat source that occur when the heat source is not producing heat.

Off cycle defrost - method in which the fan runs continuously while the system is off, so warm air melts accumulated frost from the evaporator.

Off delay timer - a timer that delays changing its contacts back to their normal position when the coil is de-energized.

Off gas - chemical gasses, typically VOC's that are the byproducts of manufacturing an item and are emitted from their sources of contaminants when their temperature is elevated. Off gassing occurs at normal room temperatures from such objects as carpeting, furniture, paneling, and other construction materials and office equipment.

Off line - any piece of equipment that is not energized. The power could be turned off automatically or manually.

Off peak periods - set times when electrical energy is sold at a reduced rate by the utility. For electric utilities these are generally nights and week ends and for gas utilities summer periods.

Off period - the period of time, other than safety shutdown, during which the burner is not firing.

Offset - a sustained deviation between the actual control point and the set point under stable operating conditions. Offset is caused by load changes affecting the system. Offset is also referred to as drift, deviation or droop.

Offset duct sections - duct sections that have the same cross sectional measurements at either end but are configured to route the duct around, over, or under an obstruction.

Offset pipe wrench - a type of pipe wrench used for work in confined areas.

Offsets - short distance changes in the direction of tubing, commonly made by using elbow fittings.

Ohm - a unit of measurement of electrical resistance. The resistance of a circuit in which a potential difference of one volt produces a current of one ampere. Symbol is "R".

Ohmmeter - an instrument for measuring electrical resistance in ohms.

Ohm's Law - the mathematical relationship between the voltage, current, and resistance of a resistive electric circuit discovered by George Simon Ohm. It is stated as voltage equals current times resistance. The mathematical relationship stating that it takes one volt to push one amp through one ohm of resistance.

Oil - a liquid lubricant; a heavy liquid fuel.

Oil binding - physical condition when an oil layer on top of refrigerant liquid hinders it from evaporating at its normal pressure/temperature condition.

Oil bleed - a bypass through or around a valve or other component in a refrigeration system that aids in returning oil to the compressor.

Oil burner - a burner for burning fuel oil. Also see atomizing oil, rotary oil, and pressure atomizing gun type burners.

Oil burner relay - a special, multi-purpose control used with oil burners. The device controls the operation of the oil burner and also acts as a safety to prevent operation in the event of malfunction. Also known as a primary control or controller.

Oil check valve - a check valve installed between the suction manifold and the crankcase of a compressor to permit oil to return to the crankcase but to prevent the exit of the oil from the crankcase upon starting.

Oil, compressor lubricating - a highly refined lubricant made especially for refrigeration compressors. May be synthetic or mineral based.

Oil cooler - a heat exchanging coil for dissipating heat from lubricating oil.

Oil, entrained - oil droplets carried by high velocity refrigerant gas.

Oil equalizer - a pipe connection between two or more pieces of equipment made in such a way that the pressure, or fluid level, in each piece is maintained equally.

Oil filled capacitor - a run capacitor that is filled with a dielectric oil designed to dissipate internal heat.

Oil filter - a device located to prevent the passage of foreign material through it to do harm to those things downstream. These types of filters maybe a refrigeration oil filter to prevent damage to the bearing surfaces or an fuel oil filter to prevent damage to the pump or nozzle.

Oil free air compressor - an air compressor used for all consumer users from light to heavy duty and most commercial applications as well.

Oil groove - a channel in bearings or other closely machined part that permits the flow of lubricant.

Oil level - the level of the oil in a compressor crankcase at which oil must be carried for proper lubrication. This is determined by the manufacturer of the compressor.

Oil level regulator - a needle valve and float system located on each compressor of a parallel compressor system. It senses the oil level in the compressors crankcase and adds oil if necessary. It receives its oil from the oil reservoir.

Oil logged - clogged or saturated with oil.

Oil loop - a loop placed at the bottom of a riser to forced oil to travel up the riser due to increased velocity of a smaller area.

Oil lube air compressor - an air compressor used for heavier duty use that has cast iron cylinder sleeves and aluminum crankcase and head for heat dissipation.

Oil passages - lubrication holes drilled in the crankshaft, connecting rods, and other components of a pump or compressor.

Oil ports - access ports on electric motors that provide a means of adding oil to the motor bearings.

Oil pressure gauge - a device used to show the pressure of oil developed by the pump.

Oil pressure safety cutout - a motor protection device

Oil pump - a device that provides the source of power for force feed lubrication systems in refrigeration systems or the oil pump in a fuel oil fired furnace.

Oil, refrigeration - specially prepared and blended oil used in refrigeration systems. The oil must be dry, otherwise, moisture will condense out and freeze in the metering device and cause the refrigeration mechanism to fail. Oil is classified as to a viscosity value of 150, 300 or 500 SSU.

Oil reservoir - that area in the base of the oil separator where oil is accumulated prior to its return to the compressor.

Oil return line - the line that carries the oil collected by the oil separator back to the compressor crankcase.

Oil ring - a loose ring, the inner surface of which rides a shaft or journal and dips into a reservoir of lubricant from which it carries the lubricant to the top of the bearing by its rotation with the shaft.

Oil rings - an expanding ring(s) mounted in grooves of the piston, designed to prevent oil from moving into the compressor clearance space.

Oil safety switch - an operating safety control that senses the net oil pump pressure differential between the compressor crankcase and the oil pump discharge. It is wired in series with the compressor contactor and will shut off during periods of low oil pressure. Usually in a manual reset controller.

Oil separator (boiler) - a device that removes oil from the exhaust steam before it enters the open feedwater heater.

Oil separator (refrigeration) - a device piped into the hot gas discharge line that is used to remove oil from the gaseous refrigerant.

Oil sight glass - a glass "bull's eye" in the compressor body that permits visual inspection of the compressor oil level.

Oil sludge - usually a thick, slushy substance formed by contaminated oils.

Oil still - a device used to recover refrigerant from oil by distillation.

Oil trap - a means of accumulating minute droplets of oil and accelerating their return to the compressor. Usually consists of a U-bend at the lowest point in the suction line. Could also be improperly installed refrigerant suction lines.

On cycle - the period when the equipment, especially refrigeration equipment, is in operation.

On delay timer - a timer that delays changing the position of its contacts when the coil is energized.

On off control - a type of control in which the controlled device is moved, from one of two fixed positions to the other, by action of the controller. For example, a valve that is moved from its open to shut position or a relay whose contacts are either open or closed.

On off valve - see 2 position valve.

On line - any piece of equipment that is energized. This could be accomplished by automatic or manual means.

On site - in the field or at the construction site. Any portion of the project that must be designed, constructed, or fabricated at the construction site. This is referred to as on site production or construction.

One pipe fitting - a specially designed tee for use in a one pipe hot water system to connect the supply or return branch into a circuit. These fittings cause a portion of the water flowing through the circuit to pass through the terminal unit.

One pipe supply riser, steam - a pipe that carries steam to a heating unit and which also carries the condensate from the heating unit. In an up-feed riser, steam travels upwards and the condensate travels downward while in a down-feed both steam and condensate travel down.

One pipe system, hot water - a hot water heating system in which one pipe serves both as a supply main and also as a return main. The heating units have separate supply and return pipes but both are connected to the same main.

One pipe system, oil - an oil supply line that is used to being the supply of fuel oil from the tank to the burner. This system has no return line.

One pipe system, steam - a steam heating system consisting of a main circuit in which the steam and condensate flow are in the same pipe. There is one connection to each heating unit that serves as both the supply and the return.

One time fuse - generic term used to describe a class "H" nonrenewable cartridge fuse, with a single element.

One time safety valve - a pressure relief valve that, after causing the system to release its pressure, must be replaced as it will not reset itself.

One ton of cooling - the removal of heat from the conditioned space at the rate of 12,000 Btu/h per hour.

Opacity - the state or quality of being opaque; smoke density is measured in Ringelmanns. Smoke opacity that exceeds a preset level will cause a stack smoke detector to pull in an alarm relay.

Open blown - loose fill insulation pneumatically installed in an unconfined attic space.

Open circuit - a circuit that does not provide a complete path for the current to flow. A noncontinuous circuit. An interrupted electrical circuit that stops the flow of electricity.

Open circuit voltage - (1) the voltage across a photovoltaic cell in sunlight when no current is flowing. (2) The voltage measured across the output terminals of a transformer. In the case of a 24 volt transformer, the output may be as high as 27-28 volts but when loaded to full va rating the output voltage would be 24 volts. Also known as (VOC) volts open circuit voltage.

Open compressor - term used to indicate an external drive compressor. Not fully hermetic.

Open damper purge - during pre-purge, the damper opens to high fire position and the burner motor drives air through the combustion chamber to remove any potentially explosive fuel or fumes. The damper closes to low fire position for ignition trials, and opens again to high fire position for the run period.

Open delta circuit - a type of 3 phase service connection that uses only two transformers to provide 240 volt three phase power and 120 or 240 volt single phase power.

Open display case - a commercial refrigerator designed to maintain its contents at refrigerating temperatures even though the contents are in an open case.

Open drip proof motor - a type of motor that has an open frame and where drops of liquid falling at a 15° angle will not affect its operation.

Open end adjustable wrench - similar in shape to an open end wrench, but has adjustable jaws.

Open end wrench - a wrench with open jaws to fit over a nut or bolt of a fixed size from either end. They are used in close quarters where the box end or socket will not allow.

Open feedwater heater - a feedwater heater in which steam and feedwater come into direct contact (mix) to raise the temperature of the water.

Open loop - a control system in which the controller controls the actuator, in a definite sequence, and the results are not reflected back to the controller. For example, an outside air controller on the intake side of a steam coil that operates a steam valve to allow a definite quantity of steam to enter the coil in accordance with outside air temperatures. The open loop is seldom used in air conditioning control systems today.

Open loop heat pump - heat pump system that uses the water in the earth as the heat transfer medium and then expels the water back to the earth in some manner.

Open loop water source - a source of ground water such as a pond, lake, or well, from which water is supplied to a geothermal heat pump.

Open loop water system - a piping system that conveys fresh water, or is open to the atmosphere at any point.

Open piping system - piping system which has an opening to the atmosphere at one or more locations.

Open type compressor system - a refrigerating system that uses a belt drive compressor or a coupling driven compressor.

Open web joist - steel joists built up out of light steel shapes with an open latticed web. Also known as bar joist or trussed joist.

Open winding - a broken wire in the motor winding that is not touching the motor shell or a motor winding that has an open circuit, preventing the flow of current.

Operating control - a control (other than a safety control or interlock) which starts, stops, and regulates burner firing according to load demand. It may also actuate auxiliary equipment.

Operating costs - the total cost of operating and maintaining an air conditioning system including energy, maintenance, and depreciation.

Operating cycle - the complete sequence of operations required in the normal functioning of an item of equipment. For a flame safeguard control, the operations performed between a call for heat and satisfaction of the call for heat (when the operating set point is reached). Also see timer cycle.

Operating differential - the overall swing (high point minus low point) of a controlled variable that is necessary to activate a controller.

Operating mode - a specific condition or state of the controlled devices in a system that allows it to accomplish a specific objective. Hydronic heating systems can have several operating modes, including off, on, priority domestic water heating, heat purging, etc.

Operating pressure - actual pressure at which the system works under normal conditions. This pressure may be positive or negative (vacuum).

Operating point - the point on a graph where the pump curve intersects the system resistance curve. This point indicates the flow rate at which the system will operate.

Operating range - the region between the maximum fuel input and minimum fuel input in which the burner flame can be maintained and stable.

Operating recorder - a machine that makes a graphical record for diagnosis of equipment performance, temperature, pressure, voltage, or any other measurable condition, in a given operating time.

Operating speed - the switching speed of a transistor expressed as either the turn on time, the turn off time, the rise time, or fall time. The amplifying speed is expressed in terms of the maximum frequency of the input signals that can be satisfactorily amplified. The operating speed of the device is dependent mainly upon the system capacitance. This specification implies both switching and amplifying speeds of the transistor.

Operational amplifier - type of amplifier used in integrated circuits that is characterized by very high voltage gains and differential inputs.

Operative temperature - the uniform temperature of a radiantly black enclosure in which an occupant would exchange the same amount of heat by radiation plus convection as in the actual non-uniform environment. Operative temperature is numerically the average, weighted by respective heat transfer coefficients (hc and hr), of the air (ta) and mean radiant temperatures (tr).

Operative temperature = (hcta + hrtr)/hc + hr). At air speeds of 0.4 meters per second (80 fpm) or less and temperature less than 50° C (120° F), operative temperature is approximately the simple average of the air and mean radiant temperatures and equal to the adjusted dry bulb temperature.

Operator - one of two functional components in a controlled device. The operator converts a signal from the controller into a useful local action. Also see final control element.

Opposed blade damper - a flap or louver type damper in which alternate blades rotate in opposite directions. This type provides an equal percentage flow characteristic successive equal increments of rotation produce equal percentage increases in flow. This is particularly useful for throttling applications where accurate control at low flow rates is required.

Optical flame detector - a detector that responds to light (visible or otherwise) emitted by a flame. Also see rectifying photocell, infrared and ultraviolet flame detectors.

Optimal climate normals - a forecasting tool that predicts temperature and precipitation on the basis of continued, observed average deviations from the norm for a given season, during the last 10 years for temperature, and the last 15 years for precipitation. The OCN emphasizes long term trends.

Optimum air supply - the quantity of air that will give the greatest thermal efficiency under actual conditions. With perfect mixing of fuel and air, the optimum air supply is equal to the chemically correct amount of air. Also see air fuel ratio.

Optimum operative temperature - temperature that satisfies the greatest possible number of people at a given clothing and activity level.

Optoelectronics - a technology that employs devices that will operate optically as well as electrically.

Orbit - the path along which electrons travel around the nucleus of an atom.

Orbiting - term used to describe the form of motion the compresses refrigerant gas between the two scroll components in a scroll compressor.

Ordinance - a statue or regulation, especially one enacted by a city government.

Organic - pertaining to or derived from living organisms. Chemical compounds that contain carbon and hydrogen, with or without oxygen, nitrogen, or other elements. Volatile organic compounds vaporize at room temperature and pressure. They are found in many indoor sources, including many common household products and building materials.

Organic compounds - chemicals containing carbon. Volatile organic compounds vaporize at room temperature and pressure. They are found in many indoor sources, including many common household products and building materials.

Orifice - an opening or window, such as in an aperture disc or iris slide. In a burner, the opening in a cap, spud, or other device whereby the flow of fuel is limited and through which the fuel is discharged to the burner.

Orifice cap (hood) - a movable fitting having an orifice which permits adjustment of the flow of fuel by the changing of its position with respect to a fixed needle or other device.

Orifice, insert - an orifice that drops into the pilot burner head and is held in place with a separate threaded nut.

Orifice, oil metering - a small hole in the pick up tube of the accumulator used to ensure oil return to the compressor in small easily handled quantities.

Orifice plate - opening where low side pressure begins. Sometimes referred to as a piston or bullet type metering device.

Orifice spud - a removable plug or cap containing an orifice that permits adjustment of the fuel flow, either by substitution of a spud with a different sized orifice or by the motion of a needle with respect to it.

Orifice tube - a metering device consisting of a restricting tube with inlet and outlet screens.

Orificing - a procedure used to reduce the field of view of a detector by means of an opaque plate or card with a selected sized opening. This is frequently required to reduce refractory interference to acceptable levels.

Original equipment manufacturer - same as the original part.

O-ring - sealing device used between parts where there may be some movement.

Orsat test kit - a gas analysis apparatus in which the CO_2 or O_2 gaseous constituents are measured by absorption in separate chemical solutions.

Oscillation - fluctuation, instability. A single swing of a swinging object. The variation between maximum and minimum values, as with an electrical current.

Oscillator - a device used to change DC voltage into AC voltage. Produces a waveform that has positive and negative pulses of equal voltage and time.

Oscilloscope - a fluorescent coated tube that visually shows an electrical sine wave.

Out contacts - those relay contacts that complete circuits when the relay coil is de-energized. Also referred to as normally closed or out contacts.

Out of phase - the condition in which two components do not reach their positive or negative peaks at the same time.

Out position - those relay contacts that complete circuits when the relay coil is de-energized. Also referred to as normally closed or out contacts.

Outage - flame failure in heating equipment.

Outdoor air - air that is brought into the ventilation system from outside the building, and, therefore, not previously circulated through the system.

Outdoor air ventilation - "fresh" outdoor air brought into a building to aid the indoor air quality.

Outdoor coil - heat pump component that is either an evaporator or a condenser, depending upon whether the system is heating, cooling or defrosting.

Outdoor design temperature - the outdoor temperature on which design heat losses are based.

Outdoor exposure test - a test where plastic pipe placed in service or stored so that it is not protected from the elements of normal weather conditions, i.e., the sun's rays, rain, air and wind to determine aging and deterioration. Exposure to industrial and waste gases, chemicals, and engine exhausts, are not considered normal outdoor exposure.

Outdoor reset control - a control method that increases the water temperature in a hydronic system as the outdoor temperature drops. These controls may be purchased with different reset ratio valves to suit specific application needs.

Outdoor thermostat - a control device used in conjunction with a regular indoor thermostat to stage the heating elements (or auxiliary heat) in a two or more stage heating system.

Outgassing - the emission of gases by materials and components usually during exposure to elevated temperature or reduced pressure.

Outlet - (1) a point at which current may be supplied to a dedicated load. (2) That part of a valve through which fluid discharges.

Outlet box - a box used to terminate a cable or conduit. Electrical connections are made within the box. A variety of covers and plates are available to close the box.

Outlet, ceiling - a round, square, rectangular, or linear air diffuser located in the ceiling, which provides a horizontal distribution pattern of primary and secondary air over the occupied zone and induces low velocity secondary air motion through the occupied zone.

Outlet, slotted - a long narrow air distribution outlet, comprised of deflecting members, located in the ceiling, sidewall, or sill, with an aspect ratio greater than 10, designed to distribute supply air in varying directions and planes, and arranged to promote mixing of primary air and secondary room air.

Outlet, vaned - a register or grille equipped with vertical and/or horizontal adjustable vanes.

Outlet velocity - the average air velocity emerging from a grille or diffuser.

Output - the power, capacity, duty, performance, net refrigeration or energy that a machine produces.

Output signal - a signal produced in response to a given input.

Outrigger - a structural member for supporting a roof or floor beyond the walls, in the direction perpendicular to the joists.

Outside air - external air. Atmosphere exterior to refrigerated or conditioned space. Ambient or surrounding air.

Outside air opening - any opening used as an entry for air from the outdoors.

Outside design temperature - the outdoor temperature on which design heat losses are based.

Outside diameter - the largest diameter of a tube or pipe. Abbreviated OD.

Outside stem and yoke valve (OS&Y) -this valve shows the positions of the stem whether it is open or closed. Used as boiler main steam stop valves.

Oven - a compartment or receptacle for heating, baking, or drying by means of heat. Also see industrial oven.

Overall coefficient of heat transfer (thermal transmittance) - the time rate of heat flow through a body per unit area, under steady conditions, for a unit temperature difference between the fluids on the two sides of the body.

Overcharge - to fill a system with refrigerant beyond its design capacity.

Over current - a condition that exists on an electrical circuit when the normal current is exceeded. Over currents take on two separate characteristics - overloads and short circuits.

Over current device - also referred to as an over current protective device. A form of protection that operates when current exceeds a predetermined value. Common forms of over current devices are circuit breakers, fuses, and thermal overloads elements found in motor controllers.

Overfed evaporator - an evaporator that is operating with a lower amount of superheat than desired. The expansion device is feeding too much refrigerant to the coil.

Overfeed stoker - a coal firing system that burns coal in suspension and on grates. Also known as a spreader stoker.

Over fire air - air introduced over the fire to aid in complete combustion. Used mostly when burning soft coal that has a high volatile content with its own feedwater pumps, fuel system, and draft fans.

Over fire draft - the draft measured over the flame within the combustion chamber.

Overhead glazing - a glazing surface that has a slope of less than 60° from the horizontal plane.

Overhead system - any steam or hot water system in which the supply main is above the heating units. With a steam system the return must be below the heating units; with water, the return may be above the heating units.

Overload - can be classified as an over current that exceeds the normal full-load current of a circuit. Also characteristic of this type of over current is that it does not leave the normal current carrying path of the circuit - that is, it flows from the source, through the conductors, through the load, back through the conductors, and to the source again.

Overload heater - an electrical device used in motor starters that will shut off the motor if the motor load exceeds acceptable amperes.

Overload protector - a device, either temperature, pressure, or current operated, which will stop operation of unit if dangerous conditions arise. Can either be low or line voltage, pilot or direct duty.

Overload relay - a thermal device that opens its contacts when the current through the heater coil exceeds the specified value for a specified time.

Over range - signal amplitudes or frequencies above the specified limits of the test instrument. For example, the uncertainty specification for the 10 volt range of an instrument with a 100% over range applies up to 20 volts.

Overrate firing - firing a boiler at an input rate in excess of its rated capacity. The rated capacity of the boiler is based upon its heat transfer surface area.

Override - a manual or automatic action taken to bypass the normal operation of a device or system.

Overshoot - the greatest amount a controlled variable deviates from its desired value before stabilizing, after a change of input.

Oversized - having more heating or cooling capacity than the minimum needed.

Oxidation - literally, oxidation is a reaction in which a substance combines with oxygen provided by an oxidizer or oxidizing agent. Based on modern atomic theory, oxidation is a reaction brought about by an oxidizing agent in which atoms, molecules, or ions lose electrons. An oxidation reaction may occur even when oxygen is not present. However it may be defined, an oxidation reaction is always accompanied by an offsetting (balancing) reduction reaction in which (1) oxygen is removed from a compound, or (2) atoms, molecules, or ions gain electrons.

Oxidize - a corrosive chemical reaction caused by exposure to oxygen gas, like rust (iron oxide) or copper oxide (which forms on or inside copper tubing).

Oxidizer - a material that yields oxygen readily to stimulate the combustion (oxidation) of organic matter.

Oxidizing agent - a chemical or substance that brings about an oxidation reaction. The agent may (1) provide the oxygen to the substance being oxidizes (in which case the agent has to be oxygen or contain oxygen); or (2) it may receive electrons being transferred from the substance undergoing oxidation.

Oxidizing flame - a lean flame or fire resulting from combustion of a mixture containing too much air (or too little fuel). This kind of flame produces an oxidizing atmosphere, which tends to oxidize materials placed in it.

Oxy-acetylene torch - a type of torch that produces a flame by the combustion of acetylene and oxygen. It consists of a torch handle with tip, regulators, hoses, and oxygen and acetylene tanks. When these two gases are mixed together in the proper ratio, a very hot flame is produced.

Oxygen (O_2) - an elemental gas that comprises approximately 20.8% of the atmosphere by volume. Oxygen is the lesser quantity of air that is necessary in the combustion of any fuel. When found in large quantity in flue gases, it is an indication of excess air being introduced into the unit. Oxygen has a molecular weight of 32.00, a specific volume of 12.1 cubic feet/lb., and a boiling temperature of -297.4° F.

Oxygen deficient - a mixture of gases containing too little oxygen to adequately support life. OSHA standard set at 19.5%.

Oxygen depletion safety shutoff system (ODS) - a system designed to act to shut off the gas supply to the main and pilot burners if the oxygen in the surrounding atmosphere is reduced below a predetermined level.

Oxygen enriched - a mixture having oxygen concentration greater than air. Normally any mixture of gases having a oxygen value over 21%.

Oxygen regulator - an adjustable device used to maintain the cylinder pressure and the working pressure that flows to the torch handle and tip.

Ozone (O_3) - triatomic oxygen; a pale bluish gaseous form of allotrope oxygen derived or formed from diatomic oxygen by silent discharge of electricity or exposure to ultraviolet radiation. It is unstable, powerful bleaching, poisonous oxidizing agent with a pungent penetrating odor.

Ozone depletion - a condition that happens when ultraviolet radiation in the stratosphere breaks CFC and HCFC refrigerants into their atomic elements ---- chlorine, fluorine and hydrogen atoms. Chlorine atoms react with and destroy stratospheric ozone, which protects earth's human and other life forms from the sun's harmful ultraviolet radiation.

Ozone depletion potential number - an ODP number is assigned to each refrigerant indicating the risk to the ozone layer caused by the chlorine in refrigerants.

Ozone depletion weight - a number assigned to a chemical representing the ability of the substance to destroy the stratospheric ozone layer. The ozone depletion weight is the same as the ozone depletion potential (ODP).

Oxygen diffusion barrier - a layer of special compounds or metal within the cross section that significantly reduce the rate at which oxygen can diffuse through a polymer tube.

Ozone layer - the outermost layer of the earth's atmosphere, (stratosphere) which absorbs ultraviolet light from the sun and shields the lower layers and the earth from harmful rays. The thinning of the ozone layer is believed to be caused by chlorine. Chlorofluorocarbons (CFC's) contain chlorine and, when released to the atmosphere, may cause deterioration of the ozone layer.

Ozone test - exposure of material to a high concentration of ozone to give an accelerated indication of oxidation in normal environments and in proximity to ozone producing apparatus.

-P-

Pags - polyalkylene glycols

Par - parallel.

PB - push button.

PCB - polychlorinated biphenyl.

Pd - pressure drop.

PE - pneumatic electric or polyethylene.

PEL - permissible exposure limit.

PEX - polyethylene cross linked tubing.

Pf - power factor.

PFC - perfluorocarbon.

Ph - phase.

pH - potential of hydrogen ions.

PID - proportional integral derivative.

PLC - programmable logic control.

Pm - preventative maintenance.

POA - pressure operated altitude.

Poe - Polyolester oil

Pos - positive.

Pot - potential.

Ppb - parts per billion.

PPE - personal protective equipment.

Ppm - parts per million.

PPT - parts per trillion.

Pri - primary.

Prot - protection.

PSC - permanent split capacitor

PSI - pound per square inch.

PSIA - pounds per square inch absolute.

PSID - pounds per square inch differential.

PSIG - pounds per square inch gauge.

PTC - positive temperature coefficient.

PVC - polyvinyl chloride.

PVE - Polyvinyl ether

Pb - barometric pressure in Hg

PS - static pressure in H_2O

Pt - total pressure in H_2O

Pv - velocity pressure in H_2O

Packaged boiler - a steam or hot water boiler in which the pressure vessel, furnace, and controls are designed, assembled, wired, and shipped as an integral unit. Packaged boilers are normally compact and efficient, delivering about one boiler horsepower (Bohp) for every five feet of heating surface. They are available in both fire tube and water tube types. Fire tube packaged boilers range in size from about 500,000 to 30 million Btu/h (10 to 600 Bohp); water tube packaged boilers range in size from about 12.5 Million to 75 million Btu/h (250 to 1500 Bohp). Also called a steam generator.

Packaged multizone units - a packaged heating, ventilating, air conditioning unit that simultaneously maintains separate temperatures on a hot and cold deck. Each zone within the building then mixes air from these two decks to maintain space temperature.

Packaged terminal air conditioner (PTAC) - an electrical heating and cooling system often found in the guest rooms of hotel/motels. The unit is installed through the wall of the room.

Packaged terminal heat pump (PTHP) - a PTAC capable of using the refrigeration system in a reverse-cycle or heat pump mode to provide heat.

Packaged units - heating and cooling systems that includes controls, air moving devices, and other accessories to provide complete performance as desired by the designer.

Packing - sealing device consisting of soft material or one or more mating soft elements. Reshaped by manually adjustable compression to obtain or maintain a leak proof system.

Packing gland - this holds packing or seals in place on valves and pumps to minimize leakage.

Packless valve - a valve that uses a flexible membrane, rather than stuffing material, to seal the stem.

Pan construction - a concrete floor or roof construction in which a prefabricated form (pan) is used repeatedly, giving the underside of the construction a waffle-like appearance.

Pane - a framed sheet of glass.

Panel - either an opaque glazing material or the operating portion of a sliding glass door.

Panel box - an electrical junction box containing fuses and switches.

Panel filter - this is the most common type of filter used in the comfort air conditioning industry today. Panel filters achieve what is generally categorized as low efficiency filtration. This typically represents the minimum acceptable quality of filtration for comfort conditioning applications. As the name implies, these filters have a flat, rectangular shape. A number of standard rectangular sizes are available, typically in depths of one or two inches. Panel filters are designed in both throw away and permanent. The throw away type usually has cardboard frames, with either metal or plastic perforated screens for the entering and leaving air sides. The filtering media can be made from any of several types of fiber fill. Spun fill fiber and hemp filter being two common types. An adhesive coating can be applied to the filter media while the filter is being made to increase its particle holding capacity. Panel filters are installed on the intake side of the fan upstream of the coil(s).

Panel heating - a heat delivery method in which elements or water pipes are embedded in the ceiling, floors or walls of the heated space.

Panel power failure relay - a relay that sheds non-critical loads during a power failure to the load management panel. Used to avoid setting a demand peak when power is lost to the load management panel.

Panel radiator - a heating unit placed on, or flush with, a flat wall surface and intended to function essentially as a radiator. Do not confuse with panel heating.

Panel systems - a heating system in which the ceiling or floor serves as the terminal unit.

Panel board - a single panel, or a group of panel units, designed for assembly in the form of a single panel; including buses, and with or without switches and/or automatic Over current protective devices for the control of light, heat, or power circuits of small individual as well as aggregate capacity; designed to be placed in a cabinet or cutout box placed in or against a wall or partition and accessible only from the front.

Paper capacitor - capacitor in which the dielectric is made of waxed paper.

Parabolic collector - curved solar collector with a lens that concentrates the solar rays onto a focal point.

Paraffinic oil - a refrigeration mineral oil containing some paraffin wax, which is refined from eastern USA crude oil.

Parallel blade damper - a flap or louver type damper in which all blades rotate in the same direction. This type provides a fairly linear flow characteristic - the flow is nearly proportional to the damper shaft rotation. This is particularly useful in mixing applications where two different flow rates must be combined.

Parallel circuit - an electrical circuit in which the identical voltage is presented to all components, with current dividing among the components according to the resistances or the impedances of the components.

Parallel compressor - many compressors piped in parallel and mounted on a steel rack. The compressors are usually cycled by the microprocessor.

Parallel connection - a type of circuit in which there are two or more separate paths or branches for flow; a circuit in which all the components are not in series.

Parallel direct return piping - a hydronic distribution system consisting of two or more branch piping paths that begin from a common source pipe and end at a common return pipe. The first branch connected to the supply pipe is also the first to connect to the return pipe.

Parallel flow - in heat exchange between two fluids, hottest portion of one meeting coldest portion of the other.

Parallel ground loop - in geothermal heat pump systems, the term used to describe the ground loop configuration that provides multiple paths for the heat exchange fluid to flow through.

Parallel piping - when two or more branch piping paths originate from a common starting point and end at a common ending point.

Parallel primary loop - a primary loop in a primary/secondary system having two or more crossovers from a common supply to a common return. A pair of closely spaced tees is installed in each crossover.

Parallel pumps - two or more pumps that are mounted to a common inlet header and discharge into a common outlet header.

Parallel reverse return piping - a hydronic distribution system consisting of two or more branch piping paths that begin from a common supply pipe and end at a common return pipe. The first branch connected to the supply pipe is the last to connect to the return pipe.

Parallel shifting - moving a reset line up or down without changing its slope to compensate for internal heat gain, or losses not proportional to the temperature difference between inside and output.

Parameter - a constant or variable quantity, the value of which is of interest in some given situation, or which determines the operating characteristics of some system.

Parapet wall - a low guarding wall or part of any wall above the roof line.

Part number - the manufacturer's unique identification number for any part, usually marked on the part itself.

Part wind start - method of motor starting that energizes only a portion of the motor windings during start-up to reduce the amperage draw on the motor, as well as to prolong the life of the contacts on the motor starter. Once the motor is up and running the run windings are introducted to the active electric circuit.

Partial premix burner - a gas burner in which part of the combustion air is mixed with the gas before they are introduced into the combustion chamber, usually by venturi action. The required secondary air reaches the combustion chamber through spaces between the burner heads. Examples are inspirating, spreader, and tunnel burners.

Partial pressure - a condition where two or more gases occupy a space and each one creates part of the total pressure. Also see Dalton's Law of partial pressure.

Partial reset - limiting the lower limit of supply temperature produced by an outdoor reset control. Partial reset is often done to prevent a conventional boiler from operating with sustained flue gas condensation.

Partial storage systems - a thermal storage system that stores only a portion of the energy needed by the load, for a 24 hour period, during the preceding off peak period.

Partial vacuum - any pressure less than atmospheric.

Partially halogenated HCFC - when not every hydrogen atom in hydrogen molecule is replaced with chlorine or fluorine atoms.

Particle count - concentration expressed in terms of the number of particles per unit volume of air.

Particles - matter in which solid or liquid substances exist in the form of aggregated molecules. Airborne particulates typically range in size from 0.01 to 20 micrometers.

Particulate matter - small, separate pieces of an airborne material. Dusts, fumes, smokes, mists, and fog are examples. Generally, anything that is not a fiber and has an aspect ratio of 3 to 1 and range in size from 0.01 to 100 microns in diameter.

Partition - an interior wall subdividing the tower into cells or into separate fan plenum chambers. Partitions may also be selectively installed to reduce windage water loss.

Parts per million (ppm) - the normal term used to express the ratio of parts of vapor or gas per million parts of air by volume at 25° C and 1 torr pressure. One part per million is equal to 0.0001%.

Pascal (Pa) - a small unit of pressure. One pound per square inch equals 6,895 Pascal's. The force per unit area.

Pascal's Law - states that the pressure applied to any part of a fluid that the force is transmitted equally in all directions.

Pass - (1) horizontal run that flue gases take through a fire tube in a boiler before existing the breeching. (2) The number of times that water flows through a water cooled shell and tube condenser. A four pass condenser coil would be more efficient than a two pass.

PASS - acronym used to remember the steps when using a fire extinguisher. They are (1) pull the key, (2) aim the extinguisher, (3) squeeze the handle and (4) sweep the bottom of the fire.

Passive recovery - a system dependent (using the refrigeration systems compressor) method of recovering refrigerant from mechanical refrigeration systems in which the gauge manifold is connected to the suction and discharge lines, using service valves or line piercing valves. Refrigerant from the system migrates to the recovery cylinder and condenses to a liquid inside the cylinder.

Passive solar design - the use of non-moving parts of a building to provide heat or cooling or to eliminate certain parts of a building that cause inefficient heating or cooling.

Passive solar system - solar systems that use the walls of the building as the collector and the heat storage medium.

Passive ventilation - building ventilation system where fans are not used to move air. Airflow is caused by wind or by the buoyancy of warmer air.

Passivity - the tendency of certain metals to develop coatings of corrosion that protect the metal from further corrosion.

Path - the part of the electrical system through which electrons travel from a source to a load, such as the electrical wiring used in a building or a piece of HVAC equipment.

Patio door - sliding glass doors, often used for access to a deck or terrace.

Pattern - refers to the uniform spray configuration of the fuel oil by the oil nozzle for an atomizing type gun burner. Nozzle patterns can be solid, semi-hollow, or hollow.

PCB - polychlorinated biphenyl. A family of compounds used as a heat transfer medium. PCB's accumulate in tissue, are environmentally hazardous, and are believed harmful to human health.

Peak - the maximum sine wave amplitude reached during an electrical cycle. Maximum load or performance.

Peak cooling load - worse case cooling requirement derived from the heat gain calculation.

Peak demand - the greatest amount of kilowatts needed during a demand interval.

Peak heating load - worse case heating requirement derived from the heat loss calculation.

Peak hold - a feature in some digital multimeters that retain the highest reading in a series of measurements.

Peak let through current - the instantaneous value of peak current let through by a current-limiting fuse, when it operates in its current-limiting range.

Peak load - the maximum load carried by a system or a unit of equipment over a designed period of time.

Peak to peak voltage - the value of AC sine wave voltage from positive peak to negative peak.

Peak voltage - the maximum positive or negative value of AC sine wave voltage. The maximum instantaneous voltage. To convert from peak voltage to average voltage use the formula $Eav = 0.637 \cdot Epk$. To convert from peak voltage to effective voltage use the formula $Eeff = 0.707 \cdot Epk$.

Pecking relay - see checking relay.

Pedestal - any concrete shape (pier), cylinder, rectangular, etc., which supports a piece of equipment such as a vessel or pump.

Pellet fuel-burning appliance - an appliance that utilizes solid fuel in the form of pellets in a closed combustion system. The pellets consist of a blend of wood waste products, (sawdust, chips), mixed with resin as binders and compressed into small pellets. The pellets are fed into the appliance in a regulated manner to provide consistent burning.

Peltier effect - where direct current is passed through two adjacent metals one junction will become cooler and the other will become warmer. This is the basis of thermoelectric refrigeration.

Penetrating oil - special lubricant that penetrates tight spaces to free rusted or locked parts.

Penetration - this is the depth of a weld into the base metal.

Pentagon - a five sided figure with each side forming a 72° angle.

Penthouse - an enclosed structure above the roof of a building, other than a roof structure or bulkhead, occupying not more than thirty-three and one third (33 1/3) per cent of the roof area. Houses equipment for elevator, ventilation or air conditioning, or other mechanical or electrical systems serving the building.

Percent - amount equal to 1/100 of a number.

Percent authority - the readjustment of a controller that determines the effect of the reset signal of the second transmitter as a percentage of the first transmitter.

Percent conductivity - the conductivity of a material expressed as a percentage of that of copper.

Percent humidity - the pounds of water actually held in air divided by the pounds of water when the same air is fully saturated.

Percentage regulation - the percentage of decrease in output voltage from a power supply from no load to full load conditions.

Perculator - the section in an absorption refrigeration system through which heated solution from the generator is pumped to the separator by the buoyancy of the entrained refrigerant gas.

Perfect combustion - the act of burning with exactly the correct proportion of fuel to air so all of the fuel and oxygen are burned. Also referred to as stoichiometric burning and theoretical combustion.

Perfect gas - a hypothetical gas obeying the relation $PV = RT$.

Perfect vacuum - theoretical state in which all atmospheric pressure is removed.

Perfect washer - theoretical achievement of 100% moisture saturation in an air washer.

Performance - output or efficiency.

Performance chart - a table showing output or efficiency under various load conditions.

Performance factor - capacity of unit in Btu/h in relation to power consumption in watts. Efficiency of a unit based on Btu's per watt. Same as EER.

Perimeter - the outer boundary of a geometric figure.

Perimeter air duct distribution system - an air conditioning system in which the radiators or registers for heating and cooling are located at the outer walls of the conditioned space. The register outlets may be located in either the floor or the ceiling.

Perimeter circuits - radiant floor heating circuits that are routed near the exposed perimeter of a room.

Period (time) - the time required to complete one alternating current cycle; time = 1/frequency.

Perm - the unit of permeance. A perm is equal to 1 grain per (sq ft) (hr) (inch of mercury vapor pressure difference).

Permalloy - a combination of nickel and iron which has a high magnetic characteristics.

Permanent magnet - a material which has its molecules aligned and has its own magnetic field. A bar of metal that has been permanently magnetized.

Permanent split capacitor motor (PSC) - a single phase induction motor which has a capacitor continuously in the run (main) windings. Unlike the split phase or capacitor start type motors there is no centrifugal switch to disconnect the run winding. This type of motor has low starting torque and is best suited for direct drive applications such as fans, blowers and some compressor motors. PSC motors are much more efficient than shaded pole motors.

Permanently lubricated motors - motors that typically will not require lubrication during their expected useful lives.

Permeability - water vapor permeability is a property of a substance that permits passage of water vapor, and is equal to the permeance of a 1 inch thickness of the substance. When permeability varies with psychrometric conditions, the spot or specific permeability defines the property at a specific condition. The permeability value of air is 1.0. Permeability is measured in perm-inches.

Permeability, electrical - the ability of a material to conduct magnetic lines of force as compared to air. The ability of a material to magnetize or demagnetize.

Permeance - (1) a measure of the ability of moisture to pass through a material. Permeance is measured in perms. (2) the property of a plastic that describes its resistance to appreciable changes in characteristics with time and environment.

Permeation - the process of gas seeping through the walls of refrigerant hoses.

Permissible exposure limit - is the concentration level established by the Occupational Safety and Health Administration (OSHA). Without qualification, the PEL implies a time weighted average (TWA) for an 8-hour work shift in a 40-hour work week. Consistent data include similarly-defined occupational exposure limits set by manufacturers such as the allowable exposure limit (AEL), the threshold limit value (TLV), and the workplace environmental exposure level (WEEL).

Permissive rules - these are requirements allowed but not required. Terms such as "shall be permitted" and "shall not be required" are used when a code statement is permissive.

Permit - an official document or certificate issued by the authority having jurisdiction authorizing performance of a specified activity.

Permit required confined space - a confined space that has a specific health and safety hazards associated with it.

Perpendicular - a line at a right angle (90°) to another line.

Perpetual inventory - an inventory system that results in the inventory account being maintained at its proper balance throughout the period and usually makes use of a subsidiary record account of each item in stock. A line item entry for each product lists units received, units sold, and the current balance of the product in inventory.

Perpetual motion - the motion of a hypothetical machine that, once started, would operate indefinitely by creating its own energy.

Personal protective equipment (PPE) - devices or clothing worn to help isolate a worker from direct exposure to hazardous materials. Examples include gloves, respirators, safety glasses, or ear plugs.

Perspective drawing - is a pictorial drawing showing an object as it appears to the eye.

Petcock - a small, manually operated faucet or valve used for draining unwanted or excess liquid or gas from the main system, or for tapping a small amount for testing.

Pete's port - the access port that permits a field technician to take pressure and/or temperature readings in the water loops of geothermal heat pump systems. The device is self sealing, which prevents water leakage from the loop.

Petroleum - see crude oil.

PEX - tubing an acronym for cross linked polyethylene tubing.

PEX-AL-PEX tubing - composite tubing having inner and outer layers of PEX and a center layer of aluminum.

pH - a term based on the hydrogen ion concentration in water, which denotes whether the water is acid, alkaline, or neutral. A pH (potential hydrogen) value of 8 or more indicates a condition of alkalinity; of 6 or less, acidity. A pH of 7 means the water is neutral.

Phase, electrical - a particular stage or point of advancement in an electrical cycle. The fractional part of the period through which the time has advanced measured from some arbitrary point usually expressed in electrical degrees, where 360° represents one cycle. Abbreviation for phase is Ø.

Phase, physical - a phase is any physically and chemically homogeneous quantity of matter. Can be in either a solid, liquid, or vapor phase.

Phase angle - the difference, in degrees, between two AC wave forms.

Phase loss monitor - motor protection device for polyphase motors that measures current flow to detect phase loss.

Phase relationship - the relationship between voltage and the current in an AC circuit. If the peaks of voltage and current are reached at the same time, the current is said to be in phase with the voltage. If these peaks occur at different times, the current is said to be out of phase with the voltage.

Phase shift - a change in phase of a voltage or current after passing through a circuit or cable.

Phase winding - an auxiliary winding, used for starting, which is out of phase with the run winding.

Phial - term sometimes used to denote the power element sensing bulb on a thermostatic expansion valve.

Phillips screw - a screw head configuration utilizing two crossed slots for extra leverage.

Phosgene - a poisonous gas formed when halide refrigerants are burned or exposed to an open flame.

Phosphorus - a chemical element, atomic number 15, used as a dopant in making n-silicon.

Photocell - see rectifying photocell flame detector.

Photoconductive - a substance that will conduct electricity when a light is shown on the P-N junction and that will block the flow of current when no light is present.

Photoconductive cells - those devices whose resistance varies in accordance to the amount of light directed on them.

Photodiode - a semiconductor component that will allow or restrict current flow in response to the application of light.

Photoelectricity - the physical action where an electrical flow is generated by light waves.

Photoelectrochemical cell - a special kind of photovoltaic cell in which the electricity produced is used immediately within the cell to produce a useful chemical product, such as hydrogen. The product material is continuously withdrawn from the cell for the direct use as a fuel or as an ingredient in making other chemicals, or it may be stored and used subsequently.

Photochemical reaction - a chemical reaction caused by light or ultraviolet radiation.

Photodetectors - those devices that convert light energy into electrical energy.

Photodiodes - those P-N junction diodes whose reverse current will vary according to the amount of light shining on them.

Photon - a particle of electromagnetic energy found in solar radiation.

Photostatically - the method by which the molecular formation of an element changes due to light.

Phototransistor - a transistor whose base collector responds to light to produce an electrical current used to energize the working circuit through a transistor. A much larger amount of current is produced than when a photodiode is used.

Photovoltaic - the generation of electrical potential across a cell by the conversion of light energy to electrical energy.

Photovoltaic array - an interconnected system of photovoltaic modules that functions as a single electrical producing unit. The modules are assembled as a discrete structure, with common support or mounting.

Photovoltaic cell - a device that converts light directly into electricity. A solar photovoltaic cell, or solar cell, is designed for use in sunlight. All photovoltaic cells produce direct current (DC).

Photovoltaic collector - a photovoltaic module or array which receives sunlight and converts in into electricity.

Photovoltaic material - a type of material that will develop voltage across its surface when a light is shown on its surface.

Photovoltaic thermal system - a photovoltaic system which, in addition to converting sunlight into electricity, collects the residual heat energy and delivers both heat and electricity in useable form. Also known as a total energy system.

Physical states - the three forms in which most substances can exist. These are solid, liquid, and gas.

Pi - the ratio of a circle's circumference to its diameter, approximately 3.1416.

Pick up allowance - that portion of the gross boiler output that is allowed for warming up the heating system and for taking care of the heat emission from a normal amount of piping. Also known as pickup loss and pickup load.

Pick up tube - the liquid drainage tube in a tank; especially a liquid receiver.

Pick up voltage - the back electromagnetic force at which a potential relay (start relay) has a sufficiently high enough voltage to open its normally closed contacts.

Picocurie (pCi) - a unit for measuring radioactivity, often expressed as Picocuries per liter (pCi/L) of air.

Picofarad - one micro-microfarad. Used in extremely small capacitors. Symbolized by μμf or pf.

Pictorial diagram - a wiring diagram that shows the actual internal wiring of a unit, much like a picture taken of a control panel. It is also called a line or label diagram.

Picture window - a large stationary (non-ventilating) window which is designed for a maximum view without obstruction. This type of window uses only fixed units. This type of window is best for a clear view and energy efficiency.

Pier - a column designed to support a concentrated load.

Piezoelectric crystals - property of quartz crystal that cause it to vibrate when a high frequency of 500 KHz or higher voltage is applied.

Piezoelectric crystals - certain materials which develop a potential difference between two faces of a crystal when it is mechanically distorted by pressure.

Piezoelectric effect - the generation of electrical potential across a given material by application of pressure across the material.

Piezo resistive sensing - a method of sensing pressure by measuring the change in volume (compression) of a material, such as silicon, in response to the pressure applied by a fluid.

Pig tail - a flexible, usually heavy duty, electrical lead attached to a component or appliance.

Pilaster - flat square column attached to a wall and projecting about a fifth of its width from the face of the wall.

Pillow block bearing - a bearing with a built in resilient support for a rotating shaft.

Pilot - a small burner (or a flame, smaller than the main burner flame) that is used to light off the main burner or burners. Also see continuous (constant, standing), expanding, intermittent, interrupted, and proven pilot.

Pilot, aerated - a pilot burner in which primary air is mixed with the gas before the gas leaves the burner ports.

Pilot and main flame supervision - the pilot and main burner flame are supervised simultaneously by one detector.

Pilot bleed relay - a relay system in some pneumatic controllers that translates the movement of the sensing element into a changing pressure to be fed to the actuator.

Pilot burner assembly - a combination gas pilot burner and flame rod flame detector; may also include an ignition electrode for spark ignition of the pilot, or a thermocouple adapter. Also called a rectification pilot or flame rectifier pilot.

Pilot control - a valve arrangement used in an evaporator pressure regulator to sense the pressure in the suction line and to regulate the action of the main valve.

Pilot control, external - a method by which the internal connection of the pilot is plugged and an external connection is provided to make it possible to use an evaporator pressure regulator as a suction stop valve as well.

Pilot device - a control component designed to control small amounts of current. Pilot devices are used to control larger control components such as reversing valves, contactors, motor starters or control relay schemes.

Pilot duty - a rating applied to inductive loads (usually a solenoid or coil) with a maximum power factor of 35%. Allowable inrush can be up to 10 times the pilot duty rating (running current).

Pilot duty rating - the contact rating of a relay designed to switch the coil of electromagnetic devices such as solenoids, relays, contactors or motor starters. The VA ratings of the controlling contacts must be equal to or greater than the sealed VA, not the inrush VA of the controlled coil to prevent excessive contact wear or contact welding.

Pilot duty relay - relays that have contacts made of gold flash over silver or fine silver, shaped and sized for low level current switching. Do not use pilot duty relays in applications requiring power duty relays.

Pilot flame - a small flame used to ignite a main flame.

Pilot flame establishing period - the period of time in ignition trails during which the flame safeguard control permits the pilot valve to be open before the flame detector is required to detect the pilot flame.

Pilot generator - a series of thermocouples together in one unit that generates the voltage required to operate a gas valve. Typical millivolt output is 250, 500 and 750.

Pilot light - the flame that ignites the main burner on a gas furnace.

Pilot link - a jumper screw or jumper wire which bypass a normally open load relay contact in the flame detection circuit of some flame safeguard controls. If removed, it allows the control to be used with a continuous pilot (safe start check is not applicable).

Pilot, non-aerated - a pilot burner in which no air is mixed with the gas before the gas passes through the burner ports. This pilot uses only secondary air.

Pilot only - see early spark termination.

Pilot operated valves - a piloted operated valve (suitable for two position applications only) is one that utilizes the difference between upstream and downstream pressure acting upon a diaphragm or piston to move the valve disc. The valve actuator needs to exert only the force necessary to open or close a small pilot port admitting fluid to the diaphragm or piston.

Pilot positioner - a pneumatic balancing relay used in conjunction with a valve operator or damper operator to assist the controller to accomplish a precise movement of the valve or damper operator being controlled. It has a separate source of supply air and can vary the pressure applied to the valve or damper operator over the full range of pressure, as needed, to achieve the required valve or damper position.

Pilot safety switch - a normally open temperature actuated switch that closed when it senses the high temperature produced by a pilot flame. If there is no flame, the switch should return to the open position.

Pilot safety valve - a valve that will close when pilot flame failure occurs shutting off the gas supply to the main and pilot burner. Usually an integral part of the gas valve.

Pilot solenoid - controls the flow of refrigerant by controlling the pressure to the equalizer of the thermostatic expansion valve.

Pilot stabilization period - a timed interval starting with the proof of the pilot and ending with the opening of the main fuel valve(s). This period permits the pilot to be established before permitting the inrush of main fuel.

Pilot, standing - a continuously burning pilot that does not operate a pilot safety device.

Pilot supervision - detecting the presence or absence of a pilot flame.

Pilot switch - a control used in conjunction with gas burners. Its function is to prevent operation of the burner in the event of pilot failure.

Pilot, thermoelectric - a safety pilot that uses a thermoelectric generator to convert heat energy into electrical energy for use in operating gas safety controls.

Pilot trail - see pilot flame establishing period.

Pilot tubing - a small diameter tube that goes from the gas valve (if 100% shut-off) or the B-cock (if non-100% shut-off) to the pilot assembly. Usually made of ¼' or ⅜" aluminum.

Pilot turndown test - a test to determine if the smallest pilot capable of holding in the flame relay is large enough to safely light off the main burner.

Pilot valve - an automatic safety shutoff valve that controls the fuel input to the pilot.

Pilotstat - coil of wire that generates a magnetic field to hold that gas valve open.

Pin - denoting an electrical terminal, usually in a connector. Normally a smaller termination than a lug.

Pinch off tool - a device used to press the walls of tubing together until fluid flow ceases.

Pinch technology - an energy analysis tool that uses vector analysis to evaluate all heating and cooling utilities in a process. Composite curves created by adding the vectors allow identification of a "pinch" point, which is the best thermal location of a heat pump.

Pipe - a seamless tube conforming to the particular dimensions commercially known as standard pipe sizes; usually copper, steel, PVC, aluminum, or brass.

Pipe bending - forming the pipe sections into predetermined radii by hot or cold bending procedures. Pipe sections are often bent to a five diameter radius, though sharper radii of three pipe diameters are sometimes required. Bending to the radius of 6, 10 or 15 times the diameter of the pipe can be accomplished.

Pipe cutter - a tool with one to four cutting wheels used to cut tubing or pipe.

Pipe die - a special tool used to cut threads on the outside of a pipe.

Pipe dope - a paint-like compound with a sealant added to it, applied to male threads only, used to lubricate the threads and to prevent leakage.

Pipe, equivalent length - the resistance of valves, controls, and fittings to flow, expressed as equivalent length of straight pipe for the convenience in calculating pipe sizes.

Pipe fabrication - production of sections, configurations, or assemblies of pipe and various pipe components such as valves and fittings. Fabrication includes a variety of processing, welding, forming, shaping, heating, cleaning and machining.

Pipe fittings - components used to join sections of pipe. These include, but are not limited to, tees, elbows, reducers, and caps.

Pipe, friction loss - the flow resistance of liquid through the pipe. It is dependent on pipe size and flow quantity. Expressed in feet of head (ft. hd.) per 100 foot of pipe.

Pipe intersection - pipe fabrication drawing that exhibits the necessary views and developments to construct pipe shapes where two or more pipes come together at a variety of angles.

Pipe lane system - system where piping runs north to south at one elevation and east to west at another.

Pipe line heater - an electric heater attached to the fuel oil line in order to maintain proper fuel oil temperature (viscosity) for moving the fuel oil through the line.

Pipe nipple - a short piece of pipe (not longer than 12") threaded at each end.

Pipe rest - supports long lengths of pipe that would otherwise cause the vise stand to tip over.

Pipe reamer - removes metal burrs that appear on the pipe after a pipe cutter has been used.

Pipe size - a term that, unless otherwise specified for pipes fewer than 12 inches in diameter. Refers to the approximate inside diameter of the pipe.

Pipe threader - a tool specifically made to establish the number of threads per inch and the taper required for each pipe size. May be either hand or power threaders.

Pipe support detail drawing - small detailed drawing showing design, location, and other information for the construction of a pipe support.

Pipe support system - complete arrangement of pipe supports (hangers, anchors, guides, snubbers) that (1) hold the weight of the system with a minimum safety factor of 3; (2) permit thermal and seismic movement; (3) damper vibration caused by mechanical equipment; and (4) maintain a safe stress limit.

Pipe thimble - sleeve used where pipes pass through walls or floors or are embedded in concrete or masonry.

Pipe thread - thread that is tapered 1/32 to 1/16" for every inch of thread.

Pipe vise - special tool used to securely hold steel pipe in position for cutting, threading or assembly.

Pipe volume - total interior volume, expressed in cubic measurement, which a cylindrical object can hold. The volume can be found by multiplying pi times the inside radius squared by he length of pipe (area of cross-section times length).

Pipe wrench - an adjustable wrench with a movable jaw and a stationary jaw, both having serrated teeth to grab onto round objects such as pipe; used to tighten or loosen pipe and fittings. Also called a Stillson wrench.

Pipeline schedule - list of all pipelines on a project giving line size, specifications, origin-termination, insulation, painting, and fabrication data, and other pertinent information.

Piping - general term used for the piping systems or for pipe and fittings in pipelines.

Piping and pick up allowance - heat stored in the boiler and piping during warm up and after shutdown and therefore deduced from gross boiler output since it is not available to the terminals.

Piping element - any component through which flow occurs.

Piping loss - heat loss through pipe walls and insulation.

Piping service - piping carrying cooling, heat, pressure, and other media that will not be part of the final product.

Piping system - all piping, valves, and fittings from the outlet of the point of delivery from the supplier to the outlets of the equipment shut-off valves.

Piping topology - the interconnected geometry formed by a specific arrangement of piping, fittings, and valves. Examples of piping topology include 2 pipe reverse return, multizone/multicirculator, and so forth.

Piston - a close fitting part that moves up and down in the cylinder or a type of non-adjustable metering device.

Piston displacement - volume displaced by piston as it travels the full length of its stroke.

Piston pin - that part of a compressor that connects the connecting rod to the piston.

Piston ring - a circular metal ring that fits around the outside of a piston, filling the space between the piston and the cylinder wall to prevent leakage in either direction.

Pitch - (1) the angle or slope of a fan blade. (2) The distance from one thread to point on the next thread. (3) The slope of a pipe used to enhance removal of waste products.

Pitch pocket - a pitch filled sheet metal retainer fitted around a pipe or similar object protruding through a roof to provide protection against water leakage.

Pitch, roof - the slope of a roof, usually expressed as a ratio of vertical rise to horizontal run.

Pitot tube - the standard pitot tube, which is used in conjunction with a suitable manometer, provides a simple method of determining the air velocity in a duct. The double concentric tube construction consists of a 1/8" OD inner tube which is concentrically located inside of a 5/16" OD outer tube. The outer static tube has 8 equally spaced, 0.04" diameter holes around the circumference of the outer tube, located 2 1/4" back from the nose or open end of the pitot tube tip. At the base end, or tube connection end, the inner tube is open ended as at the head, and the outer tube has a side outlet tube connector perpendicular to the outer tube and directly parallel with and in the same direction as the head end of the pitot tube. Both tubes haven a 90 degree radius bend in them located near the measuring end to allow the open end of the inner impact tube to be positioned so that it faces directly into the air stream when the main shaft of the pitot tube is perpendicular to the duct and the side outlet static pressure tube outlet connector is pointed in a parallel direction with the airflow facing upstream. Measurement of total pressure air stream can be made by connecting the inner tube outlet connector to one side of a manometer or draft gauge. Static pressure may be made by connecting the outer tube side outlet connector to one side of a manometer or draft gauge. Velocity pressure may be made by connecting both the inner and the outer tube connectors to opposite sides of a manometer or draft gauge. This instrument when used with a draft gauge, manometer, or micro manometer is a most reliable and rugged instrument and its use is preferred over any method for field measurement of air velocity, system total air, minimum outdoor air, maximum return air quantities, fan static pressure, fan total pressure and fan outlet velocity pressures where such measured quantities may be required.

Pitting - the resulting rough surfaces on electrical contacts on contactors, switches, and relays resulting from the heat generated from an overcurrent condition through the contacts. Pitting causes an increase in circuit resistance as well as reduction in the voltage supplied to the circuit load.

Pittsburgh lock - a corner locking longitudinal seam where a right angle flange is inserted into a formed pocket and the pocket flange flattened over the right angle flange.

Pivot - a point, shaft, etc. on which something turns; to turn as if mounted on such a point.

Placard - an easily seen tag or label that usually indicates warning or caution.

Planar diffusion - a method by which all junctions of a diode are in a single plane when the manufacturing process is complete.

Plane radiant temperature - the uniform temperature of an enclosure in which the incident radiant flux on one side of a small plane element is the same as in the existing environment.

Planned defrost - shutting the compressor off with a timer so that the space temperature can provide the needed defrost.

Plans - views of horizontal planes, showing components in their horizontal orientation. A set of construction drawings.

Plant efficiency - net electrical energy output (not generating plant auxiliaries) plus thermal energy used divided by fuel input to the plant.

Plastic range - the range on temperature between the solidus and liquidus points of a solder.

Plastomer - type of plastic liner used in ball valves to ensure tight shut-off.

Plate - one of the electron storing electrodes of a capacitor.

Plate condenser - a type of condenser used in refrigerators where the tubing is attached to a plate of sheet metal and is cooled by static air.

Plate evaporator - a flattened, solid cooling coil through which no air circulates.

Play - looseness between connected parts.

Pleated filters - extended surface filters.

Plenum - an air compartment mixing chamber which is part of a distributing system, to which one or more ducts are connected. It may be attached to, or be an integral part of, the supply outlet or the return air inlet of a furnace.

Plenum chamber - an air compartment or chamber maintained under pressure (either negative or positive) to serve other parts of an air distribution system.

Plenum chamber, water tower - the enclosed space between the drift eliminators and the fan in induced draft towers, or the enclosed space between the fan and the fill in forced draft towers.

Pliers - a hand tool that hold, cut, bend or remove objects. The name of the pliers usually tell what job it is intended for.

Pliable - easily bent, flexible.

Plot plan - plan that can be divided into various types, including master and unit plot plans. The master plot plan is the general plan that shows the overall location of equipment and structures, both existing and to be built. The unit plot plan is a smaller drawing showing only the equipment location.. In general, neither type is dimensional, though some companies do dimension the unit plot plan with coordinates.

Plug cock - see cock.

Plug, electrical - the end of a conductor that is designed to be inserted into an electrical receptacle.

Plug flow - a flow regime where the flow is predominately in one direction and contaminants are swept along with the flow.

Plug, valve - a term used to describe a disc that uses a solid, shaped plug rather than a skirted guide to characterize flow. In some valves the valve may serve as a guide.

Plug tap - the plug tap is used after the taper tap has threaded. It tapers back three of four threads before the entire threading is reached. A plug tap is used to do most of the cutting. It can be used to thread a blind hole.

Plug type fuse - a round Edison based screw in fuse. Typical sizes are 15 amps (blue), 20 amps (orange) and 30 amps (green).

Plumb - exactly level.

Plumbing system plan - these plans show the layout for the piping system that supplies the hot and cold water, the sewage disposal system, and the location of plumbing fixtures. Plans for smaller homes may include the entire plumbing system on one drawing, usually on the floor plan. But, for complex structures, separate plans for each system may be used. These plans will tell you where the main pipe and shut-off valve is located, along with such fixtures as sinks, bathtubs, and toilets.

Plume - the effluent mixture of heated air and water vapor (usually visible) discharged from a cooling tower.

Plunger type - type of solenoid that is generally used with relays that use double break contacts. The coil is surrounded by the iron core that has an opening in it for the shaft of the armature to pass.

P-N junction - where the P region and the N region are joined and formed into a single crystal.

P-N junction diode - a type of semiconductor device that allows the electrons to flow through it in only one direction.

Pneumatic - operated by or filled with compressed air.

Pneumatic actuator - a standard pneumatic actuator moves toward the advanced position as the branch line pressure increases and toward the retarded position as the branch line pressure decreases.

Pneumatic atomizing humidifier - air pressure is used to break up the water into tiny droplets and to disperse them.

Pneumatic electric relay - an air actuated device used to make or break electrical contacts in connection with the operation of the control system.

Pneumatic electric transducer - device that converts a pneumatic value to an equivalent electrical value such as volt or milliamp.

Pneumatic timer - a device that uses the displacement of air in a bellows or diaphragm to produce a time delay.

Pocket lock - see government local.

Point, critical - the point at which liquid and vapor have identical properties.

Point of access - all areas used by employees for work related passage from one area to another.

Point of duty - a statement of the air volume flow rate and static or total pressure at a stated density and is used to specify the point on the system curve at which a fan is to operate.

Point of no pressure change - point in a hydronic hot water system where the expansion tank is connected to the piping system. Pressure at this location cannot be affected by circulator operation.

Point of operation - used to designate the single set fan performance values which correspond to the point of intersection of the system curve and the fan pressure volume curve.

Point of rating - a statement of fan performance values which correspond to one specific point on the fan pressure volume curve.

Point, triple - state point at which three phases of given substance, (i.e., solid, liquid, and gas) exists in equilibrium.

Pointer flutter - pressure pulsations that cause a gauge pointer to swing above and below the actual pressure reading.

Points - teeth around the opening or a box end or socket wrench that grip the edges of the nut or the bolt head.

Poison, Class A - a DOT term for extremely dangerous poisons such as poisonous gases or liquids of such a nature that a very small amount of the gas or vapor of the liquid mixed with the air is dangerous to life. Examples would include phosgene, cyanogen, hydrocyanic acid, and nitrogen peroxide

Poison, Class B - a DOT term for liquid, solid, paste, or semi-solid substances other than Class A poisons or irritating materials that are known or presumed on the basis of animal tests to be so toxic to man as to afford a hazard to health during transportation.

Polar - molecules that have a positively charged end and a negatively charged end, each of which attracts its opposite.

Polarity - the direction of electrical current flow direction or a magnetic charge (north or south).

Polarize - to give polarity, temporary or permanent, to a conductor or ferrous material.

Pole - one set of electric contacts either in an automatic device or a manual switch. Electric devices such as relays, contactors, switches, and breakers can be purchased with one or more poles.

Pole pieces - the cores of the electromagnetic coils mounted in the stator of direct current motors.

Pole shoe - that part of a rotating device, such as a motor, that is used to hold the field coils in their proper positions around the armature.

Poles - two or more stationary electromagnets positioned at opposite sides of a circle inside the motor. They have opposing magnetic polarity.

Poles of a switch - the number of independent electrical circuits that can be simultaneously passed through a switch or relay.

Polish - to produce a highly finished or polished surface by friction using very fine abrasive.

Pollen - an airborne irritant produced by plants.

Pollutant - unwanted particulate matter in the air that often lead to unwanted health related problems. A contaminant that decreases the quality of indoor air.

Polyalkylene glycol (PAG) - a very hygroscopic (moisture absorbing) refrigeration lubricant used with HFC refrigerants. Often used in automotive air conditioning systems with HFC refrigerants. PAG's are incompatible with chlorine and have high molecular weights. This was the first generation of oil used with HFC refrigerants.

Polybutylene - a plastic used for pipe material that has excellent creep resistance as well as high resistance to stress cracking. Recommended for earth loops.

Polychlorinated biphenyl (PCB) - dielectric fluid used in capacitors and transformers that is very toxic. Use of PCB's in transformers and capacitors is strictly regulated by the EPA.

Polycrystalline silicon - silicon that has solidified at such a rate that many small crystals (crystallites) were formed. The atoms within a single crystal are symmetrically arrayed, whereas in crystallites they are jumbled together.

Polycyclic organic matter - by-products of wood combustion found in smoke and considered to be a health hazard.

Polyethylene - a family of insulating materials derived from polymerization of ethylene gas. They are basically pure hydrocarbon resins, with excellent dielectric properties. A plastic used for tubing for ground loop systems, irrigation systems, cold water lines, and heat pump piping.

Polymer - a material having molecules of high molecular weight formed by polymerization of lower molecular weight molecules.

Polyolester oil (POE) - a wax free, ester-based synthetic oil lubricant used with refrigeration systems. They are compatible with all refrigerants types but designed specifically for HFC refrigerants. They are miscible with mineral oil and alkylbenzene type oils. These oils have been used for years as jet engine lubricants.

Polyphase - more than one phase, usually three phase.

Polyphase generation - electric power production using a generator that rotates three different conducting loops at the same time.

Polyphase motor - an electrical motor designed to be used with three phase electrical current.

Polystyrene - plastic used as insulation in some refrigerator cabinet structures.

Polyurethane - any synthetic rubber polymers produced from the polymerization of an HO and NCO group from two different compounds. Used in insulation and molded products.

Poly vinyl chloride (PVC) - plastic pipe used in pressure applications for water and gas as well as for sewage and certain industrial applications.

Poly vinyl ether (PVE) - a refrigeration oil used with hydrofluorocarbons air conditioning systems. PVE type oils have a superior performance rating over POE with respect to the resistance of capillary tube blockage.

Ponded roof - a flat roof designed to hold water that acts as a cooling device.

Pop type safety valve - valve with a predetermined popping pressure. Commonly found on steam boilers.

Popping pressure - predetermined pressure at which a safety valve opens and remains open until the pressure drops.

Porcelain - ceramic china-like coating applied to steel surfaces. Also used as insulators for ignition leads on fuel oil fired equipment.

Port - an orifice or opening in a burner head through which fuel or an air fuel mixture is discharged for ignition.

Port, inlet - opening connected to the upstream side of a fluid system.

Port (valves) - is the flow controlling opening between the seat and the disc when the valve is wide open. It does not refer to body size or end connection size. Standard valve ports are the sizes normally used in the valve. Reduced ports have flow controlling areas equal to the valve port area of the valves of the same body size as the reduced port size.

Port loading - the input rate of a gas burner per unit of port area, obtained by dividing input rate by total port area. Usually expressed in terms of Btu's per hour per square inch of port area.

Portable dolly - a small platform with four wheels and which heavy objects can be placed and moved.

Portable ladder - a type of ladder that can be readily moved and carried.

Portable service cylinder - container used to store refrigerant. Two most common types are disposable and refillable.

Posistor - a thermally sensitive resistor that has a positive temperature coefficient of resistance. Also see thermistor.

Positioning controller - a control, usually modulating, that regulates air and fuel going to a boiler.

Positive - a non-negative greater than zero value that has a deficiency of electrons.

Positive acting - in pneumatic control devices this is an abrupt change in the branch line pressure of a controller from zero to 15 psig. This causes either a full open or closed condition.

Positive electrical charge - the result of electrons leaving an atom.

Positive displacement compressor - a compressor that raises vapor pressure by decreasing its volume.

Positive displacement pump - a pump where the pumping action is created by moving chambers or pistons. The flow rate of this pump is almost the same at any pressure level. Generally, self-priming and should never be operated dry, because of internal wearing of rubber parts. As discharge flow is restricted (higher pressure or head) drive horsepower requirements increases. A relief device should be provided on the discharge line to prevent over pressure and damage to the pump motor if discharge line is closed off or severely restricted. The most common positive displacement pump types are: diaphragm, gear, flexible impeller, rotary screw, roller or vane, piston, jet pump, and deep well submersible pump.

Positive positioning relay - an auxiliary device that can be fitted to a damper or valve actuator. It can position the actuator accurately with respect to the signal pressure from the controlling instrument, regardless of the load.

Positive pressure - A condition that exists if more air is supplied to a room than is exhausted. Greater than atmospheric pressure. Also see furnace pressure.

Positive pressure head - driving force used to transfer fluids through pipes. Expressed in feet of water or pounds per square inch.

Positive temperature coefficient thermistor (PTC) - electronic thermistor that increases in resistance as temperature increases. Uses to provide soft start assistance to a permanent split capacitor motor.

Post - two pin-like objects on either side of a threader. The posts fit through openings in the change plate that hold them securely. Moving the post will change the position of the dies.

Post purge - a period of time after the run period during which the burner motor (blower or fan) continues to run, driving all the products of combustion and any unburned fuel from the combustion chamber, and supplying air to burn fuel being purged from the fuel line downstream from the safety shutoff valve.

Pot - short for potentiometer.

Potable water - water that is suitable for human consumption. Domestic water.

Potassium permanganate - used in carbon filters to help reduce unpleasant odors.

Potential - electrical force, measured in volts. Potential may be positive or negative.

Potential difference - the electrical force which moves, or attempts to move, electrons along a conductor or resistance.

Potential electrical - electrical force which moves, or attempts to move, electrons along a conductor or resistance.

Potential energy - the energy present in a substance or energy present because of position.

Potential head - the energy created by or of position, measured by work possible in dropping a vertical distance.

Potential relay - a switching device used with hermetic motors that breaks the circuit to the start capacitor and/or start winding after the motor has reached approximately 75% of its running speed. It contains a normally closed electrical switch which is operated by back EMF generated across the start windings. Care must be taken in the installation process because this relay is position sensitive.

Potential transformer - a voltage transformer. The voltage supplied to a primary coil induces a voltage in a secondary coil according to the ratio of the wire windings in each of the coils.

Potentiometer - an electromechanical device consisting of a resistive element with a terminal at each end, and a third terminal connected to the wiper contact. As the wiper moves along the element, it changes the resistance in each leg (portion of the element between each end terminal and the wiper). Thus the electrical input or output can be changed mechanically. Also see auxiliary and feedback potentiometers.

Pot type burner - a non-pressurized burner, in which fuel must evaporate to support combustion.

Pound foot - basic unit of measurement of torque. The number of pound feet is calculated by multiplying the applied force (in pounds) times the radius of the turning circle (in feet).

Pound force - force applied to a 1 pound mass. Has an acceleration of 32.173 ft/s2.

Pound per square inch (**PSI**) - a measure of pressure.

Pound per square inch absolute (**PSIA**) - pressure exerted on a surface one square inch and equaling gauge pressure plus atmospheric pressure.

Pound per square inch gauge (**PSIG**) - pressure exerted on a surface one inch square and measured using a gauge without considering atmospheric pressure. Also referred to as gauge pressure.

Pour point - the lowest or minimum temperature at which an oil or distillate oil is observed to flow, when cooled under conditions prescribed by test method ASTM D97. The pour point is 3° C (5° F) above the temperature at which the oil in a test vessel shows no movement when the container is held horizontally for 5 seconds.

Poured gypsum underlayment - a pourable slurry consisting of gypsum cement, sand, and other admixtures that cures into a hard subflooring. Poured over hydronic tubing that has been stapled to the subfloor.

Power - the amount of work per unit of time. Source or means of supplying energy. In electricity the symbol is P and its units are watts and kilowatt. Formula is P = volts x amps. There are two types of power:

Apparent power - the product of measured AC voltage and current in which voltage and current reach their peak values at different times in a cycle.

True power - the average power consumed by a circuit (or component) during one full cycle of alternating current. Measured directly with a wattmeter.

Power burner - a gas burner in which combustion air is blown in by a motor driven fan. Also called a forced draft burner. Examples are aspirating, gun type, mechanical premix, and zero governor burners. Also see mechanical draft burner.

Power circuit - that portion of the total circuitry allocated to the distribution of primary AC power.

Power consumption - power used multiplied by time, i.e., kilowatt hour.

Power duty relay - relays that have contacts made of silver cadmium oxide and a very high surface resistance. Power duty relays rely on a make-break arc process to break down the resistance of deposits formed on the contacts. The cleaning process while required on power duty contacts will destroy pilot duty contacts in a short period of time. Also the light loads (amperage) of pilot duty contacts will not provide the cleaning process required on power duty contacts causing non-conductive deposits to form resulting in a failed relay. For these reasons, do not use pilot duty relays in place of a power duty relay.

Power element - the sensitive fluid filled feeler bulb, capillary tube and diaphragm element of a temperature operated TXV control.

Power equation, 1Ø - volts times amps times power factor.

Power equation, 3Ø - volts times amps times power factor times 1.7321.

Power factor (PF) - the ratio of the true or real power as measured by a wattmeter (kW) to the apparent power as measured by a voltmeter and an ammeter (kVa) for any load and time. Generally expressed as a decimal or percentage. The cosine of the angle between voltage applied and the current reading. For example 2000 watts is being supplied to a motor, but it is only using 1800 watts. Power factor would be 1800/2000 or .9 or 90% efficient.

Power factor charge - a utility charge for "poor" power factor. It is more expensive to provide power to a facility with a poor power factor which is usually less than 80% efficiency.

Power factor correction - the act of installing capacitors on the utility service's supply line to improve the power factor of the building.

Power loss - power consumed by resistance in a circuit. Voltage drop.

Power rating - the rating of a device that indicates the amount of current flow and voltage drop that can be permitted.

Power relay - generally rated up to 30 amperes and are intended for use on direct switching of small motors and heating applications that do not have a high cycle rate.

Power source - a source of electrical energy. It can be a battery, generator, piezo-crystal, and thermocouple or from which an electrical load draws its electrical energy.

Power supply - the voltage and current source for an electrical circuit. A battery, a utility service, and a transformer are power supplies.

Power transistor - used to connect the load to the line in a relay designed to control a DC load.

Power tube - see ultraviolet power tube.

Power unit - that part of the control instrument that causes the instrument to function.

Power vent - furnaces with integral means of venting products of combustion using an induced draft fan motor.

Powerpile - a multitude of single thermocouples wired in series that generates a normal open circuit voltage (VOC) of 500 to 750 millivolts. Also referred to as a thermopile.

Ppm - parts per million. The unit of concentration in solutions.

Precharged lines - refrigerant tubing that comes from the manufacturer with a refrigerant charge sealed within it.

Precipitate - to condense as with vapor or gasses, and cause free water to form. To separate a dissolved substance out from a solution.

Precipitator, electrostatic - a device that removes particles from the air by electrically charging them and then collecting them on a plate of the opposite charge.

Precision - an indication of the accuracy of the measured value.

Precision resistor - a resistor that has a high degree of accuracy.

Precooler - a cooler used to remove sensible heat before shipping, storing, or processing.

Precooler condenser - used to cool the refrigerant prior to entering the main condenser.

Predictive maintenance - a type of condition based maintenance emphasizing early prediction of failure using non-destructive techniques such as vibration analysis, thermography, and wear debris analysis.

Prefilter - a wide mesh filter for trapping large particles before they reach and clog the fine mesh, high efficiency filter.

Preheat - a process of raising the temperature of outdoor air before incorporating it into the rest of the ventilating system. Used when large amounts of very cold outdoor air must be used.

Preheat flame - this is the flame of the torch tip that preheats the base metal to be cut, soldered, or brazed.

Preheat valve - a wheel valve used to regulate gas flow on an oxyacetylene cutting torch attachment.

Preheated air - air heated prior to its use for combustion; frequently the heating is done by hot flue gases.

Preignition interlock - an interlock that proves a condition is proper for burner start-up and operation through prepurge, prior to ignition trials. It usually proves the main fuel valve(s) is closed. It may also be a damper position indicator (low or high fire proving switch), fuel pressure switch, or oil preheater controller. Also see valve closed interlock.

Premix burner - a gas burner in which the gas and air are mixed before they are introduced into the combustion chamber. Usually, the gas and air are both fed into a fan or blower, and the mixture is then blown into the combustion chamber. A premix burner burns with a short, hot flame with high heat release. Examples are mechanical premix, fan mix, and zero governor burner.

Premixer - a device to mix gas and air before delivery to a burner, such as an aspirator, an inspirator, or a fan mixer.

Prepurge - a period of time after burner start-up (call for heat) during which the burner motor (blower or fan) runs to expel the air of the combustion chamber breeching prior to ignition trials. This removes any unburned fuel so only the incoming fuel will be present for ignition.

Prescriptive design - design of a living unit or building of the same size and occupancy type as the proposed design that complies with the prescriptive requirements of this standard. The prescriptive design includes specified assumptions concerning shape, orientation, HVAC, and other system design features. The prescriptive design is used to generate the compliance requirement for the annual energy cost method.

Prescriptive requirements - specified values or rules representing the requirements that must be must in order to achieve compliance with the standard.

Pressure - force per unit area; usually measured in pounds per square inch (psi), or by the height of the column of water or mercury that the force will support (in feet, inches, or millimeters). Also see absolute, atmospheric, furnace, gauge, normal fuel supply, and static pressure.

Pressure, absolute - pressure figured from a perfect vacuum. The sum of atmospheric and gauge pressures.

Pressure atomizing gun type burner - the simplest, and probably the most common, mechanical atomizing oil burner. Light oil under pressures of up to 150 pounds per square inch is discharged through a nozzle or orifice producing a cone shaped spray. Action of the nozzle, plus the impact of a rotating airstream from a fan, mixes the oil and air. Also see atomizing oil burner.

Pressure atomizing oil burner - synonymous with mechanical atomizing oil burner. See atomizing oil burner and pressure atomizing gun type burner.

Pressure, atmospheric - the amount of pressure exerted by the atmosphere at any specific location. Usually figured to be 14.696 psi (14.7) at sea level.

Pressure, back - the pressure encountered on the return side of the system.

Pressure burner - a burner in which an air and gas mixture under pressure is supplied, usually at 0.5 to 14 inches water column.

Pressure controller - a controller which monitors the pressure of steam, air, gases, or liquids, and operates to keep the pressure within predetermined limits. It may operate as a pressure switch (on-off), or it may be a proportioning controller. Also see pressure switch and proportioning controller.

Pressure, crankcase - pressure in the crankcase of a reciprocating compressor.

Pressure, cracking - the pressure at which a pressure operated valve begins to pass fluid.

Pressure, critical - vapor pressure corresponding to the critical state of the substance at which the liquid and vapor have the same properties.

Pressure dependent VAV box - has a air damper whose position is controlled directly by the zone thermostat. The actual air flow at any given damper position is a function of the air static pressure within the duct.

Pressure differential - difference in pressure between any two points of a piping system.

Pressure differential controller - instrument that controls pressure differential between two pipes, vessels, or areas.

Pressure differential valve - a valve that senses a pressure differential and opens or closes when a specific pressure differential is reached.

Pressure, discharge - the pressure against which the compressor must deliver the refrigerant vapor.

Pressure drop - is the difference between the pressures at two points. Also resistances to flow caused by a restriction, like friction, vertical lift and accessories.

Pressure drop, critical - flow through a valve increases with increased pressure drop until a critical value is reached. Therefore, any drop in excess of the critical pressure drop should be avoided to ensure noise free operation and a minimum of valve wear. In formula form, hcritical = 50% x P1 (absolute upstream pressure).

Pressure drop type distributor - refrigerant distributor that is configured in a manner that causes a pressure drop through the device, which in turn increases the velocity of the refrigerant as it flows through the device. This ensures that the liquid and vapor entering the evaporator are well mixed, resulting in even distribution to all of the evaporator circuits.

Pressure enthalpy heat diagram - a graph of refrigerant pressure, heat and temperature properties. See Mollier's diagram.

Pressure equalizing - allowing the high side and low side pressures of a refrigeration system to become equal on the off cycle. This reduces starting load when beginning the new cycle.

Pressure, gauge - pressure above atmospheric.

Pressure gauges - an instrument that measures pressure in psi (pounds per square inch). Primary tool for service personnel for servicing and troubleshooting the system. Also call manifold gauges.

Pressure, head - force caused by the weight of a column of fluids. Expressed in feet, inches, or psi.

Pressure, heat diagram - a graph representing a refrigerant's pressure, heat, and temperature properties. Also known as a Mollier's diagram.

Pressure, hydrostatic - the normal force per unit area that would be exerted by a moving fluid on an infinitesimally small body immersed in it if the body were carried along with the fluid.

Pressure imposing element - any device or portion of the equipment used for the purpose of increasing the pressure upon the refrigerant.

Pressure independent VAV box - has an air damper whose position is controlled on the basis of measured air flow. The setpoint of the airflow controller is, in turn, reset by a zone thermostat. A maximum and minimum airflow is set in the controller, and the box modulates between the two according to room temperature.

Pressure, limiter - a device which remains closed until a certain pressure is reached and then opens and releases fluid to another part of the system.

Pressure limiting device - a control that automatically stops the compressor at a predetermined pressure.

Pressure loss - indicates how much total pressure is required to move air through a register.

Pressure motor control - a device that opens and closes an electrical circuit as pressures change to desired pressures.

Pressure null switch - a pressure switching device with a single pole double throw switching. The switch is activated by a diaphragm that senses pressure differential. The SPDT switch has a floating contact that may be positioned against either pole or can be positioned for no circuit at all. Used with non-spring return damper motors.

Pressure operated altitude valve (POA) - a device that maintains a constant low side pressure independent of altitude of operation.

Pressure, operating - the pressure at which a system is operating.

Pressure, partial - pressure attributable to each gas when the gases are a mixture.

Pressure rating - the estimate maximum pressure that the medium in the pipe can exert continuously with a high degree of certainty that failure of the pipe will not occur.

Pressure reducing station - where higher pressure steam is reduced in pressure for plant processes.

Pressure reducing valve - a diaphragm equipped valve in the make up water line of a hot water system that protects the system from city water pressures higher than the working pressure of the boiler.

Pressure regulator - a device used to maintain a constant pressure in a fuel supply line regardless of the flow. It cannot maintain a pressure greater than its inlet pressure.

Pressure regulator, dead end lock up - a pressure regulator that will maintain a pressure slightly above adjusted pressures under static conditions when no gas is flowing and will protect the downstream controls, if properly sized.

Pressure regulator, evaporator - an automatic pressure regulating valve that is mounted in the suction line between the evaporator outlet and the compressor inlet. Its purpose is to maintain a predetermined pressure and temperature in the evaporator.

Pressure regulator, lock up type - under static pressure conditions of no gas flow, outlet pressure will rise somewhat above adjusted pressure but should not rise to line pressure.

Pressure regulator, non-lock up type - under static pressure conditions, when no gas is flowing, outlet pressure will rise to line pressure.

Pressure relief valve - a safety device that opens at a preset point to protect a boiler or hot water tank from excessive pressure.

Pressure, saturation - at a given temperature, the pressure at which a liquid and its vapor or a solid and its vapor can coexist in stable equilibrium.

Pressure, static - measure of the potential energy in a fluid. The force necessary to overcome the resistance to air flow presented by ducts, dampers, filter, etc.

Pressure, suction - the pressure in the low pressure side of the refrigeration system.

Pressure switch - a switch that monitors the pressure of steam, air, gases, or liquids, and breaks a circuit when the pressure either rises or falls to a preset valve. It is also used as a limit or interlock with a flame safeguard control to shut down the burner if the pressure exceeds high pressure switch or falls below low pressure switch a preset value. It may reset automatically when the pressure returns to normal, or it may require manual resetting.

Pressure, system - the pressure which overcomes the total resistances in a system. It includes all losses as well as useful work.

Pressure tap - opening used to check pressure.

Pressure tanks - used on geothermal heat pump systems to prevent the submerged well pump from operating continuously. Water is stored in the pressure tank to provide supply water to the heat exchanger.

Pressure tap plug - threaded plug used to prevent leakage of material after taking pressure reading.

Pressure temperature relationship - the constant, predictable relationship between the pressure and temperature of a given liquid and gas mixture under saturated conditions. An increase in pressure results in a temperature increase. A decrease in temperature results in a pressure decrease.

Pressure test - an operation performed to verify the integrity of piping or tubing following its installation or modification.

Pressure, total - sum of the static pressure and the velocity pressure at a given point.

Pressure transducer - a pressure sensitive device located in the piping of a refrigeration system that will transform a pressure signal to an electronic signal. The electronic signal will then feed a microprocessor.

Pressure tube - a small line carrying pressure to the sensitive element of the pressure controller.

Pressure, vapor - the pressure exerted by a vapor. If a vapor is kept in confinement over its liquid so that the vapor can accumulate above the liquid, the temperature being held constant, the vapor pressure approaches a fixed limit called the maximum, or saturated, vapor pressure, dependent only on the temperature and the liquid. The term vapor pressure is sometimes used as synonymous with saturated vapor pressure.

Pressure, velocity - force necessary to move the mass of air to the required speed, a measure of the kinetic energy in a fluid.

Pressure vessel - the steel drum or collector where generated stream or hot water is accumulated.

Pressure water valve - a device used to control the water flow which responds to head pressure of the refrigerating system.

Pressurestat - an automatic switch that opens or closes in response to a change in pressure.

Pressuretrol - Honeywell® trademark. A "pressure controller" often used to identify the control used to limit the pressure in a steam system.

Pressurize - to introduce refrigerant or inert gas into a system to check for leaks.

Pressurized furnace - furnace that operates at slightly above atmospheric pressure.

Preventative maintenance - scheduled inspection and replacement of short lived components in order to avoid untimely and expensive replacement on an emergency basis.

Primary - the input windings of a transformer.

Primary air - the combustion air introduced into a burner that mixes with the fuel before it reaches the combustion chamber. Usually expressed as a percentage of the air required for complete combustion.

Primary air inlet - the opening or openings through which primary air is admitted into a burner.

Primary cell - a cell that cannot be recharged.

Primary coil - a tube and finned circular coil that contains a water glycol solution which surrounds the igniter and burner. This coil is used in a water glycol gas forced air furnace.

Primary control - a flame safeguard control that starts the burner in the proper sequence, proves that the burner flame is established, and supervises the flame during burner operation. It causes safety shutdown on failure to ignite the pilot or main burner or on loss of flame.

Primary controls - in air conditioning and refrigerating systems, those controls that are used to start and stop the cycle when certain temperatures or pressures are reached.

Primary distribution feeder - an electrical line supplying power to a distribution circuit, usually considered to be that portion of the primary conductors between the substation or point of supply and the center of distribution.

Primary element - that portion of the controller that feels the change in the controlled variable and initiates the change in energy that causes the controller to operate.

Primary heat exchanger - that portion of the furnace were combustion takes place. Common heat exchanger conductor materials are steel, aluminum and cast iron.

Primary loop - a main distribution circuit tied to the system's heat source, and supplying several secondary circuits.

Primary side - the side of a transformer connected to the line voltage.

Primary system - converts energy from electricity or fuel and is the source of hot or chilled water.

Primary voltage - the voltage of the circuit supplying power to a transformer is called the primary voltage, as opposed to the output voltage of a load supply voltage that is called the secondary voltage. In power supply practice, the primary is almost always the high voltage side of the transformer and the secondary is the low voltage side.

Primary winding - the coil of a transformer to which the alternating current source voltage is applied. The primary winding has fewer turns than the secondary winding on step up transformers and more turns than the secondary winding on step down transformers.

Prime - a charge of liquid required to begin pumping action of centrifugal pumps when liquid source is lower than pump. The prime may be held in the pump by a foot valve on the intake line or a valve or chamber within the pump.

Prime mover - a system that supplies the mechanical energy to rotate an electrical generator.

Prime surface - in describing a finned tube, the tube is the prime surface, and the fins are the secondary surfaces.

Priming, boiler - the discharge of steam containing excessive quantities of water in suspension from a boiler, due to violent bubbling or boiling. Priming occurs in a boiler when the water level is so high that water mixes with steam.

Priming, oil - the act of bleeding the air from the oil lines or the pump mechanism.

Primer - strong organic solvent, preferably tetrahydrofuran, used to dissolve and soften the joint surfaces in preparation for and prior to the application of solvent cement. Primer is usually colored purple.

Printed circuit card - a circuit card in which the interconnecting wires have been replaced by conductive strips painted or etched onto an insulating board similar to a phonograph.

Prints - a term used to refer to a plan or set of blueprints.

Priority load - a control strategy used in multiload systems in which some loads are temporarily turned off while the priority load is operating.

Proactive maintenance - a type of condition based maintenance emphasizing the routine detection and correction of root cause conditions that would otherwise lead to failure. Such root causes as high lubricant contaminant, alignment and balance, refrigerant charge and airflow are the most critical.

Probe - any device that senses a condition that is remote from the meter or instrument and relays the effect back to the instrument.

Process changes - changes in the condition of a process such as temperature, flow, or pressure.

Process steam - steam used in the plant for manufacturing purposes.

Process tube - a tube that extends from the compressor or filter drier of a hermetic system. It's used to gain access to the sealed system.

Process water - any water used in production or directly in the process line itself, such as a water quench tank.

Products of combustion - gases that are formed as a fuel is burned in the furnace.

Program relay - a relay that provides multiple switching actions in an electric control system, either in a fixed time sequence or in a variable sequence, depending upon signals received from two or more controllers.

Programmable logic control (PLC) system - a solid state control system that is programmed to automatically control equipment or an industrial process.

Programmable thermostat - a room thermostat that can be set to automatically operate at different temperature settings at different times.

Programmer - see programming control.

Programming control - a flame safeguard control that adds a timing function to the primary control to sequence additional burner functions such as pre-purge, post-purge, timed trials for pilot and main burner flame, and firing rate switching.

Programming terminal - used to load or amend a program in a programmable controller. Same as a loading terminal.

Propane - an easily liquefiable hydrocarbon gas. Propane is one of the components of raw natural gas, and is also derived from petroleum refining processes. Chemical formula is C_3H_8, maximum burning speed is 32 inches per second, a specific gravity of 1.5 and ignition temperature of 900° F. Refrigerant 290 (propane) is used for low temperature applications. Propane has a molecular weight of 44.10, a specific volume of 8.5 cubic feet/lb., and a boiling point of -43.67° F.

Propane gas torch - a gas torch that has a throw away tank and is easy to use. The flame adjusts easily and can be used for many operations.

Propeller fan, axial - an air moving device in which the airflow is parallel or axial to the shaft on which the propeller is mounted. These fans have good efficiency, near free air delivery and are used primarily in low static pressure, high volume applications. As static pressure is increased, horsepower increases and CFM decreases. Usually mounted in a venturi, ring, or other housing featuring simple construction and low cost.

Properties, thermodynamic - basic qualities used in defining the condition of a substance, such as temperature, pressure, volume-enthalpy, entropy.

Propagation - the spread of flame through a flammable mixture from a source of ignition.

Proportional action - an output signal changing in proportion to the amount of change in the controlled or measured variable (contract to two-position).

Proportional actuator - an actuator that can position a damper or valve anywhere between fully open or fully closed in proportion to changes in the controlled medium. When this action is done automatically, it is called modulation. Also see actuator.

Proportional band - the change in the controlled variable required to move the controlled device from one its extreme limits of travel to the other. It is normally used in conjunction with recording and indicating controllers and is expressed in percent of the chart or scale range.

Proportional control - a control response where the value of the output signal is directly proportional to the error between the target value and the actual (measured) value of the controlled variable.

Proportional control action - a control action in which a manipulated variable is proportionate to any deviation from the set point.

Proportional integral control (PI) - a control response where the value of the output signal is based on the error between the target value and the actual (measured) value of the controlled variable, as well as how long that error has existed.

Proportional integral derivative control (PID) - a control response where the value of the output signal is based on the error between the target value and the actual (measured) value of the controlled variable, how long that error has existed, and the time rate of change of error.

Proportional plus automatic reset action - a combination of proportional action and a response which continually resets the control point back toward the set point to reduce the offset.

Proportional plus rate action - a combination of proportional action and a response that precedes the normal proportional response. The combined response is proportional to the rate of change or speed with which the controlled variable deviates from the set point.

Proportional reset - a method of producing a reduced supply temperature in a hydronic distribution circuit by supplying a manually set mixing device with a "hot" fluid temperature that itself is reset by an intelligent mixing device.

Proportioning - can be applied to the maintenance of the ratio between fuel and air supply throughout the operating range of the burner.

Proportioning chemical feed pump - pump that can be adjusted to feed chemicals to a boiler over a 24 hour period.

Proportioning controller - a controller that includes a potentiometer to permit the device it controls to be set anywhere between its extreme positions. When this action is done automatically in response to changes in the controlled medium, it is called modulation. Also see modulating controller.

Proposed design - design of the living unit or building to be constructed. The design takes into account all qualities, details, and characteristics of the building that significantly affect the use of energy, such as construction, geometry, orientation, exposure, materials, equipment, and renewable energy sources.

Propylene glycol - an anti-freeze fluid cooled by a primary (phase change) refrigerant, which then is circulated by pumps throughout the refrigeration system to absorb heat.

Protection saddles - saddles used on insulated pipes that require free longitudinal movement.

Protector, circuit - an electrical device that will open an electrical circuit if excessive electrical conditions occur.

Protectorelay - Honeywell® trademark. See primary control.

Protocol - rules concerning the format and timing of messages transmitted between two communicating devices.

Proton - a particle in the center of an atom that has a positive electrical charge.

Protractor - an instrument used to measure degrees and angles of a circle.

Prove - to establish by measurement or test the existence of a specified condition, such as flame, level, flow, pressure, or position.

Proved ignition - ignition by an energy source that is supervised by a flame safeguard control that must detect the presence of energy for ignition prior to permitting the main fuel valve(s) to open.

Proved pilot - a pilot flame that is supervised by a flame safeguard control that must detect the presence of the flame prior to permitting the main fuel valve(s) to open.

Proved spark ignition - direct spark ignition that must be proved before the main fuel valve(s) can open; usually used with small oil burners. Also see direct spark and proved ignitions.

Proven combustion air - see airflow switch.

Proven pre-purge - system in which the programming control includes a provision for preventing burner operation until satisfactory airflow is established during pre-purge.

Proving pilot - sighting the pilot through the scanner to verify that the pilot is lit.

Proximate analysis - provides information regarding moisture content, volatile matter, fixed carbon, and ash content of coal.

Pry bar - a pry bar is used for leverage, to move heavy objects that would usually be hard to do alone. Pry bars are used to move, lift, tilt, and pull objects.

Psi - pounds per square inch. A measure of pressure.

Psia - pounds per square inch absolute and used for pressure measurement. Readings expressed in psia are 14.7 psi higher than readings expressed in psig.

Psig - pressure expressed in pounds per square inch gauge.

P-silicon - silicon containing a minute quantity of impurity, or dopant, such as boron, which provides insufficient electrons to exactly complete the crystal structure. There is no electrical imbalance, however.

Psychrometer - an instrument having both a dry and a wet bulb thermometer. It is used to determine the relative humidity in a space. Most instruments have a indexed scale to allow direct conversion from the temperature readings to the percentage of relative humidity.

Psychrometric chart - graphic representation of the properties of moist air, each point representing a specific condition with respect to temperature and humidity, including the relationship of wet and dry bulb temperatures, specific and relative humidity, enthalpy, specific volume and density, moisture content, and dew point temperature.

Psychrometric measurement - the measurement of temperature, pressure and humidity using the psychrometric chart.

Psychrometry - the branch of physics relating to the measurement or determination of atmospheric conditions, particularly regarding the moisture mixed with the air.

P-trap - fitting named for the shape into which the tubing is bent.

P/T plugs - pressure/temperature plugs that allow entrance into the system by the gauge stems of thermometers and pressure gauges without draining or removing the pressure of the system.

P type impurities - impurities having less than four valence electrons that donate the holes to the semiconductor crystal.

P type material - semiconductor material that has only 3 valence electrons, leaving a hole in its lattice structure, resulting in it having a net positive charge.

Pull box - a metal box at a sharp corner in conduit used to pull wires through the conduit.

Pull down - an expression indicating the action of removing refrigerant from all or part of the refrigeration system.

Pull in voltage - the minimum voltage value that causes the relay armature to seat on the pole face.

Pull out torque - the maximum torque that a motor produces without stalling. Also called breakdown torque.

Pull through - describes an air handling system that draws air over the heat exchanger before the air enters the fan itself.

Pull up torque - the minimum external torque developed by an alternating current motor during the period of acceleration from rest to the speed at which breakdown torque occurs. For motors that do not have a definite breakdown torque, the pull-up torque is the minimum torque developed up to rated speed.

Pulley - a wheel with a adjustable or fixed "V" groove. When attached to a drive and drive members, the pulley provides a means for driving the load.

Pulsating DC - a DC voltage that is not in "straight line" or "pure" DC form, such as DC voltage produced by a battery.

Pulsation - a panting of the flames in a boiler or furnace, indicating cyclic and rapid changes in the pressure in the combustion space.

Pulse - term referring to one cycle of ignition and combustion of a gas-air mixture in a pulse combustion furnace.

Pulse combustion process - repeated ignition of a gas and air mixture in a high efficiency gas fired furnace.

Pulse furnace - furnace that has a "tuned" (resonant) combustion chamber. Part of the energy normally lost through the flue is returned to the combustion chamber to start the next "pulse" of combustion. This process burns various fuels in small, discrete fuel-air mixture volumes in a very rapid series of combustions.

Pulse width modulation - a control algorithm in which the length of the on cycle varies based on the error between the target value and the actual (measured) value of the controlled variable.

Pulsing relay - see checking relay.

Pulverized coal - coal that has been pulverized to the consistency of talcum powder and that is highly explosive.

Pulverizing mill - grinds coal to the consistency of talcum powder before it is delivered to the furnace, where it burns in suspension.

Pump - any one of various machines that force gas or liquid into, or draw it out of, something as by suction or pressure.

Pump, centrifugal - pump that produces fluid velocity and converts it to pressure head.

Pump controller - starts and stops a feedwater pump, depending on the water level in the boiler.

Pump curve - a graph indicating the head energy added to the fluid by the pump versus the flow rate of the fluid through the pump. Such curves are provided by pump manufacturers.

Pump down - the act of using the compressor or a pump to reduce the pressure in the low side of the system and putting it in the high side.

Pump down control system - a control system that closes a solenoid to allow the compressor to pump all of the refrigerant from the low side of the refrigeration system into the high side. This system is used on large air conditioning systems and some commercial refrigeration systems.

Pump down cycle - operation of a refrigeration system in which refrigerant is removed from the low pressure side of the system after each run cycle to avoid high suction pressures at start up.

Pump efficiency - the ratio of the rate of head energy transferred to a flowing fluid divided by the rate of mechanical energy input to the shaft of the pump.

Pump, fixed displacement - a pump in which the displacement per cycle cannot be varied.

Pump head - the difference in pressure on the supply and intake sides of the pump created by the operation of the pump.

Pump, jet - a pump system that uses high pressure water from the pump through a jet venturi to pick up additional water from the supply system. This pump handles three to five times as much water as is supplied by the system.

Pump, reciprocating single piston - a pump having a single reciprocating (moving up and down) piston.

Pump, screw - pump having two interlocking screws rotating in a housing.

Pump, short cycling - when the capacity of the pump is too large for the system, causing very short "on" and "off" cycles.

Pump, submersible - a pump that is located below the level of water in the well. Uses multistage impellers to produce the pressure to lift water up the pipe. The motor is cooled by the flow of water through the pump.

Pump, suction - a pump that depends on negative pressure in the inlet to lift water from the earth. Actually, the pressure difference between the atmospheric pressure and pump inlet pressure forces water up the well casing. The maximum practical operating height is 15 feet.

Pumping water level - the level to which the water in a well drops when the pump is removing the full load water quantity.

Punch - a tool with a flat surface used to forcefully drive a metal part, especially a pin.

Pure capacitive circuit - the current in this type of circuit is limited by the voltage of the charged capacitor. Current leads voltage in this type of circuit.

Pure inductive circuit - a circuit in which the current lags behind the voltage by 90°. There is no true power or watts in this circuit.

Pure refrigerant - a single component fluid that does not change composition when boiling or condensing.

Pure resistive circuit - similar to a DC circuit in having true power or watts equal to the voltage times the current. In this circuit the voltage and current are in phase with each other.

Purge - eliminating a substance from a pipe or furnace by flushing it out with another substance, as in purging a furnace of unburned gas by blowing air through it. Any method which removes foreign matter. Also see post-purge and pre-purge.

Purge cart - an installation tool that uses a high capacity pump to purge air from a system, filter fluid, and help add antifreeze to a geothermal heat pump system.

Purge extender - an auxiliary timer used with a programming control to lengthen the pre-purge period at high fire in order to meet Factory Insurance Association (FIA) requirements. Some programming controls have optional plug in purge extenders with field selectable timing.

Purge period - that period before ignition and after burner shutdown where explosive combustibles are removed.

Purge timer - an auxiliary timer used with a primary control to add pre-purge capability. It may be an optional plug in device with preset timing.

Purge unit - a small condensing unit attached to a chiller that removes water vapor and noncondensable gases (usually air) from the system.

Purge valves - a specialized valve or combination of valves that allows easy forced water purging of a hydronic system or a portion thereof.

Purger - device for removing noncondensable gas from refrigerant condensers or for removing low concentration liquor from absorption system evaporators.

Purging - the releasing of compressed gas to the atmosphere through some part for the purpose of removing contaminants from that part or parts.

Purging hose - a hose through which gas, air, water, or other liquid is forced through to cleanse away unwanted matter.

Purple peeper - Honeywell® trademark; these detectors use a low power sensing tube called an aquadag tube, and require a rectification type flame signal amplifier.

Push button switch - a switch that can be opened or closed by pressing buttons on the switch. Push button switches come with a wide variety of purposes and labeling.

Push rod - any metal rod that transmits force from one point to another.

PVC - plastic pipe and fittings made from polyvinyl chloride and used for cold water supply or drain lines.

PVDF - a type of pipe that contains crystalline, high molecular weight polymer of vinylidene fluoraide, containing 59% fluorine by weight.

Pyrolysis - a chemical decomposition or breaking apart of molecules produced by the action of heat.

Pyrometer - an instrument for measuring high temperatures. A pyrometer measures higher temperatures than a mercury thermometer. The range of temperature these typically measure is from -100 to 1999° F.

Pyronometer - a device used to measure solar insulation in Btu per square foot per hour.

-Q-

Q - as used in heat loss, the value used to indicate the amount of outside air entering the building per linear foot of infiltration. It is expressed in cubic feet per hour.

Q point - the quiescent operating point. The direct current bias values of amplifier voltages and currents.

Quad - one quadrillion (1015) British thermal units. A commonly used measure of very large quantities of energy.

Quadrant - air volume damper.

Qualified person - one who has skills and knowledge related to construction practices and operation of equipment and has received safety training on the hazards involved.

Quality - weight fraction of the vapor in a vapor-liquid mixture.

Quality of steam - term used to express the moisture content present in saturated steam. The quality of steam affects the Btu content of the steam.

Quality of wet vapor - fraction by weight of vapor in mixture of liquid and vapor.

Quenching - altering the characteristics of metal by heating, then quickly submerging in a cooling agent.

Quenching flame - a reduction in temperature whereby a combustion process is retarded or stopped.

Quick burst test - the internal pressure required to burst a pipe or fitting due to an internal pressure build-up, usually within 60 to 90 seconds.

Quick closing valve - a valve that requires a one quarter turn to be fully open or closed.

Quick connect terminal - a common solderless type of electrical terminal with push-on connection.

Quick disconnect fittings - a mechanical fitting device used on refrigerant hoses that seal automatically when removed from an appliance.

Quick opening characteristic - is the relationship between the stem travel, expressed in per cent of travel, and the flow of the fluid through the valve, expressed in per cent of full flow.

Quick opening valve - a valve that opens completely by fast action, either manually or automatically controlled. A valve requiring one quarter turn or less is considered to be a quick opening valve.

-R-

R - the value for electrical resistance, used to indicate the resistance to the flow of heat through a material or assembly of materials.

Ra - reverse acting or return air

Rcr - relative climate ratio.

Reg - regulator.

Rel - recommended exposure limit.

Resil - resilient.

Rev - reversible.

Rf - radio frequency.

Rh - relative humidity.

Rheo - rheostat.

Rms - root mean square.

Rpm - revolutions per minute.

Rs - rotary scale.

Rs232 - a type of information transfer bus used to connect serial devices such as printer or modem to a computer.

Rtd - resistance temperature detector.

R-10, Tetrachloromethane (carbon tetrachloride) - a methane based refrigerant. Chemical formula is CCL_4.

R-11, Trichloromonofluoromethane - a A1 group low pressure CFC synthetic chemical refrigerant which is suitable for either low or high temperature applications. R-11 is stable, nonflammable, and nontoxic. Leaks may be detected by using a torch, soap, or electronic leak detector. R-11 will absorb most of the moisture that may have entered the system when the system was open to the atmosphere, also used as a cleaning fluid. Boiling point is 74.87° F. Chemical formula is CCL_3F. Refrigerant cylinder color code is orange. Pantone matching system number is 021.

R-12, Dichlorodifluoromethane - a A1 group low pressure CFC synthetic chemical refrigerant which was used in small air conditioners as well as domestic and commercial refrigeration systems. Has a boiling point of -21.62° F. Leaks may be detected by using a torch, soap, or electronic leak detector. Water is slightly soluble in R-12, and the resultant solution has a corrosive effect on most metals. Therefore, it is very important to make sure all the moisture is out of an R-12 system before charging it. Chemical formula is CCL_2F_2. Refrigerant cylinder color code is opaque white. No pantone matching system number. R-12 was once a popular refrigerant for refrigeration systems. It can no longer be produced in the United States and many other countries.

R-12B1, Bromochlorodifluoromethane - a methane based refrigerant. Chemical formula is $CBrCLF_2$. Halon 1211 has a boiling point of 24.8° F.

R-12B2 - Dibromodifluoromethane. Chemical formula is CBr_2F_2. A methane based refrigerant.

R-13, Chlorotrifluoromethane - a A1 group high pressure CFC refrigerant that has a boiling point of -114.6° F. Because of its high pressure and low critical temperature, it is unsuitable for ordinary applications. However, it is an excellent refrigerant in low temperature systems at temperatures of -100° F or more below zero. It is nonpoisonous, nonirritating, nonflammable, and has no corrosive effect on metals. Chemical formula is $CCLF_3$. Refrigerant cylinder color code is light blue (sky). Pantone matching system number is 2975.

R-13B1, Bromotrifluoromethane - a A1 group low temperature BFC refrigerant with a boiling point of -71.95° F. Therefore, it is not suitable at such low temperatures as is R-13. An advantage of R-13B1 over R-13 is that its condensing pressure is not as high and it has a higher critical temperature. Chemical formula is $CBrF_3$. Also known as Halon 1301. Refrigerant cylinder color code is pinkish red (coral). Pantone matching system number is 177.

R-14, Tetrafluoromethane - a A1 group PFC refrigerant having a boiling point of -198.3° F. Chemical formula is CF_4. Refrigerant cylinder color code is yellow brown (mustard). Pantone matching system number is 124. Also known as carbon tetrafluoride.

R-20, Trichloromethane (chloroform) - a methane based refrigerant. Chemical formula is $CHCL_3$.

R-21, Dichlorofluoromethane - a A1 group HCFC methane based refrigerant whose pressure and volume characteristics make it suitable for use only in centrifugal compressors. It is a slightly higher pressure refrigerant than R-11. Therefore, it does not require as high a suction vacuum and will produce more refrigeration than R-11 for the volume of refrigerant pumped. R-21 is nonpoisonous, nonirritating, and nonflammable. Boiling point is 47.8° F. Chemical formula is $CHCL_2F$.

R-22, Monochlorodifluoromethane - a A1 group low pressure HCFC synthetic chemical refrigerant which was developed for refrigeration installations that need low evaporating temperatures. R-22 is stable, nontoxic, non-corrosive, nonirritating, and nonflammable. Water is soluble in R-22, so driers and desiccants should be used to remove moisture from systems using this refrigerant. Leaks may be detected with soap, torch, or electronic leak detector. Has a boiling point of -41.36° F. Glide value of 0° F, ODP of 0.055 and a GWP of 0.34. Chemical formula is $CHCLF_2$. Refrigerant cylinder color code is light green. Pantone matching system number is 352.

R-22B1, Bromodifluoromethane - a methane base refrigerant. Chemical formula is $CHBrF_2$.

R-23, Trifluoromethane - a A1 Group HFC refrigerant that has a boiling point of -115.7° F. Chemical formula is CHF_3. Refrigerant cylinder color code is light blue/gray. Pantone matching system number is 428. Compatible lubricant is polyol ester.

R-30, Dichloromethane (methylene chloride) - a B2 group methane based refrigerant having a boiling point of 104.4° F. Chemical formula is CH_2CL_2.

R-31, Chlorofluoromethane - a HCFC refrigerant. Chemical formula is CH_2CLF.

R-32, Difluoromethane (methylene fluoride) - a A2 group HFC methane based refrigerant that has a boiling point of -61.1° F. Chemical formula is CH_2F_2.

R-40, Chloromethane (methyl chloride) - a B2 group refrigerant that was an earlier refrigerant which was widely used in commercial systems and has a boiling point of -11.6° F. Chemical formula is CH_3CL. In its pure state, R-40 has no effect on most metals. However, it should never be used with zinc or aluminum, for in contact with these metals, it breaks down to form a spontaneously combustible gas. If moisture gets into an R-40 system, an electrolytic action will take place which will dissolve natural rubber, which therefore cannot be used for gaskets or packing. R-40 is toxic and slightly asphyxiating. It can cause drowsiness, mental confusion, nausea, and even death.

R-41, Fluoromethane (methyl fluoride) - a methane based refrigerant having a boiling point of -109.1° F. Chemical formula is CH_3F.

R-50, Methane - a A3 group refrigerant having a boiling point of -258.7° F. Chemical formula is CH_4.

R-110, Hexachloroethane - a ethane based refrigerant. Chemical formula is CCL_3CCL_3.

R-111, Pentachlorofluoroethane - a ethane based CFC refrigerant. Chemical formula is CCL_3CCL_2F.

R-112, 1,1,2,2 Tetrachloro-1,2-difluoroethane - a ethane based CFC refrigerant. Chemical formula is CCL_2FCCL_2F.

R-112a, 1,1,1,2 Tetrachloro-2,2-difluoroethane - a ethane based CFC refrigerant. Chemical formula is CCL_3CCLF_2.

R-113, 1,1,2 Trichloro-1,2,2-trifluoroethane - a A1 group CFC synthetic chemical ethane based refrigerant which is nontoxic and nonflammable. Used in low capacity centrifugal chiller packaged units. Operates with very low system pressures, high gas volumes. Also used as an intermediate in the manufacture of specialty lubricants. Boiling point is 117.63° F. Chemical formula is CCL_2FCCLF_2. Refrigerant cylinder color code is dark purple (violet). Pantone matching system number is 266.

R-113a, 1,1,1 Trichloro-2,2,2-trifluoroethane - a A1 group CFC synthetic chemical ethane based refrigerant. Chemical formula is CCL_3CF_3.

R-114, 1,2-Dichloro-1,1,2,2-tetrafluoroethane - a A1 group CFC ethane based refrigerant having a boiling point of 38.8° F. Chemical formula is $CCLF_2CCLF_2$. Refrigerant cylinder color code is dark blue (navy). Pantone matching system number is 302.

R-114a, 1,1-Dichloro-1,2,2,2-tetrafluoroethane - a A1 group CFC ethane based refrigerant. Chemical formula is CCL_2FCF_3.

R-114B2, 1,2-Dibromo-1,1,2,2-tetrafluoroethane - a A1 group CFC ethane based refrigerant. Chemical formula is $CBrF_2CBrF_2$.

R-115, Chloropentafluoroethane - a A1 group CFC refrigerant having a boiling point of -38.4° F. R-115 has a glide value of <1.0° F, ODP of 0.330 and a GWP of 3.75. It is usually used in systems that require temperatures between 0° F and -60° F. The pressure characteristics of R-502 are similar to R-22, but the lower condensing temperature permits better lubrication because of the increased viscosity of the oil at the lower temperature. Leaks may be detected with soap, torch, or electronic leak detector. R-502 is used only with reciprocating compressors and is often used for frozen foods storage cases, ice cream freezers, and in some frozen-food processing plants. Refrigerant cylinder color code is light purple (lavender). Pantone matching system number is 251. Chemical formula is $CCLF_2CF_3$.

R-116, Hexafluoroethane - a A1 group ethane based refrigerant having a boiling point of -109.1° F. Chemical formula is CF_3CF_3. Refrigerant cylinder color code is dark gray (battleship). Pantone matching system number is 424.

R-120, Pentachloroethane - a ethane based refrigerant. Chemical formula is $CHCL_2CCL_3$.

R-123, 2,2-Dichloro-1,1,1-trifluoroethane - a B1 group HCFC ethane based refrigerant developed for low pressure centrifugal chiller applications. Boiling point is 82.17° F. Compatible lubricant oil is alkylbenzene or mineral oil. Chemical formula is $CHCL_2CF_3$. Ozone depletion potential is 0.015 and the greenhouse warming potential is 0.02. Refrigerant cylinder color code is light blue gray. Pantone matching system number is 428.

R-123a, 1,2-Dichloro-1,1,2-trifluoroethane - a B1 group HCFC ethane based refrigerant. Chemical formula is $CHCLFCCLF_2$.

R-124, 2-Chloro-1,1,1,2-tetrafluoroethane - a A1 group HCFC ethane based refrigerant that has a boiling point of 8.26° F. Chemical formula is $CHCLFCF_3$. Refrigerant cylinder color code is deep green (DOT green). Pantone matching system number is 335.

R-124a, 1-Chloro-1,1,2,2-tetrafluoroethane - a A1 group HCFC ethane based refrigerant. Chemical formula is CHF_2CCLF_2.

R-125, Pentafluoroethane - a A1 group HFC refrigerant that has a boiling point of -55.43° F. Chemical formula is CHF_2CF_3. Refrigerant cylinder color code is medium brown (tan). Pantone matching system number is 465.

R-133a, 2-Chloro-1,1,1-trifluoroethane - a ethane based refrigerant. Chemical formula is CH_2CLCF_3.

R-134 - a refrigerant having a boiling temperature of -9.4° F. Chemical formula is CHF_2CHF_2. Refrigerant cylinder color code is light blue.

R-134a, 1,1,1,2-Tetrafluoroethane - a A1 group HFC refrigerant developed for refrigeration systems and as a replacement for R-12. Boiling point is -15.08° F. Glide value of 0.0° F, ODP of 0.0 and a GWP of 0.28. Chemical formula is CH_2FCF_3. Compatible lubricant oil is polyol ester. Refrigerant cylinder color code is light blue (sky). Pantone matching system number is 2975.

R-140a, 1,1,1-Trichloroethane (methyl chloroform) - a ethane based refrigerant. Chemical formula is CH_3CCL_3.

R-141b, 1,1-Dichloro-1-fluoroethane - the leading substitute for CFC-11 in rigid foam blowing insulation applications such as: construction (commercial, residential and public) appliances and transport vehicles. Boiling point is 89.6° F. Chemical formula is CCL_2FCH_3. An HCFC refrigerant.

R-142b, 1-Chloro-1,1-difluoroethane - a A2 group HCFC refrigerant having a boiling point of 14.4° F. Chemical formula is $CCLF_2CH_3$.

R-143a, 1,1,1-Trifluoroethane - a A2 group HFC refrigerant having a boiling point of -53.0° F. Chemical formula is CF_3CH_3.

R-150a, 1,1-Dichloroethane - a ethane based refrigerant. Chemical formula is $CHCL_2CH_3$.

R-152a, 1,1-Difluoroethane - a A2 group HFC refrigerant having a boiling point of -13.0° F. Chemical formula is CHF_2CH_3.

R-160, Chloroethane (ethyl chlororide) - toxic refrigerant seldom used now. Boiling point is 54.32° F. Chemical formula is CH_3CH_2CL.

R-161, Ethyl fluoride - a refrigerant having a boiling temperature of -34.8° F. Chemical formula is CH_3CH_2F.

R-170, Ethane - a A3 group low temperature application refrigerant. Boiling point is -127.85° F. Chemical formula is CH_3CH_3.

R-216ca, 1,3-Dichloro-1,1,2,2,3,3-hexafluoropropane - boiling point is 96.24° F. Chemical formula is $CCLF_2CF_2CCLF_2$. A CFC refrigerant.

R-218, Octafluoropropane - a A1 group PFC refrigerant having a boiling point of -34.1° F. Chemical formula is $CF_3CF_2CF_3$.

R-227ca - chemical formula is $CHF_2CF_2CF_3$. Boiling point is 1.4° F.

R-227ea - chemical formula is CF_3CHFCF_3. Boiling point is -0.9° F.

R-E227ca2 - chemical formula is $CHF_2CF_2OCF_3$. Boiling point is 26.2° F.

R-231 - a HCFC refrigerant.

R-232 - a HCFC refrigerant.

R-233 - a HCFC refrigerant.

R-234 - a HCFC refrigerant.

R-235 - a CFC refrigerant.

R-236ca - chemical formula is $CHF_2CF_2CHF_2$. Boiling point is 41.2° F.

R-236cb - chemical formula is $CH_2FCF_2CF_3$. Boiling point is 29.5° F.

R-236ea - chemical formula is CHF_2CHFCF_3. Boiling point is 43.9° F.

R-236fa, Hexafluoropropane - a A1 group HFC refrigerant. Chemical formula is $CF_3CH_2CF_3$. Boiling point is 30.0° F.

R-241 - a HCFC refrigerant.

R-242 - a HCFC refrigerant.

R-243 - a HCFC refrigerant.

R-244 - a HCFC refrigerant.

R-245ca - chemical formula is $CH_2FCF_2CHF_2$. Boiling point is 77.9° F.

R-245cb, 1,1,1,2,2-Pentafluoropropane - a propane based refrigerant. Chemical formula is $CF_3CF_2CH_3$. Boiling point is -0.9° F.

R-245fa - chemical formula is $CHF_2CH_2CF_3$. Boiling point is 59.5° F.

R-E245cb2 - chemical formula is $CH_3CF_2OCF_3$. Boiling point is 84.0° F

R-251 - a HCFC refrigerant.

R-252 - a HCFC refrigerant.

R-253 - a HCFC refrigerant.

R-254cb - chemical formula is $CH_3CF_2CHF_2$. Boiling point is 30.6° F

R-261 - a HCFC refrigerant.

R-262 - a HCFC refrigerant.

R-C270 - chemical formula is $CH_2CH_2CH_2$. Boiling point is -28.3° F

R-271 - a HCFC refrigerant.

R-290, Propane - a A3 group HC low temperature application refrigerant. Boiling point is -43.73°F. Chemical formula is $CH_3CH_2CH_3$.

R-C316, 1,2-Dichloro-1,2,3,3,4,4-hexafluorocyclobutane - chemical formula is $C_4CL_2F_6$.

R-C317, Chloroheptafluorocyclobutane - chemical formula is C_4CLF_7.

R-C318, Octafluorocyclobutane - a A1 group PFC refrigerant having a boiling point of 21.5° F. Chemical formula is C_4F_8.

R-CE318 - chemical formula is $CF_2\ CF_2CF_2OCF_2$. Boiling point is 30.7° F

R-400 - a A1/A1 group CFC refrigerant that is a mixture of (by weight) 60% R-12 and 40% R-114. Chemical formula is $CCL_2F_2CCLF_2CCLF_2$. Boiling point is -9.2° F.

R-401A, Chlorodifluoromethane/Difluoroethane/Chlorotetrafluo roethane - a A1/A1 group HCFC zeotrope ternary blend of (by weight) 13% HFC-152a, 34% HCFC-124 and 53% HCFC-22 having a boiling point of -27.6° F. Glide value of 9.3° F, ODP of 0.030 and a GWP of 0.22. Chemical formula is $CHCLF_2CH_3CHF_2CHCLFCF_3$. Compatible lubricant is polyol ester or alkylbenzene. Refrigerant cylinder color code is pinkish red (coral). Pantone matching system number is 177. Also known as MP-39.

R-401B, Chlorodifluoromethane/Difluoroethane/Chlorotetrafluo roethane - a A1/A1 group HCFC zeotrope ternary blend of (by weight) 11% HFC-152a, 28% HCFC-124, and 61% HCFC-22. Boiling point is -30.5° F. Glide value of 8.8° F, ODP of 0.035 and a GWP of 0.24. Chemical formula is $CHCLF_2CH_3CHF_2CHCLFCF_3$. Compatible lubricant oil is polyol ester or alkylbenzene. Refrigerant cylinder color code is yellow brown (mustard). Pantone matching system number is 124. Also known as MP-66.

R-401C - a A1/A1 group HCFC zeotrope ternary blend (by weight) of 33% R-22, 15% R-152a, and 52% R-124. Chemical formula is $CHCLF_2CH_3CHF_2CHCLFCF_3$. Boiling point is -19.1° F. Refrigerant cylinder color code is aqua. Also known as MP33 and 52.

R-402A, Pentafluoroethane/Propane/Chlorodifluoromethane - a A1/A1 group HCFC zeotrope ternary blend of (by weight) 60% R-125, 2% R-290, and 38% R-22. Chemical formula is $CHF_2CF_3CH_3CH_2CH_3CHCLF_2$. Boiling point is -56.6° F. Glide value of 2.8° F, ODP of 0.020 and a GWP of 0.63. Compatible lubricant oil is polyol ester or alkylbenzene. Refrigerant cylinder color code is light brown (sand). Pantone matching system number is 461. Also known as HP-80.

R-402B - a A1/A1 group HCFC zeotrope ternary blend of (by weight) 38% R-125, 2% R-290, and 60% R-22. Glide value of 2.9° F, ODP of 0.030 and a GWP of 0.49. Chemical formula is $CHF_2CF_3CH_3CH_2CH_3CHCLF_2$. Boiling point is -53.3° F. Refrigerant cylinder color code is green brown (olive). Pantone matching system number is 385. Also known as HP-81. Compatible lubricant oil is polyol ester or alkylbenzene.

R-403A - a A1/A1 group HCFC zeotrope ternary blend of (by weight) 5% R-290, 75% R-22, and 20% R-218. Chemical formula is $CH_3CH_2CH_3CHCLF_2CF_3CF_2CF_3$. Boiling point is -58.0° F.

R-403B - a A1/A1 group HCFC zeotrope ternary blend of (by weight) 5% R-290, 56% R-22, and 39% R-218. Chemical formula is $CH_3CH_2CH_3CHCLF_2CF_3CF_2CF_3$. Boiling point is -57.1° F.

R-404A, Pentafluoroethane/Trifluoroethane/Tetrafluoroethane - a A1/A1 group HFC long term, non-ozone depleting replacement for R-502 in low and medium temperature commercial refrigeration systems. Chemical formula is $CHF_2CF_3CH_3CF_3CH_2FCF_3$. A zeotrope ternary blend of (by weight) of 44% R-125, 52% R-143a, and 4% R-134a. Boiling point (bubble point temperature) is -51.8° F. Glide value of <1.0° F, ODP of 0.0 and a GWP of 0.94. Compatible lubricant oil is polyol ester. Also known as FX-70 and HP62. Refrigerant cylinder color code is orange. Pantone matching system number is 021.

R-405A - a A1/A1 group HCFC zeotrope ternary blend of (by weight) 45% R-22, 7% R-152a, 5.5% R-142a and 42.5% C-318. Chemical formula is $CHCLF_2CH_3CHF_2CH_3CCLF_2C_4F_8$. Boiling point is -17.1° F. Compatible lubricant is polyol ester or alkyl benzene.

R-406A - a A1/A2 group HCFC zeotrope ternary blend of (by weight) 55% R-22, 4% R-600a, and 41% R-142b. Chemical formula is $CHCLF_2CH(CH_3)_3CH_3CCLF_2$. Boiling point is -26.3° F. Compatible lubricant oil is mineral or alkylbenzene.

R-407A - a A1/A1 group HFC zeotrope ternary blend of (by weight) 20% R-32, 40% R-125, and 40% R-134a. Glide value of 8.7° F, ODP of 0.0 and a GWP of 0.49. Boiling point is -49.9° F. Refrigerant cylinder color code is lime green. Chemical formula is $CH_2F_2CHF_2CF_3CH_2FCF_3$.

R-407B - a A1/A1 group HFC zeotrope ternary blend of (by weight) 10% R-32, 70% R-125, and 20% R-134a. Glide value of 5.3° F, ODP of 0.0 and a GWP of 0.70. Boiling point is -53.1° F Refrigerant cylinder color code is cream. Chemical formula is $CH_2F_2CHF_2CF_3CH_2FCF_3$.

R-407C, Difluoromethane/Pentafluoroethane/Tetrafluoroethane - a A1/A1 group HFC long term, non-ozone depleting zeotrope ternary blend replacement for HCFC-22 in various air conditioning applications, as well as in positive displacement refrigeration systems. It is a ternary blend of (by weight) of 23% R-32, 25% R-125, and 52% R-134a. Boiling point at atmospheric pressure is -46.6° F (bubble point temperature). Glide value of 8.0° F, ODP of 0.0 and a GWP of 0.37. Chemical formula is $CH_2F_2CHF_2CF_3CH_2FCF_3$. Compatible lubricant oil is polyol ester. Refrigerant cylinder color code is medium brown. Pantone matching system number is 471.

R-407D - a A1/A1 group HFC zeotrope ternary blend of (by weight) 15% R-32, 15% R-125, and 70% R-134a. Boiling point is -39.1° F. Refrigerant cylinder color code is dark brown (chocolate). Compatible lubricant oil is polyol ester. Chemical formula is $CH_2F_2CHF_2CF_3CH_2FCF_3$

R-407E - a A1/A1 group HFC zeotrope ternary blend of (by weight) 25% R-32, 15% R-125, and 60% R-134a.

R-408A - a A1/A1 group HCFC zeotrope ternary low ozone depletion potential, blend of (by weight) 7% R-125, 46% R-143a, and 47% R-22. Chemical formula is $CHF_2CF_3CH_3CF_3CHCLF_2$. Boiling point is -46.3° F. Glide value of 1.0° F, ODP of 0.026 and a GWP of 0.73. Refrigerant cylinder color code is medium purple. Pantone matching system number is 248. Compatible lubricant oil is polyol ester or alkyl benzene.

R-409A-Chlorodifluoromethane/Chlorotetrafluoroethane/Chlorodifluoroethane - a A1/A1 group HCFC zeotrope ternary refrigerant that is a blend of (by weight) 60% R-22, 25% R-124, and 15% R-142a. Chemical formula is $CHCLF_2CHCLFCF_3CH_3CCLF_2$. Boiling point is -32.4° F. Bubble point is -29.6° F. Glide value of 12.0° F, ODP of 0.050 and a GWP of 0.30. Compatible lubricant oil is alkylbenzene or polyol ester. Ozone depletion potential is 0.05 and greenhouse warming potential is 0.3. Refrigerant cylinder color code is medium brown (tan). Pantone matching system number is 465. Also known as FX-56.

R-409B - a A1/A1 group HCFC zeotrope binary blend of (by weight) 65% R-22, 25% R-124, and 10% R-142b. Boiling point is -33.9° F. Also known as FX-57.

R-410A - a A1/A1 group HFC zeotrope binary blend of (by weight) 50% R-32 and 50% R-125. Compatible lubricant oil is polyol ester. Chemical formula is $CH_2F_2CHF_2CF_3$. Boiling point is -60.5° F. Glide value of <1.0° F, ODP of 0.0 and a GWP of 0.44. Also known as AZ-20. Refrigerant cylinder color code is rose. Pantone matching system number is 507.

R-410B - a A1/A1 group HFC zeotrope binary blend of 45% R-32 and 55% R-125. Chemical formula is $CH_2F_2CHF_2CF_3$. Boiling point is -60.3° F. Compatible lubricant is polyol ester. Also known as Suva 9100.

R-411A - a A1/A2 group HCFC zeotrope refrigerant that is a ternary blend of (by weight) 1.5% R-1270, 87.5% R-22, and 11% R-152a. Boiling point is -37.5° F. Chemical formula is $CH_3CH{=}CH_2CHCLF_2CH_3CHF_2$. Refrigerant cylinder color code is dark purple (violet).

R-411B - a A1/A2 group HCFC ZEOTROPE refrigerant that is a ternary blend of (by weight) 3% R-1270, 94% R-22, and 3% R-152a. Boiling point is -42.9° F. Chemical formula is $CH_3CH{=}CH_2CHCLF_2CH_3CHF_2$. Refrigerant cylinder color code is teal with red collar.

R-411C - a A1/A1 group HCFC zeotrope ternary blend of (by weight) 3% R-1270, 95.5% R-22, and 1.5% R-152a.

R-412A - a A1/A2 group HCFC zeotrope group refrigerant ternary blend of (by weight) 70% R-22, 5% R-218, and 25% R-142a. Boiling point is -37.3° F. Chemical formula is $CHCLF_2CF_3CF_2CF_3CH_3CCLF_2$.

R-413A - a A1/A2 group HFC zeotrope group A1/A2 refrigerant ternary blend of (by weight) 9% R-218, 88% R-134a, and 3% R-600a. Chemical formula is $CF_3CF_2CF_3CH_2FCF_3CH(CH_3)_3$.

R-414A - a A1/A1 group HCFC zeotrope group A1/A2 refrigerant ternary blend of (by weight) 51% R-22, 28.5% R-124, 4% R-600a, and 16.5% R-142b. Compatible lubricant oil is polyol ester or alkylbenzene.

R-414B - a A1/A1 group HCFC zeotrope refrigerant ternary blend of (by weight) 50% R-22, 38% R-124, 1.5% 600a, and 9.5% R-142b. Compatible lubricant is polyol ester or alkylbenzene.

R-415A - a A1/A2 group HCFC zeotrope ternary blend of (by weight) 5% R-23, 80% R-22, and 15% R-152a.

R-416A - a A1/A1 group HCFC zeotrope ternary blend (by weight) 59% R-134a, 39.5% R-124, and 1.5% R-600. Compatible lubricant oil is polyol ester.

R-417A - a A1/A1 group HFC/HC zeotrope ternary blend (by weight) 46.6% R-125, 50% R-134a and 3.4% R-600 (Butane). Compatible oil is mineral, polyol ester or alkylbenzene. Ozone depletion potential is 0. Boiling point is -43.3° F. Temperature glide is 10° F.

R-422A - a A1 group HFC refrigerant that is a blend of 85.1% R-125, 11.5% R-134 and 3.4% R-600A (isobutene).

R-500 - a A1 group CFC low pressure synthetic chemical refrigerant which is a azeotropic mixture of 73.8% by weight R-12 and 26.2% by weight R-152a. Refrigerant cylinder color code is yellow. Boiling point is -28.3° F. Chemical name is dichlorodifluoromethane. Chemical formula is $CCL_2F_2CH_3CHF_2$. Water is highly soluble in R-500, and you must be extremely careful to keep water out of the system. Leaks may be detected with soap, torch, or electronic leak detector. Compatible lubricant is polyol ester. Refrigerant cylinder color code is yellow. Pantone matching system number is 109. This refrigerant can no longer be manufactured in the United States or many other countries.

R-501 - Boiling point is -42.5° F.

R-502 - a A1 group low pressure CFC synthetic chemical refrigerant for use in low temperature refrigeration systems. An azeotropic mixture of 48.8% by weight R-22 and 51.2% by weight

R-5025 - boiling point is -49.8° F. No chemical name.

R-503 - a A1 group CFC refrigerant which is a azeotropic mixture 40.1% by weight of R-23 and 59.9% by weight of R-13. Boiling point is -127.7° F. Refrigerant cylinder color code is blue green (aquamarine). Pantone matching system number is 3268. Chemical formula is CHF_3CCLF_3.

R-504 - a refrigerant that is a azeotropic mixture of R-32 and R-115. Boiling point is -71.0° F. No chemical formula.

R-505 - Boiling point is -22.0° F.

R-506 - Boiling point is 9.9° F.

R-507 - a refrigerant that is a azeotropic mixture of R-143a and R-125. Glide value of <1.0° F, ODP of 0.0 and a GWP of 0.98. Compatible lubricant oil is polyol ester. Also known as AZ-50.

R-507A - a A1/A1 group HFC, non-ozone depleting refrigerant. It is a azeotrope binary blend of (by weight) 50% R-125 and 50% R-143a. Chemical formula is $CHF_2CF_3CH_3CF_3$. Boiling point is -52.1° F. Compatible lubricant is polyol ester. Also known as AZ-50. Refrigerant cylinder color code is blue green (teal). Pantone matching system number is 326.

R-508A - a A1/A1 group PFC refrigerant which is a azeotrope binary blend of (by weight) 39% R-23 and 61% R-116. Boiling point is -125.34° F. Compatible lubricant oil is polyol ester. Chemical formula is $CHF_3CF_3CF_3$.

R-508B - a A1/A1 group PFC refrigerant which is a azeotrope binary blend of (by weight) 46% R-23 and 54% R-116. Chemical formula is $CHF_3CF_3CF_3$. Boiling point is -125.28° F. Refrigerant cylinder color code is dark blue (navy). Pantone matching system number is 302. Compatible lubricant oil is polyol ester. Also known as Suva 95.

R-509A - a A1/A1 group HCFC refrigerant which is a azeotrope binary blend of (by weight) 44% R-22 and 56% R-218. Boiling point is -52.8° F. Chemical formula is $CHCLF_2CF_3$ CF_2CF_3.

R-600, Butane - a A3 group HC low temperature application refrigerant that is sometimes used as a fuel. Boiling point is 31.1° F. Chemical formula is C_4H_{10}.

R-600a, 2-Methyl propane (isobutane) - a A3 group HC refrigerant having a boiling point of 10.89° F. Chemical formula is $CH(CH_3)_3$.

R-601 - Chemical formula is $CH_3CH_2CH_2CH_2CH_3$. Boiling point is 97.2° F.

R-601a - Chemical formula is $(CH_3)_2CHCH_2CH_3$. Boiling point is 82.0° F.

R-601b - Chemical formula is $(CH_3)_4C$. Boiling point is 49.1° F.

R-610, Ethyl ether - boiling point is 94.3° F. Chemical formula is $C_2H_5OC_2H_5$.

R-611, Methyl formate - a Group 2 low temperature application refrigerant. Boiling point is 89.2° F. Chemical formula is $HCOOCH_3$.

R-630, Methyl amine - boiling point is 19.9° F. Chemical formula is CH_3NH_2.

R-631, Ethyl amine - boiling point is 61.88° F. Chemical formula is $C_2H_5NH_2$.

R-702n, Hydrogen (normal) - boiling point is -423.0° F. Its symbol is H_2.

R-702p, Hydrogen (para) - boiling point is -423.2° F. Chemical formula is H_2.

R-704, Helium - boiling point is -452.1° F. Its symbol is He.

R-717, Ammonia - a B2 group popular refrigerant that is colorless in both the liquid and vapor states, but has a strong irritating odor. It is used for either industrial refrigeration or absorption systems. When ammonia combines with water, a caustic is formed which can burn the skin. This is especially true on any moist portion of the body, such as the eyes, nose, mouth, or any sweaty surface. It is flammable and can be explosive if mixed with air. As a liquid, it is lighter than water. As a vapor, it is lighter than air. Ammonia has no effect on metals if dry. However, it will attack copper and bronze in the presence of moisture. Ammonia will not corrode iron or steel. It has no effect on good lubricating oil, but may sludge badly if there is moisture in the system. The odor will give you immediate warning of a leak, and will form white smoke when brought in contact with a sulfur candle or burning sulfur stick. Ammonia is one of the cheapest refrigerants available. Boiling point is -28.0° F. Refrigerant cylinder color code is black. Chemical formula is NH_3.

R-718, Water - a A1 group refrigerant having a boiling point of 212.0° F. Chemical formula is H_2O.

R-720, Neon - boiling point is -410.9° F. Its symbol is Ne.

R-728, Nitrogen - boiling point is -320.4° F. Its symbol is N_2.

R-729, Air - boiling point is -317.8° F. No chemical formula.

R-732, Oxygen - boiling point is -297.3° F. Its symbol is O_2.

R-740, Argon - boiling point is -302.6° F. Its symbol is Ar.

R-744, Carbon dioxide - a A1 group refrigerant that sublimes at -109.2° F. Chemical formula is CO_2.

R-744a, Nitrous oxide - boiling point is -129.1° F. Chemical formula is N_2O.

R-764, Sulphur dioxide - a B1 group refrigerant once commonly used as a refrigerant. Chemical formula is SO_2. Cylinder color code is black. Boiling point is 14° F.

R-1112a, 1,1-dichloro-2,2-difluroethane - chemical formula is $CCL_2{=}CF_2$.

R-1113, 1-Chloro-1,2,2-trifluoroethene - chemical formula is $CCLF{=}CF_2$.

R-1114, Tetrafluoroethene - chemical formula is $CF_2{=}CF_2$.

R-1120, Trichloroethylene - boiling point is 189.0° F. Chemical formula is $CHCL{=}CCL_2$.

R-1130, 1,2-Dichloroethene (trans) - chemical formula is CHCL=CHCL. Boiling point is 118.0° F.

R-1132a, 1.1-Difluoroethene (vinylidene fluoride) - chemical formula is $CF_2{=}CH_2$.

R-1140, 1-Chloroethene (vinyl chloride) - a B3 group refrigerant. Chemical formula is $CHCL{=}CH_2$.

R-1141, 1-Fluoroethene (vinyl fluoride) - chemical formula is $CHF{=}CH_2$.

R-1150, Ethene (ethylene) - an A3 group refrigerant having a boiling point of -154.7° F. Chemical formula is $CH_2{=}CH_2$.

R-1180, Dichloroethylene - boiling point is 118.0° F. Chemical formula is CHCLCHCL.

R-1270, Propene (propylene) - an A3 group HC refrigerant having a boiling point of -53.86° F. Chemical formula is $CH_3CH{=}CH_2$.

R-value - the resistance to heat flow expressed in units of square foot hour degree F/Btu. R-value equals the number 1 divided by the U value. Higher R-value numbers indicate greater insulating values.

Race - area on which bearings ride.

Raceway, bearing - the inner or outer ring that provides a contact surface for the balls or rollers in a bearing.

Raceway, electrical - a protected runway or enclosure for conductors. A continuous channel, conduit, or conduit bodies for holding conductors or cables. Raceways include, but not limited to, rigid metal conduit, rigid nonmetallic conduit, intermediate metal conduit, liquid tight flexible conduit, flexible metallic tubing, flexible metal conduit, electrical nonmetallic tubing, flexible metal metallic tubing, under floor raceways, cellular concrete floor raceways, cellular metal floor raceways, wire ways and bus ways.

Rack system - many refrigeration compressors piped in parallel and mounted on a steel rack. The compressors are usually cycled by a microprocessor.

Radial commutator - the electrical contact surface on a rotor which is perpendicular to the shaft center line.

Radial fan - a blower that has straight blades that are, to a large extent, self cleaning, making them suitable for various kinds of material handling and grease laden air. Wheels are of simple construction and have relatively narrow blades. They can withstand the high speeds required to operate at higher static pressures (up to 12") but usually are noisier than forward curved or backward inclined blowers.

Radial piping system - a perimeter heating system employing ducts or pipes that extend from the supply plenum to each register that radiate from a central heat source.

Radiant baseboard - an extruded aluminum plate with integral tubes that is fastened to the base of walls, and releases the majority of its heat output as thermal radiation.

Radiant head burner - a raw gas burner with a head (or heads) constructed of refractory material. The raw gas is ejected from spuds into an opening in the head(s). Combustion heats the head(s) to incandescence, providing a large amount of radiant heat that tends to increase the speed of flame propagation and keep the flame shorter. Also see raw gas burner.

Radiant heat - heat that passes through air heating solid objects that in turn heat the surrounding area.

Radiant heat transfer - radiant heat transfer occurs when there is a large difference between temperatures of two surfaces that are exposed to each other, but are not touching.

Radiant heating - a heating system in which only the heat radiated from panels is effective in providing the heating requirements. The term radiant heating is frequently used to include both panel and radiant heating.

Radiant losses - heat lost from a building through radiation from the walls.

Radiant panel - a heat emitter which exchanges heat through tubing grids installed in floors, walls, or ceilings that releases 50 percent or more of its heat output as thermal radiation.

Radiant superheater - a nest of tubes that the saturated steam passes through to acquire heat.

Radiant tube type heater - an indirect fired heating system in which heating is accomplished by radiation or convection from tubular elements. The tubes are gas tight, heat resistance, open at one or both ends, and capable of withstanding explosion pressure from ignition of fuel air mixtures. Each tube has an inlet and/or burner arrangement where combustion is initiated, a suitable length where combustion occurs, and an outlet for the combustion products formed.

Radiation - transmission of heat in a straight line through space. The process in which energy in the form or rays is emitted from atoms and molecules as they undergo internal change. Also see infrared (ir) and ultraviolet (uv) radiation; and visible light.

Radiation cooling - a solar system being developed to provide cooling. In effect, the solar system is reversed and the warm liquid is pumped through the night radiators facing north to cool the water. The water is then used the next day for cooling.

Radiation, equivalent direct (EDR) - the amount of heating surface expressed in square feet which will deliver 240 Btu/hr for steam, and 150 Btu/hr for hot water systems operating at design conditions.

Radiation shield - a separate panel or panels interposed between heating surfaces and adjacent objects to reduce heat transmission by radiation. Also see liner.

Radiation, thermal - the transmission of energy by means of electromagnetic waves of very long wavelength. Radiant energy of any wavelength may, when absorbed, become thermal energy and result in an increase in the temperature of the absorbing body.

Radiated noise - the sound power level transmitted through the duct wall, expressed in decibels (dB).

Radiator - a heating unit located within the room to be heated and exposed to view. A radiator transfers heat by radiation to objects it can "see" and by conduction to the surrounding air which in turn is circulated by natural convection. A radiator is also called a convector, but the term radiator has long been established.

Radiator valve - a valve installed on a terminal unit to manually control the flow of water through the unit.

Radioactivity - process by which certain nuclides, in a spontaneous disintegration in which energy is liberated, form new nuclides. This process is accompanied by the emissions or radiation, such as alpha particles, beta particles, and gamma ray photons. Natural radioactivity is exhibited by more than 15 naturally occurring isotopes. Artificial radioactivity can be induced in certain elements under controlled conditions.

Radionuclide - an unstable nucleus of an element that decays or disintegrates spontaneously, emitting radiation.

Radius - the length of a line running from the center of a circle to the outside perimeter of the circle. Half the diameter.

Radius diffusion - the horizontal axial distance an air stream travels after leaving an air outlet before the maximum stream velocity is reduced to a specified terminal level.

Radius of a circle - a straight line drawn from the center to the extreme edge of a circle.

Radon (radon-222) - a chemically inert radioactive gas that is odorless and colorless. It is formed in the decay of uranium that undergoes radioactive decay by emitting an alpha particle. The radon decay products (also called radon daughters or progeny) can be breathed into the lung where they continue to release radiation as they further decay. Radon can enter buildings through cracks in concrete floors and walls, floor drains and sumps.

Raggle - a groove cut in masonry, especially one that receives the upper edge of a flashing above a roof. A manufactured building unit provided with a groove into which metal flashing can be fitted. To cut a raggle --- as in stone.

Rain leader - these carry rainwater from the roof to an underground or a ground level disposal conduit.

Ram air - air forced through the condenser due to the rapid movement of the vehicle along the highway.

Range - the difference between the minimum and maximum points of the control in which the control will function properly.

Rangeability - the ratio of the maximum controllable flow to the minimum controllable flow through a valve. For example, a valve with a rangeability of 50 to 1 and having a total flow capacity of 100 gallons per minute (gpm) full open can accurately control a flow as low as 2 gpm.

Range (cooling range) - in a water cooling device, the difference between the average temperature of the water entering the device, and the average temperature of the water leaving it.

Range hold - in a autoranging meter, the ability to select a specific range instead of allowing the meter to select the range.

Rank - refers to the hardness of coal.

Rankin scale - temperature scale using Fahrenheit divisions, with absolute zero at 0° R. The freezing point of water is 492° R (460° F +32° F), and the boiling point of water is 672° R
(460° F + 212° F). The zero point on this scale is -460° F.

Rapid oxidation - a reaction between the fuel, oxygen, and heat that is known as rapid oxidation or the process of burning.

Rasp file - the rasp file is used for cutting wood and for very soft metals.

Ratchet clause - a ratchet clause in a power company contract states that the high peak demand during an interval of one month will establish the peak demand charge for possibly the next 11 months.

Ratchet flare nut box wrench - used where there is limited space in which a wrench handle may be moved.

Ratchet handle - used with socket wrench; enables user to apply a force in one direction for any fractional part of a revolution and return the handle to its original position without moving the socket.

Rate - input into an appliance.

Rate action - a condition where the controller senses the rate of change of temperature and provides an immediate change of output to minimize the eventual deviation.

Rate of combustion - the amount of fuel that is being burned in the furnace or appliance per unit of time.

Rate of rise detector - a device that responds when the temperature rises at a rate exceeding a predetermined value.

Rate of shear - the difference between the velocities along the parallel faces of a fluid element divided by the distance between the faces.

Rate time - the time in minutes that rate action response precedes normal proportional action response.

Rated input - acceptable input to a furnace based on compliance with ANSI standards. The input is shown on the rating plate that is affixed to each furnace.

Rated load amps - the rated load current for a hermetic refrigerant motor compressor is the current resulting when the motor compressor is operated at the rated load, rated voltage, and rated frequency of the equipment served.

Rated pressure - the maximum pressure that the actuating components of the switch in contact with the media can withstand continuously and/or repeatedly without risk of permanent damage.

Rated voltage - that maximum voltage at which an electrical component can be operated for extended periods without undue degradation or safety hazard.

Rating - the designated limit of operating characteristics of a machine, apparatus or device based on definite conditions. Load voltage, frequency, and other operating characteristics may be given in the rating.

Ratio - the comparison of one set of values to another set of values. Mathematically, the relationship between two quantities expressed as the quotient of one divided by the other.

Raw gas burner - a gas burner in which raw gas is ejected into the combustion chamber from multiple ports. Only secondary air is used for combustion; no mixing occurs before the gas is ejected. The ports are arranged to result in maximum mixing of gas and air. Examples are radiant head, ring, and high pressure gun burner.

Raw water - untreated water from wells or city water lines. In ice making, any water used for ice making except distilled water.

RD50 - the concentration that results in 50% decrease in respiratory rate, normally measured in mice. A maximal effect generally occurs in less than 30 minutes. The RD50 response to R-717 (ammonia) is reported to take approximately two minutes.

Reactance - a measure of the combined effects of capacitance and inductance on an alternating current. The amount of such opposition varies with the frequency of the current. The reactance of a capacitor decreases with an increase in frequency: the opposite occurs with the inductance. Symbol is "X" and is expressed in units of ohms.

Reactive circuit - an AC circuit which has the property of inductance or capacitance.

Reactive power (var) - the "unused" power of an AC circuit has inductance or capacitance, which is absorbed by the magnetic or electrostatic field of a reactive circuit.

Reactivity - a description of the tendency of a substance to undergo chemical reaction either by itself or with other materials with the release of energy. Undesirable effects such as pressure buildup, temperature increase, or the formation of noxious, toxic, or corrosive byproducts may occur because of the reactivity of a substance to heating, burning, direct contact with other materials, or other condition in use or in storage.

Readily accessible - capable of being reached easily and quickly for operation, adjustment, and inspection.

Real power - the energy part of apparent power. It is the rate of supply of energy, measured commercially in kilowatts. The product of real power and length of time is energy measured by watt hour meters.

Real time - pertains to the actual time during which a physical process transpires.

Ream - to scrape the inside edge of the end of tubing or a pipe in order to remove any loose bits of metal or any lip that formed during cutting.

Reamed - description of tubing or pipe that has been scraped with a tool to remove burrs from a cut surface.

Reamer - a cutting tool with a series of teeth or sharp cutting edges; used to enlarge a hole.

Rear header - these are found on straight tube watertube boilers. They are connected to the front header by water tubes.

Rebar - a steel bar used in concrete construction to provide additional strength. Size is measured in 1/8th's. A #6 rebar would be 3/4" in diameter, a #4 rebar would be 1/2" diameter.

Recalibration - periodic adjustment of a gauge or meter to ensure accuracy.

Receding dies - dies that are placed into slots of a threader that are backed by tapered posts. As the threader is rotated clockwise, the dies move deeper into sockets to create proper thread taper.

Receiver - a storage chamber for liquid refrigerant in a mechanical refrigeration system. An internal dip tube ensures that only 100 percent liquid refrigerant leaves the device. The bottom part of a condenser coil may also be used as a receiver.

Receiver, auxiliary - an extra vessel used to supplement the capacity of the receiver when additional storage volume is required.

Receiver controller - a device that controls the final control devices such as valves or pneumatic actuators in response to air pressure signals. These signals may be supplied by pneumatic transmitters or gradual switches which supply a 3-15 psi signal over a given range.

Receiver-drier - a cylinder in a refrigeration system for the storing of liquid refrigerant and which also has desiccant.

Receiver heating element - an electrical resistance mounted in on around the liquid receiver used to maintain the high side pressure when ambient temperatures fall below preset standards.

Receptacle - a point along an electrical circuit to which an cord plug is attached for the purpose of using the current supplied by that circuit. A single receptacle is a single contact device with no other contact device on the same yoke. A multiple receptacle is two or more contact devices on the same yoke.

Receptacle outlet - an outlet where one or more receptacles are installed.

Reciprocal - the inverse.

Reciprocating - the action in which the motion is back and forth in a straight line.

Reciprocating compressor - a compressor whose piston or pistons move back and forth in their cylinders.

Reciprocating feedwater pump - a positive displacement pump used to pump liquids.

Recirculated air - return air passed through the conditioner before being again supplied to the conditioned space.

Recirculating system - term used to describe a water cooled system that utilizes a cooling tower to cool the heat laden liquid before returning it to the heat exchanger.

Recirculation, water tower - describes a condition in which a portion of the tower's discharge air re-enters the air inlets along with the fresh air. Its effect is an elevation of the average entering wet bulb temperature compared to the ambient.

Reclaim - to remove refrigerant in any condition from an air conditioning or refrigeration system and then transport the recovered refrigerant to a dedicated refrigerant reclamation center to be reprocessed to new specifications (ARI 700), which may include distillation. Reclamation requires that the reprocessed refrigerant be certified through a chemical analysis.

Reclaimed refrigerant - refrigerants reprocessed to the same specifications as for new refrigerants by means including distillation. Such refrigerants have been chemically analyzed to verify that the specifications have been met. Reclaiming usually implies the use of processes or procedures that are available only at a reprocessing or manufacturing facility.

Reclamation - restoring to usefulness.

Recombination - the process in which a free electron fills an empty hole.

Recommend - should.

Recommended exposure limit - is the highest allowable air-borne concentration that is not expected to injure a worker. It may be expressed as a ceiling limit or as a time weighted average (TWA), usually for 10 hour work shifts.

Recording ammeter - an electrical instrument that uses a pen to record the amount of current flow on a moving paper circle or strip chart.

Recording thermometer - an electrical instrument that uses a pen to record the amount of heat intensity on a moving paper circle or strip chart.

Recording voltmeter - an electrical instrument that uses a pen to record the intensity of the electromotive force (voltage) on a moving paper circle strip chart.

Recover - to remove refrigerant in any condition from a system and store it in an external refillable department of transportation (DOT) container. Recovery does not provide for any cleaning or filtration and without testing or processing in any way.

Recovered heat - this heat comes from internal heat sources. It is used for space heating, domestic or service water heating, air reheat in air conditioning, process heating in industrial applications, or other similar purposes. Recovered heat may be stored for later use where practical.

Recovered refrigerants - refrigerants removed from a system in any condition without necessarily testing or processing them. Because it is illegal and harmful to the environment to release ozone depleting refrigerants to the atmosphere, used refrigerants are recovered from systems being repaired, relocated or retired.

Recovery cylinders - refillable, heavy duty, certified pressure vessels used in the recovery of refrigerant. The color code for these cylinders is a yellow top with a gray body and must meet Department of Transportation (DOT #39) specifications.

Recovery efficiency - the percentage of refrigerant in an appliance that is recovered by a piece of recycling or recovery equipment.

Rectangle - a geometric figure with opposite sides equal in length and each corner forming a 90° angle.

Rectification - the conversion of alternating current to unidirectional or direct current. The process of changing AC into DC. Also see flame rectification.

Rectification pilot - see pilot burner assembly.

Rectification system - a flame proving system which operates on a DC flame signal provided by a rectification type flame detector.

Rectification type flame detector - ones that uses the rectification principle to indicate the presence of a flame.

Rectifier - an electrical device that allows current to flow through it only in one direction. Used to convert alternating current into direct current.

Rectifying flame rod - a metal or ceramic rod used as a flame detector in a detection system using the flame rectification principle. Also see flame rectification.

Rectifying photocell flame detector - a detector that responds to the visible light (wavelength between 0.4 and 0.8 micron) emitted by a flame. It consists mainly of a high vacuum photocell that rectifies alternating current to direct current when exposed to visible light. The cathode of the photocell is coated with caesium oxide or another material that emits electrons when light strikes it.

Recuperative coil - secondary coil used to extract the latent heat from combustion gases.

Recycle (controls) - the process of going through another operating cycle. For a flame safeguard control, automatically going through another timer cycle after a shut down in an attempt to restart the burner.

Recycle (refrigerant) - to remove refrigerant in any condition from an air conditioning or refrigeration system and then reduce contaminants through oil separation, filtration, and other methods. This term usually applies to procedures implemented at the job site or at a local service shop. Recycled refrigerant may or may not be suitable for reuse.

Recycle interlock - a running interlock used with a primary or programming control which will shut down the burner if an improper condition occurs. The control will recycle; it will restart the burner if the condition corrects itself. The interlock is usually a combustion air controller (air flow switch). See running interlock.

Recycled refrigerants - refrigerants from which contaminants have been reduced by oil separation, removal of noncondensable gases, and single or multiple passes through devices that reduce moisture, acidity and particulate matter, such as replaceable core filter driers. These procedures usually are performed at the field site or in a local service shop.

Recycling - to clean refrigerant for reuse by oil separation or single or multiple passes through devices that reduce moisture, acidity, and particulate matter.

Recycling control - a flame safeguard control which recycles following a shut down. It will restart the burner if conditions causing the shut down have been corrected.

Recycling unit - machines used to remove most contaminants from refrigerants withdrawn from a system and make the refrigerant suitable for reuse.

Red iron oxide - rust that forms when oxygen from the air and moisture combine.

Reduced voltage start - motor starting method by which a lower voltage is supplied to a motor during startup to reduce the amperage draw of the motor, as well as to prolong the useful life of the starter's contacts.

Reducer - a fitting that joins two tubes of unequal diameter.

Reducing coupling - a fitting designed to change from one pipe or tubing size to another.

Reducing elbow - a 180 degree elbow that returns the pipe to the same direction from which it started.

Reducing flame - a rich flame or fire resulting from combustion of a mixture containing too much fuel and too little air. This kind of flame produces a reducing atmosphere.

Reducing plenum duct system - a duct system configuration used to maintain a constant air velocity in the truck line as air moves through it. Usually not found in residential applications.

Redundancy - is the duplication deliberately designed into a system to provide substitution in case of equipment or control failure.

Reed valve - a thin flat tempered steel plate fastened at one end.

Reed switch - two or more highly conductive reeds encapsulated in a glass enclosure. Energy from a magnetic field will cause the reed contacts to open or close simultaneously.

Reefer - a refrigerated compartment.

Re-entrainment - situation that occurs when the air exhausted from a building is immediately brought back into the system through the air intake and/or other openings in the building envelope.

Re-expansion - in reciprocating compressors, the process by which the pressures of the clearance vapor is reduced to the level of that in the suction line.

Re-expansion line - curve on indicator diagram, representing the pressure-total volume relationship of clearance fluid during the initial portion of the return stroke of the piston, prior to the opening of the suction valve.

Reference junction - the cold junction in a thermocouple circuit that is held at a known temperature, usually 0° C (32° F).

Reference lines - these are the solid lines on a blueprint or plot plan to show that an imaginary cut has been made at a point and that, in order to see that point in detail, you must look at another drawing. The arrow indicates the direction from which the section is viewed. The letters and numbers in the circle tell what section and where it will be found.

Reference voltage - the constant voltage used as a reference point to which another voltage is compared.

Reflected radiation - is sunlight that is reflected from the surrounding trees, terrain, or buildings onto a surface exposed to the sky.

Reflective glass - glass with a metallic coating applied onto or into the glass surface to reduce solar radiant energy and visible light transmission.

Refractory - heat resistance material used to line furnaces, kilns, ovens, and combustion chambers.

Refractory block - a piece of refractory material molded with a conical or cylindrical hole through its center and mounted so that the flame fires through this hole. The block helps to maintain continuous combustion, and reduces the probability of flashback and blow off. Also called a burner block, refractory tile, or burner refractory.

Refractory hold in test - see hot refractory hold in test.

Refractory override test - see hot refractory saturation test.

Refractory saturation test - see hot refractory saturation test.

Refrigerant - a substance used in a refrigeration system to absorb heat in the evaporator by changing from a liquid to a gas and releases this heat from the condenser when the substance returns to a liquid state.

Refrigerant blend - two or more refrigerants blended or mixed together to make another refrigerant. Blends can combine as either Azeotropic or zeotropic blends.

Refrigerant charge - the quantity of refrigerant that is required for proper operation of a closed cycle refrigeration system.

Refrigerant control - a device (either manual or automatic) which meters refrigerant and maintains a pressure differential between the high and low pressure side of the mechanical refrigeration system while unit is in operation.

Refrigerant conversion - to modify an existing piece of equipment to successfully accommodate an alternate refrigerant. That is to replace refrigerant, refrigerant oil, make control adjustments, replace filter driers, etc.

Refrigerant cycle - the direct expansion refrigeration cycle consists of compressor, condenser, metering device (expansion valve or capillary), and evaporator and interconnecting tubing.

Refrigerant distributors - used to distribute the refrigerant simultaneously to several smaller tubes piped in parallel.

Refrigerant dye - coloring agent that can be added to the refrigerant to help locate refrigerant leaks in a system. Not a highly recommended practice.

Refrigerant, flammable - any refrigerant which will burn when mixed with air, such as ethyl chloride, methyl chloride, and the hydrocarbons.

Refrigerant flammability classifications - there are three refrigerant flammability classifications as dictated by the American Society of Heating, Refrigerating and Air Conditioning Engineers (ASHRAE). They are Class 1 that has "no flame propagation", Class 2 that has a "moderately flammable" (heat of combustion < 8,174 Btu/lb) rating and Class 3 has a "highly flammable" (heat of combustion > 8,174 Btu/lb) rating.

Refrigerant loop - the heat pump's refrigerant system, which exchanges energy with the fluid in the ground loop and the air side of the system.

Refrigerant migration - the condensation of refrigerant vapor at the coldest point in the system during the off cycle.

Refrigerant receiver - a refrigerant system component installed in the liquid line. It's designed to make space for liquid refrigerant flow due to the closing action of a self-regulating metering device.

Refrigerant reclaim - to reprocess refrigerant to new product specifications. This will require chemical analysis, and usually refers to the processes available at a reprocessing or manufacturing facility.

Refrigerant recovery - to remove refrigerant from the system (or a section) and place in a cylinder. No testing is required.

Refrigerant recycling - to clean refrigerant for reuse by reducing moisture, acidity, and foreign matter. Usually applies to procedures at job sits or local service shops.

Refrigerant retrofit - to furnish with new components or major parts that weren't available when a device or system was first manufactured as part of a changeover to an alternate refrigerant.

Refrigerant safety classifications - groupings that indicate the toxicity and flammability classes made up of a letter (A or B) that indicates the toxicity class, followed by a number (1, 2 or 3) that indicates the flammability class. Refrigerant blends are similarly classified, based on the compositions at their worst cases of fractionation, as separately determined for toxicity and flammability. In some cases, the worst case of fractionation is the original formulation. Class 1 flammability indicates refrigerants that do not show flame propagation in air when tested by prescribed methods at specified conditions. Classes 2 and 3 signify refrigerants with lower flammability and higher flammability, respectively; the distinction depends on both the lower flammability limit (LFL) and heat of combustion. Class A and B toxicity signify refrigerants with lower toxicity and higher toxicity, respectively, based on prescribed measures of chronic (long term, repeated exposures) toxicity.

Refrigerant, secondary - any volatile or nonvolatile substance in an indirect refrigerating system that absorbs heat from a substance or space to be refrigerated and rejects this heat to the evaporator of the refrigerating system.

Refrigerant tables - tables that show the properties of saturated refrigerants at various temperatures.

Refrigerant trap - a device placed at the outlet of the evaporator coil when the air handler is located below the condensing unit. This trap is designed to facilitate proper oil return to the compressor.

Refrigerant velocity - the movement of gaseous refrigerant required to entrain oil mist and carry it back to the compressor.

Refrigerating capacity - the rate at which a system, compared with the cooling effect produced by the melting of ice, can remove heat. Usually stated in tons or Btu per hour.

Refrigerating effect - the amount of heat a given quantity of refrigerant will absorb in changing from a liquid to a gas at a given evaporating pressure.

Refrigerating effect, condensing unit - the rate of heat removal by the refrigerant assigned to the condensing unit in a refrigerating system; this is equal to the product of the mass rate of refrigerant flow produced by the condensing unit and the difference in the specific enthalpies of the refrigerant vapor entering the unit and the refrigerant liquid leaving the unit.

Refrigerating medium - any substance whose temperature is such that it is used, with or without a change of state, to lower the temperature of other bodies or substances below the ambient temperature.

Refrigerating system, absorption - a refrigerating system in which the refrigerated gas in the evaporator is taken up in an absorber and released in a generator upon the application of heat.

Refrigerating system, central plant - a system with two or more low sides connected to a single, central high side.

Refrigerating system, chilled water - an indirect refrigerating system employing water as the circulating liquid.

Refrigerating system, compression - a refrigerating system in which the pressure imposing element is mechanically operated.

Refrigerating system, direct expansion - a refrigerating system in which the evaporator is in direct contact with the refrigerated material or space or is located in air circulating passages communicating with such spaces.

Refrigerating system, mechanical - a system where the evaporator coil produces cooling by absorbing heat from the surrounding air, raising the refrigerant to its boiling point and causing it to vaporize. The superheated vapor flows through the condenser, which condenses it into a liquid and gives off heat picked up in the evaporator coil. Then the liquid flows to the metering device, where it expands (lowering its temperature and pressure) to start the cooling cycle all over again.

Refrigerating system, multiple - a refrigerating system using the direct method in which the refrigerant is delivered to two or more evaporators in separate rooms or refrigerators.

Refrigerating system, single package - a complete factory made and factory tested refrigerating system in a suitable frame or enclosure, which is fabricated and shipped in one or more sections and in which no refrigerant containing parts are connected in the field.

Refrigeration - in general, the process of transferring heat from a place where heat is not wanted to a place that is unobjectionable. Removing heat from an enclosed space and maintaining that space at a temperature lower than its surroundings.

Refrigeration cycle - a process during which a refrigerant absorbs heat at a relatively low temperature and rejects heat at a higher temperature.

Refrigeration oil - a specially refined, chemically stable, dry, light oil that lubricates the compressor.

Refrigeration service wrench - ratcheting service tool to open and close service valves. Typical internal opening sizes include ¼", 5/16" and 3/8".

Refrigeration system - combination of interconnected refrigerant containing devices in which the refrigerant is circulated for the purpose of extracting heat to produce cooling.

Refrigeration system, absorption - a heat operated, closed refrigeration cycle in which a secondary fluid (the absorbent) absorbs a primary fluid (the refrigerant) that has been vaporized in the evaporator.

Refrigeration tables - tables that show the properties of saturated refrigerants at various temperatures.

Refrigeration, ton - the removal of heat at a rate of 200 Btu per minute, 12,000 Btu per hour or 288,000 Btu per 24 hour period. These figures come from melting 2,000 pounds of ice (one ton) in a 24 hour period. 2000 pounds times the latent heat of ice (144 Btu's per pound) equals 288,000 Btu's per day.

Refrigerator - a container and means for cooling it, such as a domestic refrigerator, or a large container such as a storage refrigerator, service refrigerator, etc.

Refrigerator, commercial - a general category referring to any of the many types of refrigerators used commercially. Included would be walk-ins, reach-ins, refrigerated display cases both service and self-service of all types, which are used by business establishments.

Refrigerator, electric - a completely self-contained unit consisting of an insulated cabinet, evaporator, condenser, and an electric compressor.

Refrigerator, gas - a refrigerator motivated by thermal energy of burning gas.

Regenerated sound - sound generated by air turbulence within ductwork. It is greatest in the low frequencies and increases with air velocity.

Register, air - a grille that is equipped with a damper or control valve, which directs air in a non-spreading jet. A combination grille and damper assembly.

Register, burner - the porting, grille work, or damper arrangement through which air is introduced into the combustion chamber around the burner ports or nozzles. It may also control the direction and velocity of the air stream for efficient mixing with the incoming fuel. A combination grille and damper arrangement covering on an air opening or end of an air duct. Also see air register.

Register, electronic - a short-term digital electronic storage circuit. The capacity of which is usually one computer word or byte.

Regulation - a measure of the amount of voltage change that occurs in the output of a generator due to the changes in the load.

Regulator - a device that functions to keep a quantity or quantities at a certain value or between certain limits. Also see atmospheric (zero governor), draft, gas pressure, and pressure regulators.

Regulator, suction pressure - an automatic valve or control device designed to maintain the pressure, and therefore the temperature, in an evaporator above a predetermined minimum.

Regulator vent - the opening in the atmospheric side of the regulator housing permitting the in and out movement of air to compensate for the movement of the regulator diaphragm.

Regulatory authority - the right, as for a government agency, to control an industrial process or mechanism in agreement with a rule.

Reheat - the process of adding heat to air to maintain the correct temperature after it has previously been cooled to some specified dew point to control humidity.

Reheat system - a system that distributes cooled air to one or more zones and reheats the cooled air to meet the comfort requirements of other zones.

Reinforced sheath - the outermost covering of a cable that has a cable sheath constructed in layers with a reinforcing material, usually a braided fiber, molded in place between layers.

Reinforcing bar - see rebar.

Relative heat gain - a measurement of the total heat gain through glazing for a specific set of conditions.

Relative humidity - the percentage of water vapor contained in a given amount of air at a specific temperature compared to the amount it could contain when 100% saturated. As air temperature changes, its ability to contain water vapor also changes. Ability to contain moisture increases with temperature. Therefore, as the temperature of a given amount of air containing a given amount of water vapor is heated, the relative humidity is lowered.

Relative measurement - a capability that compares one measurement to another. The value is the deviation of the present measurement from a previous one, expressed in absolute numbers (compares function) or as a percentage (percentage function).

Relay - an electromechanical device with contacts that open and/or close when its coil is energized or de-energized in response to a change in the conditions of the electrical circuit. The operation of the contacts effect the operation of other devices in the same circuit or in other circuits. Also see balancing, checking, flame detector, flame, and load relays.

Relay, changeover - a control used in some heat pumps to operate the reversing valve when changing from heating to cooling or vice-versa.

Relay, control - an electromagnetic device that changes the position of normally open or normally closed contacts when its coil is energized.

Relay chatter - noise due to the rapid opening and closing of relay contacts.

Relay, impedance - a control that prevents compressor restart, after any safety control shut down, until manual reset is established. Sometimes called a lock out relay.

Relay, lock-out - a relay used in conjunction with an automatic reset high pressure control to cause manual reset interruption of the unit operation when high head pressure is encountered. Reset of the system is from a remote location by interruption of power to the circuit. Also called an impedance relay.

Relay, magnetic - solenoid operated relay or contactor; a switching relay that utilizes an electromagnet (solenoid) and an armature to provide the switching force.

Relay, oil burner - dual purpose control that monitors the operation of the oil burner and acts as a safety in the event of a flame failure.

Relay socket - the base a relay plugs into, and at which external wires leading to the relay are terminated.

Relay, start - a magnetic switch that disconnects the start capacitor after the motor reaches rated speed. Terminals are 1 & 2 for the normally closed contacts and 2 & 5 for the relay coil. Also known as a potential relay.

Relay, thermal - a switching relay in which a small heater warms a bimetal element that bends to provide the switching action.

Relay, thermal overload - a thermal device that opens its contacts when the current through a heater coil exceeds the specified value for a given time.

Relay, time delay - a device for preventing short cycling of the compressor after safety shut down by interrupting the circuit for several seconds before restoring the circuit. A resistive load.

Reliable - the degree of safety that indicates a reasonable freedom from hazards will be maintained during continuous or extended operation.

Relief opening - the opening in a draft hood to permit ready escape to the atmosphere of flue products from the draft hood in event of no draft, back draft or stoppage beyond the draft hood, and to permit inspiration of air into the draft hood in the event of a strong chimney updraft.

Relief valve - a valve that opens at a designed pressure and bleeds a system in order to prevent a buildup of excessive pressure which might damage regulators and other instruments. Also see vent valve.

Relight - immediate restoration of ignition, without pre-ignition purging or interlock checks, after loss of the main burner flame. If the main burner flame is not re-established, safety shut down occurs. Also called ignition return.

Reluctance - a force working against the passage of magnetic lines of force (flux) through a magnetic substance.

Remediation - removal of contaminant sources and elimination of the cause and effect chain to prevent an IAQ problem from recurring.

Remote bulb - a part of the thermal expansion valve. The remote bulb assumes the temperature of the suction vapor at the point where the bulb is secured to the suction line. Any change in the suction vapor superheat at the point of bulb application tends to operate the valve in a compensating direction to restore the superheat to a predetermined valve setting.

Remote bulb thermostat - a control whose sensing element is located separate from the mechanism it controls.

Remote control circuit - an electric circuit that control another circuit at a remote point.

Remote power element control - a device with its sensing element located apart from its operating mechanism.

Remote reset system - a system, designed to re-establish an electrical circuit which permits starting of the compressor from a remote location, such as the room thermostat, after a safety shut down. The remote reset is used as a convenience to the homeowner.

Remote system - a refrigeration system which has its condensing unit located outside and separated from the evaporator. Sometimes referred to as a split system.

Remote temperature set point - ability to set a temperature control point for a space from outside the space. Often used in public areas.

Rendering - a perspective or elevation drawing of a project with artistic delineation of materials, shades and shadows.

Renewable energy sources - renewable energy sources of energy (excluding minerals) are derived from: (1) incoming solar radiation, including but not limited to, natural day lighting and photosynthetic processes (2) energy sources resulting from wind, waves and tides, lake or pond thermal differences and (3) energy derived from the intense heat of the earth, including nocturnal thermal exchanges.

Renewable fuse - a fuse in which the element, typically a zinc link, may be replaced after the fuse has opened, and then reused. Renewable fuses are made to class H standards.

Repeatability - the ability of a controller or interlock to maintain a constant set point characteristic.

Repeating cycle - a cycle that can continue indefinitely without loss of fuel or refrigerant. In air conditioning, the refrigerant experiences a number of physical changes as it flows through the system but finishes the cycle in the same physical state as when it started. During the process, the amount of refrigerant in the system will not change.

Repetitive accuracy - the ability of a pressure or flow switch to operate repetitively at its set point under consistent conditions.

Replacement package - parts included in a single package intended for replacement of a device.

Repulsion start induction motor - a motor which develops starting torque by the interaction of rotor currents and a single phase stator field.

Reserve capacity - the amount of generating capacity a central power system must maintain to meet peak loads.

Reset - a process of automatically adjusting the control point of a given controller to compensate for changes in the outdoor temperature. The hot deck control point is normally reset upward as the outdoor temperature drops. The cold deck control point is normally reset downward as the outdoor temperature increases. To return a control or valve to its original position.

Reset control - a device for maintaining a pressure or temperature, usually in the heating or cooling medium, which is automatically changed as some other condition changes; that is the controller is automatically reset to a different control point.

Reset line - a straight line having a specific slope on a graph of supply water temperature versus outdoor air temperature.

Reset ratio - the ratio of change in outdoor temperature to the change in control point temperature. For example, a 2:1 reset ratio means that the control point will increase 1° for every 2° change in outdoor temperature.

Residential - applications performed where individual families live. Usually considered to be individual houses.

Residual - a remainder; a remaining force as in magnetism; left over.

Residual magnetism - the magnetism that remains around a material after the magnetizing force has been around.

Residual oils - oils that are too heavy to be evaporated in any normal evaporation and distillation process and so are left over from that process.

Residual velocity - the velocity that must be kept up in the occupied zone. It generally ranges from 20 to 70 fpm.

Residue - a substance left over at the end of a process.

Resilient mount - the mounting of a motor on flexible rubber feet or mounts to prevent the transmission of vibration to the base.

Resistance, electrical - the difficulty electrons encounter moving through a conductor or substance. This causes heat in the carrier. The opposition to current flow. Measured in ohms. Symbol is "R".

Resistance formulas –

1/U = overall resistance of the wall or partition (air to air).

1/K = resistance of any homogeneous component whose K value is known, per inch of thickness.

1/C = resistance of a material, homogeneous or not (surface to surface)

1/f = film or surface resistance.

1/a = air space resistance.

Resistance heater - a unit that uses electric conductors to supply heat. The amount of heat depends upon the conductor's electrical resistance.

Resistance start induction run motor - one type of split phase motor that uses the resistance of the start winding to produce a phase shift between the current in the start winding and the current in the run winding.

Resistance temperature detector (RTD) - these are wire windings or thin film serpentines that exhibit changes in resistance with changes in temperature. While metals such as copper, nickel and nickel-iron are often used, the most linear, repeatable and stable RTD's are constructed from platinum. Platinum RTD's, due to their linearity and unmatched long term stability, are firmly established as the international temperature reference transfer standard. Thin wire platinum RTD's offer performance matching for all but reference grade wire wounds at improved cost, size and convenience.

Resistance, thermal - a measure of the ability to retard the flow of heat. The R value is the reciprocal of a heat transfer coefficient as expressed by R = 1/U.

Resistive load - an electrical load that is characteristic of not having any significant inrush current and converts electrical energy directly into heat. When a resistive load is energized, the current rises instantly to its steady state value, without first rising to a higher value. Power factor is one. Current and voltage are "in phase" with each other.

Resistivity, thermal - the reciprocal of thermal conductivity.

Resistor - a component of an electric circuit that act to oppose current flow by virtue of its electrical resistance.

1. Composition resistor - a resistor made of carbon particles mixed with a doping material and provided with terminal leads to connect the resistor (usually cylindrical in shape) into a circuit. The resistance value is determined by the relative amounts of carbon and doping material used.
2. Fixed resistor - a resistor with one essentially constant resistance value (except for changes in temperature variations).
3. Wire wound resistor - a resistor made by winding wire around a non-conducting cylinder. The resistance value is determined by the diameter and number of turns of wire.

4. Variable resistor - a resistor constructed so that the value of resistance may be varied over a range, usually by mechanical means.

Resolution - the magnitude of an input signal equal to a digit change in the least significant digit in the display. Smallest value an instrument can measure.

Resonate - to maximize or minimize the amplitude or other characteristic of the response of a given system or device.

Respirable particles - particles that can penetrate into and be deposited in the nonciliated portion of the lung. Respirable particles are from 0.5 to 10 micrometers in diameter. This size range varies from one standards or reporting body to another, and may depend on the shape of the particle as well as the size of the particle.

Respiration - the heat generated by perishable food products while in storage.

Response - the reaction of a device or system to a control signal.

Response time - see flame failure response time.

Restore - to energize a load that has been shed.

Restrictor - a device for producing a deliberate pressure drop or resistance in a line by reducing the cross sectional flow area.

Retaining wall - a wall designed to resist the lateral displacement of soil or other material.

Retentivity - the ability of a material to retain magnetism after a magnetizing force has been removed.

Retort - the space below the grates of an underfeed stoker.

Retrofit - term used in describing reworking an older installation to bring it up to date with modern equipment or to meet new code requirements.

Retrofit guidelines - guidelines intended to make the transition from a CFC mineral oil system to a system containing an alternative refrigerant and its appropriate oil.

Return air - air from the conditioned space that is returned to the conditioning equipment for reconditioning. The air may return to the system through a series of ducts, or through plenums and airshafts.

Return air control - a control strategy in which a controller receives its signal from a temperature sensor in the return air stream from a large area.

Return air plenum - an unoccupied space within the insulated building envelope through which air flows back to the space conditioning system from the space or spaces. Return plenums are normally immediately above a ceiling, and below an insulated roof or the floor above. The return air temperature is usually within a few degrees of space temperature.

Return bend - u-shaped fitting used to reverse the direction of a pipe run. Also called a turn bend.

Return branch - piping used to return water from a terminal to the circuit main pipe.

Return, dry - a return pipe in a steam heating system that carries both water of condensation and air. The dry return is above the level of the water line in the boiler in a gravity system.

Return ducts - ductwork that connects the conditioned space with the heating and cooling equipment so that this return air may be reconditioned again.

Return grille - located in the occupied space, the decorative cover for the return duct system. The air filter may be located just behind this grille.

Return intake - an opening through which air is removed from the conditioned space.

Return line - the line that returns water (condensate) back to the boiler or return tank. Also see return piping.

Return mains - the pipes that return the heating medium from the heating units to the source of heat supply.

Return manifold - a metal or thermoplastic chamber to which the return side of one or more heating distribution circuits are attached.

Return piping - the piping system that carries water from the terminal units back to the boiler.

Return plenum - the section of the return duct that is connected directly to the air handler.

Return pump - condensate or vacuum pump, returns condensate back to the boiler or storage tank.

Return stroke - the retraction of the actuator shaft as a result of a signal from the controller decreasing below the opposing force of the spring.

Return tapping - the opening in a boiler into which the pipe used for returning condensate or water to the boiler is connected.

Return water - the water leaving the heat exchange surface. This water typically returns to the water tower or well from which it originally came.

Return well - a well for the return of water after it has went through the exchanger of the geothermal heat pump.

Return, wet - that part of a return main of a steam heating system that is filled with water of condensation. The wet return usually is below the level of the water line in the boiler, although not necessarily so.

Returnable cylinder - refillable heavy duty certified pressure vessel cylinder.

Reverse acting - the output signal changing in the opposite direction the controller or measured variable changes. An example would be when an increase in the controlled or measured variable results in a decreased output signal.

Reverse biased - when negative voltage is connected to the anode and positive voltage is connected to the cathode resulting in an open connection with no current.

Reverse cycle - the ability to direct the hot gas flow into the indoor or outdoor coil in a heat pump system to control the system for heating or cooling purposes.

Reverse cycle defrost - a method of heating the evaporator coil for defrosting purposes by using a valve to direct hot gas from the compressor to the evaporator.

Reverse return system - a 2 pipe hot water heating system in which the water from the several heating units is returned along paths, arranged so that all radiator circuits of the system are practically of equal length. This system is piped in which the first terminal unit supplied with hot water is the last one to return to the boiler and vice versa.

Reversible motor - a motor which can be operated in either direction.

Reversing relay - a relay that permits a SPDT thermostat to control heating and cooling equipment by forming an interlock that prevents mutual operation.

Reversing valve - an electrically actuated pressure operated slide valve used to change the direction of refrigerant flow in a heat pump application depending upon whether heating or cooling is desired. Also known as a four way valve.

Reversing valve solenoid - an electrical solenoid coil that is attached to the reversing valve that receives a signal from the room thermostat to change the position of the reversing valve itself. The solenoid coil only moves the pilot pin carrier in the reversing valve The actual movement of the slide mechanism is achieved when the pilot pin carrier moves to allow refrigerant pressure to create a pressure differential across the valve enough to move it.

Revolutions per minute - a measure of rotary speed; the number of rotations in a 60 second period; abbreviated as RPM.

Reynolds numbers - a unitless number that can be used to predict whether flow is laminar or turbulent.

Reznor hook - a bent piece of wire with two pointed ends that are driven in a wooden joist to hold pipe, tubing, or conduit.

Rheostat - a type of variable resistor with one fixed terminal and a movable contact whereby the resistance through the rheostat can be adjusted to various levels.

Ribbon - a thin sheet of crystalline or polycrystalline material, such as silicon, produced in a continuous process by withdrawal from a molten bath of the parent material.

Ribbon cable - a cable consisting of two or more conductors laid parallel in one plane and held in place by some means.

Rich mixture - an air fuel mixture in which an excess of fuel is supplied in proportion to the amount of air available. Also see fuel rich.

Ridge - the metal covering for a roof peak that extends to either side a sufficient distance to permit adequate flashing.

Rigid conduit - steel conduit made to the same dimensions as standard pipe.

Rigid mount motor - electric motor that is bolted metal to metal to a frame. This type of motor will transmit noise.

Rigid pipe - piping material that is not intended to be bent. This pipe is usually copper or steel.

Right angle - an angle that is formed by a line that is perpendicular to another line. 90° angle.

Right hand motor rule - the rule applied to find the direction of motion of the rotor conductors of a motor.

Ring burner - a gun type, raw gas burner in which jets of gas are directed into whirling airstreams from orifices around the periphery of a ring at the mouth of the burner. Also see gun type and raw gas burners.

Ring lubrication - a system of lubrication in which the lubricant is supplied to the bearing by an oil ring.

Ring sticking - freezing of a piston ring in its groove in a piston engine or reciprocating compressor due to heavy deposits in the piston ring zone.

Ring valve - a flat, doughnut shaped valve.

Ringelmann scale - unit measure of smoke opacity; one Ringelmann equal 20% smoke density, and 5 Ringelmanns equal 100% smoke density.

Rings - circular metallic elements that ride in the grooves of a piston and provide compression sealing during combustion. Also used to spread oil for lubrication.

Rise - the measured difference between the entering air temperature (return) and the leaving air temperature (supply).

Riser - a vertical tube, pipe, or duct that carries fluid in any form from a lower to a higher level.

Riser, electrical - the bundle of wires or cables which run vertically between floors of a building.

Riser height - the vertical distance from the top of a tread or platform/landing to the top of the next higher tread or platform/landing.

Riser valve - a valve or device used to manually control the flow of a substance in vertical piping systems.

Riser, water tower - piping which connects the circulating water supply line from the level of the base of the tower or the supply header, to the tower's distribution system.

Rising stem valve - a valve whose stem rises when the valve is opened.

Rising stem with inside screw valve - probably the most common valve found in smaller sizes. The stem turns and rises on threads inside the valve body. The position of the stem indicates the position of the valve disc. The stem extends beyond the bonnet when the valve is open and, therefore, requires more headroom.

Rising stem with outside screw valve - a valve having the stem threads outside of the valve body in both the open and closed position. Stem threads are not subject to corrosion, erosion, sediment, and galling from extreme temperature changes caused by elements in the line fluid flow. Since the valve stem is outside the valve body, it is subject to physical damage when the valve is in the open position. This type of stem is well suited for steam and high temperatures, high pressure water service applications. The rising stem valve requires more headroom than a non-rising stem. The position of the valve stem indicates the position of the valve disc.

RMS value - the effective value of an alternating current sine wave which is calculated as the square root of the average of the squares of all the instantaneous values of the current through-out one cycle. Rms alternating current is that value of AC voltage that will produce as much power when connected across a resistor as a like amount of DC voltage.

Rocker switch - a pivoted switch that is operated by pressing the protruding half whereupon the flush half protrudes.

Rod and tube - the rod and tube are each made of a different metal. The tube has a high expansion rate and the rod has a low expansion rate.

Rod hanger - type of hanger used extensively to carry vertical downward loads. It allows limited horizontal pipe movement in all directions.

Roller bearing - an anti-friction bearing comprising rolling elements in the form of rollers or balls.

Rollout switch - a heat sensitive protective device that opens the circuit if flame migrates away from the burner box.

Romex - type non-metallic (NM) sheathed cable.

Roof - the roof slab or deck with its supporting members, not including vertical supports.

Roof flanges - these are placed around stacks where they pass through the roof to prevent leakage.

Roof mounted - an application where the unit is mounted on a platform designed to distribute the weight of the unit over as wide an area of the roof as possible.

Rooftop unit - HVAC system placed on a roof and connected to ducts that supply conditioned air to the area below it.

Room - separate partitioned area that may or may not require thermostatic control.

Room air temperature profile - a graph showing how room air temperature varies from floor to ceiling.

Room heater - a gas stove or appliance considered by ANSI to be a heating appliance. This heater will have an efficiency rating.

Room heating load - the rate at which heat must be added to an individual room to maintain interior comfort.

Room reset of discharge - method of zone control in which a controller receives its input signal from two temperature sensors, one in the zone and one in the discharge air stream.

Room side heat flux - the rate of heat output per unit of area from the room side of a radiant panel. It is usually expressed in Btu/hr/square foot.

Room temperature unit - a device that measures room and temperature and communicates with a controller in another location.

Room thermostat - an adjustable, temperature-operated switch, mounted in a room, that turns heat delivery on and off to regulate room temperature.

Root - the bottom of the V formed by adjoining screw threads.

Root mean square (RMS) - the square root of the average of the square of the instantaneous values of current or voltage. The value assigned to a level equal to that of a direct voltage or current producing the same power dissipation through a pure resistance. For example, the RMS value of voltage line to neutral in a home is 120 volts. During each electrical cycle, the voltage rises from zero to a peak value (120 x 1.4142 = 169.7volts), back through zero to a negative peak value then back to zero. The RMS value is .707 of the peak value (169.7 x .707 = 120 volts). The root mean square value is what a person reads on an ammeter or voltmeter.

Rosin flux - a rosin based flux containing an additive that increases wetting.

Rotary - circular, spinning or turning motion.

Rotary blade compressor - a mechanism for pumping fluid by revolving blades inside a cylindrical housing.

Rotary compressor - a mechanism for pumping fluids by revolving vanes using rotating motion.

Rotary oil burner - a burner in which atomization is accomplished by feeding oil to a rapidly rotating cup or to spinning plates. Centrifugal force throws off the oil in droplets at the periphery of the cup or plates. Also called a centrifugal atomizing oil burner. There are two basic types:

Horizontal rotary oil burner - oil is pumped through a horizontal, hollow, drive shaft to a spinning cup. The axis of the cup is also horizontal.

Vertical rotary oil burner - the oil tank is mounted higher than the burner, and oil flows downward by gravity to the spinning plates.

Rotary screw - a screw-shaped rotor, turning within a flexible stator, usually of rubber. Progressing cavities between screw and stator carry the fluid and used for pressures up to 75 psi. Can handle abrasive mixtures or slurries, at slower speeds.

Rotate - to turn or revolve around a common point.

Rotating field - the magnetic field found in electric motors where the poles keep shifting to cause motor rotation.

Rotating unit - alternately shedding and restoring loads assigned to a specific channel so that they will not be shed continuously. Also called alternating channels.

Rotation - direction in which motor shaft rotates; CW = clockwise; CCW = counterclockwise; REV = reversible; rotation can be changed. Motor manufacturer may specify rotation as viewed facing the shaft or the end of motor opposite the shaft. Always double check the direction of shaft rotation.

Rotor - the rotating part of an electrical generator or motor.

Rotor bars - bars that are part of the rotor in a squirrel cage induction motor.

Rotor slip - condition produced by a weakening magnetic field due to series impedance, eventually causing the motor speed to decrease.

Rough opening - the opening built into a frame or wall to receive a window or door unit. It is formed with a header along the top that is supported by jack studs on each side. The rough sill at the bottom is supported by cripples. The rough opening is generally greater in width and height than the window or door unit.

Rough sill - the horizontal framing piece, usually a 2 x 4, which forms the bottom of then rough opening. It is toe nailed into the jack studs and supported by cripples.

Roughing in - installation of all concealed plumbing pipes including all plumbing and/or electrical work done before setting of fixtures or finishing.

Run - a length of material that has branch connections and side outlets. May be a duct run which carry either warm or return air or a pipe run carrying a liquid fluid.

Run capacitor - an electric device that is used to momentarily store electrons and create a second phase in the start winding circuits of single phase motors. This type of capacitor is designed to stay in the circuit whenever the motor is running as a means of heat dissipation.

Run factor - the percentage of the time the unit can be expected to operate to handle the heating load during the coldest month and the cooling load during the warmest month. Also known as run fraction.

Run out - length of duct between the main truck and the diffuser in a warm air system.

Run period - the period of time after ignition trials and before the operating set point is reached during which the main burner is firing.

Run winding - the electrical winding of a motor which has current flowing through it during normal operation. This winding draws the highest operating current.

Rung - a horizontal circuit path run between vertical voltage lines on a ladder wiring diagram.

Running current - the steady state current drawn from the line when the motor is running at its rated speed, or the armature of a solenoid or coil is in its closed position.

Running load amps (RLA) - amount of current a motor draws after it has been energized and permitted to reach its normal operating revolutions per minute.

Running interlock - an interlock which proves a condition is proper for burner operation to continue through pre-purge, ignition trials, and the run period. It is used with a primary or programming control that will shut down the burner if an improper condition occurs. The interlock may be a combustion air controller (air flow switch), a damper position indicator (low or high fire proving switch), or a fuel pressure switch. It can be a recycling type, or it may require manual reset. Also see lockout and recycle interlocks.

Running time - amount of time a machine is run per hour or per a 24 hour period.

Rupture disc - a safety device that will automatically rupture at a predetermined pressure.

R value - a value indicating the thermal resistance of the material. The greater the R value, the slower heat will conduct through the material, all other conditions being equal.

-S-

SC - shading coefficient.

SCBA - self contained breathing apparatus

SCFM - standard cubic feet per minute.

SCR - silicon controlled rectifier.

SEER - seasonal energy efficiency ratio.

Shpg - shipping.

Slv - sleeve.

SMLR - screening multiple linear regression.

SMT - soil moisture tool.

SNM - special cable designed for use in hazardous locations.

SO_2 - the chemical symbol for sulfur dioxide.

Sp - static pressure or set point.

Spd - speed.

SPST - single pole single throw.

SSF - Saybolt seconds furol.

SSOV - safety shut off valve.

SSU - Saybolt seconds universal.

Ssv - suction service valve.

Stel - short term exposure limit.

Sw - switch.

Syn - synchronous.

S type cord - a 600 maximum volt cord that has a temperature range of -40° F to +140° F. It is a heavy duty, general purpose cord used for machine tools, motors, drills, and power tools.

Saddle - U-shaped piece of metal that provides support or reinforcement to pipes or tubing. It can also be used to establish a sloped run. Horizontal vessels or other pieces of equipment are supported on pedestals by means of thick steel saddles welded to the vessel which allow for the anchoring of equipment to the concrete pedestal. Saddles may also be used with guide rollers.

Saddle flange - flange usually curved and riveted or welded to fit a boiler, tank, or other vessel and receive a threaded pipe. Also called a tank or boiler flange.

Saddle valve - a valve body so shaped so that it may be clamped around a pipe or tubing to gain access without having to void or drain the water distribution system.

SAE flare fitting - tube flared to Society of Automotive Engineer (SAE) standards.

Safe - as applied to a building, means free from danger or hazard to the life, safety, health or welfare of persons occupying or frequenting it, or of the public, and from danger of settlement, movement, disintegration, or collapse, whether such danger arises from the method or materials of its construction or from equipment installed therein, for the purpose of lighting, heating, the transmission or utilization of electric current, or from its location or otherwise.

Safe lighting valve - a manual gas valve control that when turned to the pilot position and held down, permits gas flow to the pilot assembly only but not the main burner(s). Once the pilot is safely established the knob is turned to the "on" position for normal operation.

Safe start check - a feature of a flame safeguard control to prevent a burner start if a flame is detected, or if a condition simulating a flame exists in the flame detection circuit.

Safety can - approved container of not more than five gallon capacity. It has a spring closing lid and spout cover. It is designed to relieve internal pressure safety when exposed to fire.

Safety cap - this is a cap of a cylinder that protects the valves and regulator.

Safety control - any switch that does not operate during normal operation of the system, but shuts the system down if some variable gets outside of acceptable limits.

Safety control circuit - a circuit involving one or more safety controls or controllers.

Safety control/controller - an automatic control or interlock (including relays, switches, and other auxiliary equipment used to form a safety control system) which is intended to prevent unsafe operation of the controlled equipment.

Safety device - a control that is added to detect a dangerous condition and stop an action or take a new action to remedy a problem.

Safety factor - the ratio of extra strength or capacity to the calculated requirements to ensure freedom from break down and ample capacity.

Safety ground - the wire (usually green or bare) that connects the equipment grounding conductor to the earth ground.

Safety head - a type of spring loaded compressor cylinder head that allows for lifting and expansion in case liquid enters the cylinder and the piston attempts to compress it.

Safety limit controller - see limit.

Safety motor control - an electrical device used to open a circuit if the temperature, pressure, and/or the current flow exceeds safe conditions.

Safety plug - a device which will release the contents of its container above normal pressure conditions and before rupture pressures are reached.

Safety shut down (lock out) - the action of shutting off all fuel and ignition energy to the burner by means of safety control or controls such that restart cannot be accomplished without manual reset. Also see lockout switch.

Safety shut down test - a series of tests for a given flame safeguard control to make certain that the control will cause safety shut down for all the conditions that it should. For all flame safeguard controls, safety shut down should occur if a flame is not established or if it goes out. Some flame safeguard controls should cause safety shut down for other conditions as well, such as opening of a pre-ignition interlock during pre-purge, opening of a lockout interlock anytime after a few seconds, or detection of a flame (or a condition simulating a flame) during pre-purge.

Safety shutoff valve (SSOV) - a valve that is automatically closed by the safety control system or by an emergency device to completely shut off the fuel supply to the burner. The valve may be opened automatically, or it may have to be manually opened. Also see fluid power; and first stage oil, main fuel, manual reset, pilot, second stage oil, and solenoid valves.

Safety switch - see lockout switch.

Safety thermostat - a thermostat that limits the temperature of an appliance.

Safety valve - a quick opening safety device used for the fast relief of excessive pressure in a pressure vessel. For a steam boiler it opens at a preset pressure to allow excessive steam to escape.

Safety valve blowdown - drop in pressure between popping pressure and reseating pressure (usually 2 psi to 8 psi below popping pressure).

Safety valve capacity - measured in pounds of steam per hour safety valves can discharge.

Sag - the distance deviation, expressed in inches per lineal foot of flexible duct, from a horizontal or inclined plane between suspension points on installed flexible duct.

Sail switch - an airflow switch with a metal or plastic sail which is inserted into the air stream. The switch is actuated when the velocity of the air striking the sail reaches a specified value.

Salient pole - that type of field pole which projects toward the armature.

Salt brine - a solution of water and some type of salt with a freezing point below 32°F.

Sampling - the process of obtaining a sequence of values of a signal. Sampling may be at regular or intermittent intervals.

Sampling assembly - a hand-held rubber squeeze pump used to force samples of flue products through various testing devices.

Sampling rate - the number of times that a particular data channel is sampled by a commutator in one second.

Sandblast - to blow sand at high velocity with compressed air against objects to clean them.

Sanitary - as applied to a building, means free from danger or hazard to the health of persons occupying or frequenting it or to that of the public, if such danger arises from the method or materials of its construction or from any equipment installed therein for the purpose of lighting, heating, ventilating, or plumbing.

Sanitizer - one of three groups of anti-microbials registered by EPA for public health uses. EPA considers an antimicrobial to be a sanitizer when it reduces but does not necessarily eliminate all the microorganisms on a treated surface. To be a registered sanitizer, the test results for a product must show a reduction of at least 99.9% in the number of each test microorganism over the parallel control.

Sargol joint - joint in which a lip is provided for welding to make the joint fluid tight. Mechanical strength is provided by bolted flanges.

Sarlun - an improvement of the Sargol joint.

Sash - the operating portion of a hung or horizontal sliding window. A single assembly of stiles (sides) and rails (top and bottom) made into a frame and holding glass.

Sash crack - sum of all perimeters of all ventilators, sashes, or doors based on overall dimensions of such parts expressed in feet (counting two adjacent lengths of perimeter as one).

Satellite compressor - the compressor on a parallel compressor system that is dedicated to the coldest evaporators.

Saturated air - air that contains the maximum water vapor it can hold at a given temperature. Air at 100% relative humidity.

Saturated conditions - term used to indicate the boiling or condensing point of a substance, which is dictated by a specific combination of temperature and pressure.

Saturated liquid - a fluid heated to a point where any additional heat will cause it to begin to evaporate.

Saturated pressure - the force in a pressure vessel that matches the temperature of a certain contained gas at a condition where any removed heat would cause condensation and added heat would cause evaporation.

Saturated steam - is steam that contains no moisture. It is saturated with heat, since additional heat will raise the temperature above the boiling point and the removal of heat will result in the formation of water.

Saturated vapor - vapor at the same pressure and temperature as the liquid it contacts.

Saturated temperature - the temperature at which a liquid turns to a vapor or a vapor turns to a liquid.

Saturation - the state of being completely filled or charged. A condition existing when a substance contains the maximum amount of another substance that it can hold at that particular pressure and temperature.

Saturation pressure - boiling point of a liquid.

Saturation temperature - the temperature at which boiling or condensation can take place.

Saybolt seconds universal (SSU) - a measurement of oil viscosity. The number of seconds it takes oil at a standard temperature (130° F for lighter oils and 210° F for heavier oils) to flow through a hole of a standard size and fill a 60 milliliter flask.

Scaffold - any elevated platform that is used for supporting workers, materials or both.

Scale - (1) a measurement band on a test instrument. (2) any film or oxidation deposit formed on a metal surface immersed in water. (3) when used in blueprint reading, it is a reference to the size of the drawing compared to the size of the object or building, which the drawing represents.

Scale drawing - a drawing usually considerably reduced in size from the actual or designed object, construction site or building, but which is draw to scale.

Scale forming salts - salts such as calcium carbonate and magnesium carbonate that wen in solution tend to form a hard, brittle scale on hot surfaces.

Scale trap - a pipe fitting that separates scale from water.

Scaling - the formation of lime and other deposits on the water side surfaces of heat exchangers.

Scanner - a collective term applied to all flame radiation detectors. These detectors monitors and proves pilot and main flames of the heating appliance.

Scavenger pump - a mechanism used to remove fluid from a sump or container.

Scfm - cubic feet per minute of standard air (.075 lb/ft3) at 70° F at sea level.

Schedule - a pipe size system (inside diameters and wall thicknesses) originated by the iron pipe industry whereby the larger the number the thicker the wall.

Schedule, drawing - a detailed tabulation of components, items or parts to be finished.

Schedule number - approximate value of the expression 100 x p/s, where "p" is the service pressure and "s" is the allowable stress in psi.

Schematic diagram - a pictorial representation of an electrical circuit showing components, lines and connections, in symbolic form and in relationship to each other.

Schottky barrier - a cell barrier established at the interface between a semiconductor, such as silicon, and a sheet of metal.

Schrader valve - valves that use a valve core, like a tire-valve stem, to gain access to a sealed system.

Scientific notation - the use of "powers of 10" to simply large and small numbers.

Score - to make a scratch mark or line to show a starting point or cutting line.

Scotch boiler - a horizontal, fire tube boiler consisting of a cylindrical steel shell with one or more cylindrical, internal combustion chambers located (generally) in the lower portion, and with a bank or banks (passes) of tubes attached to both end closures. The combustion chamber(s) and tubes are completely surrounded by water. A scotch boiler is described by a number of passes, and whether it is a dry back boiler (for stationary service) or a wetback boiler (for marine or stationary service). Also see dry back, wet back, fire tube and multi-pass boilers.

Scotch yoke - a mechanism used to change reciprocating motion into rotary motion or vice-versa. Used to connect the crankshaft to piston in a refrigeration compressor.

Scratch coat - the first coat of plaster, which is then scratched to provide the bonding surface for the second coat.

Screening multiple linear regression - a tool that partitions the forecast into high and low frequency temperature-precipitation elements. It uses the same data a CCA, but is applied to forecasts of single seasons rather than multi-season patterns, as done with CCA.

Screw compressor - a rotary compressor that employs two intermeshed, counter rotating screws for positive displacement.

Screw extractor - a tool used to remove broken bolts, screws, etc. from holes.

Screw pump - see screw compressor.

Screw threads - a helical ridge of uniform section formed inside of a hole (as in a nut) or on the outside of a fastener (as in a screw or bolt).

Screwdriver - are hand tools used for turning screws. The length of a screwdriver is measured from the tip of the blade to the beginning of the handle. Screwdrivers come on eight basic designs: standard, phillips, clutch head, allen, stubby, offset, tork, and starting.

Screwed end - pipe or fitting that is joined by threaded connections.

Screwed flange - flange that is attached to a pipe by a screwed connection.

Screws - threaded fasteners used to obtain a good hold and provide a means for repeatedly removing fastened parts.

Scroll type compressor - one that involves the mating of two spiral coils (scrolls) to form a series of crescent shaped pockets. During compression, one scroll remains stationary while the other scroll orbits around the fixed scroll.

Scrubber - a device that removes undesirable gaseous elements from the flue gas.

Scupper - an opening in a wall or parapet that allows water to drain away.

SE type cord - a 600 maximum volt cord that has a temperature range of -58° F to +221° F. It is a heavy duty, all temperature cord used for construction equipment motors, and outdoor lighting.

Seal, bellows - metal bellows used in a shaft seal, or in place of packing for valves. Also used in long pipelines instead of gaskets to compensate for expansion of the line with temperature changes.

Seal, shaft - see shaft seal.

Seal air - the air supplied to any device at a significantly higher pressure than the surrounding air for the specific purpose of excluding contaminations by the surrounding atmosphere.

Seal, compressor - a device to prevent the leakage of refrigerant gas at the point where the crankshaft must pass through the crankcase. Usually takes the form of two finely finished surfaces separated only by a thin film of oil and maintained under pressure by means of a spring.

Seal leak - the escape of oil/or refrigerant at the junction where the shaft exists the housing.

Seal of a trap - the vertical distance that gas must travel through the water in the trap before it can bubble up into the living space.

Sealed - tightly closed; not open to the atmosphere. Having a non-leaking union between two elements.

Sealed bellows sensor - a temperature sensor with a vapor, gas, or fluid which changes in pressure and volume as temperature changes.

Sealed combustion - a design for a combustion type heat source in which all air required for combustion is drawn through sealed venting material from the outside of the building. All exhaust gases are also routed through sealed venting material to the outside of the building.

Sealed unit - a hermetic welded motor compressor assembly that has the motor and compressor inside a sealed dome or housing. The pressures can be accessed by saddle or schrader valves.

Sealed VA - the VA after the controlled device has been operated or energized. The sealed VA of an electro-magnetic device will be different for all devices depending upon the design of the particular device. The sealed VA can be obtained from the manufacturer.

Sealtight - a brand of weatherproof flexible conduit.

Seam weld - longitudinal seams formed by welding. Heavy gauge metal (18 gauge or heavier) is more easily welded than lighter gauge metal because lighter gauge metals tend to burn through before it reaches proper welding temperatures.

Seamless - piercing and rolling of a solid billet or cupping a plate to form pipe.

Seasonal efficiency - the consummate efficiency of a heating system over a 12 month period considering flue losses, electrical consumption, warm up and cool down cycles, pilot losses and fuel consumption, when compared to the design loss of the building and the annual degree day value for the geographical area of the building.

Seasonal energy efficiency ratio (SEER) - total cooling output of an air conditioner during its normal annual usage period for cooling (in Btu) divided by the total electric energy input during the same period as determined by conditions.

Seasonal peak - the maximum demand placed on the utility's capacity resulting from seasonal factors. Some utilities have summer peaks, some winter peaks, some both.

Seasonal performance factor (SPF) - the measure of the efficiency of heating equipment over the length of the heating season. It is the ratio of the heat energy output compared to its electrical energy input over an entire heating season.

Seat - that portion of a valve mechanism against which the valve presses to effect shutoff.

Seat, front - that part of a refrigeration valve that forms a seal with the valve button when the valve is in the closed position.

Second Law of Thermodynamics - heat will flow only from the material at a higher temperature to a material at a lower temperature.

Second stage oil valve - in oil burners with 2 firing rate levels (stages), the automatic safety shutoff valve which opens to admit more fuel when the higher combustion rate is required. The second stage may be optional.

Secondary air, combustion - air that is introduced at the point of combustion or air which is not mixed with the fuel before it enters the combustion chamber.

Secondary air, discharge - the air surrounding an outlet that is captured or entrained by the initial outlet discharge air stream.

Secondary cell - a cell that can be charged by applying DC voltage from a battery charger.

Secondary circuit - an individual distribution piping circuit with its own circulator that receives heat from the building's primary circuit.

Secondary control - safety device used to shut off the system if certain unsafe conditions exist.

Secondary distribution system - a lower voltage, alternating current system which connects the secondaries of distribution transformers to the customers load devices.

Secondary fluid - an anti-freeze fluid cooled by a primary (phase change) refrigerant, which is then circulated by pumps throughout the refrigeration system to absorb heat.

Secondary refrigerating system - a refrigerating system in which the condenser is cooled by the evaporator of another or primary refrigerating system.

Secondary side - the side of a transformer connected to the loads it is operating. The secondary voltage could be higher or lower than the voltage being supplied to the primary side of the transformer

Secondary surface - a coil surface, such as a fin, that is not in direct contact with the refrigerant.

Secondary voltage - the output, or load supply voltage of a transformer or sub-station. Can be higher or lower than the applied (primary) voltage.

Secondary water - hot or chilled water distributed to terminal units.

Secondary winding - the coil of a transformer into which voltage in induced; energy is delivered to the load circuit by the secondary winding.

Section - a representation of an object as it would appear if cut by an imaginary plane, showing the internal structure.

Sectional boiler - see cast iron sectional boiler.

Sectional header boiler - see water tube boiler.

Sections - the cast iron assemblies that are joined together to form a sectional boiler block.

Sediment - particles of foreign matter present in boiler water.

Seebeck effect - when two different adjacent metals are heated, an electric current is generated.

Seebeck EMF - the open circuit voltage caused by the difference in temperature between the hot and cold junctions of a thermocouple circuit.

Seimic pipe support - pipe support hanger, restraint or snubber designed to withstand the piping load that may result from an earthquake.

Seize - when two parts rub together so that they get hot and force the lubricant out of the area they "gall" and finally "freeze" or stick together.

Selective absorber surface - surface used to increase the temperature of a solar collector.

Selective energy systems - form of cogeneration in which part, but not all, of the site's electrical needs are met solely with on-site generation, with additional electricity purchased from a utility as needed.

Selector relay - relays that are used where the application requires the comparison, selection and transmission of one of two proportional signals.

Self checking circuit - see dynamic ampli-check and dynamic self check.

Self closing - as applies to a fire door or other opening protective, means normally closed and equipped with an approved device which will insure closing after having been opened for use.

Self contained air conditioning unit - a refrigerating system having all working or major parts in one unit that can be moved without disconnecting any refrigerant lines. This would include the condenser and evaporator coils, blower assembly and condenser fan assembly, and a complete set of controls.

Self contained control - a control that has the source of power, sensing element and final control mechanism contained within a single instrument.

Self contained recovery - recovery or recycling equipment that is capable of removing the refrigerant from an appliance without the assistance of components contained in the appliance.

Self excitation - the process by which a device supplies its own excitation voltage rather than having it supplied from some external source. A self excited generator supplies its own excitation voltage.

Self extinguishing - the ability of a plastic to resist burning when the source of heat or flame that ignited it is removed.

Self inductance - The property whereby a voltage is produced in a circuit (and in particular in an induction coil) wherever current changes in the circuit.

Self tapping screws - sheet metal screws with auger type tips designed to start its own hole. These screws come in various shank sizes and lengths.

Semi-automatic burner - see semi-automatic system.

Semi-automatic system - a system which requires some manual operation; a system in which the burner is started and ignited manually; purged, sequenced, and modulated automatically; and stopped manually; with certain steps and conditions supervised by safety interlocks.

Semi-conductor - a material that has electrical properties of current flow, between a conductor and an insulator, and is used to manufacture solid state devices such as diodes and transistors.

Semi-conductor fuses - fuses used to protect solid state devices.

Semi-extended plenum - a trunk duct that is extended as a plenum from a fan or HVAC unit to serve multiple outlets and/or branch ducts.

Semi-hermetic compressor - a motor compressor design that combines the benefits of the hermetic and open type compressors. This type of compressor is keep hermetic by the use of gasketed openings and fittings and is field serviceable.

Semi-solid nozzle - a fuel oil nozzle type where most of the oil is concentrated on the outer edge of the spray pattern, while a lesser concentration is found within the spray area. Also known as a semi-hollow nozzle.

Sensible heat - the heat energy which changes the temperature of a substance without changing its state. This heat is measured with a thermometer. For example, raising the temperature of water to the boiling point or reducing the temperature of air without removing moisture.

Sensible heat change - the addition or removal of sensible heat in air without changing its moisture content.

Sensible heat factor - the ratio of the sensible heat cooling load to the total heat load.

Sensible heat loss - heat lost through the process of convection.

Sensible heat ratio - the ratio of the sensible (heating or cooling) load to the total heat load.

Sensing bulb - sensing element placed and located on the suction line at the outlet of the evaporator coil to control the operation of the TXV located at the inlet of the evaporator coil.

Sensing bulb thermostat - thermostat that uses a pressurized vessel to open or close electrical contacts.

Sensing device - a pilot device that detects a quantity and converts it into an electrical signal.

Sensing element - the first system element or group of elements. The sensing element performs the initial measurement operation.

Sensing tube - an electron tube that is sensitive to ultraviolet flame detectors. Also see aquadag tube and ultraviolet power tube.

Sensitivity - the ratio of the rate of response of a controller to each unit of change of the controlled device. For example, in a pneumatic thermostat, sensitivity is the PSI change in control air pressure for each degree of temperature change felt by the sensing element.

Sensor - an element used in a detector that perceives the presence of radiant energy and converts it into electrical energy.

Separator, oil - a device used to separate refrigerant oil from refrigerant gas and return that oil to the crankcase of the compressor.

Septic tank - a private sewage system where the wastes are directed to this concrete unit in which the wastes putrefy and the liquid flows through.

Sequence - see timer sequence.

Sequence controls - group of devices that act in series or in a time order.

Sequencer - a time delay type relay that uses a heater wire or element as the method to activate the contacts to turn strip heaters (heating elements) on in stages instead of all at once.

Sequential light off - in a multi-burner system, ignition of the pilots one at a time, following each other in a predetermined order. Each pilot is in a separate valve zone (see valve zone). Each pilot must be proved before the next one can be ignited. After all pilots have been proved, all main burners are ignited in unison. (The main burners are all in a single valve zone.)

Serial number - a unique identifying number assigned to a machine or part.

Series 90 - Honeywell nomenclature designing a group of controllers and motors which are characterized by low voltage, 3 wire, modulating operation.

Series circuit - a circuit in which the components are arranged end-to-end having only one continuous path and the voltage is additive, but not the current. The same current flows through all components in the circuit and this arrangement is considered to be a voltage divider circuit.

Series flow - a flow path in which only one path exists for fluid to flow.

Series loop system - a forced hot water heating system with the terminal units connected so that all the water flowing through the circuit passes through each series connected unit in the circuit.

Series motor - a motor in which the armature and field windings are connected in series with each other across the applied voltage. Also called series wound motor. Shunt motors are characterized by high starting torque and poor speed regulation.

Series/parallel circuit - a combination of series (single path) and parallel (multiple path) circuits.

Series resistance - any sum of resistances installed in sequential order within one circuit.

Series pumps - two or more pumps installed in the same piping circuit, and having the same flow direction.

Series wound motor - a commutator motor in which the field circuit and armature circuit are connected in series.

Serious hazard - means a hazard of considerable consequence to safety or health through the design, location, construction, or equipment of a building, or the condition thereof, which hazard has been established through experience to be of certain or probable consequence, or which can be determined to be, or which is obviously such a hazard.

Serpentining - the arrangement of tubes in a coil to provide circuits of the desired length. Intended to keep pressure drop and velocity of the substance passing through the tubes within the desired limits.

Serrated - having sawlike teeth or notches along the cutting edge.

Servel system - one type of continuous operation absorption refrigerating system.

Service - all the conductors and equipment for supplying electrical energy to the dwelling.

Service compartment - a normally enclosed compartment of an assembly to which ready access must be gained by the operator for occasional operations such as "lighting up;" or the adjustment, cleaning or servicing of such parts of the equipment as air filters, blowers, motors, and controllers.

Service conductors - the portion of the electrical service system that is the conductors that carry electrical energy from the utility transformers to the entrance pipe or conduit on the building.

Service connection - a point at which fuel, atomizing medium, or power is connected to the boiler, firing equipment, or controlled devices.

Service drop - connecting wires from the power lines to the point of entry to the building.

Service entrance - the associated cables, conduit, boxes and meters used to bring electric power from the main power line to the building.

Service equipment - necessary equipment usually consisting of circuit breakers, switches, fuses, and their accessories, located near the point of entrance of the supply conductors of the building. These wires are intended to constitute the main control and means of cut-off for supplying the power to the building.

Service factor, motor - a measurement that states the percent beyond the rated load or horsepower which a motor can carry without overheating. A 1.15 Service factor means the motor can deliver 15% more than the rated horsepower without injurious over-heating. A 1.0 Service factor motor should not be overloaded beyond its rated horsepower. The ambient temperature of the environment must be within the defined limits.

Service factor, pipe - a factor which is used to reduce a strength value to obtain an engineering design stress. The factor may vary depending on the service conditions, the hazard, the length of service desired, and the properties of the pipe.

Service lateral - underground service conductors extending from the street main to the service entrance conductors.

Service maintenance - workplace activities such as constructing, installing, setting up, adjusting, inspecting and modifying machines or equipment. These activities include lubrication, cleaning or unjamming and making adjustments or tool changes, where the employee may be exposed to the expected release of energy or the energizing or startup of equipment.

Service manifold - an instrument equipped with suction and discharge/liquid pressure gauges and with hoses for connection to the suction and discharge/liquid service valves and to a refrigerant cylinder, recovery machine or vacuum pump.

Service panel - from the meter the electrical service continues to the service panel. In this box, the electric is divided down and distributed to the various circuits that will distribute it through out the application.

Service raceway - a rigid metal conduit, electrical metallic tubing, or other raceway, that encloses service entrance conductors.

Service valve - a device to be attached to the system which provides an opening for the gauges and/or charging lines. Also provides means of shutting off or opening gauge and charging ports, and controlling refrigerant flow in the system.

Serviceability - an installation is classified as serviceable when sufficient area is provided for access to control panels, electrical connections, and refrigerant lines. The better the access, the higher the serviceability of the system.

Serviceable hermetic - a hermetic unit housing the compressor and the motor that can be disassembled by use of bolts and threads.

Servo - a servo motor supplies power to a servo mechanism. A servo mechanism is a lower power device (electrical, hydraulic, or pneumatic) used to put in operation and control a more complex or powerful mechanism.

Servo mechanism - a closed-loop system which initiates an input signal with deviation from a desired condition; the signal is fed back into the control system until a continued response eliminates the signal.

Set screw - fasteners used primarily to hold pulleys and fan wheels to shafting material.

Setback - reducing the temperature setting of a setpoint control, such as a room thermostat, to conserve energy.

Setpoint - a predetermined value to which a controller or interlock is adjusted and at which it performs its intended function.

Setpoint controller - a control that attempts to maintain a preset temperature at some location in a system by controlling the operation of other devices.

Setpoint demand - a demand for boiler operation from a load that needs to be supplied with water at a specific set temperature.

Setpoint temperature - a desired temperature that is to be maintained if possible, and is set on some type of controller such as a thermostat or setpoint controller.

Severe duty - a totally enclosed motor with extra protection (shaft slinger, gasketed terminal box) to resist entry of contaminants. Used in extra duty, damp or other non-hazardous contaminated environments.

SEW-A type cord - a 600 maximum volt cord that has a temperature range of -58° F to +221° F. It is a heavy duty, all temperature cord that is UL listed for outdoor use and used for construction equipment motors, and outdoor lighting.

Shaded - glazed area which is externally protected from direct solar radiation by use of devices permanently affixed to the structure or by an adjacent building, topographical feature, or vegetation.

Shaded pole - an extra pole on a stator winding that is wrapped with a copper ring, causing it to be out of phase with the stator winding.

Shaded pole motor - a small special type of single phase alternating current induction motor that utilizes a copper shading coil to provide the necessary displacement for starting. This motor has low starting torque and is used for light starting loads. These motors have no brushes or commutators and are rated in watts as opposed to horsepower because of their small size.

Shading coefficient - the ratio of solar heat gain occurring through non-opaque portions of the glazing, with or without integral shading devices, to the solar gain occurring through an equivalent area of unshaded, 1/8 inch thick, clear, double strength glass.

Shading coil - a large copper wire, loop, or band connected around part of a magnetic pole piece to oppose a change of magnetic flux. The loop produces a tiny magnetic field that is out of phase with the main poles, which aides in the starting the motor.

Shaft - a long, thin, usually cylindrically shaped part.

Shaft, building - an enclosed space extending through one or more stories of a building, connecting vertical openings in successive floors, or floors and the roof. May also be referred to as chases.

Shaft efficiency - prime movers shaft energy output divided by its energy input in equivalent, consistent units. For a steam turbine, input can be the thermal value of the steam or the fuel value to produce the steam. For a fuel fired mover, it is the fuel input.

Shaft end - end of the motor where the rotor shaft exists.

Shaft seal - a metal, ceramic or synthetic device used to prevent leakage between the crankshaft and the housing.

Shall - term used to indicate provisions that are mandatory if compliance with the standard is claimed. In reference to the National Electric Code (NFPA 70) indicates a mandatory requirement.

Shank - the shaft of a tool, such as a screwdriver, drill bit, or nut driver.

Sharp freezing - refrigeration at temperature slightly below freezing, with moderate air movement.

Shear strength - this is the ability of a joint to withstand a parallel or right angle force.

Shear stress - frictional force overcome in sliding one "layer" of fluid along another, as in any fluid flow. The shear stress of petroleum oil or other Newtonian fluid at a given temperature varies directly with shear rate (velocity). The ratio between shear stress and shear rate is constant; this ratio is termed viscosity of a Newtonian fluid, the greater the shear stress as a function of rate of shear. In a non-Newtonian fluid -- such as a grease or polymer containing oil (e.g. multi-grade oil) -- shear stress is not proportional to the rate of shear. A non-Newtonian fluid may be said to have an apparent viscosity, a viscosity that holds only for the shear rate (and temperature) at which the viscosity is determined.

Sheath - the material, usually an extruded plastic or elastomer, applied outermost to a wire or cable. Very often referred to as a jacket.

Sheathing - the covering (usually wood boards, plywood, or wall boards) placed over exterior studding or rafters of a building providing a base for the wall or roof cladding.

Sheave - a wheel or pulley with a grooved rim, used with a belt to transfer rotary motion.

Shed - to de-energize a load in order to maintain a kW demand set point.

Sheet metal - galvanized steel that refers to metal that is formed in sheets 1/8th of an inch or less.

Shell - the outside casing of a compressor.

Shell, boiler - metal welded to the tubesheets of a boiler to form its basic structure.

Shell and coil heat exchanger - heat transfer surface that has one fluid, usually a liquid, flowing through a coil of tubing or piping that is located within a surrounding shell. Another fluid, usually a vapor, flows within the shell, facilitating the heat transfer between the two fluids. This type of heat exchanger must be cleaned chemically. A designation for heat exchangers, condensers, and chillers consisting of a tube bundle within a shell of casing.

Shell and tube flooded evaporator - a device that flows water through tubes built into a cylindrical evaporator or vice-versa.

Shell and tube heat exchanger - a heat exchanger design in which one fluid passes through a group of several parallel tubes while the other fluid passes through the spaces between these tubes.

Shell type condenser - a cylinder or receiver that contains condensing water coils or tubes.

Shell, water tower - the chimney-like structure, usually hyperbolic in cross-section, utilized to induce air flow through a natural draft tower. Sometimes referred to as a stack.

Shielded cable - a special cable used with equipment that generates a low voltage output. Used to minimize the effects of frequency "noise" on the output signal.

Shim - a thin piece of metal used for filling space, leveling, etc. Materials used for shims.

Shipping bolts - packaging bolts that stabilize heavy equipment in transit. They should be removed before operation of the equipment.

Shoe - metal piece attached (usually welded) to the underside of a pipe and resting on supporting steel. Used to reduce wear from sliding of lines subject to movement. The adding of a shoe permits insulation to be added to the pipe, and can provide elevation from the support to allow for a slight slope to the pipe.

Shop drawings - drawings and other data prepared by the contractor, or any subcontractor, manufacturer, supplier or distributor that illustrates how specific portions of the work shall be fabricated and/or installed.

Shop fabrication - fabrication completed in a shop as opposed to field construction. Shop fabrication has many advantages over field construction, offering a wider variety of production methods in a controlled environment.

Shop prime - the highest grade of the kind of sheet metal specified.

Shop prime coat - a prime coat of paint applied in the shop to protect the fabricated item during the construction period.

Short - an accidental connection between conductors of different phases or opposite polarity. Short circuits provide a low resistance path and therefore allow a very large current to flow. A short circuit can be highly destructive if the flow of current is not interrupted by a fuse or a circuit breaker. Also, to intentionally bypass part of a circuit with a jumper.

Short circuit - an abnormal connection of relatively low resistance between two points of a circuit, resulting in a flow of excess (often damaging) current between these two different polarities or phases. This condition can create a relatively large flow of current between these points, possibly causing damage. Also called a short.

Short circuit air - situation that occurs when the supply air flows to exhaust registers before entering the breathing zone. To avoid short circuiting, the supply air must be delivered at a temperature and velocity that results in mixing throughout the space.

Short circuit rating - the maximum short-circuit current an electrical component can sustain without the occurrence of excessive damage when protected with an over current protective device.

Short circuiter, automatic - a device designed to automatically short circuit the commutator bars in some forms of single phase commutator motors.

Short cycling - a system that starts and stops more frequently than it should.

Short cycling, boiler - an undesirable operating condition in which the duration of the boiler firing cycles is considered too short for acceptable efficiency or equipment life. Short cycling often occurs when a boiler is excessively oversized or connected to a highly zoned distribution system.

Short cycling timer - a cam operated motor driven, on delay timer that allows time to operate and reset short cycled contacts to their original position.

Short term exposure limit - the concentration to which workers can be exposed continuously for a short period of time without suffering adverse effects.

Short to ground - term used to describe a no resistance electrical path between a hot leg or terminal to a ground terminal or ground.

Shorted capacitor - a situation where the two aluminum plates are touching each other.

Shorted motor winding - part of an electric motor winding is shorted out because one part of the winding touches another part, where the insulation is worn or in some way defective.

Shorted to ground - a situation where the windings of a motor are touching the case. The ohmic value in this situation is zero.

Should - term used to indicate provisions that are not mandatory but which are desirable as good practice.

Shoulder nipple - a short nipple that includes a very short section of unthreaded pipe.

Shroud - a housing over the condenser or evaporator.

Shunt - any parallel connection in a circuit.

Shunt (meter) - a resistor connected in parallel with a meter movement to extend the range of current the meter will measure.

Shunt wound motor - a direct current motor in which the field circuit and armature circuit are connected in parallel.

Shut down - the total process of terminating the operation of a combustion system. Also see safety and unison shut down.

Shutoff valve - see manual shutoff and safety shutoff valves.

Shutter - a movable screen or cover that opens and closes an aperture. Also see air shutter.

Shutter control - a damper assembly and pressure operated actuator used on systems with non-unloading type hermetic compressors in an air cooled condensing unit to maintain minimum required condensing pressure during low ambient operation.

SI unit - (Le Systeme International d'Unites) metric system of measurement adopted by most technical industries throughout the world.

Side cutters - a plier like, sharp edged wire cutter.

Side lap - where flat sheets overlap or where corrugations occur in sheet metal along the side of a panel, they may be overlapped, forming a loose joint. The joint may be left free or filled with mastic materials, depending upon applications.

Side step fixed ladder - a fixed ladder that requires a person to get off at the top to step to the side of the ladder side rails to reach the landing.

Siemen - a unit of conductance equal to the inverse ohms. Siemens = 1/ohms.

Sight glass, liquid - a glass tube or window in a refrigerating system which shows the amount of refrigerant or presence of gas bubbles in the liquid line.

Sight glass, oil - a glass "bull's eye" located in the compressor crankcase permitting visual inspection of the compressor oil level.

Sight glass, steam boiler - a glass tube located on a steam boiler application to show the level of the water line externally that corresponds to the water level internal to the steam boiler.

Sighting pipe - a pipe, mounted to the front plate of a combustion chamber, through which an optical flame detector sights the burner flame.

Signal - an electrical current used to convey information either digital, analog, audio, or video.

Signal light - a light that is used to show when some electric component or circuit is energized by illuminating the light.

Silica gel - a chemical compound used as a drier, which has the ability to absorb moisture when heated, moisture is released and compound may be reused.

Silicon - a dark gray chemical element, atomic number 14 and is semi-metallic in nature. A common constituent of sand and quartz (as the oxide). Crystallizes in face centered cubic lattice like diamonds. An excellent semiconductor material.

Silicones - resins used in insulation systems, having silicone in their chemical make-up, resulting in high heat resistance and water repellency.

Silicon controlled rectifier (SCR) - an electronic semi-conductor that contains silicon. The gate of the SCR must be triggered before the device will conduct current.

Sill - the main horizontal member forming the bottom of the frame of a window or door. It is generally slanted down to the outside to shed standing water.

Silver brazing - a brazing process in which brazing alloy contains some silver as part of joining the alloy. Process is considered to be above 840° F.

Silver solder - a solder that has silver added for strength.

Simulation - a resemblance to something else, or an imitation of.

Sine wave AC current - a wave form of a single frequency alternating current. The graphical representation of all points traced by the sine of an angle as the angle is rotated through 360 degrees.

Sine wave voltage - a voltage waveform whose value at any point is proportional to the trigonometric sine of the angle of the generator producing it. A generator rotated to the 45° position would have a trigonometric sine value of 0.707 times an assumed voltage value of 100 volts would produce an output voltage of 70.7 volts.

Single circuit system - a hydronic distribution system that uses diverter tees to connect each heat emitter branch circuit to the main piping circuit.

Single cleat ladder - a ladder consisting of a single pair of side rails connected together by cleats, rungs, or steps.

Single cut file - a single cut file that has one set of parallel lines or cuts running across the face. The single cut file is used for finish surfaces.

Single duct system - an air system that uses one duct to distribute the air to each zone.

Single family house - building containing one or two living units or a townhouse, including a manufactured house (modular) but not a manufactured house (mobile home).

Single glazing - use of single panes of glass in a window or door. These units are not as efficient as double glazing.

Single pass - referring to the recovery/recycle system that removes the refrigerant from the system being repaired and passes it through a system of filter-driers and perhaps a distillation process, and then into the recovery cylinder.

Single phase - only one alternating voltage or current produced or used. The electrical power supplied to equipment or small motors, normally under 7 ½ hp.

Single phase motor - an electric motor which operates on single phase alternating current.

Single phasing - that condition which occurs when one phase of a 3 phase system opens, either in a low voltage (secondary) or high voltage (primary) distribution system. Primary or secondary single phasing can be caused by any number of events. This condition results in unbalanced loads in polyphase motors and unless protective measures are taken, causes overheating and failure.

Single phasing - a condition that occurs when one of three phases is lost.

Single pole - one set of electrical contacts that will open or close (or make) when the switch is moved from one position to the other.

Single pole breaker - circuit breaker used to disconnect the black (hot) wire on a 120 volt single phase branch circuit.

Single pole double throw switch (SPDT) - an electric switch with one pole blade and operated two contact points. The first being normally open, the second being normally closed.

Single pole single throw switch (SPST) - an electric switch with one pole blade and one contact point that opens and closes.

Single port burner - a burner where the entire air-gas mixture emits from a single port.

Single seated valve - a valve with only one seat and one disc and generally suitable for tight shutoff. Since there is nothing in a single seated valve to balance the force exerted by the fluid pressure against the disc, the single seated valve requires more force to close than a double seated valve of the same size.

Single stage air compressor - a system where air is compressed once and forced into the hose or tank, could be a single or multi-cylinder unit. Normally used for air pressures below 150 psi.

Single stage compressor - compressor having only one compressive step between inlet and outlet.

Single stage oil pump - pump used on a single pipe gravity feed system, where the supply of oil is located higher than the oil pump burner assembly.

Single thickness flare - the part of the tubing that forms the flare is the thickness of the tubing.

Single throw - a switch having only one "on" or "off" position.

Single zone system - a system that provides heating and cooling to a single space in which desired conditions are controlled throughout by a single temperature control device.

Sintered oil bearing - a porous bearing metal, usually bronze, and which has oil in pores of bearing metal.

Sinuous header - these are found on watertube boilers. Tubes are expanded, rolled, and beaded into front and rear headers.

Sinusoidal - current that varies in proportion to the sine of an angle or time function; e.g., ordinary alternating current. Stable wave form of an undistorted alternating current or voltage.

Siphon - a pigtail shaped protective device used between the steam and bourdon tube in a steam pressure gauge.

Siphoning - the flow of liquid, through a closed tube, from a higher to a lower level.

Site plan - a layout of the building to be conditioned and its location on the ground, as well as all buried utility lines, cables, pipes, and/or conduits.

SJ type cord - a 300 maximum volt cord that has a temperature range of -40° F to +140° F. It is a medium duty, general purpose cord, used for power tools, floor polishers, sanders, and portable lighting.

SJE type cord - a 300 maximum volt cord that has a temperature range of -58° F to +221° F. It is a medium duty, all temperature cord, used for blowers, motors, dryers, and outdoor lighting.

SJEW type cord - a 300 maximum volt cord that has a temperature range of -58° F to +221° F. It is a medium duty, all temperature cord, that is UL listed for outdoor use and used for blowers, motors, dryers, and outdoor lighting.

SJO type cord - a 300 maximum volt cord that has a temperature range of -40° F to +194° F. It is a medium duty cord having a jacket that resists swelling and decomposing when exposed to oil and grease. Used for motors, power tools, and portable lighting.

SJOW-A type cord - a 300 maximum volt cord that has a temperature range of -40° F to +194° F. It is the same as SJO cord but it is UL listed for outdoor use and used for motors, power tools, and portable lighting.

SJOOW-A type cord - a 300 maximum volt cord that has a temperature range of -58° F to +221° F. It is a medium duty cord having a jacket that resists swelling and decomposing when exposed to oils, acids and chemicals. It is also abrasion and flame resistant. Used for motors, power tools, lighting, blowers, and industrial equipment.

SJT type cord - a 300 maximum volt cord that has a temperature range of 0° F to +140° F. It is a medium duty, general purpose cord used for motors, power tools, lawn and garden equipment, and appliances.

SJTW-A type cord - a 300 maximum volt cord that has a temperature range of 0° F to +140 F. It is a medium duty, general purpose cord that is UL listed for outdoor use and used for motors, power tools, and lawn and garden equipment.

SJTO type cord - a 300 maximum volt cord that has a temperature range of 0° F to +140° F. It is a medium duty cord having a jacket that resists swelling and decomposing when exposed to oil and grease. Used for motors, power tools, and lighting.

Skeleton wall - a nonbearing wall supported by each story on a skeleton frame. Also known as a panel wall.

Skelp - plate prepared by forming and bending and welding into a pipe.

Skin condenser - condenser using the outer surfaces of the cabinet as the heat radiating medium.

Skirt - (1) portion of a vertical vessel that extends from the vessel to the base and foundation. (2) the bottom part of a piston.

Skirt vents - holes provided at the top of the skirt near the vessel bottom for ventilation.

Skirted guide - see characterized guide and valve skirt.

Skylight - a special window set into a roof to provide natural light.

Slab - the upper part of a reinforced concrete floor, which is carried on beams below. A platform on which the outdoor condensing unit sits.

Slab below grade - any portion of a slab floor in contact with the ground which is more than 24 inches below the final elevation of the nearest exterior gade.

Slab mounted - an installation where the unit sets on a level flat slab located high enough above the surface of the earth (grade level) to eliminate snow problems.

Slab on grade, exterior - any portion of a slab floor in contact with the ground which is less than or equal to 24 inches below the final elevation of the nearest exterior grade.

Slab on grade system - a type of radiant floor panel in which tubing is cast into a concrete slab poured on a stable soil or foundation.

Sledge hammer - a hammer used for breaking and pounding. A sledge hammer usually has a longer handle.

Sleeper - small low elevation concrete pier usually rectangular in shape and 12 to 24 inches in height used to support piping close to grade, especially large diameter pipe.

Sleepers - wooden strips, usually arranged in parallel, and secured to a subfloor.

Sleeve - a tube or tube like part that fits around another part.

Sleeve bearing - a bearing made of a sleeve bushing, not a ball or a roller bearing. Preferred where low noise level is important, as on fan and blower motors. Unless otherwise stated, sleeve bearing motors listed can be mounted in any position, including shaft-up, or shaft-down (all position mounting).

Slide type damper - a damper consisting of a single blade that moves normal (at right angle) or near normal to the flow.

Slide valve - term used to describe a valve that has an internal slide mechanism that shifts from one side to the other to change the position of the valva. A reversing valve is an example of this type of device.

Sliding armature - an armature that mounts between two slots in a contactor frame and moves up and down the slots when the contactor is energized.

Sling psychrometer - a humidity measuring device with wet and dry bulb thermometers. It is rapidly moved through the air when measuring humidity.

Slinger ring - a ring fastened around the circumference of an axial fan. The slinger ring picks up cool condensate and flings it against the face of the condenser increasing subcooling.

Slinky loop - a flattened, circular coil of plastic pipe resembling a slinky, which is used as a ground loop on geothermal heat pump systems.

Slip - the difference between the speed of the rotating magnetic field of an induction motor and the speed of the rotor under full load.

Slip joint pliers - a hand tool where the slip joint can be locked into two different positions for different applications. There are many types of slip joint pliers. Each kind has a different grip size and handle size.

Slip rings - metal rings connected to the rotating armature windings in a generator. Brushes sliding on these rings provide connections for the external circuit. Always used in pairs.

Slips and drives - components of a duct system that are used to fasten sheet metal duct sections together. These components are used for the vertical connections and S-locks are used for the horizontal connections.

Slope coil - an evaporator that sits diagonally in a plenum.

Slot - a long narrow air outlet usually comprised of deflecting members generally located in the ceiling, high sidewall or perimeter sill and designed to distribute air in varying directions and planes. Aspect ratios of length/height are generally greater than 10/1.

Slot insulation - the insulation lining the slot containing the motor windings. Also called slot liner.

Slow blow fuse - a fuse with a built in trip delay commonly used in HVAC installations due to the power surge on start up.

Slow closing solenoid valve - solenoid valve that moves from the open to the closed position slowly to prevent the condition of water hammering from happening.

Slow opening valve - valve that requires five full turns of its handwheel to be fully open or closed.

Sludge - oil or mineral waste deposit in a crankcase or boiler.

Slug - detached mass of liquid or oil that causes an impact or hammer in a circulating system.

Slugging - condition in which a mass of liquid enters compressor causing hammering. Low or no superheat measurement.

Slurry - flow medium which has solid material suspended in a liquid.

Small appliance - any of the following products that are fully manufactured, charged, and hermetically sealed in a factory with five pounds or less of refrigerant. This would include refrigerators and freezers designed for home use, room air conditioners (including window air conditioners and packaged terminal air conditioners), packaged terminal heat pumps, dehumidifiers, under-the-counter ice makers, vending machines, and drinking water coolers.

Smoke - an air suspension (aerosol) of particles, usually but not necessarily solid, often originating in a solid nucleus, formed from combustion or sublimation. Also defined as carbon or soot particles less than 0.1 micron in size which result from the incomplete combustion of carbonaceous materials such as coal, oil, tar, and tobacco.

Smoke alarm - a single or multiple station alarm that reacts in response to smoke.

Smoke damper - a listed device that is designed to resist the passage of air and smoke. The device is arranged to operate automatically, controlled by a smoke detection system, and when required, is capable of being positioned manually from a remote command station.

Smoke density - varies from clear to dark. Determined by the amount of light that passes through the smoke as it leaves the boiler.

Smoke detector - a detective device which monitors smoke density, and actuates an alarm system if the opacity exceeds a preset level. Also see stack smoke detector.

Smoke developed rating - a numerical value assigned to materials that are tested in accordance with ASTM E-84 to determine the flame spread and smoke developed ratings of the material. By limiting these parameter's the material's contribution to smoke and spread of flame is lessened. The smoke developed rating is a factor determined by a comparison of the smoke developed ratings of untreated red oak wood.

Smoke pipe - see chimney connector.

Smoke stick - a diagnostic tool used to observe air flow. Usually it consists of a chemical in a squeezable container. When squeezed it emits smoke which visibly follows air flow currents.

Smoke test - a test to determine the tightness of a chimney and for detecting any cracks in a masonry chimney flue or deterioration or breaks in the integrity of a factory built or metal chimney flue system. The procedure involves lighting a smoke bomb or building a smoky fire in a fireplace or solid fuel burning appliance, covering the chimney termination, and checking for smoke escapement through the chimney walls.

Smooth pipe - a generic term for tubing with relatively smooth inner walls. Examples include drawn copper tube, polybutylene tubing, PEX and PVC.

Snake - a flexible wire used to push or pull wires through a conduit, a partition, or other inaccessible place. Also see fish tape.

Snap action - term used to describe a control, valve, or switch that is either fully open or fully closed. Used to describe valves that do not modulate. For switches the four means of snap action is achieved by bimetallic, mercury, over center or magnetic. Also known as detent action.

Snap lock - a corner locking longitudinal seam where a right angle flange containing notches at 2 inch intervals is inserted into a performed pocket with a 180 degree open hem. When fully inserted, the right angle flange snaps behind the open hem, forming a tight fitting longitudinal seam.

Snap ring pliers - pliers used to either take off or install snap rings by either expanding or decreasing the size of the ring.

Snap switch - one with contacts that make and break quickly through mechanical linkage.

Sniffer - a common term used to describe some refrigerant leak detectors. These instruments use a geiger counter approach for leak detection.

Snubbed - a filter circuit used to dampen out transient voltages in motor control and other types of circuits.

SO type cord - a 600 maximum volt cord that has a temperature range of -40° F to +194° F. It is a heavy duty cord having a jacket that resists swelling and decomposing when exposed to oil and grease. Used for tools, cranes, hoists, and industrial equipment.

Soap bubble test - a method of detecting leaks by coating with a soap solution. Any major leak will cause bubbles to form.

Soapstone - a type of rock cut into narrow strips used as a marking device; will not break up as easily as chalk.

Socket - a device for electrical connection of a plug, bulb, etc.

Socket fitting - a fitting used to join pipe in which the pipe is inserted into the fitting. A fillet weld is then made around the edge of the fitting and the outside wall of the pipe.

Socket wrench - a wrench used to remove nuts that do not have any obstructions over them.

Sodium - a silver-white, alkaline, metallic element found in nature only in combined form. Symbol is Na.

Soffit - underside of a stair, arch, or cornice.

Soft copper - a bendable form of tubing that has been annealed. Soft copper tubing is available in three thicknesses -- Type K, L, and M.

Soft flame - a flame partially deprived of primary air such that the combustion zone is extended and inner cone is ill defined.

Soft hammer - this is a soft headed hammer made of different sorts of material, such as rawhide, rubber, wood, plastic, or lead. It is used to drive parts into place, or to separate them without damage to the surface.

Soft soldering - see soldering.

Soft start kit - a kit consisting of a "PTC" positive temperature thermistor used to provide added torque for the motor on start up. Wired in parallel with the run capacitor.

Soft tubing - copper tubing that is sold in coils of varying lengths and can be bent with relative ease.

Soil gases - gases that enter a building from the surrounding ground. Examples would be radon, volatile organics and pesticides.

Soil moisture tool - a newly implemented forecasting tool that provides an indication of surface climate on the basis of soil moisture conditions and the temperature deviation during the previous month.

Soil pipe - any pipe that carries human waste.

Soil stack - a vertical drain pipe or stack that receives discharge from water closets (toilets) or urinals.

Soil swing curve - the effect of air and solar changes on the soil temperatures from the surface to a depth of 12 feet.

Soil temperature variation - an assumed figure that is a balance between the allowable earth temperature and the depth of the pipe loop location.

Solar cell - also known as a photovoltaic cell. Device that converts solar radiation directly into electricity.

Solar collector - device used to trap solar radiation, usually having an insulated black surface.

Solar constant - the solar intensity incident on a normal surface located outside the earth's atmosphere at a distance from the sun equal to the mean distance between the earth and the sun. At sea level in July the solar intensity value is about 300 Btu per (sq ft) (hr) since about 28% is absorbed in the earth's atmosphere.

Solar energy - the energy received by the earth from the sun.

Solar energy source - source of natural day lighting and of thermal, chemical or electrical energy derived directly from conversion of incident solar radiation.

Solar energy systems - systems used to collect, convert, and distribute solar energy in forms useful within a business or residence. A passive system uses no additional energy from other sources for the distribution of the solar generated heat. An active system may use blowers, supplementary coils, etc.

Solar gain - the absorption of heat from the sun. The amount of solar radiation (measured in Btu's) received by a surface.

Solar heat - heat from visible and invisible energy waves from the sun.

Solar heat gain coefficient (SHGC) - the ratio of the solar heat gain entering the space through the glazing product to the incident solar radiation. Solar heat gain includes directly transmitted solar heat and absorbed solar radiation which is then reradiated, conducted or convected into the house.

Solar index - a number from 1 to 100 indicating the percent of household water that could have been supplied that day by a typical solar hot-water system.

Solar payback - this is the point at which the investment for a solar system has saved enough money in fuel bills to pay for itself.

Solar radiant heat - solar heated water or an anti-freeze solution is piped through heating coils imbedded in concrete in the floor or in plaster in ceilings or walls.

Solar radiation - the amount of sun's heat that must be considered in the heat gain calculations. Its primary effect is through windows.

Solar shield - the shielding or blockage of the sun's heat.

Solar transmittance - the percentage of total solar energy that glazing transmits through a window. A standard clear dualpane has a solar transmittance of 71%.

Solder - a joining process when two or more materials are connected using a filler metal whose liquidus temperature is below the melting point of the base metal.. Common solders are a metal alloy of tin and lead used to join two metal surfaces together, or to patch metal parts or surfaces.

Solder pot - a device using a low-melting solder and an overload heater sized for the amperage of the motor it is protecting. The solder will melt, opening the circuit when there is an excessive overload.

Soldering - the process of joining two metals by adhesion of a low melting temperature metal (less than 840° F).

Soldering gun - a soldering gun is used to solder electrical connections. It comes in fixed and adjustable wattages.

Sole plate - the bottom horizontal piece in a frame wall, usually single or double 2x4's. The wall is nailed into the deck or rough floor through the sole plate.

Solenoid - a coil of wire which, when current flows through it, will act as a magnet and will pull a movable iron core, or armature, to a central position.

Solenoid coil - component of a heat pump system which, if energized, will change the position of the plunger of the pilot valve. Also known as the RVS.

Solenoid valve - an automatic safety shutoff valve that is opened or closed by the action of a solenoid upon an armature attached to the valve disc, resulting in fast opening and closing times (less than one second). It is widely used for low volume duty, such as a pilot valve, or a main fuel valve in a small to medium size on off system. Also called a magnetic valve. Also see armature and solenoid.

Solid - the non-fluid form of a substance in which the motion of the molecules is restricted, giving it a definite shape and volume; form that is neither gas nor liquid.

Solid absorbent refrigeration - a refrigeration system which uses solid substance as an absorber of the refrigerant during the cooling part of the cycle and releases this refrigerant when heated during the generating part of the cycle.

Solid fuel heating - a system that uses solid natural resources such as wood or coal to provide heat.

Solid state - the electronic properties of crystalline materials (usually semiconductor in type). The interaction of light, heat, magnetic fields, and electric currents in these crystalline materials are involved in solid state devices. Less power is required to operate solid state devices and a greater variety of effects can be obtained.

Solid state device - a group of electronic components that perform jobs similar to vacuum tubes, but unlike tubes they are not electrodes enclosed in glass but rather specifically prepared solid matter.

Solid state hard starting kit - used to increase the starting torque of a permanent split capacitor motor. This kit contains a solid state relay and a AC electrolytic starting capacitor.

Solid state relay - a device or an electronic component constructed from semiconductor material, used to control either AC or DC loads. Advantages are that it has no moving parts, is resistant to vibration, and is sealed against dirt and moisture.

Solid state starting relay - actually an electronic component known as a thermistor, intended to replace the current type starting relay.

Solidify - to make or become firm, hard, or solid. To crystallize.

Solidus - the temperature at which a metal material begins to melt. Below this point the brazing or soldering material is solid.

Solubility - the ability of a substance to be dissolved by, or in, another substance.

Soluble - a substance that can be dissolved in a given liquid.

Solute - a component (originally solid, liquid, or gas) of a solution, usually present in lesser amount than the solvent.

Solution - a liquid that has another liquid or solid completely dissolved in it. A lithium bromide water solution, commonly used in absorption systems, is water with a quantity of lithium bromide dissolved in it. "Strong" and "weak" solutions are those with respectively high and low concentrations of another liquid or solid.

Solvent - (1) a substance capable of dissolving another substance, especially grease or oil. (2) A liquid component of a solution usually present in greater amount than the solute.

Solvent cement - a solvent adhesive that contains a solvent that dissolves or softens the surfaces being bonded so that the bonded assembly becomes essentially one piece of the same type of plastic.

Solvent weld - a chemical process used to join plastic pipe.

Sone - calculated sound loudness rating. One sone is the loudness of a 1,000 Hz. Signal with the sound pressure of 40 decibels.

Soot - a black substance, mostly consisting of small particles of carbon, which can result from incomplete combustion and appears as smoke. Soot gives smoke its color.

Soot blowers - used to remove soot from around tubes to increase boiler efficiency. Mostly found on watertube boilers.

SOOW-A type cord - a 600 maximum volt cord having a temperature range of -58° F to +221° F. It is a medium duty cord same as SJOOW-A cord used for motors, machine and power tools, lighting, blowers, construction and industrial equipment, cranes, and hoists.

Sound attenuation - the extent to which sound power level is reduced (attenuated) as it travels through a flexible duct, expressed in decibels (dB).

Sound power level - the acoustic power radiating from a sound source, expressed in decibels.

Sound tracer - an instrument that helps to locate sources of sound.

Sound traps - acoustical devices usually installed at the beginning of a supply duct system. These devices contain sound absorptive materials designed to reduce air distribution noise due to the air conditioning equipment. They are also used in office spaces to reduce cross talk.

Source - the part of an electrical system that supplies energy to other parts of the system, such as a battery that supplies energy for a flashlight.

Source nipple - short length of heavy walled pipe between high pressure mains and the first valve of bypass, drain, or instrumentation connection.

Sources of ignition - devices or equipment that, because of their intended modes of use or operation, are capable of providing sufficient thermal energy to ignite flammable gas-air mixtures.

South pole, magnetic - that part of the magnet into which the magnetic lines of flux flow.

SOW-A type cord - a 600 maximum volt cord having a temperature range of -40° F to +194° F. It is the same as SO cord but is UL listed for outdoor use and used for tools, cranes, hoists, and industrial equipment.

Space charge - the electrical space charge resulting from ions and electrons. The cloud of electrons around a hot vacuum tube cathode.

Space, confined - for the purpose of the National Fuel Gas Code (NFPA 54), a space whose volume is less than 50 cubic feet per 1,000 Btu per hour of the aggregate input rating of all appliances installed in that space. Rooms communicating directly with the space in which the appliances are installed, through openings not furnished with doors, are considered a part of the unconfined space.

Space heater - a heater used to raise the temperature of the space within a building to the desired comfort level for human occupancy.

Space heating - the process of adding heat to a building or room to maintain indoor comfort.

Space thermostat - a thermostat whose sensor is located in the controlled space.

Spalling - hairline cracks in boiler brickwork (refractory) due to changes in furnace temperatures.

Span - the algebraic difference between the upper and lower values of the range. A 40° F to 240° F device would have a 200° F span.

Spandrel wall - that portion of a skeleton wall above the head of a window or door.

Spanner wrench - used to make adjustments on packing and other notched nuts.

Spark arresters – screening material or a screening device attached to a chimney termination cap to prevent the passage of sparks and brands to the outside atmosphere.

Spark gap - a gap between two electrodes. When a high potential difference is imposed across the electrodes, current will jump across the gap, creating an electrical spark for the purpose of igniting the pilot.

Spark generator - a high voltage transformer which produces a spark, used to ignite gas pilots or oil burners (direct spark ignition).

Spark igniter - in a direct spark ignition system, the assembly which includes the ignition electrodes, a ceramic insulator, and a mounting bracket.

Special purpose motors - designed for a particular application. Developed when an OEM has refined the operating characteristics or construction features of the motor. Does not have standard operating characteristics or standard mechanical features.

Special tools - those tools that are not available on the open retail market.

Specific density - a statement of mass per unit of volume measurement.

Specific gravity – the ratio of the density of one substance to the density of another substance taken as a standard. (water for liquids and solid; air or hydrogen for gases)

Specific heat - in the foot-pound-second system, the amount of heat (Btu) required to raise 1 pound of a substance 1° F. In the centimeter-gram-second system, the amount of heat (calorie) required to raise 1 pound of a substance 1° C. For example, the specific heat of water is 1 and the specific heat of air is 0.24.

Specific humidity - is the actual weight of water vapor mixed in the air. This weight is expressed in pounds of water vapor per pound of dry air.

Specific volume - the volume of a substance per unit of mass. The cubic feet of air and water vapor per pound of dry air.

Specific weight - weight of a substance as compared to water.

Specifications - a written document describing in detail the scope of work, materials to be used, method of installation, and quality of workmanship, for a particular job. Usually utilized in conjunction with working drawings.

Spectacle blind - double flange unit that can be swiveled to either a blind position (closed) or an open position.

Speed (rpm) - the running speed of an AC motor is dependent upon the number of poles and the frequency of the power source.

$$RPM = \frac{120 \times \text{frequency}}{\text{number of poles}}$$

number of poles	synchronous speed	normal speed
2	3600	3450
4	1800	1725
6	1200	1075
8	900	825

Speed handle - the speed handle is used with a socket wrench when the nut or bolt is easy to get to and when loosening or tightening can be done.

Speed ratio - the ratio or relationship of input speed divided by output speed. Speed ratio is normally expressed as (x):1.

Speed reducer - a mechanical device, incorporated between the driver and the fan of a mechanical draft tower, designed to reduce the speed of the driver to an optimum speed for the fan. The use of geared reduction units predominates in the cooling tower industry, although smaller towers will utilize differential pulleys and V-belts for the transmission of relatively low power.

Speed regulation - the ability of a motor to maintain a steady speed with changes in load.

Spike - a transient voltage surge.

Spillage - combustion products flowing from the appliance air openings or draft hood relief openings due to a malfunction of the venting system.

Spiral - curved flutes with a sharp cutting edge.

Splash bar - one of a succession of equally spaced horizontal bars comprising the splash surface of a fill deck in a splash filled cooling tower. Splash bars may be flat, or may be formed into a shaped cross-section for improved structural rigidity and/or improved heat transfer capability. When flat, they are sometimes referred to as slats.

Splash filled - descriptive of a cooling tower in which splash type fill is used for the primary heat transfer surface.

Splash lubrication - a system of lubrication in which parts of a mechanism dip into and splash the lubricant onto themselves and/or other parts of the mechanism.

Splash method - a method for more efficient evaporation in which water, dropping from a higher level in a cooling tower, splashes on slots with air passing through them.

Splash system, oiling - a method of lubricating moving parts by agitating or splashing.

Splashproof motor - a motor protected from falling droplets from any angle up to 100 degrees from vertical.

Splice - connection of two or more conductors.

Spline - a series of grooves, cut lengthwise, around a shaft or hole.

Split - the difference in temperature between the temperature of the air or liquid entering a coil and the coil operating temperature.

Split circuit receptacle - a receptacle that can be connected to two branch circuits. These receptacles may also be used so that one receptacle is live at all times, and the other receptacle is controlled by a switch. The terminals on these receptacles usually have break away tabs so the receptacle can be used either as a split circuit receptacle or as a standard receptacle.

Split phase motor - a motor with two stator windings. The most common type of single phase induction motor. Equipped with a starting winding displaced in magnetic position from, and connected in parallel with the main winding. Generally, this motor has low to moderate locked rotor torque with high locked rotor amps, high break down torques, and is used on belt drive blowers, fans, pumps, etc. The start winding in this motor is disconnected by a centrifugal switch or potential relay after motor attains speed, then operates on the main or run winding.

Split rings - a group of copper bars mounted on the end of a rotor shaft of a DC motor and connected to the brushes and the rotor windings.

Split system - a refrigeration or air conditioning installation which places condensing unit outside or at a remote location from the evaporator. Also applies to heat pump installations.

Spoiler screw - a screw or bolt moved in or out of the gas jet stream in a burner to control the primary air injection.

Spontaneous combustion - occurs when combustion materials self ignite.

Spoon of a valve - an internal mechanism in a mixing valve that slides or rotates to regulate the amount of hot and cold fluid streams entering the valve.

Spot smoke tester - an instrument used to measure the free carbon content of the flue products from a fossil fuel fired unit. The result is obtained by comparison to a standard using spots of carbon or black color of varying density.

Spray angle - specific coverage that a fuel oil nozzle must conform to. It must match up with the size and shape requirements of the combustion chamber and burner. Square type heat exchangers will tend to have a wide (80°-90°) spray angle, whereas long narrow heat exchangers have a smaller (30°- 45°) spray angle.

Spray cooling - a method of refrigerating by spraying refrigerant inside of an evaporator or by spraying refrigerated water.

Spray filled - descriptive of a cooling tower which has no fill, with water-to-air contact depending entirely upon the water break-up and pattern afforded by pressure spray nozzles.

Spray pond - a means of cooling condenser water by spraying it into the cooler ambient air. The falling water is then collected in a pond surrounding the spray.

Spread - this is the maximum width of the total air stream or pattern at the point of terminal velocity. It is measured in feet.

Spreader burner - a low pressure gas burner which uses venturi action to introduce part of the air required for combustion. The venturi delivers the air fuel mixture to a large vertical port, where a spreader plate spreads the flame toward the outside wall of the combustion chamber. Also called a large port burner. Also see inspirating and partial premix burners.

Spreader stokers - spreader stokers use a combination of suspension burning and grate burning. Coal is continually projected into the furnace above an ignited fuel bed. The coal fines are partially burned in suspension. Large particles fall to the grate and are burned in a thin, fast burning fuel bed. Because this firing method provides extreme responsiveness to load fluctuations and because ignition is almost instantaneous on increased firing rate, the spreader stoker is favored over other stokers in many industrial applications. The spreader stoker is designed to burn about 50% of the fuel in suspension. Thus, it generates much higher particulate loadings than other types of stokers and requires dust collectors to trap particulate material in the flue gas before discharge to the stack. To minimize carbon loss, fly carbon re-injection systems are sometimes used to return the particles into the furnace for complete burnout. Because this process increases furnace dust emissions, it can be used only with highly efficient dust collectors.

Spring benders - springs that are designed to be slipped over tubing to completely cover the area of the bend and to prevent kinking.

Spring hanger - support that allows variations in pipe position due to changes in temperature that is often used for vertical lines.

Spring loaded check valve - a check valve that contains a spring to close the valve regardless of its mounting position.

Spring loaded relief valve - a class of pressure valves employing a preloaded spring. The valve opens when the spring tension is exceeded and automatically closes when the pressure falls below a preset value.

Spring pressure - on thermostatic expansion valves, the pressure that determines the amount of superheat that must be maintained in the evaporator. The higher the spring pressure, the higher the superheat. The spring pressure, along with the evaporator pressure, pushes to close the thermostatic expansion valve. Also referred to as the superheat spring pressure.

Spring range - the range through which the signal applied must change to produce total movement of the controlled device from one extreme position to the other.

Spring return actuator - an actuator which is automatically driven to the closed position by a spring mechanism when power is removed. Also see actuator.

SPT-1 type cord - a 300 maximum volt cord that has a temperature range of +32° F to +140° F. It is a light duty, general duty cord used for lamps, speakers and appliances.

SPT-2 type cord - a 300 maximum volt cord that has a temperature range of +32° F to +140° F. It is a light duty cord but with heavier construction than SPT-1 cord used for lamps, speakers and appliances.

SPT-3 type cord - a 300 maximum volt cord that has a temperature range of +32° F to +140° F. It is a medium duty cord but with heavier construction than SPT-2 cord used for tools, major appliances, fans, and motors.

Spud - see orifice spud.

Square - to machine, cut, or make at a right angle. A geometric figure with four equal length sides and four 90° angles. A measurement of two dimensions length and width.

Square feet of steam - refers to the rating of the radiators to be used in a steam heating system.

Square foot of heating surface or radiation - equivalent direct radiation (EDR). When applied to boilers, it is 240 Btuh or the amount of heating surface that will emit 240 Btu's per hour when supplied with 215° F steam and surrounded by air at 65° F. The equivalent square foot of heating surface may have no direct relation to the actual surface area.

Square head cock - a type of valve often used as a balancing valve. In place of the valve handle, the stem is made square. A wrench is used to adjust the valve setting.

Square mil - the area of a square one mil by one mil.

Square of fan static law - a formula that determines the power required to increase the static pressure. If you have a 1 horsepower motor and you wanted to double the static pressure in the duct system, you would need to increase to a 4 horsepower motor. (2 x 1)2

Square measure - a measurement of two dimensions --- length and width.

Squirrel cage blower - blower typically found on air handling systems connected to duct runs. These blowers have a low pressure at their inlet but a high pressure at the outlet, which helps the system overcome the resistance in the ductwork. Also called a forward curved centrifugal blower. A fan which has blades parallel to fan axis and moves air at right angles or perpendicular to fan axis.

Squirrel cage rotor - a solid rotor used in AC induction motors which has round metal conductors joined together at the ends and placed inside the laminated metal motor core.

Squirrel cage winding - a permanently short circuited winding, usually un-insulated and chiefly used in induction motors, having its conductors uniformly distributed around the periphery and joined by continuous end rings. This secondary circuit is usually called the rotor.

ST type cord - a 600 maximum volt cord that has a temperature range of 0° F to +140° F. It is a heavy duty, general purpose cord, used for motors, power tools, construction equipment, and portable lighting.

Stability - the quality of a burner that enables it to remain lit over a wide range of fuel air mixture ratios when its surroundings are at room temperature.

Stable - unchanging or not subject to change.

Stable atom - an atom that will not release electrons under normal conditions.

Stable flame - a flame envelope that retains its form and continuity over the entire range of firing rates (low fire, high fire, and modulating) for a burner.

Stack - (1) one or several flues or pipes arranged to let the products of combustion escape to the outside. A single smoke pipe or chimney. (2) The thickness of the stator. (3) a vertical drain pipe.

Stack control - see stack switch.

Stack effect - the difference in pressure in a building due to the difference between the building temperature and the outside ambient temperature. The stack effect can overpower the mechanical system and disrupt ventilation and circulation in a building.

Stack height - the height of a gravity convector between the bottom of the furnace or heating unit and the top of the outlet opening.

Stack loss - see flue gas loss.

Stack, soil - the vertical disposal pipe that connects all of the horizontal fixture branches to the building drain. When this stack handles human waste, it is referred to as a soil stack.

Stack smoke detector - a smoke detector mounted on a stack to monitor the smoke produced by a boiler, oven, or furnace. The sensor consists of a photocell mounted on the stack opposite a light source. An alarm relay pulls in whenever smoke opacity exceeds a preset level or if the light source burns out.

Stack switch - a combustion monitoring device used with oil burners. If the temperature in the vent does not attain the required temperature, ignition failure is assumed and the switch interrupts the circuit to the burner. Also known as a stack relay.

Stack thermometer - a high temperature thermometer used to measure the heat of the furnace exhaust gasses.

Stack, vent - that portion of the soil or waste stack that extends above the highest fixture and then continues up through the roof.

Stack, waste - when the vertical disposal pipe that connects all of the horizontal fixture branches to the building drain does not handle any human waste, it is referred to as a waste stack.

Stack, water tower - an extended fan cylinder whose primary purpose is to achieve elevation of the discharge plume.

Stage - a level of operating capacity for a component or system.

Stage differential - the change in temperature at the thermostat needed to turn heating or cooling equipment off once it is turned on.

Staged system - a system that has more than one mode of heating or cooling operation.

Staging interval - the minimum time period for shedding or restoring two sequential loads.

Staging thermostat - a thermostat that is designed to open and close more than one set of contacts to control several modes of heating or cooling operation.

Stagnant zone - any area or space having low air motion and stratification. Having a stagnant zone doesn't mean that the air quality is poor.

Stair rail system - a vertical barrier erected along the unprotected sides and edges of a stairway to prevent employees from falling to a lower level.

Stand alone system - any control system that performs all control functions without direction from any other unit.

Stand by - an additional machine which is used as a substitute in times of emergency.

Stand by loss - heat energy required to maintain stored water at an elevated temperature.

Stand pipe - a wet or dry fire pipe line, extending from the lowest to the uppermost story of a building or structure, equipped with a shut-off valve with hose outlets at every story.

Standard - (1) a standard documents performance requirement for specific products that appear in the code. If you examine your local code, you will note a list of approved standards for plumbing/electrical products. (2) designation of cast iron flanges, fittings, and valves suitable for a maximum working steam pressure of 125 psig.

Standard 15 (Safety Code for Mechanical Refrigeration) - this standard prescribes safeguards for design, construction, installation, and operation of refrigerating systems.

Standard 34 (Designation and Safety Classification of Refrigerants) safety group - this classification consists of a letter (A or B), which indicates the toxicity class, followed by a number (1, 2 or 3), which indicates the flammability class. Toxicity classes A and B signify refrigerant with lower and higher toxicity, respectively, based on prescribed measures of chronic (long-term, repeated exposures) toxicity. Flammability class 1 indicates refrigerants that do not show flame propagation in air when tested by prescribed methods at specified conditions. Classes 2 and 3 signify refrigerants with lower and higher flammability respectively. The distinction depends on both the lower flammability limits and heat of combustion.

Standard air - see standard conditions.

Standard air density - .075 pounds per cubic foot. Equivalent to dry air at 70° F and at sea level. 1/13.33 cubic foot.

Standard atmosphere - a condition where air is at 14.7 psia pressure at 59° F, a pressure of 29.921 inches of Hg and a relative humidity of 30%.

Standard conditions - used as a basis for air conditioning calculations. Temperature of 59° F, pressure of 29.921 inches Hg, 0 feet of attitude, 14.696 psia, specific volume of 13.33 cubic foot, specific density is .075 pounds per cubic foot, and the relative humidity of 30%.

Standard cubic feet per minute (SCFM) - the volume of air delivered by a compressor or required by a pneumatic tool.

Standard expansion tank - an expansion tank that does not have a diaphragm.

Standard marks - marking on either side of a threader indicating the exact position for the change plate. If deeper threads are required, the change plate is positioned above the standard marks. If shallower threads are required, the change plate is positioned below the standard marks.

Standard port ball valve - a ball valve design where the orifice through the ball is slightly smaller than the diameter of the pipe to which the valve is connected.

Standard weight - schedule of wrought iron pipe weights in common use.

Standard ton of refrigeration conditions - refrigerant at a rate of one ton of ice melting per 24 hours (288,000 Btu per 24 hours, 12,000 Btu per hour, or 200 Btu per hour) in a system operating at 5° F evaporating, 86° F condensing, liquid subcooled 9° F, and suction gas superheated 9° F.

Standby - the state or condition of equipment which will permit complete resumption of stable operation within a short period of time. The state of a flame safeguard control while awaiting a call for heat; power is available at the control, but the operating sequence has not yet begun.

Standby power - electric energy supplied by a utility to a cogenerator or vice-versa during either one's unscheduled outage.

Standing pilot - gas burner ignition system in which a pilot flame remains lit between cycles of on/off burner operation.

Standing pressure test - method of leak detection that involves pressurizing the system and testing the following day to determine if pressure loss has occurred.

Standing seam - an unsoldered longitudinal seam that permits limited expansion and contraction. Formed by abutting two right angle flanges where one flange is 1/4" higher than the second flange. The long flange is turned over the shorter flange 180 degrees and button punched at intervals to provide a method of fastening. In some cases a second fold is made by turning both flanges inward another 180 degrees. If the second fold is made then the button punch is omitted. The double seam is used for roof and architectural applications.

Star transformer - a 3 phase transformer that has the ends of end winding connected to a common point. The star transformer produces a balance of all hot legs to ground.

Star winding - a winding layout of some 3 phase motors in which the end of the windings are connected together. Also known as a wye connection.

Start capacitor - an electric device that is used to momentarily store electrons, creating a second phase in starting windings of single phase motors. This type of capacitor is designed to stay in the circuit only a short period of time.

Start interlock - an interlock that proves a condition is proper for burner start up. If it is open, the burner cannot be started. It may be a valve closed interlock, damper position indicator (low or high fire proving switch), fuel pressure switch, electrical start switch, or oil pre-heater controller.

Start relay - a normally closed potential relay used to either disconnect the start capacitor or start winding once the motor comes up to speed and generates sufficient back emf.

Start/stop station - a control panel at which the operator can manually turn on or turn off remote electric equipment.

Start switch - see starting switch, centrifugal.

Start winding - a winding in an electric motor used only during a brief period when motor is starting. Usually these windings are smaller in diameter and have a higher ohmic value than the run winding.

Starter - a motor energizing switch with an overload cut-out that can be reset.

Starting holding coil circuit - that circuit in a magnetic starter that is energized when the start button is pressed to close the main contacts and which remains energized as long as the main current flows through the starter to the motor.

Starting torque - the amount of turning force produced by a motor as it begins to turn from a stopped position.

Starting screwdriver - are used to help put in or take out screws. Some have twisting centers or magnetized ends. Once the screw has been started, a common screwdriver should be used.

Starting switch, centrifugal - a centrifugally operating switch usually used in connection with split phase and capacitor start motors to open or disconnect the starting winding circuit after the rotor has attained a predetermined speed, and to close or reconnect it prior to the time the rotor comes to rest.

Starved evaporator coil - an evaporator that is receiving insufficient refrigerant and developing excessive superheat.

State, gaseous - one of three states or phases of matter characterized by fastest freedom of molecules and lack any inherent fixed shape or volume.

State, liquid - one of three states or phases of matter characterized by limited freedom of molecules and by substantial incompressibility.

State, solid - one of the three states or phases of matter characterized by stability of dimensions, relative incompressibility, and molecular motion held to limited oscillation.

Static balancing - balancing without rotation or knife edge balancing.

Static charge - an electrical charge that is bound to an object. An unmoving electrical charge.

Static checkout tests - tests performed on a flame safeguard system before mounting the flame safeguard control and without operating the burner; to ensure correct wiring and proper operation of external controllers, limits, interlocks, valves, transformers, and motors.

Static discharge head - vertical distance (in feet) from pump to point of discharge.

Static electricity - electricity that results from the electron being displaced and not returning to the original atom; usually results from friction.

Static friction - the force just sufficient to initiate relative motion between two bodies under load. The value of the static friction at the instant relative motion begins is termed break-away friction.

Static head - the pressure due to the weight of a fluid in a vertical column or, more generally, the resistance due to lift.

Static head loss - reduction in pump head caused by an increase in elevation.

Static pressure - the pressure which tends to burst a pipe (if a positive pressure) or to collapse it (if a negative pressure). It is used to overcome the frictional resistance to flow through the pipe. Expressed as a unit of pressure, it may be given in either absolute or gauge pressure. It is frequently expressed in feet of water column or (in the case of pipe friction) in mil-inch of water column per foot of pipe. In an air distribution system, it refers to the pressure necessary to overcome the total resistance created by the ductwork.

Static regain - a method of duct sizing used to design supply air systems of any velocity or pressure. It is used to meet the requirements of maintaining uniform static pressure at all branch and duct outlets.

Stationary air pockets - air that accumulates at the high points of piping systems and creates air binding.

Stationary blade compressor - a rotary pump which uses blade inside pump to separate intake chamber from the exhaust chamber.

Stator - that portion of the motor on which the coils are wound.

Status indicator - a PC card, which is optional, whose function it is to give a visual readout of the operating or failure modes of a cuber at a glance.

Stay bolt - a bolt threaded through or welded at each end, into two spaced sheets of a firebox or box header to support flat surfaces against internal pressures.

Stays - bolts or other devices designed to prevent tubesheet bulging when a boiler is under pressure.

Steady state condition - a stabilized condition of a piece of heating or cooling equipment where not much change is taking place.

Steady state efficiency - (1) a measure of the carbon dioxide in the flue gases, expressed as a percentage, to determine the level of completion of the chemical reaction during combustion taken at "steady state" conditions meaning there is no further change in the reaction process. (2) the efficiency of converting fuel into heat when a heat source operates continuously with non-varying input and output conditions.

$$SSE(\%) = \frac{\text{Fuel Input - Flue Loss - Condensate Loss}}{\text{Fuel Input}} \times 100$$

For furnaces tested under the isolated combustion system (ICS) method and for outdoor furnaces, cabinet heat loss (jacket loss) must also be deducted from the energy input.

$$SSE(ICS)(\%) \frac{\text{Fuel Input - Flue Loss - Condensate Loss - Jacket Loss}}{\text{Fuel Input}} \times 100$$

Steam - water in the vapor phase or state. The vapor formed when water has been heated to its boiling point, corresponding to the pressure it is under.

Steam and water drum - the pressure vessel in a steam boiler that contains both steam and water.

Steam atomizing oil burner - a burner for firing fuel oil which is atomized by steam. It may be of the inside or outside mixing type.

Steam boiler - a closed pressure vessel in which water is converted to steam by the application of heat. They are usually available in standard sizes up to and above 100,000 lb. steam/hr. and are used for space heating applications in both new and existing systems. They for used for many auxiliary uses, such as hot water heat exchangers, absorption cooling, laundry, and sterilizers.

Steam bound - conditions that occurs when the temperature in the open feedwater heater gets too high and the feedwater pump cannot deliver water to the boiler.

Steam, dry saturated - steam at the saturation temperature corresponding to the pressure, and containing no water in suspension. Steam that is lacking suspended water particles.

Steam grid - grid of steam nozzles that release steam directly into the air.

Steam heating system - a heating system in which the heating units give up their heat to the room by condensing the steam furnished to them by a boiler or other source.

Steam jet refrigeration - a refrigeration system which uses a steam venturi to create a high vacuum (low pressure) on a water container causing water to evaporate at low temperature.

Steam quality - the amount of vapor in a mixture of liquid and vapor, expressed as a percentage by weight.

Steam, saturated - steam in contact with the water from which it is generated. May be either wet or dry.

Steam separator - device used to increase the quality of steam. Found in the steam and water drum.

Steam space - the space above the water line in the steam and water drum.

Steam strainer - used before steam traps and turbine throttle valves to remove solid impurities.

Steam, superheated - steam at a temperature higher than the saturation temperature corresponding to the pressure.

Steam trap - an automatic valve which traps air but allows condensate to pass while preventing passage of steam.

Steam turbine - used to drive boiler auxiliaries or generators in large plants.

Steam, wet saturated - steam at the saturation temperature corresponding to the pressure, and containing water particles in suspension.

Steel, fireproofed - coating of special concrete placed on steel structures to prevent the steel from buckling from the heat of a fire.

Steel joist - any secondary steel member of a building or structure made of hot or cold-formed solid or open web sections, or riveted or welded bar, strip or sheet members or slotted and expanded or otherwise deformed rolled sections.

Steel strap - simple support used for stationary and noninsulated pipelines in heating and common housing plumbing.

Stellite® - trade name for a very hard metal alloy.

Stem - that part of a service or globe valve that is turned to open or close the valve.

Stem travel - the movement of a valve stem along the axis of its stroke.

Step down transformer - a transformer that has a secondary voltage lower than its primary voltage.

Step up transformer - a transformer that has a secondary voltage higher than its primary voltage.

Stepped down voltage - a voltage decrease produced by a transformer with more windings in its primary than in its secondary.

Stepped up voltage - a voltage increase produced by a transformer with more windings in its secondary than its primary.

Stepper motor - a motor that provides precise control by operating in a series of steps, as dictated by an external temperature sensor.

Sterilizer - one of three groups of antimicrobials registered by EPA for public health uses. EPA considers an antimicrobial to be a sterilizer when it destroys or eliminates all forms of bacteria, fungi, viruses, and their spores. Because spores are considered the most difficult form of a microorganism to destroy, EPA considers the term sporicide to be synonymous with a sterilizer.

Stick tight nozzle - a nozzle designed to hold the flame on the nozzle rather than floating in front of it. This is necessary in premix burners since the rate of flame propagation is high due to the lean mixture used. Also see flame retention nozzle.

Stinger leg - a higher voltage wire from a 3 phase delta transformer.

Stippling - the process by which solar collector manufacturers use acid to ripple the glass texture. Stippled glass reduces reflection loss.

Stoichiometric air fuel ratio – describes a condition in which the reactants of a chemical reaction are present in the exact quantities, as predetermined for the chemical equation of the reaction. It describes perfect combustion when the reactants are fuel and oxygen. The air fuel ratio that results in complete burning without excess air. Also see perfect combustion.

Stoke - kinematic measurement of a fluid's resistance to flow defined by the ration of the fluid's dynamic viscosity to its density.

Stoker - a machine used to supply a furnace with coal. Also see capacity of classification of stokers.

Stoker combustion rate - this rate is expressed in terms of the number of pounds of fuel or the Btu developed per square foot of stoker grate area per hour.

Stop cock - a quick opening or closing valve usually found on gas lines.

Stop valve - a shut off valve for controlling the flow of gases and liquid. Stop valves are used to shut off the flow within a system in order to isolate portions of the system for repairs, replacement or maintenance.

Storage battery - another term used for lead acid secondary cells.

Storage capacity - the volume of fluid that may be safely stored in a vessel such as a receiver.

Stored heat - heat from external or recovered heat sources that is held in reserve for later use.

Story - the vertical dimension between successive levels of horizontal framework ties, girts, joists, or beams. Story dimensions vary depending upon the size and strength characteristics of the framework material used.

Straight blade - the straight blade (or standard) is the most popular screwdriver. There are two types of standard screwdrivers, the keystone and the cabinet bit. The main difference between them is the way the tips are ground.

Straight drill - this type of drill is the most popular and useful. It is generally lightweight and can be used in almost every location. Has a drill angle of 118°.

Straight through valve - a valve with one inlet and one outlet. A fluid entering the inlet port flows straight through to the outlet port. It can be a two position valve or a modulating valve. Also called a 2 way valve.

Straight tube boiler - see water tube boiler.

Strain - change of shape or size of a body produced by the action of a stress.

Strain relief - a device that firmly attaches the power cord to the appliance.

Strainer - a device such as a screen or filter used to retain solid particles while liquid passes through. Does not contain a desiccant material.

Strand - a single un-insulated conductor.

Stranded conductor - a conductor of several strands of solid wire twisted together. Standard cables have 7, 19, and 37 strands.

Strap on aquastat - a special type of electromechanical setpoint controller (thermostat) designed to strap directly to the outside of a pipe.

Strap type pipe wrench - a wrench that has a nylon strap attached so that it allows the strap to be tightened around the pipe and tightened without marring the pipe.

Stratification boundary - the boundary between room currents moving faster than 15 fpm and the stratification zone.

Stratification of air - a condition in which there is little or no air movement in a room. Air lies in different temperature layers.

Stratosphere - the atmosphere between 7 and 30 miles above the earth where a layer of ozone filters out harmful ultraviolet light.

Stratospheric ozone layer – the atmospheric zone above the earth, extending from six to fifteen miles above the earth's surface. It protects the earth from ultraviolet rays from the sun.

Streamliner duct fitting - where it is impossible to offset a duct around an obstruction such as small beams, plumbing and steam lines, a fitting is used in which the sides are expanded to maintain the same duct area while an open sheet metal taper is installed within the fitting. The taper should have a maximum of 20 degrees on the entering side and a maximum of 60 degrees on the leaving side.

Street elbow - an elbow fitting that is threaded internally at one end and externally at the other.

Stress - the resultant internal force that resists change in the size or shape of a body acted on by external forces. When external forces act on a body they are resisted by reactions within the body which are termed stresses.

Stress crack - a crack in piping or other component caused by age or abnormal conditions such as vibration or temperature changes.

Strike - door part of a door latch.

Striker - this is a device such as a friction lighter that is used to light the torch.

Strip chart - recording chart that records temperature, pressure, voltage, amperage or any other required variable.

Stroboscope - a device that can be turned to the speed of a rotating shaft to appear to stand still. The dial reads in RPM so the speed can be instantly noted.

Stroke - a single movement of a piston, stem, or crank arm from one end of its range to the other. For a reciprocating engine, a stroke constitutes a half revolution of the engine and equals twice its crank radius.

Structural steel member - any primary or secondary member of a building or structure consisting of a rolled steel structural shape other than cold-formed steel, light gauge steel or steel joist members.

Strut - horizontal or angled structural member used primarily for longitudinal strength.

Stubby - the stubby, or short screwdriver, is used to work in small or hard to fit places.

Stud - bolts extending from front or rear of a device, such as a motor, by which it is mounted. Often used on small, direct drive fans and blowers.

Studs - vertically set skeleton members of a partition or wall to which lath or wall surface is attached.

Stubs - short copper or copper coated steel pipes welded or brazed into openings in the housing of a hermetic compressor for attaching tubing or valves.

Stuffing box - a packing gland surrounding a shaft, steam or rod to prevent leakage.

STW-A type cord - a 600 maximum volt cord that has a temperature range of 0° F to +140° F. It is a heavy duty, general purpose cord that is UL listed for outdoor use and used for motors, power tools, construction equipment, and portable lighting.

Subbase - the part of the space temperature thermostat that is mounted on the wall and to which the interconnecting electrical wiring is attached.

Subcooled - term that describes a substance at a temperature that is below its saturation point.

Subcooled liquid - liquid cooled below the condensing temperature corresponding to the existing pressure. Liquid subcooling increases the net refrigeration effect, and reduces flash gas. For example R-22 at a pressure of 76 psig corresponds to 45° F. The temperature of the liquid line is 40° F. This condition would show 5° F of subcooling.

Subcooling - liquid temperature and pressures are directly related. Subcooling is cooling a liquid below its saturation temperature corresponding to the pressure. Subcooling is measured by comparing the temperature of the liquid line to what the liquid temperature would ordinarily be at the measured discharge pressure.

Subcooling coil - a supplementary coil in an evaporative condenser, usually a coil or loop immersed in the spray water tank, which reduces the temperature of the liquid leaving the condenser.

Sublimation - a condition where a substance changes from a solid to a gas without becoming a liquid. Skips a change of state.

Submaster controller - a controller whose set point is reset by a master controller.

Submerged vertical boiler - see fire tube boiler.

Submersible motor - a motor whose housing is designed so that the motor can run completely submerged under water.

Substance - any form of matter or material.

Substation - a station in a power transmission system in which electrical power is transformed to a conveniently used form. The station may consist of transformers, switches, circuit breakers, and other auxiliary equipment.

Subsystem - a number of systems which operate independently but are commonly linked to a higher level controller.

Suction gas - the refrigerant vapor in an operating refrigeration system found in the tubing from the evaporator to the compressor and in the compressor shell.

Suction lift - a situation where the liquid source is lower than the pump. Pumping action creates a partial vacuum and atmospheric pressure forces liquid up to the pump. Theoretical limit of suction lift is 34 feet; practical limit is 25 feet or less, depending on pump type and elevation above sea level.

Suction line - tube or pipe used to carry refrigerant gas from the evaporator outlet to the compressor inlet.

Suction line accumulator - a device (sometimes an accessory) used to prevent liquid flood back from entering the compressor.

Suction line service valve - valve installed in the suction line. It has gauge port connections for connecting the suction hose of a service manifold.

Suction manifold - a device to distribute suction gas equally from a common suction line to multiple compressors connected in parallel.

Suction pressure - refrigerant pressure on the low side of the refrigeration system. Generally speaking, it is the pressure between the metering device and the compressor inlet. Suction pressure varies depending on (1) type of refrigerant, (2) temperature, moisture, and CFM of air across the evaporator, (3) amount of refrigerant charge, (4) discharge pressure, and (5) clean or dirty evaporator coil surface. Also called back pressure, low side pressure or crankcase pressure.

Suction pressure control valve - a device located in the suction line that maintains constant pressure in the evaporator during the running portion of the cycle.

Suction riser - a vertical tube or pipe that carries suction gas from an evaporator on a lower level to a compressor on a higher level.

Suction service valve - a two way manual operated valve located at the inlet to the compressor, which controls suction gas flow and is used to service the unit.

Suction side - low pressure side of the system extending from the refrigerant metering device through the evaporator to the inlet of the compressor.

Suction temperature - the boiling temperature of the refrigerant corresponding to the suction pressure of the refrigerant.

Suction type mixer - see aspirator mixer.
Suction valve - a valve that controls the flow of refrigerant from the suction line into the compressor cylinder head.

Suction valve lift unloading - the suction valve in a reciprocating compressor cylinder is lifted, causing the cylinder to stop pumping therefore controlling the capacity of the system and reducing amp draw of the compressor.

Sulfur content - manufacturers of fuel oil try to keep the sulfur content of fuel oils as low as possible. Sulfur in fuel becomes a pollutant when the oil is burned. For a fuel oil to receive a Grade 1 rating, it can have no more than 0.5% sulfur. Grade 2 fuels can have no more than 1.0% sulfur.

Sulfur dioxide - a colorless, extremely irritating gas or liquid. Molecular weight is 64.063, specific volume of 5.9 cubic feet/lb., and a boiling point of 14.0° F. Also see R-764.

Sump, water tower - a depressed chamber either below or along-side (but contiguous to) the collection basin, into which the water flows to facilitate pump suction. Sumps may also be designed as collection points for silt and sludge to aid in cleaning.

Sun effect - solar energy transmitted into space through windows and building materials.

Superheat - the temperature of vapor above its boiling point or the difference between the temperature at the evaporator outlet and the converted temperature of the refrigerant evaporating in the evaporator. Example: with a refrigerant gas temperature of 50° F, an evaporating temperature of 40° F, the superheat would be 10° F.

Superheat switch - this is a normally closed safety switch. It stops the compressor if the system is low on refrigerant.

Superheated vapor - a gas that has been heated to a temperature that is above its boiling point as a liquid at the existing pressure. For example R-22 at a pressure of 195.9 psig is equal to the temperature reading of 100° F. Taking the temperature of the hot gas discharge line shows a temperature reading of 150° F. Thus, this condition would show a 50° F superheat reading.

Superheated steam - steam heated to a temperature higher than its vaporization temperature at its working pressure.

Superheater - a heat exchanger arranged to cool liquid going to the evaporator and using this heat to superheat the vapor leaving evaporator.

Superheater drain valve - a valve found on the superheater header outlet. This valve is used to maintain flow throughout the superheater during startup and shutdown.

Superheater header - main inlet and outlet line to and from the superheater tubes in the superheater.

Supervise - to monitor a condition requiring attention and initiate corrective action if necessary.

Supervisory control – a higher level manual or automatic control (central or network) of a normally stand alone control system.

Supervisory fuel cock - a manually operated plug cock, used for shutting off fuel, which has an electric interlock. Its primary purpose is to minimize the hazard of lighting off manually ignited multi-burner equipment.

Supplemental thermal - heat required when recovered engine heat is insufficient to meet thermal demands.

Supplementary firing - injection of fuel into an exhaust gas stream to raise its energy content (heat).

Supplementary heat - electric strip heaters (can be any other type of fossil fuel system) that are used to increase the heating capacity of a heat pump system or temper the air that is being supplied to the occupied space during the defrost cycle mode.

Supplementary power - electric energy supplied by an electric utility in addition to the energy the facility generates.

Supply - a fuel or water line flowing into a component. A water line or air duct carrying heated or cooled fluid to the conditioned space for heat transfer.

Supply air - that air delivered to each or any space supplied by the air distribution system or the total air delivered to all spaces supplied by the air distribution system, which is provided for ventilating, heating, cooling, humidification, dehumidification and other similar purposes.

Supply air ducts - ducts that carry the conditioned air to the conditioned space. May be square, rectangular, round or oval and made of metal or fiberglass.

Supply air reset - energy conserving control strategy which monitors heating and cooling loads in building spaces and adjusts discharge air temperature to the most efficient level to satify the measured load.

Supply branch - the piping used to supply heated water or air from a main, circuit main, or trunk to the terminal unit or diffuser.

Supply mains - the pipes used to distribute water from the boiler or conditioned air to the supply branches.

Supply opening - any opening through which air is supplied into a ventilated space, zone or area being heated, cooled, humidified, dehumidified or filtered. Supply openings can be sidewall, ceiling, floor, or baseboard.

Supply piping - that portion of the piping system that carries water from the boiler to the terminal units or to the point of use.

Supply plenum - the duct section that connects directly to the air handler portion of the air conditioning system.

Supply tank - a tank connected directly to the oil burning equipment.

Supply tapping - the opening in a boiler into which the supply main is connected.

Supply voltage - the energy source supplied to the controller and/or auxiliary device.

Supply water - the water that is be supplied to a heat exchange surface, such as the condenser. The water has not yet absorbed or rejected any system heat. Make-up water for a water tower application.

Supply water reset - energy conserving control strategy which automatically changes the setpoint of water supplied to a heating or cooling loop to the highest setpoint (chilled water) or lowest setpoint (hot water) to satisfy the heating or cooling load in a building space.

Surface blowdown valve - used to remove impurities from the surface of the water in a steam and water drum.

Surface condenser - a shell and tube vessel used to reduce the exhaust pressure on the outlet end of turbines or engines.

Surface film resistance - the reciprocal of the heat transfer film coefficient, in °F·ft2·h/Btu.

Surface plate - a tool with a very accurate flat surface that is used for measuring purposes and for lapping flat surfaces.

Surface reflectance - the fraction of the radiant flux falling on a surface that is reflected by it.

Surface tension - caused by impurities on the top of the water in the steam and water drum.

Surface thermostat - devices designed for mounting on and measuring the temperature of the surface, such as that of a pipe.

Surge - the modulating action of temperature or pressure before it reaches its final value or setting.

Surge, electrical - a transient variation in the current or voltage at a point in the circuit. Surges are generally unwanted and temporary.

Surge suppressor - a device that limits peak voltage to a predetermined value when voltage spikes or surges appear on the connected line.

Surge tank - a container connected to a refrigerating system which increases gas volume and reduces rate of pressure change.

Surge voltage protection - a protective circuit that is built into an electrical test instrument to protect it from large spikes that could otherwise damage it.

Survey - a detailed list of load factors and job factors that are required prior to estimating heat load calculations.

Suspension sling - used to support the drum of a HRT boiler.

Sustained pressure test - a test run a pipe where a constant internal pressure test lasts for 1000 hours.

SVT type cord - a 300 maximum volt cord that has a temperature range of 0° F to +140° F. It is a heavy duty, general purpose flexible cord, used for vacuum cleaners and office equipment.

Swage - the process of enlarging one tube end so that the end of other tube of the same size will fit within.

Swaged joint - the joining of two pieces of copper tubing by expanding or stretching the end of one piece of tubing to fit over the other piece.

Swaging tool - a tool used to enlarge a piece of tubing for solder or braze connection.

Swamp cooler - evaporative type cooler in which air is drawn through porous mats soaked with water. The efficiency of this machine is dependence upon the wet bulb temperature.

Swash plate-wobble plate - device used to change rotary motion to reciprocating motion, used in some refrigeration compressors.

Sweat fitting - a metal fitting with close tolerances that is joined by sweating.

Sweating - (1) condensation of moisture from air on cold surfaces. (2) Method of soldering in which the parts to be joined are first coated with a thin layer of solder.

Sweet water - term sometimes used to describe tap water.

Sweet water brine - water with a freezing point of 32° F.

Swing check - a check valve with an internal disc that swings on a pivot.

Swing joint - a piping joint that is made up of a 90° ell and a street ell. They are used in connecting fuel oil tanks to permit easy movement.

Swing space - room needed to turn a nut, bolt, wrench or tool.

Swing strut - pipe support component that can carry load in its axial direction while not restricting the pipe in any direction perpendicular to its axis within certain limits. A swing strut resists both tension and compression forces.

Swinging armature - an armature used in a contactor that is mounted on a line and moves up and down in a swinging motion.

Switch - a mechanical or electrical device that makes or breaks contacts to either complete or open an electrical circuit; may be automatic or manual. Also see interlock; and airflow, lockout, low oil temperature, master, pressure, sail, and valve closed indication switches. Typical switches are:

Normally open (N/O) - a contact pair which is open when the device is in the de-energized position.

Normally closed (N/C) - a contact pair which is closed when the device is in the de-energized position.

Switch, auxiliary - an accessory switch available for most damper motors and control operators that can be arranged to open or close an electric circuit whenever the control motor reaches a certain position.

Switch configuration - the terms pole and throw in combination with the terms single and double are used to describe the action of electric switches. A single pole switch has functionally one movable contact. A double pole switch has two. A single throw switch can connect to only one circuit or one fixed contact. A double throw switch can connect a common lead to either of two circuits or two different fixed contacts. These features are usually abbreviated SPST --- single pole, single throw; SPDT --- single pole, double throw, etc.

Switch, disconnect - a switch usually provided for a motor that completely disconnects the piece of equipment from the power source.

Switch, flow - a control installed in the liquid circuit. Used to prevent operation of the unit when the liquid flow is not up to the required rate.

Switch, general use - a switch intended for use in general distribution and branch circuits. It is rated in amperes, and it is capable of interrupting its rated current at its rated voltage.

Switch, general use, snap - a form of general use switch constructed so that it can be installed in device boxes or on box covers, or otherwise use in conjunction with wiring systems recognized by the NEC.

Switch, mode switch - the position selector switch on the zone thermostat that selects the type of operation of the system. These positions are cooling, off, heating, auto, and emergency heat.

Switch, motor circuit - a switch rated in horsepower that is capable of interrupting the maximum operating overload current of a motor of the same horsepower rating as the switch at the rated voltage.

Switching regulator - a system used to improve the voltage regulation of a power supply. Components are switched in and out of the system according to the need for them in improving regulation.

Switching relay - devices that operate by a variation in the conditions of one electrical circuit to affect the operation of devices in the same (or another) circuit. Used to increase switching capability and isolate electrical circuits. Also provide for electrical interlocks within a system.

Syphon seal - a corrugated metal tubing used to hold seal ring and provide leak proof connection between seal ring and compressor body or shaft.

Symbol - a letter or picture representing some part of an electric circuit, fitting, equipment, valve or component.

Symbol palette - a collection of symbols representing physical components in schematic drawings (piping and electrical).

Symptom - a characteristic of a machine's performance that indicates a malfunction.

Synchronize - to balance out combustion controls before switching to automatic operation.

Synchronous motor - an AC motor in which the average speed of normal operation is exactly proportional to the frequency of the system to which it is connected.

Synchronous speed - the speed of the rotating magnetic field set up by the stator of an induction motor. In a synchronous motor the rotor locks into step with the rotating magnetic field. To determine synchronous speed the formula cycles (Hz) x 120 divided by the number of poles is used.

Synthetic dust weight arrestance - a measurement of a filter's ability to remove synthetic dust from test air.

Synthetic hydrocarbon - oil molecule with superior oxidation quality tailored primarily out of Paraffinic materials.

Synthetic lubricant - a lubricant produced by chemical synthesis rather than by extraction or refinement of petroleum to produce a compound with planned and predictable properties.

Synthetic natural gas (**SNG**) - gas which is manufactured to emulate the characteristics of natural gas.

Synthetic rubber, neoprene - soft resilient material made of a synthetic chemical compound.

System - a set or arrangement of components so related or connected as to form a unity or organized whole; a collection of consecutive operations and procedures required to accomplish a specific objective. Also see automatic, closed circuit oil, conductivity flame rod, dead end, flame detection, flame safeguard, manual, multi-burner, rectification, and semi automatic system.

System analysis - method to evaluate tradeoffs among envelope components and heating, ventilating, and air conditioning equipment such that the building's annual energy cost does not exceed a value specified by a building that meets the prescriptive requirements of the standard.

System characteristic curve - graph used to evaluate the performance of a hydronic system by plotting the systems head in feet on the vertical axis and its flow rate on the horizontal axis.

System charge - the amount of refrigerant that is contained in an air conditioning or refrigeration system.

System dependent recovery equipment - equipment that makes use of an appliance's own components to remove refrigerant from the appliance.

System, direct return - a hydronic piping arrangement for a heating system in which the heated fluid is returned to the boiler by the shortest direct path, resulting in considerable differences in the lengths of the several circuits composing the system.

System evacuation - when using a vacuum pump, the process by which moisture is removed from an air conditioning or refrigeration system prior to the introduction of refrigerant into the system.

System, forced circulation - a heating, air conditioning, or refrigerating system in which the heating or cooling fluid circulation is effected by a fan or pump.

System, gravity circulation - a heating or refrigerating system in which the heating or cooling fluid circulation is effected by the motive head due to difference in densities of cooler and warmer fluids in the two sides of the system.

System lag - the difference in temperature between the point at which the thermostat closes and the point at which the thermostat starts to rise or fall.

System, one pipe - a hydronic piping system in which the fluid withdrawn from the supply main passes through heating and cooling units and returns to the same supply main.

System overshoot - the temperature rise of the controlled space above the desired set point of the thermostat.

System pressure regulator - a valve located between the compressor and the receiver of a refrigeration system. This valve controls the amount of liquid refrigerant that bypasses a parallel liquid receiver.

System pump down - the process by which the system refrigerant is stored in the compressor, receiver, and condenser coil in preparation for a repair on the low pressure side of the refrigeration system.

System resistance curve - a curve plotted on a graph that indicates the heat loss of a piping system versus the flow rate through it.

System temperature - the average of the temperatures of the water leaving the boiler and returning to the boiler.

-T-

TAB - testing and balancing.

TD - temperature difference.

TDR - time delay relay.

TE - totally enclosed.

TEAO - totally enclosed, air over.

TEFC - totally enclosed, fan cooled.

TENV - totally enclosed, non-ventilated.

TEWI - total equivalent warming impact.

THHN - a 90º C 600 volt wire.

THWN - a 75º C 600 volt wire.

TLC - total load change.

TLV - threshold limit value.

TP - total pressure.

TRV - thermostatic radiator valve.

TVOC - total volatile organic compound.

TWA - time weighted average.

TXV - thermostatic expansion valve.

Tachometer - a device that measures or records speed or velocity of a motor in revolutions per minute.

Tagging - procedure whereby all lines, coordinates, equipment, valves, and so forth are identified on the model by standard or special tags.

Tagout - the practice of placing tags on all operational switches, in accordance with an established procedure, to inform others that repairs are in process and may not be operated until the tagout device is removed.

Tagout device - a warning device, such as a tag, and a means of attachment.

Tail pipe - the outlet pipe from the evaporator.

Take off - duct section connected to the main truck line, providing a connection point for the branch duct. Can be oval, square or round.

Tang - a prong on a chisel or file that fits into a shaft or handle.

Tank farm - any area of a refinery or plant that is primarily used for the storage of liquid or gaseous products. Usually a group of vessels or holding tanks.

Tank, supply - separate tank connected directly or by a pump to a oil burning appliance.

Tankless water heater - a heat exchanger that heats service hot water as used and therefore does not employ a storage tank. Also called an instantaneous heater.

Tap (electrical) - (1) a method of obtaining or drawing current from a circuit. (2) on multi speed motors, wires or terminals that make it possible to select the number of poles or windings being used.

Tap (mechanical) - a device to thread the inside (female) of a pipe or fitting.

Tap drill - a drill used to form a pilot hole prior to placing tapped threads in a hole. The drill is the size of the root diameter of the tap threads.

Tap-a-line - a device used to puncture or tap a line where there are no service access. Sometimes referred to as a saddle valve or piecing valve.

Tapered - a piece of material that increases or decreases in size at a consistent rate.

Tapped - a circuit that has been cut-in by another circuit, or split in its configuration by a circuit. Also refers to multiple voltage input transformers and multi-speed motors.

Taps - on a multi-speed motor, wires or terminals that make it possible to select the number of poles being used.

Tare weight - the weight of a container or cylinder before being filled.

Target flow rate - a calculated flow rate through a piping system based on an expected rate of heat transport, and an assumed circuit temperature drop.

Target value - the desired value of a controlled variable.

T drill® - a specialized tool for fabricating branch connections in copper tubing.

Technician - anyone who performs maintenance, service, or repair operations that could reasonably be expected to release Class I or Class II substances from appliances into the atmosphere per section 608 of the EPA. Technicians may be installers, contractor employees, in-house service personnel, or owners.

Tee - a T-shaped fitting that connects three lengths of pipe or tubing at right angles to each other.

Tee joint - joint between two members located at right angles to each other forming a tee.

Tee reducing - fitting used to connect the branch line to the water supply system. This fitting is designed to accommodate the changes in pipe sizes that often occur when branch lines are taken off.

Teflon - a synthetic rubber material used for O-rings, pipe thread compound and slide valve mechanisms.

Telemetry - also called telemetering. The sensing and measuring of information at some remote location, then transmitting that information to some other location for reading the measurements. The same facilities can be used for control and other signals.

Temperature - the degree or intensity of hotness or coldness of anything as measured by a thermometer. The measurement of the intensity or speed of molecular motion. Also see ambient and ignition temperature.

Temperature, absolute - temperature expressed in degrees above absolute zero.

Temperature, absolute zero - the zero point on the absolute temperature scale, 459.69° below the zero of the Fahrenheit scale and 273.16° below the zero of the Centigrade scale.

Temperature alarm - temperature sensitive device with an alarm that warns against a rise or drop in temperature.

Temperature, ambient - the temperature of the air around the object under consideration.

Temperature coefficient – a factor used to calculate the change in indication or output of an instrument with changes in temperature. Uncompensated changes in temperature contribute to instrument uncertainty by an amount determined by the temperature coefficient.

Temperature coefficient of resistivity - the amount of resistance change of a material per degree of temperature rise.

Temperature, condensing - the temperature of the fluid in the condenser at the time of condensation.

Temperature, critical - the saturation temperature corresponding to the critical state of the substance at which the properties of the liquid and vapor are identical.

Temperature, dew point - the temperature at which the condensation of water vapor in a space begins for the given state of humidity and pressure as the temperature of the vapor is reduced. The temperature corresponding to saturation (100% humidity) for a given absolute humidity at constant pressure.

Temperature difference - the number of degrees between two temperatures; determines the speed of heat transfer from the hotter to the colder substance.

Temperature, discharge - the temperature of the gas leaving the compressor.

Temperature drop - a measurement of the difference in heat between two points of a system, such as at the furnace plenum and the outlet grille.

Temperature, dry bulb - the temperature of the air measured with an ordinary thermometer. It indicates only sensible heat changes.

Temperature, effective - an arbitrary index that combines into a single value the effect of temperature, humidity, and air movement on the sensation of warmth or cold felt by the human body. The numerical value is that of the temperature of still, saturated air that would induce an identical sensation.

Temperature, entering - the temperature of a substance as it enters a piece of equipment.

Temperature, evaporating - the temperature at which a fluid boils under the existing pressure.

Temperature, final - the temperature of a substance as it leaves a piece of equipment.

Temperature glide - a range of condensing or evaporating temperatures for one pressure. Found in blended or ternary refrigerants. The fluctuation of temperature at a given pressure in a zeotropic blend, caused by fractionation between the evaporator inlet and outlet or fractionation between the condenser inlet and outlet.

Temperature-humidity index - an index that combines dry bulb and wet bulb temperatures to estimate human response to the environment.

Temperature indicator - device used for measuring temperature. It can be locally mounted dial, remote mounted dial capillary tube, or an electric thermocouple.

Temperature limit control - a controller that prevents the temperature of a heat source from exceeding a set value.

Temperature, mean radiant - the temperature of a uniform black enclosure in which a solid body or occupant would exchange the same amount of radiant heat as in the existing non-uniform environment.

Temperature range - the specific span of temperatures within which safety or operating controls are designed to operate.

Temperature recorder - instrument that records a permanent continuous history of pipe, equipment, or air temperature.

Temperature, regulated valve - a valve that is open or closed depending upon the temperature of the fluid it controls.

Temperature rise - the amount of heat a motor generates above ambient or the temperature difference between the supply and return air temperatures.

Temperature, room - temperature of any room, as for example (1) a room in which a refrigerator is being operated or tested; (2) a room being conditioned for the comfort of occupants. Used colloquially to mean ordinary temperature one is accustomed to find in dwellings.

Temperature, saturation - the boiling point of a refrigerant at a given pressure. It is considered to be the evaporator temperature in refrigeration.

Temperature scales - The Centigrade and Fahrenheit scales derive their degree values by dividing the difference between the ice point and steam points of water as follows. Centigrade equals100 units and Fahrenheit equals 180 units. The value of a Fahrenheit degree is therefore five ninths of a Centigrade degree. The Kelvin scale, sometimes called Centigrade absolute, has its zero point at the lowest attainable temperature, 273.16 C below the triple point temperature of water and exactly 273.16 below the zero of the Centigrade scale. The Rankin scale, sometimes called Fahrenheit absolute, has its zero point at the lowest attainable temperature, 459.688 F below the triple point temperature of water and exactly 459.688 below the zero of the Fahrenheit scale. To convert Centigrade to Kelvin temperature, add 273.16 degrees. To convert Fahrenheit to Rankine temperature, add 459.688 degrees.

Temperature sensing bulb - bulb containing a fluid and bellows or diaphragm. Temperature increase on the bulb causes the bellows or diaphragm to expand.

Temperature, suction - the temperature of the gas as it enters the compressor.

Temperature swing - (1) degrees of indoor temperature change in relation to outdoor temperature change on any given day (2) the temperature difference between the low and high temperatures of the controlled space.

Temperature variation - the temperature difference between points within the same space.

Temperature, wet bulb - thermodynamic wet-bulb temperature is the temperature at which liquid or solid water, by evaporating into air, can bring the air to saturation adiabatically at the same temperature. Wet bulb temperature is the temperature indicated by a thermometer having a bulb covered with a wetted wick.

Temperature, wet bulb depression - the difference between the wet bulb and the dry bulb temperatures.

Tempered glass - glass that has been heat treated to become more than three times stronger. Also known as safety glass because it breaks into small pebbles rather than jagged shards. It should be used in sliding glass doors and windows susceptible to impact and breakage.

Temporary service stairway - a stairway where permanent treads and/or landings are to be filled at a later date.

Tensile strength - a term denoting the greatest longitudinal tensile stress a substance can bear without tearing apart or rupturing.

Tensile stress - this occurs when two forces of equal intensity act on an object, pulling in opposite directions. This affects such items as boiler plates and staybolts.

Terminal - a point on an electrical device where connections may be made. It could be a screw or quick connect device where a conductor(s) is intended to be connected.

Terminal connectors - small metal ends of varying size and shape that crimp onto the wires to make connections simpler and faster.

Terminal markings - identification letters and symbols assigned to circuitry in schematic drawings to aid in tracing the circuits.

Terminal reheat system - centralized blower and cooling system that supplies cool air to multiple zones. Each zone contains a hot water coil or electric heater that reheats the cooled supply air as determined by the zone thermostat.

Terminal unit - a radiator, conveyor or the like that heats the conditioned space by transferring heat carried by hot water or steam.

Terminal velocity - a point where discharged air from a grille reduces in velocity to 50 fpm for heating and 35 fpm for cooling. The velocity of an air stream is the highest sustained velocity in the mixed air steam at the end of the throw, measured in feet per minute. Terminal velocity is not sharply defined for all applications.

Terminate - to end an action or sequence of events such as a defrost cycle or employment.

Termination temperature - temperature at which the terminals of devices connected to the conductor will withstand.

Ternary - having three elements, parts, or divisions.

Terrazo - marble aggregate concrete that is cast in place, or precast, and ground smooth.

Test charge - an amount of refrigerant vapor forced into a refrigeration system to test for leaks.

Test cord - an electrical cord that supplies current directly to the compressor terminals for testing and service purposes.

Test dial, gas supply meter - one of the recording dials that indicates a quantity of gas per revolution of the dial.

Test light - a light provided with test leads used to test or probe electrical circuits to determine their condition.

Test points - labeled test points that provide the option of testing different parts of the electrical circuit with a volt or ohmmeter.

Tests - procedures for determining the manner which equipment is functioning, or for determining the existence, type, and location of any trouble. See hot refractory hold in, hot refractory saturation, ignition interference, ignition spark response, pilot turndown, safety shutdown, and static checkout tests.

Theoretical air - the exact amount of air required to supply oxygen for complete combustion of a given quantity of a specific fuel.

Theoretical combustion - see perfect combustion.

Therm - a quantity of heat equivalent to100,000 Btu's (25,200 Kcal). The therm is the traditional unit in which natural gas is sold.

Thermal - having to do with heat.

Thermal aging - exposure to a given thermal condition or a programmed series of conditions for prescribed periods of time.

Thermal alloying - the act of uniting two different metals to make one common metal by the use of heat.

Thermal balance point - the lowest outdoor temperature at which the heat pump can supply 100% of the heating required for the space.

Thermal break - the placement of insulation between two materials to prevent rapid heat transfer between them.

Thermal bulb - a sealed probe charged with a fluid that expands or contracts when heated or cooled; used to sense remote temperatures.

Thermal capacity - the capacity of a substance for storing heat. It is expressed in the amount of heat required to raise the temperature of a substance a given amount, or, to change its state --- such as liquid to vapor. The maximum amount of instantaneous heat that a system can produce.

Thermal comfort - when a person is surrounded by an environment whose temperature and relative humidity permit the person to lose, without conscious effort, metabolic heat at the same rate he or she produces it.

Thermal conductance (C) - time rate of heat flow through a body (frequently per unit area) from one of its bounding surfaces to the other for a unit temperature difference between the two surfaces, under steady conditions (Btu/h·ft2·°F).

Thermal conductivity (k) - the time rate of heat flow through a unit area ((one square foot) of 1 inch or one foot thick homogeneous material in a direction perpendicular to isothermal planes, induced by a unit temperature gradient. Thermal conductivity must be evaluated for a specific mean temperature and moisture content, because in most materials it varies with the temperature and moisture content. The thermal conductivity is the reciprocal of the thermal resistance. Units are Btu·in/h·ft2·°F or Btu/h·ft·°F.

Thermal cutout - an over current device that protects the circuit during excessive heating conditions.

Thermal delay relay - a relay that prevents short cycling of compressor by delaying the thermostat's call for cooling for several seconds.

Thermal efficiency - the output of a machine (expressed in Btu) divided by the input (expressed in Btu). The ratio of heat supplied from burning fuel to the heat absorbed by a fluid in a boiler or air in a furnace. This is the First Law of Efficiency.

Thermal electric - electrical energy derived from heat energy, usually by heating a working fluid that drives a turbogenerator.

Thermal emittance - the ratio of the radiant flux emitted by a body to that emitted by a blackbody at the same temperature and under the same conditions.

Thermal energy - heat. Measured in calories or Btu's.

Thermal envelope - the shell formed by all building surfaces that separates heated space from unheated space. Exterior walls, windows, and exposed ceilings are all parts of the thermal envelope of a building.

Thermal environment - those characteristics of the environment which affect a person's heat loss.

Thermal equilibrium - the condition of a system when the rate of energy loss from the system is exactly the same as the rate of energy input to the system.

Thermal expansion - the expansion of a material when subjected to heat.

Thermal head - the head produced by the difference in the weight of the heated water in the supply side of the system and the cooler water in the return side. This is the only head available to cause circulation of water in a gravity system.

Thermal internal - the tendency of a substance to resist temperature changes.

Thermal mass - material with significant heat capacity and surface area that affects building loads by absorbing or releasing heat or both due to the fluctuation of any of the following:

(a) interior temperature
(b) interior radiant conditions
(c) exterior temperature
(d) exterior radiant conditions

Thermal mass wall insulation position - (1) exterior insulation position - wall having mass thermally coupled directly to the room air and having the entire effective mass layer on the interior of an insulation layer. (2) Integral insulation position - wall having mass thermally coupled to room air and having either insulation and mass materials well mixed, as in wood (logs), or substantially equal amounts of mass material on the interior and exterior of insulation, as in concrete blocks with insulated cores. (3) Interior insulation position - wall having mass located on the exterior of the insulating material(s) or otherwise not meeting either of the above definitions for exterior or integral positions.

Thermal movement - any expansion or contraction of the piping system due to a change in temperature associated with the line material or external conditions.

Thermal overload - an overload device that senses the current draw of a load by the heat produced, that is proportional to the current draw.

Thermal overload element - the alloy piece used for holding an overload relay closed. It melts when the current draw is too great.

Thermal protector - a protective device built into motors that protects them from overheating due to overload or failure to start. Basic types are automatic reset and manual reset. With the automatic reset type, when the motor cools, the protector automatically restores power. Automatic reset protective devices should not be used where unexpected restarting would be hazardous. The manual reset types are equipped with an external button that must be pushed to restore power to the motor. Recommended where unexpected restarting could be hazardous, such as saws, conveyors, etc.

Thermal radiation - the transmission of heat from a hot surface to a cooler on in the form of invisible electromagnetic waves, that on being absorbed by the cooler surface, raise the temperature of that surface.

Thermal rating - the maximum and/or minimum temperature at which a material will perform its function without undue degradation.

Thermal relay, hot wire relay - an electrical control used to activate a refrigeration system. This system uses a wire to convert electrical energy into heat energy.

Thermal resistance, electrical - that change in the electrical resistance of a material when subjected to heat. Resistance to heat flow from conductors to outer surface of insulation or sheath in a wire or cable.

Thermal resistance (R) - a measure of the ability to retard the flow of heat. The R value is the reciprocal of thermal conductivity expressed in units of hr / ft2 /degree F per Btu. As the R value of an element or assembly increases, the transmission of heat energy through the element or assembly decreases.

Thermal resistivity - thermal resistance of a unit cube of material. The reciprocal of thermal conductivity. Units are Btu·in/h·ft2·°F.

Thermal sensation - a conscious feeling commonly graded into the categories of cold, cool, slightly cool, neutral, slightly warm, warm and hot.

Thermal shock - shock caused by a sudden rise or drop in temperature of the fluid in contact with a valve.

Thermal stress - forces built up as anchored piping undergoes changes in temperature during operation. A rise in temperature causes expansion. As the pipe cools down during downtime or between operations, the pipeline contracts.

Thermal timer - a device that will actuate one or more switches when heated bimetal warps sufficiently.

Thermal transmission - the flow of heat energy induced by a temperature difference. Heat may be transferred by conduction, convection, radiation, and mass transfer. These can occur separately or in combination, depending on specific circumstances.

Thermal transmittance (U) - overall coefficient of heat transmission or thermal transmittance (air to air) expressed in units of Btu per hour per square foot per degree F. It is the time rate of heat flow. The U value applies to combinations of different materials used in series along the heat flow path and also to single materials that comprise a building section, and includes cavity air spaces and surface air films on both sides. U = 1/Rt.

Thermal transmittance, overall (Uo) - the overall heat transmission or thermal transmittance of a gross area of the exterior building envelope, expressed in units of Btu per hour per square foot per degree F. The Uo value applies to the combined effect of the time rate of heat flows through the various parallel paths, such as windows, doors and opaque construction areas, comprising the gross area of one or more exterior building components, such as walls, floor, or roof/ceiling.

Thermal trap - a piping assembly that discourages heat migration due to natural convection when the system circulators are off.

Thermistor - a solid-state device whose electrical resistance varies with temperature. Its temperature coefficient of resistance is high, nonlinear, and negative. Also see posistor.

Thermocouple - a device for measuring temperature consisting of two electrical conductors of dissimilar metals joined at a point, called the "hot junction", and where a voltage proportional to the temperature is developed across the output. This volts open circuit (VOC) voltage is usually in the range of 18-30 millivolts. One type of thermocouple is used in pilot safety systems to prove the existence of a pilot flame. Thermocouples have very large operating temperature ranges and the advantage of very small size. However, they have the disadvantage of small output voltages, noise susceptibility from the wire loop, and a relatively high drift.

Thermocouple lead wire - similar to thermocouple wire except the degree of accuracy in temperature measurements is not as high and it is used to transmit thermocouple information to remote indicators.

Thermocouple thermometer - an electrical instrument using a thermocouple as a source of electrical flow that is connected to a milliammeter calibrated in temperature degrees.

Thermocouple wire - a 2conductor cable, each conductor employing a dissimilar metal, made up specifically for temperature measurements.

Thermodisk defrost control - an electrical switch with a bimetal disk that is controlled by electrical energy.

Thermodynamic wet bulb temperature - that temperature at which water, by evaporating into air, can bring the air to saturation adiabatically at the same temperature. The wet bulb temperature measured with an appropriate psychrometer can approach the thermodynamic wet bulb temperature.

Thermodynamics - the science that deals with mechanical action or relations of heat.

Thermodynamics, Laws of - laws upon which rest the classical theory of thermodynamics. These laws have been stated in many different, but equivalent ways. The first law: (1) when work is expended in generating heat, the quantity of heat produced is proportional to the work expended; and conversely, when heat is employed in the performance of work, the quantity of heat that disappears is proportional to the work done (Joule); (2) if a system is caused to change from an initial state to a final state by adiabatic means only, the work done is the same for all adiabatic paths connecting the two states (Zemansky); (3) in any power cycle or refrigeration cycle the net heat absorbed by the working substance is exactly equal to the net work done. The second law: (1) it is impossible for a self-acting machine, unaided by any external agency, to convey heat from a body of lower temperature to one of higher temperature (Clausius); (2) it is impossible to derive mechanical work from heat taken from a body unless there is available a body of lower temperature into which the residue not so used may be discharged (Kevin); (3) it is impossible to construct an engine that, operating in a cycle, will produce no effect other than the extraction of heat from a reservoir and the performance of an equivalent amount of work (Zemansky).

Thermoelectric - conversion of heat to electrical energy or vice-versa.

Thermoelectric refrigeration - a refrigeration mechanism that depends on the Peltier effect. Direct current flowing through an electrical junction between dissimilar metals provides heating and cooling effect depending on the direction of current flow.

Thermography - the use of infrared thermography whereby temperatures of a wide variety of targets can be measured remotely and without contact. This is accomplished by measuring the infrared energy radiating from the surface of the target and converting this measurement to an equivalent surface temperature.

Thermomastic - a type of material that has a high heat absorbing ability. Used to prevent overheating of a device when making solder or brazed type connections.

Thermometer - instrument for measuring the intensity of heat. May be dial, spirit, or electronic type. This instrument measures only sensible heat.

Thermometer bulb - the rounded, lower end of a thermometer.

Thermometer, digital - a thermometer that uses solid state circuitry and direct reading digital numbers for temperature read outs. These thermometers are easier to read and are more accurate than ordinary spirit or column type thermometers.

Thermometer, resistance - temperature sensor consisting of a metal or silicon chip in which electrical resistance is a function of the temperature. The thermistor increases or decreases its resistance as the temperature rises or falls.

Thermometer, stack temperature - a thermometer with a range of 200° F to 1,000° F used to measure the temperature of the flue products from fossil fuel type heating units.

Thermometer well - a pocket or recess in a system, designed to hold a thermometer bulb; a temperature test point.

Thermometry - the measurement of heat.

Thermomodule - a number of thermocouples used in a parallel circuit to achieve low temperatures.

Thermophotovoltaic cell - a device that concentrates sunlight on a small heat absorber made of metal or other suitable material, heating it to a high temperature. The secondary thermal radiation re-emitted by the absorber is used as the energy source for a photovoltaic cell. The cell is chosen for maximum efficiency at the wavelength of the secondary radiation.

Thermopile - a number of thermocouples connected in series in a single unit (all the hot junctions are heated together), producing a much higher voltage than the individual thermocouples, usually in the range of 250-750 millivolts. Some types may be used as sources of electrical energy.

Thermoplastic - insulation that will re-soften and distort from its formed shape by heating above a critical temperature peculiar to the material.

Thermosiphoning - the tendency of heated water to rise vertically within a hydronic piping system, even when the circulators are off.

Thermostat - an automatic controller that responds to temperature changes to maintain the temperature within predetermined limits.

Thermostat, defrost - a thermostat mounted on the outdoor coil that initiates or terminates the defrost cycle when the temperature of the coil reaches the preset cut-in/cut-out set points of the control.

Thermostat, direct acting - an instrument for activating a control circuit upon sensing a predetermined low temperature.

Thermostat, modulating - a temperature controller that employs a potentiometer winding instead of switch contacts.

Thermostat, on and off - a thermostat designed to open or close an electric circuit in response to a temperature change.

Thermostat, operating - a thermostat that controls the operating temperature of a device.

Thermostat, outdoor - a control used to limit the amount of auxiliary electric heat according to the outdoor ambient temperature to reduce electrical surges and cost.

Thermostat override - (1) continued equipment operation after the thermostat is satisfied. (2) to jumper out or bypass.

Thermostat, reverse acting - an instrument for activating a control circuit upon sensing a predetermined high temperature.

Thermostat, room - a thermostat properly located in a room so as to respond to representative room temperature and thereby control heating or cooling devices.

Thermostat, safety - a thermostat that automatically limits the temperature of a appliance.

Thermostat, sub base - when installed with a thermostat, it permits selection of function of either heating, cooling, automatic change over, emergency heat, continuous fan operation, or off position.

Thermostatic control - a device that operates a system or part of the system based on temperature changes.

Thermostat controlling element - the portion of a thermostat that reacts to temperature change by opening or closing the contacts through mechanical linkage. The two types of elements are bimetal and bulb.

Thermostatic expansion valve (TXV) - a control valve operated by temperature and pressure within the evaporator coil that controls the flow of refrigerant. The power element feeler bulb is attached to the outlet of the evaporator coil. May be internally or externally equalized.

Thermostatic motor control - a device used to control the cycling of the unit through the use of a control bulb attached to the evaporator.

Thermostatic operator - a non-electric actuator that mounts to a valve and opens and closes the valve as necessary to maintain a set temperature.

Thermostatic radiator valve - the assembly of a thermostatic operator and a radiator valve.

Thermostatic trap - a steam trap that opens in response to a decrease in temperature and closes when steam reaches it. The temperature sensitive element is usually a sealed bellows or series of diaphragm chambers containing a small amount of volatile liquid.

Thermostatic valve - a valve controlled by thermostatic elements.

Thermostatic water valve - a valve used to control the flow of water through the system that is actuated by a temperature difference. Used on units such as water cooled compressor or condensers.

Thermosyphon - the natural circulation of a gas or liquid that occurs as it is heated. The warm, lighter material rises as the cooler material falls. Eventually, the warm material arrives at the top. A solar heating system in which the storage tank is higher than the collectors.

THHN wire -a UL listed, 90º C, small diameter, general purpose, 600 volt building wire to be used as power, lighting, and control wiring. The thin high grade, flame retardant, heat, abrasion, and moisture resistance PVC with polyamide nylon jacket permits the easy use of more wires in a given size conduit. Used in dry applications only. TH = thermoplastic, H = hot applications, N = non-metallic.

Thimble - a fixed or removable ring, tube, or lining usually located in the hole where the chimney connector or vent connector passes through a wall or enters a chimney or vent.

Thin wall conduit - galvanized, lightweight conduit. Also called electro-mechanical tubing or electrical metallic tubing (EMT).

Thread - a spiral groove cut into the outside of a pipe or the inside of a pipe fitting with a similar taper. When inside and outside thread of the same size are joined together properly, they form a leak-proof connection.

Thread angle - the V-angle of the threads. The sides of the threads normally form a 60º angle.

Thread cutting oil - a lubricating oil used to keep the dies lubricated and cool, and to prevent the teeth of dies from breaking. Thread-cutting oil contains a sulfur, lard or mineral base.

Thread joint compound - non-hardening materials used on pipe to ensure a tight seal.

Threaded rod - metal rod with its entire length threaded. It can easily be cut to any length with a hacksaw. Also known as all thread rod.

Threader - a device or tool used to cut thread on a piece of pipe.

Threads per inch - measurement determined by counting the number of crests (or roots) per inch of threaded section.

Three phase current - operating by means of a combination of three alternating current circuits that differ in phase by one third of a cycle (120 mechanical degrees). This type of motor can yield high starting and breakdown torques. They are more efficient than single phase, and larger horsepower ratings can be furnished. May be delta or wye.

Three phase delta connection system - in wiring, this arrangement employs three wires. Assuming 100 amps and 1,000 volts in each phase winding, the pressure between any two conductors is the same as the pressure in the winding, and the current in any conductor is equal to the current in the winding multiplied by the square root of 3, that is 100 x 1.7321 = 173.21 amps, or 173 amps.

Three phase 4 wire system - a three phase, three wire system having a fourth wire connected to the neutral point of the source that may be grounded.

Three phase motor - an induction type of motor that has a very high starting torque and requires no special starting apparatus. The motor must be operated on 3 phase current. Rotation maybe changed by reversing any two leads.

Three piece circulator - a circulator in which the wetted parts are separated from the motor by a coupling assembly.

Three pipe system - a system where one pipe supplies chilled water, a second pipe supplies heated water and the third pipe carries the return flow. Commonly referred to as a mixing valve.

Three pole breaker - a triple pole single throw (TPST) circuit breaker used to disconnect all three hot wires by a single throw of the switch lever.

Three way diverting valve - a three way diverting valve is a valve that always have one inlet and two outlets. Fluid entering the inlet port can be diverted to either of the two outlet ports in nay proportion desired by moving the valve stem. Valves designed for diverting service can usually be used also for mixing applications.

Three way switch - a three terminal switch in which a circuit may be switched to either of the two paths.

Three way mixing valve - a three way mixing valve is a valve that has three openings. It always has two inlets and one outlet. The proportion of fluid entering each of the two inlets can be varied by moving the valve stem. Valves designed for mixing service are not generally suitable for diverting service.

Three wheel cutter - a cutter with three cutter wheels, two of which have replaced the rollers; used to cut pipe when complete rotation is impossible.

Three wire control - another name for floating control in which three wires are required between the controller and actuator.

Three wire control circuit - in this circuit three wires are typically used to operate a heating only application (R-W) and continuous fan operation (R-G). The red low voltage wire is the uninterrupted power from the transformer power side that goes to the thermostat and the white wire from the thermostats "W" terminal goes to the heating circuit terminal in the furnace. The green wire goes from the thermostats sub-base to the 24 volt coil of the indoor fan relay.

Threshold current - the symmetrical RMS available current at the threshold of the current-limiting range, where the fuse becomes current-limiting when tested to the industry standard. This value can be read off of a peak let through chart. A threshold ratio is the relationship of the threshold current to the fuse's continuous current rating.

Threshold differential pressure - the set pressure at which a differential pressure bypass valve begins to allow flow to pass through itself.

Threshold limit value (TLV) - a term used by the American Conference of Governmental and Industrial Hygienists to express the airborne concentration of a material to which nearly all workers can be exposed day after day without adverse effects. "Workers" means healthy individuals. The young, old, ill, or naturally susceptible have lower tolerances and need to take additional precautions. ACGIH expresses TLV's in three ways: (1) TLA-TWA, the allowable time-weighted average concentration for a normal eight hour workday or forty hour work week; (2) TLV-STEL, the short term exposure limit or maximum concentration for a continuous exposure period of fifteen minutes (with a maximum of four such periods per day, with at least sixty minutes between exposure periods, and provided that the daily TLV/TWA is not exceeded). (3) TLV-C, the ceiling exposure limit or the concentration that should not be exceeded even instantaneously.

Throttle - to reduce or cut off fluid flow.

Throttling - the expansion of gas through an orifice or controlled opening without the gas performing any work in the expansion process.

Throttling range - in proportional controllers the change in controlled variable required to move the controlled devices or devices from one extreme limit of travel to the other.

Throttling valve - a valve that is used for control of flow rate.

Through fixed ladder - a fixed ladder that requires a person getting off at the top to step between the side rails of the ladder to reach the landing.

Through stays - these are found on firetube boilers (HRT and scotch marine) to keep the front and rear tube sheets from bulging.

Through-the-wall - describes a type of air conditioning installation in which the unit's chassis is mounted through the outer wall of the building, with the evaporator inside, the condenser outside, and the other components in between.

Throw, air - the horizontal distance an air stream travels, after leaving a horizontal sidewall outlet, before the maximum velocity is reduced to the terminal velocity. For a perimeter outlet, throw is the vertical distance the air stream travels before the maximum velocity is reduced to the terminal velocity.

Throw, mechanical - the distance between the centerlines of the crankshaft and the connecting rod journal; the offset in a crankshaft hat is equal to half the stroke of the piston.

Throw, electrical - the number of settings of a switch or relay in which an electrical current can pass through the switch or relay contact points. Movement of the pole (switch) in an electrical device.

Throw of an outlet - measured in feet, is the distance from the center of the outlet to a point in the mixed air stream where the highest sustained velocity of the mixed air stream has been reduced to a specified level. Generally taken as the full distance from the outlet center to a position, or to a point midway between two outlets in the direction of the air path.

Thrust - the force acting lengthwise along the axis of a shaft either towards or away from it.

Thrust bearing - one that supports a load that acts along the center line of the supported component. In a motor mounted with its shaft in a vertical position, the weight of the rotor must be supported by a thrust bearing.

THWN wire - a UL listed, 600 volt, general purpose 75° C wet or dry location conductor insulated with a flame retardant, heat, abrasion, and moisture resistant PVC with polyamide nylon coating to be used as power, lighting, and control circuits. TH = thermoplastic, W = wet location use, N = non-metallic.

Thyristor - a general term used to describe a group of four layer semiconductor devices that can be pulsed into conduction, but can be turned off only by interrupting the applied voltage. SCR's and triacs are thyristors. An electronic component that has only two states of operation; on or off.

Tick rate - the frequency of a Geiger counter effect instrument makes in proximity to a leak.

Tie rod - the sections of cast iron sectional boilers are held in tight contact by means of tie rods that pass entirely through the sections.

Tight house - construction type that has insulated walls and ceilings, vapor barriers, weather stripping on doors and windows, snug doors, and fireplace damper. In regards to infiltration and building construction standards the air changes per hour is considered to be one half.

Tight shutoff - virtually no flow or leakage when a valve is in its closed position. Generally speaking, only single seated valves have tight shutoff. Double seated valves can be expected to have 2 to 5% leakage in the closed position.

Time delay fuse - a circuit interrupting device that permits momentary over current for circuits normally drawing heavy current for a short time. Used for inductive and capacitive load circuits.

Time delay relay - a relay in which there is a delay between the time the coil is energized or de-energized and the contacts open or close. Often used to control fans for greater heating or cooling efficiency.

Time of day rate - time of day rates allow the utility company to vary the power billing rate with the time of day the power is used.

Time weighted average (TWA) - the concentration for a conventional 8 hour workday and a 40 hour work week, to which nearly all workers may be repeatedly exposed, day after day, without adverse effect. Usually annotated as TWA.

Timed trial for ignition - refers to burners (usually oil) using direct spark ignition of the main burner. See main burner flame establishing period.

Timed trial for main flame - see main burner flame establishing period.

Timed trial for pilot - see pilot flame establishing period.

Timed two position action - a variation of the two position action in which the on periods are prematurely shortened. Through this may be accomplished by many means, it is usually accomplished in electric room thermostats by means of a heating element that is energized during the on cycle (for heating) and off periods (for cooling).

Timeout - on a status indicator this is a condition caused by a timing circuit that will cut the cuber off, if the ice harvest is not triggered during an ice harvest, in the amount of time determined by the preset timer adjustment.

Timer - a timing device used in a flame safeguard programming control to sequence the operation of a burner system. It consists of an AC synchronous motor driving a series of cams that open or close several sets of switch contacts at predetermined times.

Timer clock - a mechanical or electric device used to monitor the actual time of day.

Timer cycle - the time required for a complete rotation of the timer in a flame safeguard programming control, excluding the time the timer stops for high fire or low fire proving and for the run period.

Timer, defrost - a timer that operates at the same time as the refrigeration system. After a set period of operating time, the timer trips to initiate a defrost operation if the termination switch is closed.

Timer dial - a rotating disk on the end of the timer that shows what sequence period the programming control is in. It enables the operator to determine if the system is operating properly and, in some cases, to move the timer to the proper positions for check out or troubleshooting.

Timer motor - electric motor drives a cam that actuates defrost switch at specific time intervals to begin defrosting of outdoor coil when heat pump is on the heating cycle.

Timer sequence - the order in which timer functions are performed in a programming control.

Timer thermostat - a thermostat control that includes a clock mechanism. Unit automatically controls the room temperature and changes the temperature according to the time of day.

Timers - a mechanism used to control on and off times of an electrical circuit.

Tinning - the practice of coating a clean soldering iron with molten solder to facilitate further soldering.

Tin snips - heavy duty shears for cutting sheet metal.

Tip jack - a single conductor socket that connects to the F terminal of a flame signal amplifier, used for troubleshooting the amplifier and flame relay.

Tip speed - the peripheral speed (in feet per minute) of a fan or blower blade at any specified RPM. The speed of the outer edge of a fan blade or blower.

Title sheet - usually the first sheet of a set of working drawings, most often including a drawing of the plot on which the building is to be constructed.

TLV Ceiling limit - the ceiling exposure limit or concentration not to exceed at any time, even for very brief times. The ACGIH publishes a book annually that explains and lists TLV's.

Toggle bolt - threader fastener with spring loaded wings used to fasten heavier objects to a hollow section of wall. The wings are collapsed inserted into the hole in the wall where they expand to bear against the inside of the wall.

Toggle switch - an electrical circuit control that is opened or closed by flipping a projecting lever.

Tolerance - amount of variation or change allowed from a standard accuracy, especially the difference between the allowable maximum and minimum sizes of some mechanical or electrical part. For example, a 50 microfarad run capacitor with a tolerance of plus or minus 10% may be either 45 to 55 microfarad.

Ton of refrigeration - the refrigeration effect equal to the melting of one ton of ice in 24 hours. It may be expressed as 288,000 Btu's per 24 hours (2000 pounds of ice x 144 (latent heat of fusion) or 12,000 Btu/hr or 200 Btu's per minute.

Ton refrigeration unit - a unit that removes the same amount of heat in 24 hours as melting one ton of ice.

Topping cycle - cogeneration facility in which the energy input to the facility is first used to produce useful power, and the rejected heat from production is used for other purposes.

Topping off - the process of adding refrigerant to a system to bring up to full capacity.

Torch ignition - manual ignition of a pilot in a semiautomatic system, or manual ignition of the main burner in a manual system.

Torque - the forces produced by twisting or rotating equipment. Can be measured in foot-pound (ft/lbs), inch-pounds (in/lbs), ounce-feet (oz/ft), or ounce-inches (oz/in). In motors the torque values are: locked rotor torque, or starting torque and breakdown torque. There are three types of torque:

Breakdown - the maximum torque that a motor will develop with rated voltage applied at rated frequency, without an abrupt drop in speed.

Locked rotor - the minimum torque that a motor will develop at standstill for all angular positions of the rotor, with rated voltage applied at rated frequency. Also called starting torque or breakaway.

Pull up - the minimum torque developed by an alternating current motor during the period of acceleration from standstill to the speed at which breakdown torque occurs with rated voltage applied at rated frequency.

For motors which do not have a definite breakdown torque, the pull-up torque is the minimum torque developed up to rated speed.

Torque, full load - maximum torque delivered without overheating.

Torque, stall - torque developed when starting.

Torque, starting - amount of torque available, when at 0 rpm, to start and accelerate the load.

Torque wrench - a special wrench that may be used to measure the torque or pressure applied to a nut or bolt head.

Torr - a unit of pressure equal to 1/760 atmosphere or approximately 1 mm of mercury.

Torsional stress - one that resists forces tending to twist a body.

Total air - (1) the total amount of air supplied to a burner. It is the sum of primary, secondary and excess air. (2) a mixture of discharged and entrained air.

Total air rate - total mass flow of dry air per hour through a water tower.

Total charge - the amount of refrigerant needed for a system's operation, plus the refrigerant needed to fill the lines.

Total cooling capacity - the combined latent and sensible heat removal abilities of a cooling unit. It is expressed in Btu/H at standard rating conditions for air cooled units of 95° F condenser inlet air, 80° F DB, 67° F WB evaporator entering air. For water cooled units, conditions are 80° F DB, 67° WB evaporator entering air, 75° F water in, 95° F water off condenser. These conditions are specified by the Air Conditioning and Refrigeration Institute.

Total cooling effect - the difference between the total enthalpy of the dry air and water vapor mixture entering a unit per hour and the total enthalpy of the dry air and water vapor (and water) mixture leaving the unit per hour, expressed in Btu per hour.

Total current - the current that flows from the voltage source of a circuit.

Total energy management - conservation concept where a building is looked at in terms of its total energy usage, rather than analyzing the requirements of separate systems.

Total energy system - form of cogeneration in which all electrical and thermal energy needs are met by on-site systems. A total energy system can be completely isolated or switched over to a normally disconnected electrical utility system for back-up.

Total equivalent length - the sum of the equivalent lengths of all piping, fittings, valves, and other components in a piping circuit.

Total equivalent warning impact (TEWI) - a unit of measurement that assesses the total effect CFC's, HCFC's, and HFC's have on global warming.

Total force - total pressure that is acting on an area, determined by diameter and pressure.

Total head - the sum of discharge head, suction lift and friction losses.

Total heat - the sum of sensible and latent heat. Also called enthalpy.

Total heat load - all the sources of heat for a conditioned space; including infiltration, conduction, radiation and internally generated heat.

Total heating capacity - the output of a furnace or winter air conditioner in Btu/h is the total heating capacity. The winter air conditioner capacity must meet or exceed the heat loss of the conditioned space. Also known as bonnet capacity.

Total pressure - the sum of both the static pressure and the velocity pressure of a moving stream of air at the point where the measurement is taken.

Total pressure loss - friction loss of the ductwork the supply fan must overcome to deliver the required volume of air to the conditioned space.

Total R value - the overall total thermal resistance of an assembly of materials that are assembled together.

Total resistance - the total opposition to current flow of a circuit that may be found by removing the voltage source and connecting an ohmmeter across the points where the source was connected.

Total suspended particles - the weight of particulate matter suspended in a unit volume of air.

Total system charge - the amount of refrigerant that is contained in an air conditioning or refrigeration system. On package units, the total system charge is indicated on the nameplate of the unit. On split systems, the total system charge is determined by weighting the refrigerant as it is introduced into the system.

Total volatile organic compounds (**TVOC**) - the sum of organic vapors in air measured by an appropriate sampling and analysis procedure. TVOC's are difficult to measure and interpret with normal IAQ instrumentation and methods.

Total voltage - the voltage supplied by a source.

Total water rate - total mass flow of water per hour through the tower.

Totalizer - dial that determines hourly or daily flow rates.

Totally enclosed air over motor (**TEAO**) - a motor constructed with no ventilation openings that requires external airflow over the motor enclosure for cooling.

Totally enclosed fan cooled motor (**TEFC**) - a motor constructed with no ventilation openings that includes an external fan with a shroud to move cooling air over the motor.

Totally enclosed non-ventilated motor (**TENV**) - a motor constructed with no ventilation openings.

Touch test - used primarily as a troubleshooting technique. The technician can determine the general operating conditions of a system by touching the refrigerant lines and sensing their temperatures. A solid understanding of the basic refrigeration system is needed for this method to be effective.

Tower pumping head - the static lift from the elevation of the basin curb to the centerline elevation of the distribution system inlet; plus the total pressure (converted to feet of water) necessary at that point to effect proper distribution of the water to its point of contact with the air.

Tower, water - an enclosed device for evaporating cooling water by contact of air. Efficiency is determined by the wet bulb temperature of the ambient air.

Tower, water, atmospheric - one in which the air movement through the tower is dependent only upon atmospheric conditions. Also known as natural draft cooling tower.

Tower, water, forced draft - a mechanical draft water cooling tower having one or more fans located in the air entering the tower.

Tower, water, induced draft - a mechanical draft water cooling tower having one or more fans located in the air leaving the tower.

Townhouse - a living unit in which one or more walls are partition, lot line, or common walls but not containing common floor and ceiling combinations.

Toxic - poisonous, hazardous to life.

Toxicity classification - refrigerants shall be classified for toxicity to one of two classes in accordance with ASHRAE 34. A Class A refrigerant is one that has a toxicity that has not been identified at concentrations less or equal to 400 parts per million (ppm), based on data used to determine Threshold Limit Value--Time Weighted Average (TLV-TWA) or consistent indices. A Class B refrigerant is one for which there is evidence of toxicity at concentrations below 400 parts per million (ppm), based on data used to determine Threshold Limit Value--Time Weighted Average (TLV-TWA) or consistent indices.

TP card - a specially sized pocket temperature pressure card that shows the temperature pressure relationships of various refrigerants. May also be referred to as a PT card.

Tracer gases - compounds, such as sulfur hexafluoride, that are used to identify suspected pollutant pathways and to quantify ventilation rates. Tracer gases may be detected qualitatively by their odor or quantitatively by monitoring equipment.

Tracking collector - a solar energy collector that constantly positions itself perpendicularly to the sun as it rotates.

Tracks - moves over or along; traverse.

Trade names - because of pioneering development, or merely common usage, some control devices are referred to by their manufacturer's marketing names. The new control technician should not let this become confusing.

Transducer - a device actuated by power from one system and supplies power in another form to the second system. Examples are a strain gauge or thermocouple.

Transfer air - air moved from one conditioned space to another, usually through doorways and grilles. Note: By comparison, infiltration air and exfiltration air are usually considered to be transferred from, or to, unconditioned spaces.

Transformer - an electromagnetic device, with two or more coils linked by magnetic lines of force, used to increase or decrease AC voltage. With a voltage increase, current decreases and vice versa. In HVAC controls the transformer usually converts line voltage (115v) to a low voltage (24v).

Transformer rectifier - a combination transformer and rectifier in which the input an AC. May be varied and then rectified in DC.

Transient - an instantaneous voltage surge through a circuit that may be caused by lightning, static electricity or the activation of inductive loads.

Transistor - a small, solid state electronic component employing a semiconductor such as germanium or silicon to perform the functions of amplifying, switching or controlling electrical signals. The transistor has replaced the older vacuum tube in a variety of applications because of its many mechanical rigidity, long life, freedom from heater current, and ability to operate at low voltages with relatively high currents. Complex circuits utilizing many transistors and other circuit components that may be integrated into a single chip.

Transition duct section - the piece of duct section that is used to join two different duct sections together with differing cross sectional measurements

Transmission - the heat loss or gain from a building through exterior components such as windows, walls, floors, etc.

Transmission coefficient - the ratio of the solar heat gain through a glazing system to that of an unshaded single pane of double strength glass under the same set of conditions.

Transmission, heat, coefficient of - any one of a number of coefficients used in the calculation of heat transmission by conduction, convection, and radiation, through various materials and structures.

Transmission line - a system of conductors suitable for conducting electric power or signals efficiency between two or more terminals. Also known as line.

Transmission loss - the reduction of airborne sound power that is caused by placing a wall or barrier between the reverberant sound field of a source and its receiver. Transmission loss is a property of the wall or barrier.

Transmission power loss - the amount of energy lost during transmission due to electrical resistance. Transmission power loss is determined using the equation: power loss = I2R

Transmissivity - the capacity of a material to transmit radiant energy. Transmittance is the ratio of the radiant flux transmitted through a body to that incident to it.

Transmittance, thermal (U factor) - the time rate of heat flow per unit area under steady conditions from the fluid on the warm side of a barrier to the fluid on the cold side, per unit temperature difference between the two fluids. Common unit is Btu per hour/ square foot/ Fahrenheit degree.

Transmitter - a sensor (not a controller) that transmits a directly functional signal in proportion to the variable measured. This proportion is factory set and cannot be adjusted. With this control there is no ability for sensitivity adjustment.

Transom - a window directly above a door, often used to mount a window air conditioners.

Transverse - pertaining to occurrences in the direction of tower width.

Transverse duct joint - those points where one duct section is joined to a fitting or another duct section.

Trap - a vertical U-shaped offset filled with water and located beneath plumbing fixtures to form a seal against the passage of foul odors or gases.

Trapeze - pipe hanger arrangement formed by two suspended hanger rods and a crossbar that supports the pipe.

Travel coefficient - the ratio between the flow at a given valve stem position and the flow through the valve at its full open position; expressed as a decimal fraction.

Tread depth - the horizontal distance from front to back of a tread, excluding nosing, if any.

Tri cocks - usually three in number. Located on water column or on boiler itself. Under pressure the top try cock should show steam and the lower two, water. Under vacuum air will enter through openings. Try cocks are used to check water in glass and may be used to maintain proper water level if glass is broken.

Tri state control - another name for floating control in which the actuator is either opening the valve, closing the valve, or holding the valve at its current position.

Triac - a bidirectional, three lead rectifier consisting of two SCR's connected in inverse parallel. Triac is an abbreviation of a triode AC switch. It can be triggered into either forward or reverse conduction by pulses applied to the gate terminal.

Trial for ignition - refers to burners (usually gas or oil) using direct spark ignition of the main burner. See main burner flame establishing period.

Trial for main flame - see main burner flame establishing period.

Trial for pilot - see pilot flame establishing period.

Triangle - a three sided geometric figure.

Tribology - the science and technology of interacting surfaces in relative motion, including the study of lubrication, friction and wear. Tribological wear is wear that occurs as a result of relative motion at the surface.

Trichlorotrifluoroethane - complete name for R-113. A Group 1 refrigerant. Chemical compounds that make up this refrigerant are chlorine, fluorine, and ethane.

Trim - trim consists of all parts of a valve that are in contact with the flowing medium, but are not part of the valve shell or casting. Thus seats, discs, stems, packing rings are all trim components. The term trim is usually used in connection with trim materials.

Trimmer - beam framing an opening in a wood joist floor supporting the header beam.

Triode - a three element electronic device, usually a vacuum tube.

Trip out point - the point at which an electrical control will open its contacts in the event of an unsafe condition.

Triple action control - a preassembled control that provides high limit, low limit, and circulator control in hydronic systems using tankless water heaters

Triple point - the pressure temperature condition in which a substance is in equilibrium in a solid, liquid and vapor states.

Troffer - a recessed lighting unit, usually long and installed with the opening flush with the opening. Most standard size is 2 foot by 4 foot and containing 4 bulbs or lamps.

Troposphere - the lower atmospheric level that extends upward from ground level to about 7 miles and that part of the atmosphere in which most weather disturbances occur. Global warming takes place in the troposphere.

Troubleshooting - the process of observing a malfunctioning system's operation and diagnosing the cause of the malfunction.

Truck, refrigerated - a commercial vehicle equipped to maintain a temperature below atmospheric conditions.

True power - the average power consumed by a circuit (or component) during one full cycle of alternating current. Measured directly with a wattmeter.

True RMS - a circuit used for measuring the effective heating value of AC signals. The RMS value of an AC waveform equals the DC value that would produce equivalent heating in a purely resistive load. A true RMS circuit makes accurate measurements even if the waveform is not sinusoidal.

Trunk - the largest or main circuit that carries the combined capacity of two or more sub-circuits. The main duct of an air system.

Trunk line - the main run of a duct system. This run of ductwork is connected to the air handler. The take-offs are then connected to the main trunk line.

Truss - a structure composed of a combination of members, usually in some triangular arrangement so as to constitute a rigid framework.

Tube - a small diameter, flexible pipe, usually made of copper, aluminum, or plastic.

Tube and plate system - a type of radiant panel construction in which tubing is saddled with or snaps into a preformed metal plate intended to laterally disperse heat away from the tubing.

Tube brushes - used in firetube boilers to remove soot and debris from inside the tubes.

Tube bundle - the arrangement of tubular water pipes within the evaporator of a chiller or heat exchanger.

Tube, capillary - in refrigeration practice a tube of small internal diameter used as a liquid refrigerant flow control or expansion device between the high and the low sides; also used to transmit pressures from the sensing bulb of some temperature controls to the operating element.

Tube, constricted - a tube that is restricted in diameter.

Tube, finned - heat transfer tube or pipe with extended surfaces in the form of fins, discs or ribs.

Tube, finned - a heat transfer tube or pipe that has extended surfaces in the form of fins, ribs or discs.

Tube sheet - tubes are rolled, expanded, and beaded into front and rear tube sheets of HRT and scotch marine boilers and upper and lower tube sheets of vertical firetube boilers.

Tube, soft copper - deoxidized, and dehydrated, seamless, soft copper tube, thoroughly annealed to assure quality for bending and flaring.

Tube-within-a-tube - a water cooled condensing unit in which a small tube is placed inside a larger tube. Refrigerant passes through one tube while water flows through the other.

Tubeaxial fan - an air moving device where the airflow is parallel or axial to the shaft on which the propeller is mounted. The propeller is housed in a cylindrical tube or duct. This design enables duct fans to operate at higher static pressures than propeller fans. Commonly used in spray booth and other ducted exhaust systems. As static pressure is increased, horsepower increases and CFM decreases.

Tubing - a fluid carrying pipe where OD always equals tubing size.

Tubing cutter - tool used to make an accurate 90 degree cut on either hard or soft cutter.

Tubing, pre-charged - a refrigeration line set that is fitted with quick connect type fittings on each end. The line has been factory evacuated and filled with refrigerant. The lines may also be evacuated and sealed or pressurized with nitrogen.

Tubular - round gauge glass used for pressures up to and including 250 psi.

Tuned circuit - an electronic circuit, containing both inductance and capacitance, that is adjusted to resonate or operate at a specified frequency.

Tungsten load - incandescent lamps are tungsten loads. The cold resistance of a tungsten load is extremely low. This can result in inrush currents of as much as 6 times the steady state current, causing contact erosion or welding. Tungsten contact ratings are specified as amp-t or amp tungsten.

Tunnel burner - a gas burner in which the actual combustion begins in a refractory tunnel or tuyere block. It is a relatively high capacity partial premix type of gun burner, which uses the kinetic energy of the gas stream issuing from the orifice to bring in at least part of the air required for combustion. A wide range of control of flame characteristics is obtained by adjustment of the primary air, thus causing combustion to be delayed until secondary air can reach the air fuel mixture. Also see gun type and partial premix burners.

Turbidity - the degree of opacity of a fluid.

Turbine - an engine driven by the pressure of steam, water, or air against the curved vanes of a wheel.

Turbine, double flow - turbine that splits steam in two directions before passing over the turbine blades.

Turbine, single flow - turbine that utilizes steam in a single path over the turbine blades.

Turbine stages - that part of the turbine where steam gives up its energy to the turbine blades. As the steam pressure drops, the stages (blades) become larger.

Turbine, steam - prime mover of electrical generators whose power is derived from high velocity steam passing through nozzles and directed at the turbine's rotor blades.

Turbulence - fluid flow that is not parallel to the wall of the pipe or duct through which the fluid is flowing. Turbulence could be caused by rough inner walls, obstructions or directional changes.

Turbulent flow - a classification of flow where streamlines repeatedly cross over each other as the fluid flows along the pipe. Turbulent flow is the dominant type of flow in hydronic heating systems.

Turnbuckle - simple position adjustment device for use on pipe supports.

Turndown ratio - the ratio of the maximum usable flow to the minimum controllable flow through a valve; usually somewhat less than the rangeability, as the maximum usable or required flow is usually less than the total flow capacity of a valve.

Turndown range - the range of firing rates over which satisfactory combustion may be obtained. For example, the firing rate of a burner with a 4 to 1 turndown range can be varied from about 25% to 100%.

Turndown voltage - the amount of millivoltage required to just hold open the pilot safety device to allow gas to flow to the main burner. If the voltage drops below this value then the pilot safety would drop out.

Turning vanes - rigidly installed curved metal air guides places in elbows, stack heads, etc., to produce as uniform as possible air pattern through the fitting.

Turns - the number of wire loops in a coil, especially in a transformer.

Turns ratio - the ratio of the number of turns of the primary winding (Np) of a transformer to the number of turns of the secondary winding (Ns).

Tuyere - a nozzle or opening through which combustion air is introduced into a combustion chamber.

Two conductor Romex - a cable used in circuit breaking containing three wires -- black (hot), white (neutral), and bare copper (grounding).

Two mode switching - two stage switching operations such as start/stop, on/off and open/closed.

Two phase power - produced by having an alternating with two sets of coils wound 90° out of phase with each other.

Two pipe system, fuel oil - a piping system where one pipe (suction) is used for the supply, while the other is for the return of any unused oil that the burner didn't use. These systems are considered self-priming.

Two pipe system, hot water - a heating system using two separate pipes to supply and return water from the terminal units. The essential feature of a two pipe system is that each heating unit receives a direct supply of the heating medium which cannot have served a preceding heating unit. Another name for a parallel direct return or parallel return distribution piping system.

Two pole breaker - a double pole single throw (DPST) circuit breaker used for connecting a 240 volt single phase circuit. This is usually a dedicated circuit to the piece of equipment only.

Two position action - the type of action where the output signal is changed to either a maximum or minimum value with no intermediate steps (contrast to proportional).

Two position actuator - an actuator that moves a damper or valve to either of only 2 positions - on or off (open or closed), or high or low. Also see actuator.

Two position valve - a valve that is either fully open or fully closed, with no positions in between. Also called an on-off valve.

Two speed compressor - a compressor in a refrigeration system that has two different speeds to match system load capacity. This is accomplished by using different motor windings whether or not they are internal (hermetic) or external (open drive) of the compressor itself.

Two stage air compressor - a system where air is compressed in the first stage (larger diameter cylinder), cooled in the intercooler, and then compressed again in the second stage (smaller diameter cylinder). Used where air pressure is above 150 psi.

Two stage firing - a system where the combustion rate of the burner can be set at either of two fixed levels.

Two stage heating system - system that tries to match the furnaces heating capacity to the actual heating requirements by energizing the furnaces heating controls in steps or stages.

Two stage heating thermostat - control device used with a two stage heating system. If more heat is needed than the first stage can supply, the thermostat will energize the second stage of heating.

Two temperature valve - a pressure opened valve used in the suction line on multiple refrigeration systems that maintain evaporators in the system at different temperatures.

Two way valve - see straight through valve.

Two well system – a geothermal heat pump configuration in which the supply and return water piping for the system are connected to separate underground wells.

Two wire control circuit - in this circuit two wires are typically used to operate a heating only application. The red low voltage wire is the uninterrupted power from the transformer power side that goes to the thermostat and the white wire from the thermostats "W" terminal goes to the heating circuit terminal in the furnace.

Type I certification (small appliances) - this certification (under Section 608 of EPA standards) must be attained before working on small alliance equipment that is manufactured, charged and hermetically sealed with five (5) pounds or less of refrigerant. This classification includes refrigerators, freezers, room air conditioners, package terminal heat pumps, dehumidifiers, under-the-counter ice makers, vending machines and drinking water coolers.

Type II certification (high pressure appliances) - this certification (under Section 608 of EPA standards) must be attained before working on high pressure equipment that uses refrigerant with a boiling point between -50° C (-58° F) and 10° C (50° F) at atmospheric pressure. Includes 12, 22, 114, 500 and 502 refrigerants.

Type II certification (very high pressure appliances) - this certification (under Section 608 of EPA standards) must be attained before working on very high pressure equipment that uses refrigerant with a boiling point below -50° C (-58° F) at atmospheric pressure. Includes 13, and 503 refrigerants.

Type III certification (low pressure appliances) - this certification (under Section 608 of EPA standards) must be attained before working on low pressure equipment that uses refrigerant with a boiling point above 10° C (50° F) at atmospheric pressure. Includes 11, 113, and 123 refrigerants.

Type "B" vent - a gas only vent pipe that is comprised of two pieces of pipe. The inner pipe is made of aluminum and the outer is made of galvanized steel. It is round in shape. Designed for category I type furnaces only.

Type "BW" vent - a gas only vent pipe that is comprised of two pieces of pipe. The inner pipe is made of aluminum and the outer is made of galvanized steel. It is oval in shape and to be used with wall furnaces only.

Type "K" copper - a thick walled flexible tubing. Required for underground installations. Typical uses include water service, plumbing, heating, steam, gas, oil, oxygen, and applications where thick walled tubing is required. Can be used with sweat, flared and compression fittings. Available in hard or soft types.

Type "L" copper - standard tubing used for interior, above ground plumbing. Uses include heating, air conditioning and steam. Can be used with sweat, flare, or compression fittings. Available in hard or soft types.

Type "L" vent - a gas/oil vent pipe that is comprised of two pieces of pipe. The inner pipe is made of stainless steel and the outer is made of galvanized steel. Used on category I furnace appliances only.

Types of motors - classified by operating characteristics and type of power required:

1. Capacitor start motor (CS): a type of single phase induction motor that has a capacitor in series with the starting winding only during the starting operation. This type of motor has high locked rotor torque with low amp draw, high breakdown torques used on hard starting applications such as compressors, positive displacement pumps, conveyors, etc.
2. Capacitor start-capacitor run (CSR): performance and applications is similar to capacitor start induction run, the capacitor enables operation at lower current demand and improved efficiency. Used on larger single phase HP ratings.
3. Permanent split capacitor (PSC): single phase induction motor that has a capacitor continuously in the start (auxiliary) winding. Unlike the split phase or capacitor, start type motors there is no centrifugal switch to disconnect the start windings. This type motor has low starting torque and is best suited for direct drive applications such as fans, blowers, etc. PSC motors are much more efficient than shaded pole motors.
4. Shaded pole motor (SP): a special type of single phase induction motor that utilizes a copper shading coil to provide the necessary displacement for starting. Low starting torque, typically used in direct drive fans, blowers, etc.
5. Split phase motor (SP): the most common type of single phase induction motor. Equipped with a start winding displaced in magnetic position from, and connected in parallel with the run winding. The start winding is disconnected from the circuit by either a potential or start relay or a centrifugal switch. Generally, this motor has low to moderate locked rotor torque with high locked rotor amps, high breakdown torques, and is used on belt drive blowers, fans, pumps, etc.
6. Three phase: operates on three phase power (3 power lines) dependent upon designs can yield high starting and breakdown torques. They are more efficient than single phase, and larger HP ratings can be furnished. Found in all larger size motors.
7. Direct current: requires direct current supply and primarily used for adjustable speed drive applications or motor vehicles.
8. Universal motor: one than can be operated on either AC or DC, brush type. Used primarily in high speed applications such as drills, saws, etc.

-U-

U - thermal transmittance.

Uo - overall thermal transmittance.

UEL - upper explosive limit.

UV - ultra violet or unit ventilator.

U value - a coefficient expressing the thermal conductance of a composite structure in Btu's per square foot hour degree F temperature difference. The reciprocal of the resistance factor of the insulation. The lower the U value, the better the insulating value. Also known as U factor.

U.L. approved - Underwriters Laboratory approved.

U.L. group - the Underwriters Laboratories classification reflects the comparative life hazard of refrigerants in the absence of flames or surfaces of high temperatures. Group 1 is the most toxic (e.g. R-764) and group 6 is the least. This measure is used to classify refrigerants in older codes, still in effect in some jurisdictions.

U.L. labeled - see labeled.

U.L. listed - a product certification, indicating that the product meets certain safety standards set for it by Underwriters Laboratories. The UL listing mark may appear in various forms. If the product is too small for the marking to be applied to the product, the marking must appear on the smallest unit container in which the product is packaged. Not a product endorsement.

U.L. recognized - refers to a product that is incomplete in construction features or limited in performance capabilities. A "recognized" product is intended to be used as a component part of equipment that has been "listed." A "recognized" product must not be used by itself. A UL product may contain a number of components that have been "recognized."

Ultimate analysis of coal - when a more thorough analysis of coal is required, an ultimate analysis is preformed. The ultimate analysis breaks down coal into its chemical constitutes and heating and caloric values. An example for semi bituminous coal is sulfur 2.26%, hydrogen 4.14%, carbon 79.97%, nitrogen 1.26%, oxygen 4.18%, and the calorific value is 14,081 Btu/lb.

Ultimate CO_2 - the percentage of carbon dioxide in dry combustion products when a fuel (gas) is completely burned with exactly the amount of air needed for complete combustion. This is the theoretical maximum CO2 that can be obtained for a given gas in burning the gas in air.

Ultra-low - temperatures ranging from -50° F to -250° F.

Ultrasonic - a range of sound waves beyond human hearing capability, that can be used in test equipment to identify failing equipment, plumbing, or electrical systems.

Ultraviolet - invisible radiation waves with frequencies shorter than wave lengths of visible light and longer than x-ray.

Ultraviolet degradation - the degradation caused by long time exposure of a material to sunlight or other ultraviolet rays containing radiation.

Ultraviolet flame detector - a detector that responds to the ultraviolet radiation (wavelength between 0.02 And 0.4 micron) emitted by a flame. Also see minipeeper and purple peeper.

Ultraviolet power tube - a high power electron tube sensitive to ultraviolet radiation. When saturated with ultraviolet radiation, this tube delivers about 1/10 watt to the flame signal amplifier. The flame relay requires about 3/4 watt to pull in, so little power amplification is required.

Ultraviolet radiation (UV) - invisible, electromagnetic waves of high frequency (short wavelength - between 0.02 and 0.4 micron). Ultra violet radiation comprises a little less than one percent of the light spectrum.

Ultraviolet solution - a liquid added to air conditioning or refrigeration systems that aid in the leak detection process. The liquid solution glows brightly when seen under ultraviolet light.

Unadaptered person - a person who has been in a space for less than 15 seconds and has become accustomed to the odors and contaminants in the space. The 15 second definition for adaptation is from ASHRAE and does not imply that all persons will accept the surroundings in this time frame.

Unconditioned space - space within a building that is not conditioned space.

Unconfined space - a space having a volume not less than 50 cubic feet per 1,000 Btu/hr of the aggregate input rating of all appliances installed in the space. Rooms communicating directly with the space, in which the appliances are installed, through openings not furnished with doors, are considered a part of the unconfined space.

Undercharged - a refrigeration system that is short or lacking of refrigerant.

Underfeed stoker - a coal firing system that introduces the coal under the fire.

Underfed evaporator - an evaporator that is operating at a low efficiency due to a reduced amount of refrigerant being supplied to it. Also referred to as a starved evaporator.

Underground cable - cable specially designed for below grade installations. Type UF is an approved underground cable.

Undershoot - when the actual value of a controlled variable drops below the intended lower limit of operation.

Undersized - where the capacity of the machine or system cannot keep up with the load applied. An air conditioning system that runs continuously and doesn't cool the conditioned space to the required setting is said to be undersized.

Uneven parallel system - parallel compressors of unequal sizes mounted on a steel rack and controlled by a microprocessor.

Ungrounded conductor - the conductor of an electrical system that is not intentionally connected to ground. This conductor is referred to as the "hot" or "live" conductor.

Unheated space - space within a building that is not heated space.

Union - fitting used to connect two straight lengths of pipe or tubing, neither of which can be turned. A device placed in a pipeline to facilitate disassembly of the system.

Unison light-off - in a multi-burner system, ignition of all pilots at the same time. All pilots are in the same valve zone (see valve zone). When all pilots have been proved, all main burners are ignited at the same time.

Unison shutdown - in a multi-burner system, if the flame goes out or the flame signal is lost for any one burner in the valve zone, all burners in that zone are shut down automatically.

Unit - a factory-made, encased assembly of the functional elements indicated by its name, such as air conditioning unit, room-cooling unit, humidifying unit, etc.

Unit bearing - an all position mounting bearing, constructed with a long, single sleeve bearing. Used for fan duty only.

Unit cooler - a direct cooling, factory made, encased assembly including a cooling element, fan and motor, and directional outlet.

Unit heater - a self-contained, automatically controlled, vented, fuel burning device having an integral means for circulation of air. It is not intended for attaching pipes or ducts for the distribution of the heated air. A unit heater may be floor mounted or suspended.

Unit heater cooler - a self contained, packaged air conditioner for localized temperature control.

Unit ventilator - a terminal unit in which a fan is used to mechanically circulate air over the heating coil. These units are so constructed that both outdoor and room air may be circulated so as to provide ventilation as well as heat. These units may contain a cooling coil for summer operation.

Unitary cooling and heating equipment - one or more factory-made units that normally include an evaporator or cooling coil or a compressor and condenser combination, and may include a heating function as well. Where such equipment is provided in more than one assembly, the separate assemblies shall be designed to be used together.

Unitary heat pump - one or more factory-made units that normally include an indoor conditioning coil, compressor(s), and outdoor coil or refrigerant-to-water heat exchanger, including means to provide both heating and cooling functions. When such equipment is provided in more than one assembly, the separate assemblies shall be designed to be used together.

Unity - pure resistance where the phase angles for the voltage and current rise and fall at the same value. The power factor for this type of electrical circuit is 100%

Universal certification - in relation to EPA section 608, anyone that has or holds Type I, Type II and Type III certification.

Universal joint - a type of swivel for a socket wrench handle that permits reaching around objects.

Universal motor - a brush type motor that is designed to operate on either AC or DC. Used primarily in high speed applications such as drills, saws, etc.

Universal orifice - a combination fixed and adjustable orifice designed for the use of two different gases, such as LPG and natural gas.

Universal replacement part - a part that can be used as a replacement part for many different models of equipment, including that of different manufacturers.

Unloader - a device on or in a compressor for equalizing the high and low side pressures for a brief period during start-up, in order to decrease the starting load on the motor or to disengage certain cylinders of a multi-cylinders compressor to modulate the unit's cooling capacity.

Unproved ignition - ignition by an energy source assumed to be energized during the main burner flame establishing period.

Unsaturated state of air solubility - when the amount of dissolved air in the system water is low enough to enable the water to resorb any residual air trapped in other parts of the system.

Unusual sources - sources that release contaminants at rates that are significantly higher than those typically found in spaces of the same type, or sources that release unusual (foreign) contaminants for the type of space.

Unvented - an appliance that is not provided with a flue pipe or flue collar, and from which the combustion products are freely emitted into the space in which the appliance is located.

Updraft - excessively low air pressure existing at the outlet of a chimney or stack that tends to increase the velocity and volume of gases passing up the stack.

Up-feed system - hydronic system in which the supply main is located below the level of the terminal units.

Upflow furnace - a central furnace with essentially vertical airflow, that discharges air at or near the top of the furnace; as distinguished from a downflow furnace.

Upper explosive limit - the maximum concentration of a combustible gas or mixture of gases in air or oxygen that will ignite and burn (propagate a flame).

Upper zone - room volume above the stratification level created by displacement ventilation.

Upshot burner - a gas burner in which the gas is delivered vertically from below the combustion chamber, resulting in a vertical flame.

Upstream - that area that is closer to the source of fluid flow.

Urethane foam - a type of insulation that is formed in between inner and outer walls of an appliance, display case or container.

Usable temperature - the temperature or range of temperatures at which heat energy can be absorbed, rejected, or stored for use within the system.

Usage profile - a graph that shows the quantity of domestic hot water or electrical usage required at any hour of the day.

Useful oil pressure - the difference in pressure between the discharge and suction sides of the compressor oil pump.

"U" tube manometer - a low pressure measuring instrument having the measuring liquid in a "U" shaped tube where the pressure would push down on the fluid measuring material and support it on the other side.

Utility - any service necessary for the functioning of a plant such as electricity, gas, water, fire protection, sewage disposal, steam, fuel oil, instrumentation air, utility air, cooling water, drains, condensate, flare system, or blowdown system. In general, anything that is not directly associated with the primary processes in a plant can be considered a utility.

Utility air - air line used in cleanup operations, to drive motors, and other service functions.

Utility gases - natural gas, manufactured gas, liquefied petroleum gas-air mixtures or mixtures of any of these gases.

Utility rate structure - a utility's approved schedules of charges to be made in billing for utility service rendered to various classes of customers.

Utility service - the utility company. Also, the amount and configuration of voltage supplied by a supplied by a utility company. There are four main types of commercial utility services: 208 volts AC wye, 480 volts AC wye, 240 volts AC delta, and 480 volts AC delta.

Utility steam - steam that is generated and used for service in a plant.

Utility transformer - primary and secondary coils of wire that reduce (step down) the utility supply voltage for use within a facility.

Uv power tube - see ultraviolet power tube.

U value - see U factor.

-V-

V - volt, volume.

Va - volt ampere. A designation of power in terms of voltage and current.

VAC - volts alternating cycle.

VAV - variable air volume.

VDC - volts direct current.

VDT - video display terminal.

VOC - volatile organic compounds.

VOM - volt-ohm-milliammeter.

VP - velocity pressure.

VTVM - vacuum tube voltmeter.

V belt - endless belt having a cross section resembling a truncated triangle or V. Usually made of reinforced rubber, V belts can either be smooth or notched and comes in almost any length.

V block - a V shaped groove in a metal block to hold round objects.

V max - maximum (peak) voltage in an alternating current cycle.

V pulley - a pulley that is used with a V belt.

V rms - voltage root mean square; average voltage equal to the maximum (peak) voltage multiplied by 0.707.

VA rating - for transformers, the product of the output voltage times the maximum amps that can be supplied by the secondary windings.

VAC - abbreviation for volts alternating cycle. This is the unit for rating capacitors.

Vacuum - an enclosed space from which practically all air has been removed; a reduction of pressure below atmospheric pressure; a space with nothing at all in it.

Vacuum control system - in many automobile air conditioning systems, the intake manifold vacuum is used to operate dampers and controls in the system.

Vacuum controller - a controller that can be set to operate at pressures less than atmospheric pressure.

Vacuum gauge - a field tool used to measure the vacuum level in a closed system. Vacuum gauges can read the level in either inches of mercury or microns.

Vacuum heating system - a two pipe heating system equipped with the necessary accessory apparatus and controls to permit the pressure in the system to go below atmospheric pressure.

Vacuum pressure - pressure below atmospheric pressure usually measured in inches of mercury..

Vacuum pump - a special high efficiency compressor used to create high vacuums for testing or drying purposes.

Vacuumstat - a Honeywell® trademark. See vacuum controller.

Valence electrons - electrons in the outer orbit of an atom. The number of atoms in this orbit dictates whether the material is a conductor or an insulator.

Valence state - energy content of an electron in orbit about an atomic nucleus.

Valley - the trough or gutter formed by the intersection of two inclined planes of a roof.

Valve - a device used to isolate a part of a piping system by stopping flow; for controlling or regulating pressure or flow of a fluid or liquid. Capable of regulating the flow (cut-off, diversion, or modulation) of the medium moving through the pipe. Also see cock, automatic control, butterfly, diaphragm, diverting, double seated, firing rate, manual shutoff, mixing, safety shutoff, single seated, solenoid, straight through, two position, and vent valves.

Valve actuator - a device that physically moves the stem or shaft of a valve using electrical, thermal, or pneumatic input energy.

Valve automatic control - a valve combines a valve body and a valve actuator or motor. A signal from some remote point can energize the actuator either to open or close the valve, or to proportion the rate of flow through the valve.

Valve, back pressure - an automatic valve located between the evaporator outlet and the compressor inlet that is responsive to its own inlet pressure or to the vapor flow when necessary to prevent the evaporator pressure from falling below a selected value.

Valve body - the portion of the valve body through which the medium flows.

Valve body rating - the maximum pressure that a valve body can withstand. The actual body rating is the maximum safe permissible fluid pressure at a given temperature. The nominal body rating is a specified rating (often cast on the valve body) based on the specified body material and universally accepted construction characteristics determined by specifications issued by authorities having jurisdiction. It provides a convenient method of classifying the valve by pressure for identification purposes. The nominal body rating is not the same as the actual body rating. Also, the maximum rating of the valve may be limited to the value below the body rating by the packing and disc material.

Valve bonnet - the removable cover or top of the valve body (attached by screws). The bonnet assembly may include the valve trim.

Valve, cap seal - a manual valve whose seal is protected by a tightly fitting cap.

Valve, charging - a valve located on the liquid line, usually near the receiver, through which refrigerant may be charged into the system.

Valve, check - valve that allows fluid to flow in one direction only.

Valve closed indication switch - a single pole double throw switch in a valve actuator that can be wired to either make or break a circuit when the valve is in the closed position. When used with a valve with a second seat (not to be confused with a double seated valve), it constitutes a valve closed interlock. The second seat closes the valve before it reaches the end of its stem travel. The valve closed indication switch in the actuator operates only after the valve stem continues past the point at which the second seat has closed the valve. Also see valve closed interlock.

Valve closed interlock - an interlock that ensures that the automatic main fuel valve(s) is closed before allowing burner operation to start or to proceed through pre-purge; usually used as a pre-ignition interlock. Also see valve closed indication switch.

Valve closing time - the time it takes an automatic valve to close completely after it is de-energized by a flame safeguard control.

Valve, condenser shut off - a valve located in the hot gas discharge line at the inlet of the condenser.

Valve, constant pressure - a controlling device for regulating the flow of volatile refrigerant into the cooling unit, actuated by changes in pressure of the low side.

Valve, cylinder discharge - the valve in a compressor through which the gas leaves the cylinder.

Valve, cylinder suction - the valve in a compressor through which the gas enters the cylinder.

Valve, diaphragm - a form of packless valve, manually or mechanically actuated. Also a valve actuated by pressure of a motivating fluid on one side of the diaphragm that seals the motivating fluid from the flowing fluid controlled by the valve. Flowing and motivating fluids could be, but need not, be the same.

Valve, direct acting - one that closes with the admission of fluid pressure to a diaphragm and opens when pressure is released.

Valve disc - the movable portion of a valve that comes into contact with the seat as the valve closes, and which controls the flow of the medium through the valve. A contoured disc controls the flow by a shaped end, sometimes called a valve plug, and is usually end-guided at the top or at the bottom (or both) of the valve body. A V-port disc has a cylinder (called a skirt or guide) that rides up and down in the seat ring. The skirt guides the disc and the shaped openings in the skirt vary the flow area. A quick opening disc is so machined that maximum flow is quickly achieved when the disc lifts from the seat. It is either end-guided or guided by wings riding in the seat ring.

Valve, discharge - on a compressor, the valve that allows compressed refrigerant to flow from the cylinder to the discharge port.

Valve, emergency relief - a manually operated valve for the discharge of refrigerant in case of fire or other emergency.

Valve, equipment shut-off - a valve located in the piping system, used to shut off individual equipment.

Valve, expansion - a type of refrigerating metering control that maintains a pressure difference between the high and low sides in the refrigeration system. The valve is caused to operate by pressure in the low or suction side to redirect the opening of the valve on the high side.

Valve, float - a regulating valve controlled by liquid level or a valve actuated by a float in a liquid container.

Valve flow characteristic - the relationship between the stem travel, or lift, expressed in percentage, and the flow of fluid through the valve, expressed in percentage of full flow. Also see equal percentage and linear characteristics.

Valve flow meter - measures the flow of a substance by the movement of a piston in a valve caused by resistance to the flow of the substance.

Valve, gate - a valve that has a wedge shaped leaf which, when closed, seals tightly against two metal seats that are at slight angles. Usually used where flow will be completely closed or open.

Valve, globe - a compression type valve, commonly used where there is occasional or periodic use, such as lavatory connections, hose bibs, and oil shut off valves.

Valve guide - the part of the valve disc that keeps the disc aligned with the valve seat. Also see characterized guide and valve skirt.

Valve, king - a manual stop valve located between the outlet of the receiver and the inlet of the metering device.

Valve, low side float - a float valve operating by change in level of low pressure liquid, opens at low level and closes at high pressure.

Valve modulating - a valve that can be positioned anywhere between fully on and fully off to proportion the rate of flow in response to a modulating controller.

Valve, packless - a valve that does not use packing to prevent leaks around the valve stem. Flexible material is generally used to seal against leaks and still permit valve movement.

Valve, pilot - a small capacity pressure control valve used to control the action of a larger capacity control valve.

Valve plate - that part of a compressor located between the top of the compressor body and the head. Usually contains the compressor suction and discharge valves.

Valve plug - a term used to describe a disc that has a solid, shaped end (instead of a skirted guide) to characterize flow through a valve. In some valves, the plug may serve as a guide.

Valve port - the flow controlling opening between the seat and the disc of a valve. It does not refer to the body size nor to the end connection size of the valve. Regardless of the pipe size, the valve port size determines the flow.

Valve, pressure relief - a safety valve design to prevent the development of a dangerous condition by relieving either pressure, temperature, or vacuum in a hot water supply system.

Valve purge - a shut-off valve that, when opened, vents noncondensable gases.

Valve ratings - maximum fluid pressure and temperature ratings are limitations placed upon a valve by the maximum pressure and temperature to which the valve may be subjected. Determining factors may be packing, body material, disc material, or actuator limitations. Also see capacity index, close off ratings, pressure drop, rangeability, tight shutoff, travel coefficient, turndown, valve body rating, and valve flow characteristic.

Valve, reducing - a valve that reduces high inlet pressure to the required lower outlet pressure.

Valve reed - a thin piece of steel located on top of the compressor cylinder and under the valve plate. The valve reed flexes to open and close as gas is compressed inside the cylinder.

Valve, reverse acting - a valve that opens with the admission of fluid pressure to a diaphragm and closes when pressure is released.

Valve, reversing - the control used to regulate the flow of hot high pressure vapor from the compressor and the cool low pressure vapor to the compressor from the coils depending upon the direction of heat flow desired.

Valve seal over travel interlock - see valve closed interlock.

Valve seat - the stationary portion of a valve with which the movable portion (disc) comes into contact to stop flow completely.

Valve, service - a device used by service technicians to check the operating pressures, charge or evacuate a refrigeration system.

Valve, service shut-off - a valve, installed by the serving gas supplier between the service meter or source of supply and the customer piping system, to shut off the entire piping system.

Valve skirt - a valve guide that has notches or V's cut into it to characterize the flow through the valve; also known as a skirted guide or a characterized guide.

Valve, slide - the slide valve portion of the reversing valve that shifts the refrigerant flow.

Valve, solenoid - a valve actuated by magnetic action by means of an electrically energized solenoid.

Valve stem - a cylindrical shaft that is moved manually or by an actuator, and to which the disc is attached.

Valve, stop - a shut-off valve other than a valve for controlling the flow of refrigerant.

Valve, suction - a valve in a refrigeration compressor that allows vaporized gas to enter the cylinder from the suction line and prevents its return.

Valve, thermostatic radiator - a device that combines a temperature sensitive power unit with a valve. Mounts in the hot water or steam supply pipe adjacent to a radiator. May also be applied to baseboard and similar types of heaters. Requires no electricity.

Valve trim - all parts of a valve that are in contact with the flowing medium, but are not a part of the valve shell or casting. Seats, discs, stems, and packing rings are all trim items.

Valve, two position - a valve that is either full on or off with no intermediate position. Also referred to as an on-off valve.

Valve, TXV, bi-flow - a thermostatic expansion valve that is designed to provide pressure reduction and refrigerant flow control in either direction.

Valve, water regulating - valves that regulate the flow of water through the condenser while it is running. Most water regulating valves are controlled by the head pressure of the refrigeration system.

Valve zone - a section of a multi-burner system that is controlled by one valve (or one set of valves; e.g., double block and bleed). Example: "5 burners per zone" means one valve controls the gas to five burners.

Vane - thin plate or strip in the opening of a grille.

Vane ratio - in air distributing devices the ratio of depth of vane to shortest opening width between two adjacent grille bars.

Vanes - the blades mounted on a rotor inside a compressor or other rotating machine.

Vapor - the gaseous form of any substance that is usually a liquid or a solid; a gas, near equilibrium with the liquid phase of the substance, that does not follow the gas laws.

Vapor barrier - a thin plastic or metal foil sheet used in air conditioned structures to prevent water vapor from entering the insulating material. Vapor barriers are always installed on the warm side of the insulation layer.

Vapor charge bulb - a charge in a thermostatic expansion valve bulb that boils to a complete vapor. When this point is reached, an increase in temperature will not produce an increase in pressure.

Vapor charged - lines and component parts of a system that are charged at the factory. A type of refrigerant charging where only vapor or gas is introduced into the refrigeration system.

Vapor compression - the process on which the basic refrigeration cycle is based. Refrigerant vapor is compressed to increase its temperature and pressure, after which the heat absorbed into the system, can be rejected to a medium at a lower temperature.

Vapor density - the weight of a vapor or gas compared to the weight of an equal volume of air; an expression of the density of the vapor or gas calculated as the ratio of the molecular weight of the gas to the average molecular weight of air, 29. Mw of gas/29 = vapor density. Materials lighter than air have vapor densities of less than 1.0. Materials heavier than air have vapor densities greater than 1.0. All vapors and gases will mix with air, but the lighter materials will tend to rise and dissipate (unless confined). Heavier vapors and gases are likely to concentrate in low places (along or under floors; in sumps, sewers, and manholes; in trenches and ditches) where they may create a fire, explosion, or health hazards.

Vapor heating system - a two pipe steam heating system which operates at pressures at or nar atmospheric and which returns the condensate to the boiler or receiver by gravity.

Vapor line - found only in dual action heat pumps, it is the suction line in the cooling mode and the hot gas discharge line in the heating mode.

Vapor lock - a condition where liquid is trapped in a line because of a bend or improper installation that prevents the vapor from flowing.

Vapor pressure - (1) the pressure exerted by a saturated vapor above its own liquid in a closed container. (2) pressure of a confined vapor in equilibrium with its liquid at specified temperature thus, a measure of a liquid's volatility.

Vapor pressure curve - a graphic presentation of various pressures produced by refrigerant under various temperatures.

Vapor pump - another name for a refrigeration compressor. The compressor takes superheated vapor from the evaporator section and adds the heat of compression and rejects the new superheated refrigerant out to the condenser.

Vapor refrigerant charging - adding refrigerant to a system by allowing vapor to move out of the vapor space of a refrigerant cylinder and into the low pressure side of the refrigeration system.

Vapor retarder - a layer of low moisture transmissivity material (not more than 1.0 perm dry cup) placed over the warm side (in winter) of insulation, over the exterior of below grade walls, and under floors as ground cover to limit the transport of water and water vapor through exterior walls, ceilings and floors. Vapor retarding paint, listed for this application, also meets this definition.

Vapor, saturated - a vapor condition that will result in condensation into droplets of liquid as vapor temperature is reduced.

Vapor, superheated - vapor at a temperature that is higher than the saturation temperature (i.e., boiling point of water) at the existing pressure.

Vapor, water - used commonly in air conditioning terminology to refer to steam in the atmosphere.

Vaporization - the change of liquid into a gaseous state.

Vaporstat - a control with a large diaphragm that makes it highly sensitive to low pressure.

Variable - a factor or condition that can be measured or altered by a control system, such as temperature, pressure, flow, liquid level, humidity, weight, or chemical composition.

Variable air volume controller - device having electronic or pneumatic components used to regulate the volume of air in a distribute system.

Variable air volume system - a space conditioning system that maintains comfort levels by varying the volume of conditioned air to the zones served. This system delivers conditioned air to one or more zones. The duct serving each zone is provided with a motorized damper that is modulated by a signal from the zone thermostat.

Variable area flow meter - this meter measures the flow of a substance by how much resistance is created by a float or piston that changes the area (size) of the flow path.

Variable pitch pulley - a pulley that can be adjusted to provide different pulley ratios.

Variable resistor - a resistor whose resistance value can be varied between its minimum and maximum values.

Variable speed circulator - a circulator that is operated over a wide range of speeds using either an internal or external variable speed drive.

Variable speed motors - shaded pole or permanent split phase capacitive motors able to change speeds without having to be disconnected by a switch or relay. Commonly used to operate light loads such as ceiling fans and blower motors.

Varistor - a resistor that changes its resistance value with a change in voltage. Varistors are used to protect electronic circuits from transient voltage spikes.

Varnish - when applied to lubrication, a thin, insoluble, nonwipeable film deposit occurring on interior parts, resulting from the oxidation and polymerization of fuels and lubricants. Can cause sticking and malfunction of close clearance moving parts.

VARs - (**volts-amps-reactive**) unit for the measurement of reactive power. In a pure inductive circuit, the value of the voltage multiplied by the current. Also known as wattless power.

Vaulted ceilings - all ceilings where enclosed joist or rafter space is formed by ceilings applied directly to the underside of the roof joists or rafters.

Vector - a line that represents the direction and magnitude of a quantity.

Velocimeter - a instrument used to measure air velocities using a direct reading air speed indicating dial.

Velocity - the rate of motion in a particular direction in relation to time speed. Usually expressed in feet per second.

Velocity head - the mechanical energy possessed by a fluid due to its velocity.

Velocity meter - a meter used to detect the velocity of fluids, air, or water.

Velocity noise - undesirable sound created when the flow velocity through the pipe or piping component exceeds a certain value.

Velocity, outlet - the average discharge velocity of primary air being discharged from the outlet, normally measured in the plane of the opening.

Velocity pressure - in a moving fluid, the pressure capable of causing an equivalent velocity as applied to move the same fluid through an orifice such that all pressure energy expended is converted into kinetic energy.

Velocity profile - a sketch that shows how the velocity of the fluid varies from the centerline of a pipe out of the pipe wall.

Velocity recovery fan cylinder - a fan cylinder of a water tower on which the discharge portion is extended in height and outwardly flared. Its effect is to decrease the total head differential across the fan, resulting in either an increase in air rate at constant horsepower, or a decrease in horsepower at constant air rate.

Velocity reduction method - a method of duct sizing where arbitary reductions are made in velocity after each branch or outlet.

Velocity riser - a vertical pipe that is sized smaller to increase the velocity and ensure oil entrainment in refrigerant piping systems. The velocity riser (smaller-size pipe) on a double riser suction line is sized for minimum load capacity.

Velocity, room - the average, sustained, residual air velocity level in the occupied zone of the conditioned space. Usually expressed in feet per minute.

Velocity, terminal - the highest sustained airstream velocity existing in the mixed air path at the end of the throw.

Velometer - an instrument used to measure the air velocity in a duct system.

Vena contracta - the smallest area of a fluid stream leaving an orifice.

Veneered wall - a wall having a facing of masonry or other weather-resisting noncombustible materials securely attached to the backing, but not so bonded as to exert common action under load.

Vent - (1) a small hole or opening to permit passage or escape, as of a gas. (2) A passageway used to convey flue gases from gas utilization equipment, or their vent connectors, to the outside atmosphere.

Vent alarm - a device placed in series in the vent pipe of a fuel oil tank that whistles as the air is escaping from the tank as the tank is being filled. The fill person would listen to the sound and as the whistling stops they know that the tank is nearly filled and stops the flow of oil.

Vent cap - a protective cover or housing attached to the vent termination, intended for preventing downdrafts and the entry of rain, snow, hail, and vermin.

Vent condenser - removes the oxygen and other noncondensables gases in a deaerating Feedwater heater.

Vent connector - the pipe that connects a gas fired device to a gas vent or chimney. Also called breeching or flue pipe.

Vent damper - a device intended for installation in the venting system, in the outlet of or downstream of the appliance draft hood, of an individual automatically operated fuel-gas burning unit and which is designed to automatically open the venting system when the appliance is in operation and to automatically close off the venting system when the appliance is in a standby or shutdown condition. Can be electrically, mechanically or thermally actuated.

Vent free - certain gas stoves or fireplaces are not required to be vented, therefore, they are called "vent free."

Vent gases - products of combustion from gas appliances plus excess air, plus dilution air in the venting system above the draft hood.

Vent pipe - a vertical pipe used to ventilate plumbing systems and to remove odors and sewer gases from the building to the outside and to provide circulation of air to and from any part of the drainage system.

Vent stack - the upper portion of a waste or soil stack extending through the roof of a building, installed primarily for the purpose of providing circulation of air.

Vent system - a continuous open passageway from the flue collar or draft hood of a fuel burning appliance to the outside atmosphere for the purpose of removing products of combustion.

Vent terminal - the fitting at the end of a vent pipe that directs the flue gases into the outside atmosphere and keeps out rain, snow debris, and animals.

Vent valve - an electrically operated, normally open valve that allows any gas that leaks through any upstream valve(s) to escape to the outside atmosphere. Also called a relief valve. Also see bleeding and double block and bleed.

Vent valve, steam - a device for permitting air to be forced out of a heating unit or pipe and which closes against steam.

Vent valve, water - a device for permitting air to be forced out of a heating unit or pipe and which closes against water.

Vented gas categories - (a) Category I. An appliance that operates with a non-positive vent static pressure and with a vent gas temperature that avoids excessive condensate production in the vent. (b) Category II. An appliance that operates with a non-positive vent static pressure and with a vent gas temperature that may cause excessive condensate production in the vent. (c) Category III. An appliance that operates with a positive vent static pressure and with a vent gas temperature that avoids excessive condensate production in the vent. (d) Category IV. An appliance that operates with a positive vent static pressure and with a vent gas temperature that may cause excessive condensate production in the vent.

Vented wall furnace - a self-contained, vented, fuel-gas burning appliance complete with grilles or equivalent, designed for incorporation in or permanent attachment to the structure of a building and furnishing heated air, circulated by gravity or by a fan, directly into the space to be heated through openings in the casing. Such appliances should not be provided with duct extensions beyond the vertical and horizontal limits of the casing proper, except that boots not to exceed 10 inches beyond the horizontal limits of the casing for extension through the walls of nominal thickness are permitted. When such boots are provided, they shall be supplied by the manufacturer as an integral part of the appliance. This definition excludes floor furnaces, unit heaters, direct vent wall furnaces, and central furnaces.

Ventilation - the process of supplying or removing air by natural means (including infiltration, operable windows, and doors) or mechanical means (such as motor driven fans and blowers) to or from a space for the purpose of providing fresh air, and controlling air contaminant levels, humidity, or temperature within the space. For indoor air quality considerations, this term always refers to outside air and not to total air moved in and out of an occupied space. Such air may or may not be conditioned.

Ventilation air - the air required for the removal of excessive external heat surrounding an appliance, or for the removal of such heat within the enclosure in which the appliance is located. For certain spaces, transfer air from spaces with acceptable indoor air quality may also be considered ventilation air.

Ventilation rate - the rate at which indoor air enters and leaves a building. Expressed in one of two ways: the number of changes of outdoor air per unit of time (air changes per hour, or ACH) or the rate at which a volume of outdoor air enters per unit of time (cubic feet per minute, or CFM).

Ventilation system efficiency - a ratio of the rate of outdoor air supplied to the occupied zones served by a ventilation system and the actual outdoor air rate supplied by the system as measured at its outdoor air intake. The difference between the outdoor air supplied by the system and the outdoor air delivered to the occupied zones served by the system exists due to various air distribution inefficiencies including imperfect mixing in the space that is expressed by the air change effectiveness.

Ventilator - see exhaust fan.

Venting - (1) the removal of combustion products as well as process fumes to the outer air. (2) the act of releasing refrigerants to the atmosphere. The venting of ozone depleting refrigerants is prohibited by the Clean Air Act of 1990.

Venting system - a continuous open passageway from the collar or draft hood of a gas-burning appliance to the outside atmosphere for the purpose of removing flue or vent gases.

Ventricular fibrillation - the random, irregular heartbeat that often occurs when a person gets an electrical shock of 100 to 200 milliamps of current.

Venturi - a short tube with a constricted, throat-like passage that increases the velocity and lowers the pressure of a fluid passed through it.

Venturi effect - effect created as a fluid flows through a reducing orifice, creating an increase in velocity of the fluid as well as a reduction in pressure at the inlet of the device.

Venturi mixing burner - see inspirating burner.

Venturi port - any valve in which the internal port is smaller than the pipeline bore.

Verturi type distributor - a distributor that relies on the venturi effect to thoroughly mix the liquid and vapor refrigerant before distributing the refrigerant to the evaporator sections. Theses distributors have the advantages of causing little turbulence in the refrigerant and causing a minimum of overall pressure loss as refrigerant flows through the device.

Vertex - the starting point of an angle, from which two lines diverge.

Vertical firetube boiler - a one pass boiler that has fire tubes in a vertical position. Vertical firetube boilers are classified as wet-top or dry-top.

Vertical ground loop - used in geothermal heat pump systems where buried loops are installed vertically where ground space is limited.

Vertical panel radiator - a panel radiator that is taller than it is wide, and has tubes oriented in a vertical direction.

Vertical riser - a vertical portion of the suction line, usually used with an oil trap, in refrigeration system.

Vertical rotary oil burner - see rotary oil burner.

Very high pressure appliance - an appliance that uses a refrigerant with a boiling low point below 58° F at atmospheric pressure. This includes but not limited to refrigerants R-13 and R-503.

Vessel - any container used in conjunction with a piping system to hold, transform, or store the medium.

Vibration - a periodic motion of the particles of an elastic body or medium in alternately opposite directions from the position of equilibrium when that equilibrium has been disturbed. A quivering or trembling motion.

Vibration absorbers - a flexible, shock absorbing fitting installed next to or below a vibrating component to prevent the transfer of vibration through the line or base.

Vibration eliminators - soft or flexible substance or device that will reduce the transmission of a vibration.

Vibration isolators - mounting pads for vibrating machinery that prevents the transmission of the vibration to other parts of the building.

Vibration switch - a circuit control that shuts down machinery in case of excessive vibration.

Virgin refrigerant - new, original, non-recycled refrigerant.

Viscosity - the property of a liquid that results in a resistance to flow. Viscosity decreases as the temperature increases. The unit of the coefficient of viscosity is called the poise. Measured in terms of Saybolt Seconds Universal or Saybolt Furol Seconds.

Viscosity, absolute - the force per unit area required to produce unit relative velocity between two parallel areas of fluid unit distance apart.

Viscous - thick, resistant to flow. Applied to a liquid.

Viscous friction - internal friction that develop within a fluid or between the fluid and a surface due to the viscosity of the fluid. Viscous friction is responsible for loss of head energy.

Vise grips - adjustable pliers that can be locked at a desired opening.

Vise jaws - serrated jaws that are part of a clamping device (part of the pipe vise). The lower jaw is stationary while the upper jaw is adjustable to allow for holding different size pipes.

Visible light - light that can be perceived by the naked eye. Electromagnetic waves with frequencies between those of infrared and ultraviolet, with wavelengths between 0.4 And 0.8 micron.

Visible light transmittance - the percentage of visible light that is transmitted through glazing. Visible light transmittance is affected by special coatings and the number of panes of glass or film in an insulating glass unit.

Vital heat - heat generated by the ripening of fruits and vegetables in storage.

Vitreous - pertaining to a composition of materials that resemble glass, such as lavatories, electrode holders, or drain pipe.

Volatile liquid - a liquid that evaporates easily and rapidly at a low temperature and pressure.

Volatile matter - gases given off when burning coal.

Volatile organic compounds (VOC's) - compounds that evaporate from the many housekeeping, maintenance, and building products made with organic chemicals. These compounds are released from products that are being used and that are in storage. In sufficient quantities, VOC's can cause eye, nose, and throat irritations, headaches, dizziness, visual disorders, memory impairment; some are known to cause cancer in animals; some are suspected of causing, or are known to cause, cancer in humans. At present, not much is known about what health effects occur at the levels of VOC's typically found in public and commercial buildings.

Volt - the difference of electrical potential between two points of a conductor carrying a constant current of one ampere, when the power dissipated between these points is equal to one watt. A unit of electromotive force. One volt will produce a flow of one ampere through a one ohm resistance.

Volt ampere (va) - the unit of measurement of apparent power. Volts times amperes.

Voltage - a term used to indicate the electrical potential difference or electromotive force in an electrical circuit. It is voltage or electrical pressure that causes current to flow.

Voltage control - device used to provide some electrical circuits with uniform or constant voltage.

Voltage drop - a term expressing the amount of voltage loss from the original input in a conductor or load. Also referred to as IR drop.

Voltage flicker - significant fluctuation of voltage.

Voltage, nominal - a nominal value assigned to a circuit or system for the purpose of conveniently designating its voltage class such as 120/240 or 480Y/277. The actual voltage at which a circuit operates can vary from the nominal within a range that permits satisfactory operation of equipment.

Voltage rating - a rating that indicates the amount of voltage that can be safety connected to a device.

Voltage regulator - a device or circuit that maintains a constant value of voltage.

Voltage relay - a normally closed relay wired into a motor's start winding circuit. The relay remains closed upon start up due to a voltage drop in the circuit. As the motor attains rated speed, the back electromotive voltage rises and operates a magnetic relay that opens the normally closed start winding contacts that are in series with the start windings. Also referred as a start relay or a potential relay.

Voltage to ground - for grounded circuits, the voltage between the given conductor and that point or conductor of the circuit that is grounded. For ungrounded circuits, the greatest voltage between the given conductor and any other conductor of the circuit.

Voltaic cell - a cell produced by suspending two dissimilar elements in an acid solution. A potential difference is developed by chemical action.

Volt ampere - the unit of measurement of apparent power. Also the unit of electrical power. Abbreviated va.

Voltmeter - an instrument used to measure the potential difference between two specific points.

Volt-ohm-milliammeter (VOM) - a multi-function, multi-range meter that usually is designed to measure voltage, current, and resistance.

Volume - the three dimensional capacity (length, width and height) of a given space. Always expressed in cubic terms.

Volume dampers - a damper designed to control the direction and quantity of air flow within a duct system.

Volume factor - provides the number of gallons of water (or fluid) per linear foot of a piping material.

Volume of a pipe - the measurement of the space within the walls of the pipe. To find the volume of a pipe, multiply the length (or height) of the pipe by the product of the inside radius times the inside radius by Pi (3.1416).

Volume, specific - the volume of a substance per unit mass; the reciprocal of density. Units are cubic feet per pound, cubic centimeters per gram, etc.

Volumetric efficiency - term used to express the relationship between the actual performance of a compressor or vacuum pump and the calculated performance of the pump based on its displacement versus its actual pumping ability.

Volumetric efficiency, apparent - the ratio of length of suction line, on indicator card, to stroke.

Volumetric efficiency, due to cylinder heating - the ratio of the total to the apparent volumetric efficiency. Also called real or no-clearance volumetric efficiency.

Volumetric efficiency, total - the ratio of the actual volume of gas removed by the compressor or pump to the actual displacement of the compressor or pump.

Volute - the case containing the spinning rotor in a centrifugal compressor or pump.

VOM (**volt-ohm-milliammeter**) - a common test instrument that combines a voltmeter, ohmmeter and milliammeter in one case.

Vortex tube - mechanism for cooling or refrigerating that accomplishes cooling effect by releasing compressed air through specially designed openings. Air expands in rapidly spiraling columns of air that separates slow moving molecules (cool) from the faster moving molecules (hot).

Vortex tube refrigeration - refrigerating or cooling devices using the principle of vortex tubes.

Vortexing - a whirlpool action in the sump of a cooling tower or receiver.

-W-

W - symbol for watt, wattage, or width.

WOG - water, oil, or gas.

WSFO - weather service forecast office.

WSO ap - weather service office (airport)

Walk-in cooler - a large commercial refrigerated space kept below room temperature.

Wall heat capacity - effective wall heat capacity (Btu/ft2 ·°F) for purposes of calculating thermal mass performances for exterior walls. It is the sum of the products of the mass of each individual material in the wall per unit area of wall surface times its individual specific heat.

Wall stacks - these pieces of sheet metal are ducts that run vertically in the studding space of a frame wall. The available depth in a stud wall is 3 5/8" deep. The industry has standardized the size of wall stack material to be 3 ¼" deep. Standard widths of wall stack material are 8", 10", 12" and 14". The 14" wide wall stack will easily fit into a standard 16" on center stud spacing of 14 5/8".

Wall thickness - a term expressing the thickness of a layer of applied insulation or jacket.

Walls - those portions of the building envelope that are vertical or tilted at an angle of 30° or less from the vertical plane. These include above-grade where all of the exterior wall or any given story if 50% or more of the gross exterior wall area of the story is exposed to outside air or below-grade where all of the exterior walls of any given story if more than 50% of the gross exterior wall area of the story is below grade.

Warm air heating system - a warm air heating plant consists of a heating unit (fuel consuming furnace) enclosed in a casing, from which the heated air is distributed to the various rooms of building through ducts. If the motive head producing flow depends on the difference in weight between the heated air leaving the casing and the cooler air entering the bottom of the casing, it is termed a gravity system. If a an is used to produce circulation and the system is designed especially for the fan circulation, it is termed a forced air warm system.

Warm side - the side of any partition that is at the higher temperature.

Warm-up burner - a burner usually smaller than the main burner, that is ignited by another ignition source, and is used to warm up the boiler.

Warp switch - an electrical control with a bi-metal heat sensing element that changes shape when heated, causing the contacts to open or close.

Warping - bending or distortion of boiler or superheater tubes, usually caused by overheating.

Warranty - an assurance of a products proper performance.

Wash - the slight slope or beveled edge on the top surface of a masonry chimney designed to shed water away from the clay flue liner.

Washer - a flat, metal ring used to lock a nut in place, or fill a space. Can be lock, flat, star, or fender.

Waste heat - heat rejected from the building (or process) because its temperature is too low for economical recovery or direct use.

Waste pipe - any pipe that receives the discharge of any fixture except the water closet.

Waste water system - a system in which the condenser water is discharged to a sewer rather than being recycled. Wastewater systems do not recycle or reuse the water in the system.

Water boilers - these boilers are available in standard sizes from over 35,000 to over 100,000,000 Btu's per hour. Many of these boilers are in the low pressure class (15 psig and less) and are used primarily for space heating applications in both new and existing buildings.

Water box - a container or reservoir at the end of a chiller where water is introduced and contained.

Water chiller - a refrigeration system that cools water that is pumped into other parts of the system to maintain the desired condition in a specific area.

Water, circulating - in a water cooled or water cooling device, the quantity of water entering the device per unit of time.

Water closet - a toilet.

Water column - abbreviated as wc. 27.71" wc equal 1 psi. A unit used for expressing pressure. One inch of water column equals a pressure of 0.578 oz/in2 or 0.43 lb/ft. .

Water column, boiler - a column of piping, located on the upper outside of the boiler shell to maintain a water level. Consists of try cocks, water glass gauge, low water cut-off, pressure gauge and pressuretrols.

Water column blow down valve - valve on the bottom of the water column used to remove sludge and sediment that might collect at the bottom of the water column.

Water cooled condenser - a condensing unit that is cooled by water.

Water, cooling - water used for condensation of refrigerant; condenser water.

Water cooling tower - a structure that captures the water leaving the condenser of a water cooled refrigeration system and lowers its temperature so that the water can be recirculated through the condenser for additional cooling.

Water defrosting - the use of water to melt ice and frost from the evaporator during the off cycle.

Water, domestic - pure water intended for human consumption.

Water draft - the air pressure of a 0.01 to 0.03 inch water column drawn over a flame.

Water economizer - a system by which the supply air of a cooling system is cooled directly or indirectly by evaporation of water, or other appropriate fluid, in order to reduce or eliminate the need for mechanical cooling. As with an air economizer, a water economizer must be integrated into the system so that the economizer can supply a portion of the cooling concurrently with the refrigeration system.

Water feed valve - a inline valve that reduces the incoming pressure to the pressure required for the hydronic system. This valve automatically feeds water into the system to maintain the desired water pressure.

Water finding paste - a pasty substance that changes color when exposed to free water. It is smeared on a dip stick and then the stick is lowered into a fuel oil tank to determine the level (if any) of water in the tank.

Water hammer - a violent fluctuation in pressure within a pipe line caused by the sudden stopping or slowing down of flow velocity by rapidly opening or closing a faucet or valve, usually causing a loud noise.

Water heater - any heating device with open piping connections to the potable water supply system which intended to operate at a pressure not exceeding 160 psig, has a temperature not exceeding 210° F, has a capacity not exceeding 220 gallons, and has a heat input not greater than 200,000 Btu per hour.

Water jacket - in a compressor, the water used for cooling the cylinder head and/or walls.

Water leg boiler - a firebox boiler in which the water vessel extends to the base on both sides of the combustion chamber. Also see fire-box boiler.

Water level - surface or height of still water.

Water loading - circulating water rate per horizontal square foot of fill plan area of the cooling tower.

Water logged - a condition in which the captive air in an expansion tank is eventually replaced with water. Usually results in a no heat situation.

Water loops - the underground piping arrangements through which water or water/antifreeze solutions flow in geothermal heat pump systems.

Water, make-up - water supplied to replenish, as water replacing that lost by evaporation.

Water main - the principle water supply line.

Water manometer - used to measure pressure in inches of water column. There are 27.93 inches of water column in one psig.

Water regulating valve - an operating control valve that regulates the opening of the valve in relation to the high side refrigerant pressure.

Water rate - mass flow of water per square foot of fill plan area of the cooling tower per hour.

Water, raw - in ice making, any water used for ice making except distilled water.

Water, return - water off the liquid source heat pump that is returned to the source of supply.

Water, service line - the water supply line that runs from the street main to the building.

Water side - in a boiler, the tubes and vessels which contain water and steam, as opposed to the fire side.

Water source, buried closed loop - water is circulated through a closed loop pipe system to use heat transfer between the loop and the earth as a heat energy source or sink.

Water source, reservoir - a body of water, such as a large lake or stream, is used as a heat energy source or sink.

Water table - the top surface of the water in the water bearing unconsolidated formation. Source of underground water used for geothermal heat pump systems. Wells must be drilled to reach the water table, which can be up to 200-300 feet below the surface of the earth. The top surface of the aquifer.

Water tight - equipment that is constructed so that water or moisture is excluded from an enclosure.

Watertight joints - sheet metal joints formed by soldering, welding, or brazing the material to stand hydrostatic tests set for the application of the fabricated product.

Water treatment - the treatment of water with chemicals to reduce its scale forming properties or to change other undesirable characteristics.

Water-tube boiler - a boiler in which the water and steam passes through tubes surrounded by the products of combustion. Water tube boilers have capabilities for higher pressures and much greater outputs than fire tube boilers. There are two general types - straight tube and bent tube.

Straight tube boiler (also called horizontal) - the main bank of tubes is straight and on a slope of 5 to 15° from the horizontal. This type has further classifications: sectional header boiler - the tube bank is comprised of multiple parallel sections, each section having a front and rear header connected by vertical rows of tubes, all within a common drum.

Box header boiler - the tubes in the bank are connected by two inclined, rectangular headers (a front and a rear).

Cross drum boiler - a sectional header or box header boiler in which the axis of the horizontal drum is at right angles to the center lines of the tubes in the main bank.

Longitudinal drum boiler - a sectional header or box header boiler in which the axis of the horizontal drum is parallel to the center lines of the tubes in the main bank (in a horizontal plane).

Bent tube boiler - consists of two or more drums, requiring the tubes to be bent outward radially near their ends in order to interconnect the drums. The multiple drums result in very high outputs (over one million pounds of steam per hour, or 1250 million Btu/h), so this type of boiler is usually found in power plant installations. One version, called a low head boiler, has three drums with relatively short tubes in a vertical plane.

Water vapor - water contained in atmospheric air, actually, it is low temperature superheated steam.

Water vapor permeance - the rate of water vapor transmission by diffusion per unit area of a body between two specified parallel surfaces, induced by unit vapor pressure difference between the two surfaces.

Water vapor permeability (u) - the rate of water vapor transmission by diffusion per unit area of flat material of unit thickness induced by unit vapor pressure difference between two surfaces, under specified temperature and humidity conditions.

Water vapor resistance (Z) - the reciprocal of permeance.

Water vapor retarder - material or construction that adequately impedes the transmission of water vapor under specified conditions. Water vapor retarders have a water vapor permeance of less than 1.0 perm when tested in accordance with ASTM E 96-92.

Water wall - vertical or horizontal tubes found in the furnace area of water tube boilers that lengthen the life of the refractory.

Water wall blow down valve - an approved valve used to remove sludge and sediment from water walls and water wall headers.

Water white - a low iron-content glass used in solar heat collection panels.

Waterside explosion - see explosion.

Watt - the unit of electrical power. In a purely resistive alternating current circuit or a direct current circuit, one watt of power is consumed by a load when one volt causes a flow of one ampere through the load. Watts equal volts time's amps. Also equal to one joule per second and to 1/746 of one horsepower.

Watt hour - a unit of electrical energy expenditure. It is the equivalent of a 1 watt bulb operating for one hour of time.

Watt hour meter - a meter that indicates the instantaneous rate of power consumption of a device or circuit. Used by the electrical power companies to monitor consumers electrical consumption.

Watt meter - an instrument used to measure electrical power.

Watt second - the energy related to one watt acting for one second; one joule.

Watts Law - the mathematical relationship of power to voltage and current. In a DC circuit or a purely resistive AC circuit, power equals voltage times current.

Watts output - torque (ounce inches) x speed (rpm) divided by 1,352.

Wave form - a graphic representation of the different voltage values of an electrical current over a period of time.

Wavelength - the distance between corresponding points on two successive waves of a periodic waveform. The quotient of the velocity of propagation divided by the frequency.

Wax - an ingredient in many lubricating oils that may separate out if cooled sufficiently. Wax may plug the expansion valve and reduce heat transfer of the coil.

Weather proof - electrical devices designed to be protected from the weather.

Weather stripping - pliable material that seals cracks around windows, doors and openings to prevent infiltration.

Weather tight joint - sheet metal joints formed in such a manner as to resist moisture penetration when exposed to maximum wind and rain (or snow) conditions occurring during the most severe weather conditions normally encountered in the area of the installation.

Wedge - disk shape for certain types of gate valves.

Weep hole - a small opening in a wall or window member, through which accumulated condensation or water may drain to the building exterior, as from the base of the cavity wall, a wall flashing, or a skylight.

Weight switch - an electrical control used in an icemaker. Switch is actuated by a water filled canister and reset when the canister empties.

Weight type alarm whistle - an alarm whistle that signals high or low water by the gain or loss of buoyancy of weights in water within the water column.

Weight valve - a valve used with high side float refrigerant metering system to prevent evaporation in the line between the float and the evaporator when the system is in the off cycle.

Weir port - particular shape of a valve body interior.

Welding - the connection of metal objects by partially melting and fusing them together.

Well casing - a perforated, non-corrosive pipe (PVC or steel) inserted into a well to keep the walls from collapsing. This allows both the insertion of a submersible pump and free flow of water into the well.

Well casing screen - a section of well casing made of stainless steel screen located below the pumping level of the well to allow water entry into the casing.

Well water source, closed loop - a system that removes water from the earth by means of a drilled or bored well and returns it to the earth by means of a separate drilled or bored well.

Well water source, open loop - a system that removes water from the earth by means of a drilled or bored well and returns the water to the earth through a separate disposal system.

Western union splice - a standard splice made by twisting two wires together.

Wetback boiler - a scotch, fire tube boiler with a steel baffle at the back of the furnace to direct the products of combustion from the combustion chamber to the second pass. The baffle covers the rear end of the furnace and tubes, and is completely water cooled. The boiler gets its name from the wall of water between the baffle and the rear end of the boiler. Also see fire tube and scotch boilers.

Wet base boiler - a boiler design in which the boiler sections totally surround the combustion chamber.

Wet bulb - device used in the measurement of the relative humidity. Evaporation of moisture lowers the temperature of the wet bulb compared to the dry bulb in the same atmosphere.

Wet bulb depression - the difference between the wet and dry bulb temperatures. This indicates the % of relative humidity.

Wet bulb hygrometer - see psychrometer.

Wet bulb temperature - the lowest temperature that will register on a standard thermometer having its bulb enclosed in a wet wick over which air is circulated. This reading can be plotted with the dry temperature on a psychrometric chart to determine the relative humidity. The wet bulb reading is always lower than the dry bulb temperature unless the relative humidity is 100%. Then they are the same. The instrument used to obtain the wet and dry bulb reading is a sling psychrometer.

Wet bulb thermometer - a thermometer whose bulb is covered with a piece of water soaked cloth. The lowering of temperature that results from the evaporation of water around the bulb indicates the air's relative humidity.

Wet cell battery - a cell or connected group of cells that convert chemical energy into electrical energy by reversible chemical reactions.

Wet compression - the mixing of liquid refrigerant with the suction vapor entering the compressor. This causes the compressor's discharge to be saturated instead of superheated.

Wet contacts - contact points that are closed with a liquid mercury bulb.

Wet filter - an air filter saturated with water or oil to increase its filtering ability.

Wet heat - heating system using either hot water (hydronic) heat or steam heat.

Wet location - installations underground or in concrete slabs or masonry in direct contact with the earth, and locations subject to saturation with water or other liquids, such as vehicle washing areas, and locations exposed to weather and unprotected.

Wet return - that part of a return main of a steam heating system which is completely filled with water or condensate.

Wet rotor circulator - a circulator in which the motor's armature and its impeller are an integral unit that is surrounded and cooled by the fluid passing through the circulator.

Wet steam - when steam contains particles of water that have not been evaporated, it is say to be wet.

Wet vent - a vent in which a portion of the piping used to vent the system is the drainage pipe through which the waste flows.

Wetted hole - the portion of the well below the static water level in the well.

Wetted surface cooling tower - an application where water in a cooling tower is spread over a wetted surface while air is passed over it to enhance evaporation.

Wetting - the ability of a solder or brazing alloy to spread and adhere to a solid base metal. Wetting is reduced when contaminants such as oil, grease, or oxides are present.

Wheatstone bridge - a bridge with four main arms containing a resistor in each arm. With known values for three resistors, the value for the unknown resistor can be computed.

Wheeling - use of the transmission facilities of one system to transmit gas or power for another system.

White noise - white noise is noise of wide frequency range in which the amplitude of the noise is essentially the same in all frequency bands.

Wholesaler - the middleman between the manufacturer and retailer. One who sells to the retailer in whole quantities for resale to the consumer.

Wick - a component part of a humidifier that increases the surface area of the water, increasing the efficiency of the humidifier.

Wild leg - the >190 volt source produced by one line in a closed delta connection. Also known as the bastard leg or stinger leg.

Wind chill factor - the still air temperature that would have the same chilling effect on human skin as the combination of the recorded wind speed and ambient temperature.

Wind effect - the increase in evaporation rate due to air travel over a water surface.

Wind load - the total force or load imposed upon a structure by a wind blowing against its surface.

Windage - water lost from the water tower because of the effects of wind currents. Sometimes referred to as blowout.

Winding - one or more wire turns that form a continuous coil or conductive path.

Window size - always expressed as width first, then height.

Window unit - a common term used when referring to air conditioners that are placed in a window.

Wiper - the moving contact that makes contact with a terminal in a stepping relay or switch. In a potentiometer, the contact that moves along the element, dividing the resistance according to its mechanical position.

Wire - a conductor, bare or insulated.

Wire, drawing - restriction of area for a flowing fluid, causing a loss in pressure by (internal and external) friction without loss of heat or performance of work; throttling.

Wire gauge - refers to the system of sizing wire by its diameter; starting with 0000 or 000 as the largest diameter and going up to size 40 or above for the smallest diameter.

Wire glass - ¼" clear or obscure glass having a layer of diamond or square pattern wire mesh embedded in the glass lite.

Wire nut - a connector for wires, tightened like a screw or nut.

Wire stripper - a tool that removes the insulation from wire.

Wiring diagram - a diagram showing electrical devices by symbols and their interconnections. Types are schematic, pictorial (or line), and installation.

Wiring sub-base - a panel, providing terminals for connecting external devices (controllers, interlocks, etc.), on which the thermostat or flame safeguard control is mounted.

Within sight - see in sight.

Within sight from - see in sight.

Wobble plate-swash plate - type of compressor designed by General Motors to compress gas and having pistons with piston motion parallel to the crankshaft.

Woodruff key - a device used to align and hold a flywheel, pulley or other rotating mechanism on a shaft. Often half moon shaped.

Work - the result of energy expended. The transfer of energy from one object to another, measured in force multiplied by distance, or foot pounds.

Work, effective - net mechanical energy required by, or load imparted to, piston of compressor.

Work, indicated - work equivalent of indicator card area for a reciprocating compressor or engine.

Work hardening - the tendency of metal to become stiff and brittle when hammered or bent.

Work order - the written authorization to service or install equipment.

Working alone - performance of any work by an individual out of audio or visual range of another individual for more than a few minutes. No other person is aware of the individual working alone, the nature of the work being done, or the time period the individual expects to work. A worker working on a roof-top piece of equipment should not perform practices known to be hazardous. Always work under conditions where the availability of emergency aid is compatible with the nature of the hazard and the degree of exposure.

Working drawings - drawings intended for use by a contractor, subcontractor or fabricator for a building project, containing the necessary information to manufacture or erect an object or structure.

Working level - one working level of radon - 222 daughters will give out alpha particle energy of 1.3 x 105 meV in one liter of air upon decay to lead - 210.

Working level month - exposure to 1 wl for a working month of 173 hours.

Working pressure - (1) a test pressure at which fuel gases are used. Working pressure is reduced from the higher cylinder pressure through the use of a regulator. (2) the allowable upper pressure limit at which a pipe or piping component can safely be used.

Working range - the desired controlled or measured variable values over which a system operates.

Working voltage - the maximum continuous voltage a capacitor will withstand. Also referred to as the voltage rating.

Wound induction motor - a three phase induction motor used for large air conditioning units. Stator winding is the same a squirrel cage stator however, the rotor contains wound coils of wire instead of squirrel cage rotor bars, and will contain as many poles as there are stator poles.

Wound rotor induction motor - one which the secondary circuit consists of a polyphase winding or coils whose terminals are either short circuited or closed through suitable circuits.

Wrist pin - a retaining peg that joins a connecting rod to a piston.

Wrought copper - copper that has been hammered or formed by pressure into a desired shape.

Wrought fittings - fittings joined by sweat connections to either hard-drawn or soft copper tubing. Wrought fittings are usually made of copper but sometimes of brass.

Wrought iron - iron of low carbon content useful because of its toughness, ductility, and malleability.

Wye connection - also referred to as the star connection. A wye system is a four wire connection made by joining one end of each of three windings. Wye systems provide three single phase 120 volt circuits for light residential loads, and can also be used to provide a 208 volt three phase source when three of the four conductors are used. When used with motors, the windings can be arranged in series or parallel, depending on the voltage at which the motor is to be used.

Wythe - in reference to masonry chimneys, a course, a thickness, or a continuous vertical section of masonry separating flues in a chimney.

-X-

X - symbol for reactance.

X axis - a horizontal line.

-Y-

Y axis - a vertical line.

Yellow flame burner - see long flame burner.

Yellow tips - the appearance of yellow tips in an otherwise blue flame, indicating the need for additional primary air.

Yield strength - the minimum stress that a material will start to physically deform without further increase in load.

Yoke - that part of the flaring tool that fits over the block.

Yoke vise - a vise is used for holding pipe while it is being worked on. The yoke vise has an upper and lower serrated jaw and operates on a hinge allowing it to be opened and closed to hold pipe. A chain vise has a bottom jaw and a chain that loops around the pipe and holds the pipe securely while allowing it to be tightened.

-Z-

Z - symbol for impedance.

Zener diode - a diode that has a constant voltage drop when operated in the reverse direction. Zener diodes are commonly used as voltage regulators in electronic circuits.

Zeolite - a resin chemical material used to de-mineralize water for softening water.

Zeotrope refrigerants - refrigerant blends that change volumetric composition and saturation temperatures as they evaporate or condense at constant pressures. Zeotropes have a temperature glide as they evaporate and condense. Zeotrope and non-azeotrope mean the same thing. Zeotropic is derived from the Greek words zein (to boil) and tropos (to change).

Zerk fitting - grease fittings outfitted with a one way check valve that is used to inject grease for lubricating a bearing.

Zero (absolute pressure and temperature) - in absolute measuring systems, the absence of a condition, e.g., heat or pressure. In non-absolute systems, an arbitrary starting point for measurement.

Zero adjustment - a mechanical means by a dial, screw, or button for setting a meter pointer on the zero mark of the meter scale.

Zero crossing network - a network designed to insure that a relay, or other device, turns on only when the AC line voltage is at or near zero of the sine wave.

Zero gas - gas at atmospheric pressure (zero gauge pressure), usually supplied by a zero governor.

Zero governor - a regulating device that is normally adjusted to deliver gas at atmospheric pressure (zero gas) within its flow rating. Also called an atmospheric regulator.

Zero governor burner - a power burner in which gas is supplied at atmospheric pressure (zero gas) maintained by a zero governor. A fan, or blower, blows the air through a venturi, where venturi action pulls the gas into the air stream. The quantity of gas drawn into the air stream is a direct function of the air velocity, so the air fuel ratio will remain constant regardless of the air volume. The firing rate is controlled by air adjustment alone. Also see aspirating, forced draft, and premix burners.

Zero ice - trade name for dry ice.

Zero potential - having zero or no voltage.

Zero resistance - indication of continuity, or a complete path for electron movement from one point to another.

Zinc - the coating that is put on sheet metal to make it galvanized.

Zinc fume fever - caused by inhalation of zinc oxide fumes and characterized by flu-like symptoms.

Zone - specific area to be heated or cooled at a specified temperature and controlled by one thermostat. A space within a house with a distinct pressure compared to other pressure zones. Zones within a building may require different indoor design temperatures based on their ultimate use. Also see valve zone.

Zone, comfort average - the range of effective temperature over which the majority (80% or more) of adults feel comfortable.

Zone, comfort extreme - the range of effective temperatures over which one or more adults feel comfortable.

Zone controls - controls used to maintain each specific area or zone within a building at a desired condition. Can be hydronic or forced air.

Zone reheat - a system where zone heaters tempers or reheat air distributed from the centrally located system.

Zone system - for a hydronic, electric or forced air system where more than one thermostat is used. This permits independent control of room air temperature at more than one location.

Zone valve - controls used to maintain each specific area or zone within a building at a desired temperature or condition.

Zoning - the practice of dividing a building into small sections for heating and cooling control. Each section is selected so that one thermostat can be used to determine its requirements.

Zoning, mixed air - hot and cold air are mixed in just the right proportions to maintain the desired zone temperature. The air is channeled into a heating section (hot deck) and cooling section (cold deck) and then mixed.

Zoning, multiple unit - a separate heating cooling unit is used for each zone.

Zoning, volume - has just one central heating and cooling unit with ducts leading to each zone.

Appendix

Appendix Table of Contents

Common Industry Abbreviations for Associations

AABC - Associated Air Balance Council.

ABC - Associated Builders and Contractors.

ABMA - American Boiler Manufacturers Association.

ACCA - Air Conditioning Contractors of America.

ACEC - American Consulting Engineers Council.

ACEEE - American Council for an Energy Efficient Economy.

ACGIH - American Conference of Governmental Industrial Hygienists - an organization of professionals in governmental agencies or educational institutions engaged in occupational safety and health programs. ACGIH develops and publishes recommended occupational exposure limits for chemical substances and physical agents.

ADC - Air Diffusion Council.

ADDA - American Design Drafting Association.

ADI - Air Distribution Institute.

AEE - Association of Energy Engineers.

AFFI - American Frozen Food Institute.

AGA - American Gas Association.

AGAL - American Gas Association Lab.

AGC - Associated General Contractors of America.

AGMA - American Gear Manufacturers Association - composed of member companies who manufacture speed reducers, enclosed gear drives, open gearing, and gear type shaft couplings. They establish standards for the design and application of gear products. The combined experience of company members and technical members assures gear users that products will perform satisfactorily when built, selected, and tested in accordance with AGMA standards.

AHAM - Association of Home Appliance Manufacturers.

AIA - American Institute of Architects.

AEE - Association of Energy Engineers.

AIEE - American Institute of Electrical Engineers.

AIHA - American Industrial Hygienists Association.

AIPE - American Institute of Plant Engineers.

AMCA - Air Movement and Control Association - a nationally recognized association that establishes standards, tests and certifies the performance of air moving devices.

AAMA - American Architectural Manufacturers Association

ANSI - American National Standards Institute - an organization that coordinates voluntary standards to promote to interchangeability of parts and products. ANSI does not itself develop standards. Rather, it facilitates development by establishing consensus among qualified groups. ANSI ensures that the guiding principles --- consensus, due process, and openness are followed. A privately funded organization that identifies industrial/public national consensus standards and coordinates their development. Many ANSI standards relate to safe design/performance of equipment and safe practices and procedures.

API - American Petroleum Institute.

ARI - Air Conditioning and Refrigeration Institute - an association for refrigeration and air conditioning manufacturers. ARI does research and produces educational materials and sets standards.

ARL - Applied Research Laboratories.

ARW - Air Conditioning and Refrigeration Wholesalers.

ASA - Acoustical Society of America, American Standards Association, American Subcontractors or the Association or American Supply Association.

ASHRAE - American Society of Heating, Refrigerating and Air Conditioning Engineers.

ASI - American Standards Institute.

ASES - American Solar Energy Society.

ASME - American Society of Mechanical Engineers - an organization to which all engineers can belong. Through ASME, engineers can keep current on new procedures in the engineering field.

ASPE - American Society of Plumbing Engineers or American Society of Professional Estimators.

ASSE - American Society of Sanitary Engineers or American Society of Safety Engineers.

ASTM - American Society for Testing and Materials - An organization that devises consensus standards for materials characterization and use.

AWS - American Welding Society.

BOCA - Building Officials and Code Administrators.

BOMA - Building Owners and Managers Association International.

CABO - Council of American Building Officials.

CAGI - Compressed Air and Gas Institute.

CBAC - Commercial Building Air Conditioning Center.

CDA - Copper Development Association.

CEC - California Energy Commission.

CERI - Clean Energy Research Institute.

CGA - Canadian Gas Association.

CI - Combustion Institute.

CPSC - Consumer Products Safety Commission.

CRMA - Commercial Refrigerator Manufacturers Association.

CSA - Canadian Standards Association and/or Cryogenic Society of America Inc - a private accredited test lab that writes and tests to its own standards. Provincial Law mandates that all products sold for consumer use must have CSA or equivalent approval. This law is enforced by the Canadian Utilities. CSA evaluates products and underwrites that they are safe per their tests. The Canadian Standards Association develops safety and performance standards in Canada for electrical products, similar to but not always identical to those of Underwriters Laboratories in the U.S.

CSI - Construction Specifications Institute.

CTI - Cooling Tower Institute.

DOE - Department of Energy.

DOT - Department of Transportation - a federal agency charged with the regulation and control of all hazardous materials shipped through commercial means.

EASA - Electrical Apparatus Service Association.

ECI - Evaporative Cooling Institute.

EEA - Earth Energy Association.

EEI - Edison Electric Institute.

EIA - Environmental Information Association.

EIC - Environmental Industry Council.

EJMA - Expansion Joint Manufacturers Association.

EMA - Environmental Management Association.

EPA - Environmental Protection Agency - an agency of the U.S. government charged with the responsibility of protecting the environment and enforcing the Clean Air Act of 1990. The EPA regulations require the use of environmentally safe procedures in the use of refrigerants.

EPRI - Electric Power Research Institute.

FAA ap - Federal Aviation Administration (airport).

FDA - U.S. Food and Drug Administration - the FDA is the federal agency that establishes requirements for the labeling of food and drugs to protect consumers.

FEMA - Federal Emergency Management Agency.

FSEC - Florida Solar Energy Center.

Factory Mutual - the identification of the Factory Mutual system and the Factory Mutual Engineering Corporation, an established United States approval agency for gas detection instrumentation. Traceable to National Bureau of Standards.

GAMA - Gas Appliance Manufacturers Association.

GEA - Geothermal Energy Association.

GRC - Geothermal Resources Council.

GRI - Gas Research Institute.

HA - Heating Alternatives, Inc.

HBI - Healthy Buildings International.

HEI - Heat Exchange Institute.

HI - Hydraulic Institute or Hydronics Institute.

HUD - U.S. Housing and Urban Development

HVI - Home Ventilating Institute.

IAEI - International Association of Electrical Inspectors.

IAPMO - International Association of Plumbing and Mechanical Officials.

IARW - International Association of Refrigerated Warehouses.

IBCA - Industry Bar Code Alliance.

IBEW - International Brotherhood of Electrical Workers.

IBR - Institute of Boiler and Radiator Manufacturers.

ICBO - International Conference of Building Officials.

ICC - Interstate Commerce Commission - a government body that controls the design and construction of pressure containers.

ICRA - International Compressor Remanufacturers Association.

IDEA - International District Energy Association.

IDHCA - International District Heating and Cooling Association.

IEEE - Institute of Electrical and Electronics Engineers.

IES - Illuminating Engineering Society or Institute of Environmental Sciences.

IEQA - Indoor Environmental Quality Alliance.

IFCI - International Fire Code Institute.

IFMA - International Facility Management Association.

IGCC - Insulating Glass Certification Council.

IGSHPA - International Ground Source Heat Pump Association.

IGT - Institute of Gas Technology.

IHACI - Institute of Heating and Air Conditioning, Inc.

IFH - Industrial Health Foundation.

IIAR - International Institute of Ammonia Refrigeration.

IIE - Institute of Industrial Engineers.

IKECA - International Kitchen Exhaust Cleaning Association.

IMACA - International Mobile Air Conditioning Association.

IOA - International Ozone Association.

ISA - Instrument Society of America.

ISB - Institute for Sick Buildings.

ISO - International Organization for Standardization.

ITSAC - International Thermal Storage Advisory Council.

MACS - Mobile Air Conditioning Society.

MANA - Manufacturers Agents National Association.

MCAA - Mechanical Contractors Association of America.

MHANA - Masonry Heater Association of North America.

MICA - Midwest Insulation Contractors Association.

MSCA - Mechanical Service Contractors of America.

MSSVFI - Manufacturer's Standardization Society of the Valve and Fittings Industry, Inc.

NAAMM - National Association of Architectural Metal Manufacturers.

NAC - National Asbestos Council.

NACE - National Association of Corrosion Engineers.

NADCA - National Air Duct Cleaners Association.

NAFA - National Air Filtration Association.

NAFEM - North American Association of Food Equipment Manufacturers.

NAHB - National Association of Home Builders.

NAIMA - North American Insulation Manufacturers Association.

NAM - National Association of Manufacturers.

NAOHSM - National Association of Oil Heating Service Managers

NAPE - National Association of Professional Engineers or National Association of Power Engineers.

NAPHCC - National Association of Plumbing Heating Cooling Contractors.

NARI - National Association of the Remodeling Industry.

NAS - National Academy of Sciences.

NASM - National Association of Service Managers.

NATT - National Association of Trade and Technical Schools.

NAWD - National Association of Wholesaler-Distributors.

NAWIC - National Association of Women in Construction.

NBBPVI - National Board of Boiler and Pressure Vessel Inspectors.

NBFU - National Board of Fire Underwriters.

NCAC - National Council of Acoustical Consultants.

NCIAQ - National Coalition on Indoor Air Quality.

NCPWB - National Certified Pipe Welding Bureau.

NCRSA - National Commercial Refrigeration Sales Association.

NCSBCS - National Conference of States on Building Codes and Standards.

NEBB - National Environmental Balancing Bureau.

NEC - National Electric Code.

NECA - National Electrical Contractors Association.

NEDA - National Environmental Development Association.

NEHA - National Environmental Health Association.

NEIC - National Energy Information Center.

NEMA - National Electrical Manufacturers Association - this organization sets standards for motors, controls, and most electrical machines in the United States.

NEMI - National Energy Management Institute.

NESA - National Energy Specialist Association.

NFPA - National Fire Protection Agency or National Fluid Power Association - an international standards making organization dedicated to the protection of people from fires and electric shock. The NFPA is responsible for developing and writing the National Electric Code, Sprinkler Code, Life Safety Code and well as other codes.

NFRC - National Fenestration Rating Council

NFSA - National Fire Sprinkler Association.

NGA - National Geothermal Association.

NGWA - National Ground Water Association.

NHRAW - Northamerican Heating Refrigeration Air Conditioning Wholesalers.

NIACA - National Insulation and Abatement Contractors Association.

NIBS - National Institute of Building Sciences.

NIOSH - National Institute of Occupational Safety and Health.

NIST - National Institute of Standards and Technology - a national institute that helps set standards and technology for mobile air conditioning systems.

NRCA - National Roofing Contractors Association.

NSC - National Safety Council.

NSF - National Sanitation Foundation.

NSPE - National Society of Professional Engineers.

NSPI - National Spa and Pool Institute.

NSWMA - National Solid Wastes Management Association.

NUCA - National Utility Contractors Association.

NWSA - National Welding Supply Association.

OPEI - Outdoor Power Equipment Institute.

OSHA - Occupational Safety and Health Administration – passed a law through Congress, December 31, 1970, to maintain occupational safety and to protect persons in places of employment.

PAFTT - Program for Alternative Fluorocarbon Toxicity Testing - a testing agency formed by several refrigerant manufacturers to study and make toxicity tests on new refrigerants.

PATMI - Powder Actuated Tool Manufacturers' Institute.

PDI - Plumbing and Drainage Institute.

PEC - Piping Education Council.

PEMAEC - Process Equipment Manufacturers' Association and Environmental Council.

PFI - Pipe Fabrication Institute.

PLCA - Pipe Line Contractors Association.

PMAA - Petroleum Marketers Association of America.

PMI - Plumbing Manufacturers Institute or Project Management Institute.

PPFA - Plastic Pipe and Fittings Association.

PPI - Plastic Pipe Institute.

PSC - Piping System Council.

RETA - Refrigerating Engineers and Technicians Association.

RPA - Radiant Panel Association.

RRF - Refrigeration Research Foundation.

RSES - Refrigeration Service Engineers Society.

SAE - Society of Automotive Engineers.

SAVE - Society of American Valve Engineers.

SBCCI - Southern Building Code Congress International.

SDMA - Spiral Duct Manufacturers Association.

SEIA - Solar Energy Industries Association.

SES - Standards Engineering Society.

SFPE - Society of Fire Protection Engineers.

SMACNA - Sheet Metal and Air Conditioning Contractors National Association.

SME - Society of Manufacturing Engineers.

SPE - Society of Petroleum Engineers or Society of Plastics Engineers.

SPED - Society of Piping Engineers and Designers.

SPI - Society of the Plastics Industry.

SRCC - Solar Rating and Certification Corporation.

STI - Steel Tank Institute or Steel Tube Institute.

STLE - Society of Tribologists and Lubrication Engineers.

SWE - Society of Woman Engineers.

TEMA - Tubular Exchanger Manufacturers Association.

TIEQ - Total Indoor Environmental Quality Coalition.

TSARC - Thermal Storage Applications Research Center.

UAJAPPFI - United Association of Journeymen and Apprentices - Plumbing/Pipe Fitting Industry.

UBC - Uniform Building Code - The UBC is designed to be compatible with the International Conference of Building Officials (ICBO), International Association of Plumbing and Mechanical Officials (IAPMO), and the National Fire Protection Association (NFPA), which make up the enforcement tools of a jurisdiction.

UL - Underwriters Laboratories - an independent organization that sets safety standards for motors and other electrical equipment. A laboratory that tests devices and materials for compliance with the standards of construction and performance established by the laboratory and with regards to their suitability for installation in accordance with the appropriate standards of the National Board of Fire Underwriters.

UMC - Uniform Building Code -

UPC - Uniform Plumbing Code

USCEA - United States Council for Energy Awareness.

USDA - United States Department of Agriculture.

USECRE - United States Export Council for Renewable Energy.

USMA - United States Metric Association.

VI - Vibration Institute.

VMAA - Valve Manufacturers of America.

VRA - Valve Re-manufacturers Council.

WDA - Wholesale Distributors Association.

WHO - World Health Organization.

WOHMA - Waste Oil Heating Manufacturers Association.

WQR - Water Quality Research.

WSC - Water Systems Council.

Business Terminology

Account - a record of similar transactions under one title in the ledger.

Account balance - the difference between the sum of the debit postings and the sum of the credit postings in an account.

Accounting - the act of recording, classifying, and summarizing in a significant manner and in terms of money, transactions, and events which are, in part at least, of a financial character, and interpreting the results thereof.

Accounts payable - a term applied to the aggregate of amounts owned to the creditors of a business.

Accounts receivable - a term applied to the aggregate of amounts due from the customers of a business.

Accrual accounting - this is the form of accounting legally required of most small businesses. In this form of accounting, both cash and credit transactions are recorded as they occur.

Accrual basis - a basis of record-keeping and a method of accounting whereby all income earned and expenses incurred, whether prepaid or unpaid, are identified with specific periods of time; distinguished from "cash basis."

Accruals - the recognition of income and expenses as they occur even though they are not received or paid for until a later period.

Accrued expenses - the liability for expenses incurred, but not paid, at a given date.

Accrued income - the asset arising from income earned, but not received, at a given date.

Accrued liabilities - amounts representing liabilities incurred, but not paid, at a given date.

Affected employee - an employee whose job requires them to operate or use a machine or equipment that is being serviced or maintained under lock-out/tag-out conditions. Also, an employee whose job requires them to work in the area where servicing or maintenance is being performed.

Aging accounts receivable - preparing a schedule of accounts receivable with thee accounts listed according to age. Each account is then examined and a decision made as to which are probably uncollectible.

Asset - property which had money value.

Authorized capital stock - the maximum number of shares of stock which a corporation may issue under its charter.

Balance sheet - a statement of the financial condition of a business at a certain date showing its assets, liabilities and owner's equity.

Bank draft - a check drawn by a bank on another bank, payable to a third party.

Bankruptcy - a legal procedure for formally liquidating a business, carried out under the jurisdiction of courts of law.

Bill of lading - a written document issued by a carrier, constituting a receipt for the goods shipped, and a contract to deliver the goods at a specified place to a named person or his order.

Board of directors - persons elected by the stockholders to manage the corporation's affairs.

Bond - an interest bearing written promise of a corporation, generally secured, to pay a fixed amount to holders at a designated future time, usually three years or more.

Budget - a managerial plan of proposed operations to accomplish financial objectives.

Capital investment - an expenditure for an investment whose returns are expected to extend beyond one year.

Corporation - an artificial being, created by the state through a charter to engage in a particular kind of business.

Cost evaluation - a determination of the total cost of owning a piece of equipment for a specific period of time. Includes initial cost of equipment and accessories, cost of operation, cost of maintenance and/or repair, cost of financing, etc., all normalized to a specific point in time.

Credit memo - a notice to a buyer of a credit granted by a seller.

Creditor - one to whom a debt is owned.

Current assets - cash and other assets expected to be converted into cash within operating cycle, usually one year or less.

Current liabilities - liabilities representing the short term debt of a business, expected to fall due within a year or less.

Customer relations - the ability to gain customer confidence and trust by effectively and promptly serving their wants and needs.

Direct cost - those costs which may readily be attached or directly related to the unit of production.

Emergency plan - a document that details procedures, exit routes, and assembly areas for facility personnel in the event of an emergency.

Employee - is a worker whose services are hired by an employer, and from whose pay state and federal taxes are deducted.

Entrepreneur - one who assumes the risk and management of a business.

Estimate - a prediction of the expected cost for equipment or repair work.

Estimated tax - four times a year, a self employed person must make a payment to the internal revenue service based on what he or she expects to earn that year.

Excise tax - an internal tax such as the tax levied on those who produce chlorofluorocarbons (CFCs).

Expense - a present or past expenditure defraying a current cost of operating or doing business.

Face amount - the sum of money stated on a check, bond, note or other instrument, exclusive of interest accumulations.

Financial leverage - the use of other people's money to increase the return on the investment of the owners.

Financial statement - a balance sheet, statement of income or any other presentation of financial data.

Fiscal year - a twelve month accounting period ending with the last day of any month.

Fixed assets - assets that are for the most part permanent in nature and whose useful life is more than one year.

FOB - free on board - a shipping term meaning that the parts or material ordered are free of shipping charges to the point specified. Beyond that point, the purchaser must pay for shipping

General journal - the record used for recording business transactions not recorded in special journals.

General ledger - the record containing the accounts in which are summarized all the transactions of a business.

Gross profit - the excess of net sales over the cost of goods sold.

Indirect costs - all of the costs which can not be directly attached to the unit of production.

Intangible assets - assets having no physical nature but having value derived from the rights conferred upon their owner.

Interest - the charge made for the use of money.

Inventory - itemized list of goods, property, etc.; stock listing, store of goods, etc.

Invoice - an itemized statement of charges for billing.

Journal - a book of original entry where transactions are initially recorded.

Ledger - a group of accounts where transactions are recorder from the journal.

Liabilities - creditor claims on the assets of a business.

List price - a published price, subject to trade discounts.

Long term liabilities - obligations not due and payable for a comparatively long period, normally more than a year.

Maturity date - the date on which an obligation and/or debt becomes due.

Mechanic's lien - a legal claim to ownership of repaired equipment if payment for the repair is not made.

Merchandise inventory - a list of goods on hand for resale and its money value.

Merchandise turnover - turnover - the number of times the average merchandise inventory is sold during the financial period.

Net profit - the excess of income over costs and expenses of a business.

Net worth - the excess of assets over creditor claims against the assets. Assets minus liabilities equal net worth.

Net working capital - current assets less current liabilities.

Notes payable - amounts owed because of promissory notes given to creditors.

Notes receivable - a promissory note which is a formal written promise from a customer to pay a definite sum at a specific future time.

Organization costs - the costs of organizing a corporation (e.g., legal fees, charter costs.) The sum of these is considered an intangible asset of the company.

Overhead - all of the indirect costs which can not be directly related to a unit of production and which must be included in the cost of the sale by some allocation method.

Owner's equity - the net worth of a firm or company. The amount of the total assets of the company and which belong to the owners.

Paid up capital - the total of the par or stated value received from stockholders for capital stock issued to them.

Par value - the face value placed on capital stock or other securities.

Parent corporation - a corporation holding more than 50 percent of the voting stock of another corporation and thus controlling the activities and resources of the owned firm.

Partnership - a relationship between two or more persons, based upon a written, oral or implied agreement, to combine their resources and skill in a joint enterprise and to share profits and losses jointly.

Payback - total investment divided by one year's saving. This gives the number of years it will take to recover the investment with no regard to interest rates or taxes.

People skills - one of the most misunderstood applications of the HVAC industry. These are the skills that are necessary to work well with and for others.

Petty cash - a fund of cash that is established to make minor disbursements and avoids writing checks for small amounts.

Physical inventory - a listing of merchandise on hand, determined by actual count, weight or measurement, and the pricing thereof.

Profit and loss statement - a financial statement showing the income and expenses of a business for the period and the profit (or loss) resulting from such dealings.

Profit margin - the difference between cost and selling price. Also called mark-up.

Profit, net income - the excess of revenues over expenses.

Promissory note - a written promise by the maker to pay a certain sum of money on a fixed or determined date to the order of a designed person.

Proposal - a formal, written cost estimate or bid.

Proprietary - sole ownership of property, a business, an item of labor, or an object that extends legal ownership rights.

Purchase order - a business form that authorizes an employee to make a purchase on open account.

Revoke - to cancel or take back a permission or contract.

Return on investment (ROI) - the percentage of the earning on the total assets of a corporation.

Sales - an account that records the sale of merchandise by a credit entry for the selling price of the goods.

Share - one of the units of equal value into which each class of the of the capital stock of a corporation is divided.

Small business - any business entity (including a sole proprietorship, corporation, partnership or other legal entity) which is owned and operated independently from all other businesses, which has the purpose of making a profit, and which has fifty or fewer employees, or which has a million dollars or less per year in gross sales.

Sole proprietorship - a business owned entirely by one person.

Special journals - journals used to record transactions of a particular kind such as a sales journal, purchase journal or a cash receipts journal.

Stock certificate - printed evidence of ownership in a corporation.

Stock holder - the owner of one or more shares of the capital stock of a corporation.

Subsidiary - a firm which is owned and controlled by another corporation.

Computer Terminology

Back up - to make a copy of data or programs from a hard drive to floppy disks, CD's, DVD's or other storage media.

BASIC - an acronym for Beginners All-purpose Symbolic Instruction Code. A high level, English like programming language used for general applications.

Bit - abbreviation for binary digit. A unit of information equal to one binary decision, or the designation of one of two possible and equally likely values or states (such as 0 or 1) of anything used to store or convey information.

Byte - a group of eight contiguous bits that occupy a single memory location in a microcomputer or computer.

Computer - series of electrical components that accept inputs from an operator and control outputs.

Computer aided drafting (CAD) - system by which a drawing can be made with a computer and associated software for preparing technical drawings such as floor plans and schematics.

Cross hair - the crossed vertical and horizontal lines on a CAD graphic display unit that represents the cursor position or the vertical and horizontal lines of an input device like a digitizing puck.

Data byte - the eight-bit binary number that the microprocessor chip will use in an arithmetic or logical operation or store in memory.

Diskette - a magnetic recording medium that functions the same as a disk, except the media is flexible and is similar to a 45 RPM record. Used only in floppy disk drives.

Encode - to use a code, frequently one composed of binary numbers, to represent individual characters or groups of characters in a message. To change from one digital code to another. If the codes are greatly different, the process is called code conversion.

Floppy disk - a type of disk drive, in a computer that uses a storage medium made of a flexible nature. Sometimes called a flexible disk. Also see diskette.

Internet accessibility - the ability of a control system to send or receive information from a remote site using the Internet.

Logic - a decision making capability of computer circuitry.

Megabyte - one million bytes of memory

Memory - any device that can store logic 0 and logic 1 bits in such a manner that a single bit or group of bits can be accessed and retrieved.

Microcomputer - a computer system based on a microprocessor that contains all memory and interface hardware necessary to process software.

Microprocessor - electrical component consisting of integrated circuits that may accept information, store it, and control and output device.

Random access memory - a particular type of memory-type building block that is commonly available in metal-oxide-semiconductor integrated circuit form and sometimes containing a given number of flip-flops. Bits of binary information can be written into or taken away from any of the flip-flops.

Read only memory (ROM) - a semiconductor memory from which digital data can be repeatedly read out, but cannot be written into as in the case for read/write memory.

Read write memory - a semiconductor memory into which logic 0 and logic 1 states can be written (stored) and read out again (retrieved).

Software - a term used to describe all computer programs, whether in machine, assembly, or high-level language.

Medical Terminology

Abdominal thrust - an upward push to the abdomen given to clear the airway of a person with a complete airway obstruction.

Acidosis - a condition of decreased alkalinity of the blood and tissues. Symptoms may include sickly sweet breath, headache, nausea, vomiting, visual disturbances. Usually the result of excessive acid production. Tissues and central nervous system functions are disturbed.

Acute exposure - exposure of short duration, usually to relatively high concentrations or amounts of material.

Acute health effect - an adverse effect on a human or animal body, with symptoms developing rapidly.

Acute lethality - the death of animals immediately or within 14 days after a single dose of or exposure to a toxic substance.

Acute toxicity - adverse health effects resulting from brief exposure to a chemical.

Allergen - any substance or vapor that causes a person to react with skin eruptions or eyes and nose irritations.

Allergic reaction - a reaction, sometimes severe, to certain substances such as the poison in insect bites and stings. These situations may cause hives, severe swelling, redness, difficult breathing, loss of consciousness, or death.

Allergic rhinitis - inflammation of the mucous membranes in the nose that is caused by an allergic reaction.

Analgesia - loss of sensitivity to pain.

Anesthesia - loss of sensation or feeling.

Anesthetic EC50 - the concentration of a substance that caused the temporary loss of ability to perceive pain and other sensory stimulation to 50% of test animals, normally measured for 10 minutes exposures. EC50 refers to the effective concentration for 50% of specimens.

Anosmia - loss of sense of smell.

Anoxia - a lack of oxygen in blood or tissues.

Antidote - a remedy to counteract a poisons toxic effects. It may act to eliminate, absorb, or neutralize the poison.

Arterial bleeding - bleeding from an artery, characterized by a bright red, spurting flow.

Asbestosis - chronic lung disease caused by inhaling airborne asbestos fibers.

Asphyxia - lack of oxygen and thus interference with the oxygenation of the blood. Asphyxia can lead to unconsciousness.

Asphyxiant - a vapor or gas that can cause unconsciousness or death by suffocation (lack of oxygen). Most simple asphyxiants are harmful to the body only when they become so concentrated that they reduce the available oxygen in the air (normally about 21%) to dangerous levels (18% or lower). Some are chemical asphyxiants like carbon monoxide (CO) or cyanide, that reduces the blood's ability to carry oxygen. Other examples are CO_2, N_2, H_2, methane, and hydrocarbon gases.

Asphyxiation - a condition that causes asphyxia or suffocation caused by a lack of oxygen in the blood. Asphyxiation is one of the principal potential hazards of working in confined spaces.

Ataxia - a loss of muscular coordination of gait or movement.

Blood borne - carried by the blood. Certain diseases such as hepatitis are blood borne.

Blood pressure - the force of the circulating blood pushing against the walls of the blood vessels.

Building related illness (BRI) - term that refers to a discrete, identifiable disease or illness. Illness can be traced to a specific pollutant or source within a building.

Carcinogen - a chemical that has been demonstrated to cause cancer in animals and, therefore, is considered capable of causing cancer in humans. Findings are based on the feeding of abnormally large quantities of a chemical to test animals or by the application of concentrated solutions to the animal's skin. A chemical is considered to be carcinogen if (1) it has been evaluated by the International Agency for Research on Cancer and found to be a carcinogen or potential carcinogen; (2) it is listed as a carcinogen in the annual report on carcinogens published by the National Toxicology Program; (3) it is regulated by OSHA as a carcinogen; (4) if one positive study has been published.

Cardiac arrest - the condition when the heart stops beating. Cardiopulmonary resuscitation must be given promptly to keep blood flowing to the brain and the cells of the body.

Cardiac arrhythmia - an irregularity in the rhythm of the heart's beating. The heart may speed up quickly then slow down for no apparent reason.

Cardiac emergency - a life threatening condition when the heart is not functioning properly such as during a heart attack or a cardiac arrest.

Cardiac sensitization - an acute effect in which the heart is rendered more sensitive to the body's own catecholamine compounds or administered drugs, such as epinephrine, possibly resulting in irregular heart beat (cardiac arrhythmia), that could be fatal.

Cardiopulmonary resuscitation - an emergency procedure for a person who is not breathing and whose heart has stopped beating (cardiac arrest). The procedure involves a combination of chest compressions and rescue breathing.

Cardiovascular disease - a disease characterized by the gradual clogging of the blood vessels by fatty substances. These conditions are associated with heat attack, stroke, high blood pressure, and diabetes.

Carotid artery - the artery that supplies blood to the head.

Carotid pulse - the beat that is felt at the side of the neck when the carotid artery is pressed. Located between the windpipe and the neck muscle, the carotid pulse is checked to determine the presence or absence of a heartbeat.

Chest compression - a procedure for manually circulating blood in a person whose heart has stopped beating. It involves pressing up and down on the lower half of the breastbone. CPR is a combination of chest compressions and rescue breathing.

Chronic exposure - continuous or intermittent exposure extending over a long time period, usually applies to relatively low material amounts or concentrations.

Chronic health effect - an adverse effect on a human or animal body with symptoms that develop slowly over a long time period and persist or that recur frequently.

Chronic toxicity - adverse or chronic effects resulting from repeated doses of or exposures to a substance over a relatively prolonged period of time such as months, years, or decades. Ordinarily used to denote effects noted in experimental animals.

Cyanosis - a dark purplish coloration of the skin and the mucous membrane due to deficient oxygenation of the blood.

Epidemiology - a science that deals with the incidence, distribution, and control of diseases in a population.

Fibrosis - the formation of fibrous tissue, as in a reparative or reactive process to particulates, in excess of amounts normally present in the lung tissue walls. This reduces the oxygen and CO_2 exchange efficiency.

First aid - immediate measures that can be taken by the victim or others in order to reduce or eliminate the potential effects of a chemical exposure or other injury.

Frost bite - when skin freezes.

Gangrene - death of tissue combined with putrefaction.

Hypersensitivity diseases - diseases characterized by allergic responses to animal antigens. The hypersensitivity diseases most clearly associated with indoor air quality are asthma, rhinitis, and hypersensitivity pneumonitis. Hypersensitivity pneumonitis is a rare but serious disease that involves progressive lung damage as long as there is exposure to the causative agent.

Hypersensitivity pneumonitis - a group of respiratory diseases, including humidifier fever that involves inflammation of the lungs. Most forms of hypersensitivity pneumonitis are thought to be caused by an allergic reaction triggered by repeated exposures to biological contaminants.

Immediately dangerous to life and health (IDLH) - a condition that poses a threat of exposure to airborne contaminants when that exposure is likely to cause death or immediate or delayed permanent adverse health effects or prevent escape from such environment. The revised and added IDLH's of 1994 are based on additional toxicity criteria and data. Most fire codes use the IDLH, based on definitions from the Standards Completion Program (SCP) as a criterion for ventilation rates and emergency discharge treatment. Standard 15 also uses the SCP IDLH as one of several criteria to determine refrigerant quantity limits for occupied spaces.

Inflammation - a series of reactions produced in the tissues by an irritant, injury, or infection characterized by redness and swelling caused by an influx of blood and fluids.

Ingestion - the taking in of a substance through the mouth.

Inhalation - the breathing in of a substance in the form of a vapor, gas, fume, mist, or dust.

Legionnaire's disease - disease named after an outbreak of the illness at an American Legion Convention in July of 1976.

Malignancy - abnormal mass of new tissue growth that serves no function in the body and that threatens life or health.

Metabolic rate - the rate of energy production of the body. Metabolism, which varies with activity, is expressed in met units. One met is defined as 58.2 W/m2 (18.4 Btu/h ft2) which is equal to the energy produced per unit surface area of a seated person at rest. The surface area of an average man is about 1.8 m2 (19 ft2).

Metabolism - the process of change some chemicals go through after absorption by the body.

Multiple chemical sensitivity - a term used by some people to refer to a condition in which a person is considered to be sensitive to a number of chemicals at very low concentrations. There are a number of views about the existence, potential causes, and possible remedial actions regarding this phenomenon.

Nausea - a tendency to vomit; a feeling of sickness in the stomach.

Psychosocial factors - psychological, organizational, and personal stressors that could produce symptoms similar to poor indoor air quality.

Respiratory system - the breathing system, including the lungs and air passages (trachea or windpipe, larynx, mouth, and nose), as well as the associated nervous and circulatory supply.

Sick building syndrome (SBS) - term that refers to a set of symptoms that affect a number of building occupants during the time they spend in the building and diminish or go away during periods when they leave the building. Cannot be traced to specific pollutants or sources within the building.

Silicosis - a condition of massive fibrosis of the lungs causing shortness of breath because of prolonged inhalation of silica dusts.

Stethoscope - an instrument used to detect sounds.

Area of Circles in Square Feet

Diameter in inches	Square foot area	Diameter in inches	Square foot area
1	.0054	1.5	.0123
2	.0218	2.5	.0341
3	.0491	3.5	.0668
4	.0873	4.5	.1105
5	.1364	5.5	.1650
6	.1964	6.5	.2305
7	.2673	7.5	.3068
8	.3491	8.5	.3940
9	.4418	9.5	.4923
10	.5454	11	.6600
12	.7854	13	.9218
14	1.069	15	1.227
16	1.396	17	1.576
18	1.767	19	1.969
20	2.182	21	2.405
22	2.640	23	2.885
24	3.142	25	3.409
26	3.687	27	3.976
28	4.276	29	4.587
30	4.909	31	5.241
32	5.585	33	5.940
34	6.305	35	6.611
36	7.069	37	7.467
38	7.876	39	8.296
40	8.727	41	9.168
42	9.621	43	10.08
44	10.56	45	11.04
46	11.54	47	12.05
48	12.57	49	13.10
50	13.64	51	14.19
52	14.75	53	15.32
54	15.90	56	17.10
58	18.35		

Capacity of Pipe in Cubic Feet of Gas per Hour

Based on a 0.5 psig or less and a Specific Gravity of 0.60.

Nominal Iron Pipe Size "	Internal Pipe Size	Length of pipe, feet							
		10	20	30	40	50	60	70	80
⅜	.493	72	49	40	34	30	27	25	23
½	.622	132	92	73	63	56	50	46	43
¾	.824	278	190	152	130	115	105	96	90
1	1.049	520	350	285	245	215	195	180	170
1.25	1.380	1050	730	590	500	440	400	370	350
1.50	1.610	1600	1100	890	760	670	610	560	530
2	2.067	3050	2100	1650	1450	1270	1150	1050	990
2.5	2.469	4800	3300	2700	2300	2000	1850	1700	1600
3	3.068	8500	5900	4700	4100	3600	3250	3000	2800
4	4.026	17500	12000	9700	8300	7400	6800	6200	5800

Color Code for Resistors

Color	Digits	Multiplier	Tolerance
Black	0	1	± 20%
Brown	1	10	± 1%
Red	2	100	± 2%
Orange	3	1,000	± 3%
Yellow	4	10,000	--
Green	5	100,000	± 5%
Blue	6	1,000,000	± 6%
Violet	7	10,000,000	± 12.5%
Gray	8	0.01	± 30%
White	9	0.1	± 10%
Gold	--	0.1	± 5%
Silver	--	0.01	± 10%
No color	--	--	± 20%

Concentrations

1,000,000 ppm	= 100%
10,000 ppm	= 10.0%
10,000 ppm	= 1.0%
1,000 ppm	= 0.1%
100 ppm	= 0.01%
10 ppm	= 0.001%
1 ppm	= 0.0001%
1,000 ppb	= 1 ppm
100 ppb	= 0.1 ppm
10 ppb	= 0.01 ppm
1 ppb	= 0.001 ppm
1,000 ppt	= 0.001 ppm
100 ppt	= 0.0001 ppm
10 ppt	= 0.00001 ppm
1 ppt	= 0.000001 ppm
1,000 ppt	= 1 ppb
100 ppt	= 0.1 ppb
10 ppt	= 0.01 ppb
1 ppt	= 0.001 ppb

Multiplying factor	Prefix	Symbol
1,000,000,000,000 = 10^{12}	tera	T
1,000,000,000 = 10^{9}	giga	G
1,000,000 = 10^{6}	mega	M
1,000 = 10^{3}	kilo	k
100 = 10^{2}	hecto	h
10 = 10^{1}	deka	de
0.1 = 10^{-1}	deci	d
0.01 = 10^{-2}	centi	c
0.001 = 10^{-3}	milli	m
0.000001 = 10^{-6}	micro	μ
0.000000001 = 10^{-9}	nano	n
0.000000000001 = 10^{-12}	pico	p

Cubic Feet of Airflow per minute

Btu/Hr = 1.08 x ΔT ° x CFM (Based on .075ft^3 air density)										
Btu/Hr Output	**30°**	**40°**	**50°**	**60°**	**70°**	**80°**	**90°**	**100°**	**110°**	**120°**
30,000	926	694	556	463	397	347	308	278	253	231
35,000	1080	810	643	540	463	405	360	324	295	270
40,000	1234	925	741	617	529	463	412	370	337	309
45,000	1389	1042	833	694	595	521	463	417	379	347
50,000	1543	1157	926	772	661	579	514	463	421	386
55,000	1698	1273	1019	849	728	637	566	509	463	424
60,000	1852	1389	1111	926	794	694	617	556	505	463
65,000	2006	1505	1203	1003	860	752	669	602	547	502
70,000	2160	1620	1296	1080	926	810	720	648	589	540
75,000	2315	1736	1389	1157	992	868	772	694	631	579
80,000	2469	1852	1481	1235	1058	926	823	741	673	617
85,000	2623	1968	1574	1311	1124	984	875	787	715	656
90,000	2778	2083	1667	1389	1190	1042	926	833	758	694
95,000	2932	2199	1759	1466	1257	1100	977	880	800	734
100,000	3084	2315	1852	1543	1323	1157	1029	926	842	772
105,000	3241	2431	1944	1620	1389	1215	1080	972	884	810
110,000	3395	2546	2037	1698	1455	1273	1132	1019	926	849
115,000	3549	2662	2130	1775	1521	1331	1183	1065	968	887
120,000	3704	2778	2222	1852	1587	1389	1235	1111	1010	926
125,000	3858	2894	2315	1929	1653	1447	1286	1157	1052	965

Decibel Ratings & Hazardous Time Exposure of Common Noises

Typical Level (Decibels)	Example	Dangerous Time Exposure
0	Lowest sound audible to human ear.	
30	Quiet library, soft whisper.	
40	Quiet office, living room, bedroom away from traffic.	
50	Light traffic at a distance, refrigerator, gentle breeze.	
60	Air conditioner at 20 feet, conversation, sewing machine.	
70	Busy traffic, office tabulator, noisy restaurant. At this decibel level, noise may begin to affect hearing if you are constantly exposed.	Critical level begins
80	Subway, heavy city traffic, alarm clock at two feet, factory noise.	More than 8 hours
90	Truck traffic, noisy home appliances, shop tools, lawnmower. As loudness increases, safe time exposure decreases.	Less than 8 hours
100	Chain Saw, boiler shop, pneumatic drill. Exposure may be dangerous at 100 db, and with every 5 db increase, the "safe time" is cut in half.	Less than 2 hours
120	Rock band concert in front of speakers, sandblasting, and thunderclap. At 120 db, exposure can injure the ear.	Immediate danger
140	Gunshot blast, jet plane at 50 feet. Noise at 140 db may cause actual pain in the ear.	Any length of exposure time is dangerous
180	Rocket launching pad. Without ear protection, noise at this level causes irreversible damage.	Hearing loss inevitable

SUPPLY DUCT SIZES

CFM	(Based on a pressure drop equal to 0.10 in. WG/100 ft.)		
200	10 x 6	8 x 8	6 x 10
300	12 x 6	10 x 8	8 x 10
400	16 x 6	12 x 8	10 x 10
500	18 x 6	14 x 8	11 x 10
600	22 x 6	16 x 8	12 x 10
700	18 x 8	14 x 10	12 x 12
800	20 x 8	16 x 10	13 x 12
900	22 x 8	18 x 10	14 x 12
1000	24 x 8	20 x 10	16 x 12
1200	28 x 8	22 x 10	18 x 12
1400	32 x 8	24 x 10	20 x 12
1600	26 x 10	22 x 12	
1800	30 x 10	24 x 12	
2000	32 x 10	26 x 12	

RETURN DUCT SIZES

CFM	(Based on a pressure drop equal to 0.08 in. WG/100 ft.)		
200	10 x 6	8 x 8	6 x 10
300	14 x 6	10 x 8	6 x 10
400	20 x 6	14 x 8	6 x 10
500	24 x 6	16 x 8	6 x 10
600	20 x 8	16 x 10	6 x 12
700	22 x 8	18 x 10	6 x 12
800	24 x 8	19 x 10	6 x 12
900	26 x 8	20 x 10	6 x 12
1000	30 x 8	22 x 10	6 x 12
1200	34 x 8	26 x 10	6 x 12
1400	30 x 10	24 x 12	6 x 12
1600	35 x 10	26 x 12	
1800	37 x 10	30 x 12	
2000	42 x 10	32 x 12	

RETURN AIR GRILLE SIZE

(Based on approximately 350 ft. min with effective grille face area of 70 %.)

CFM	Size in inches
100	10 x 6
125	12 x 6
170	12 x 8
145	14 x 6
200	14 x 8
245	24 x 6
335	24 x 8
310	30 x 6
425	30 x 8
250	12 x 12
375	12 x 18
575	18 x 18
750	18 x 24
1000	24 x 24

Efficiencies

Cooling Efficiency SEER	Approximate Annual Operating Cost							
6	200	300	400	500	600	700	800	900
7	175	260	345	430	515	600	690	780
8	150	225	300	375	450	525	605	685
9	135	200	270	335	400	465	540	610
10	120	180	240	300	360	420	485	550
11	110	165	220	275	330	385	440	500
12	100	150	200	250	300	350	405	455
13	90	135	185	230	280	325	370	415
14	85	130	170	215	260	300	345	385
15	80	120	160	200	240	280	320	360
16	75	115	150	190	225	265	300	340
17	70	105	140	175	210	250	285	320

To use this chart: Estimate the approximate cooling efficiency of your current air conditioner. Then estimate your current annual cooling cost, based on your total electric bill, eliminating the cost of other electrical use appliances. Use the chart to estimate how much you can save with a higher efficiency system. For example --- if your current air conditioning unit has a SEER of 6 and your annual operating cost is $600.00, the cost to operate a new 14 SEER system will be approximately $260.00, giving you an annual savings of more than $340.00.

Efficiencies

Heating Efficiency AFUE	Approximate Annual Operating Cost						
60	400	500	600	700	800	900	1000
65	365	460	550	640	735	825	915
70	340	425	510	595	675	760	845
75	315	395	470	550	630	710	785
80	295	365	440	515	585	660	735
90	255	320	385	450	515	580	640
95	240	305	365	425	485	545	605

To use this chart: Estimate the approximate heating efficiency of your current furnace. Then estimate your current annual heating cost, based on your total fuel bill, eliminating the cost of other fuel use appliances. Use the chart to estimate how much you can save with a higher efficiency system. For example --- if your current furnace is 60% efficient and your annual operating cost is $700.00, the cost to operate a new 93% AFUE system will be between $425.00 and $450.00, giving you an estimated annual savings of approximately $250.00.

Electrical Formulas

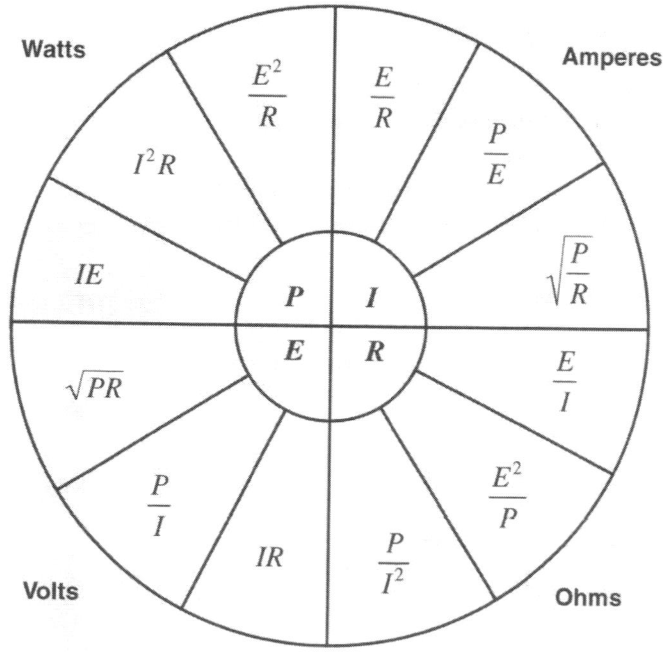

Ohm's Law states that the current in an electric circuit is directly proportional to the applied voltage, and inversely proportional to the circuit resistance.

Quantity	Symbol	Unit	Abbreviation
Capacitance	C	Farad	F
Charge	Q	Coulomb	C
Current	I	Ampere	A
Electromotive force	E	Volt	V
Force	F	Newton	N
Frequency	f	Hertz	Hz
Impedance	Z	Ohm	Ω
Inductance	L	Henry	H
Power	P	Watt	W
Reactance	X	Ohm	Ω
Resistance	R	Ohm	Ω
Torque	T	Pound-foot	lb/ft
Work	W	Joule	J

Capacitive reactance

$$X_c = \frac{1}{2\pi fC}$$

X_c	= capacitive reactance in ohms
f	= frequency in hertz
C	= capacitance in farads
2π	= 6.28, or 2 x π (pi = 3.14)

Capacitors (connected in parallel)

$$C_{total} = C_1 + C_2 + C_3 + \ldots C_n$$

Capacitors (connected in series)

$$C_{total} = \frac{1}{1/C_1 + 1/C_2 + 1/C_3 + \ldots 1/C_n}$$

Capacity

Btu/hr = GPM x 500 x ▲T (of the water)

Quantity of sensible heat = 1.08 x CFM x ▲T

Quantity of latent heat = .68 x CFM x ▲Grains of moisture

Total heat = 4.5 x CFM x ▲Enthalpy

Current imbalance in three-phase circuits

$Deviation_{MAX}$ = maximum deviation from average current

E_{AVG} = average current

$$\% \text{ current imbalance} = \frac{Deviation_{MAX}}{I_{AVG}} \times 100$$

Efficiency of any device

$$\% \text{ efficiency} = \frac{\text{output}}{\text{input}} \times 100$$

Impedance

$$Z = \sqrt{R2 + (Xc - Xc)2}$$

Z	= impedance in ohms
R	= resistance in ohms
X_c	= capacitive reactance in ohms
X_L	= inductance in ohms

Inductive reactance

$$X_L = 2\pi fL$$

X_L	= inductance reactance in ohms
f	= frequency in hertz
L	= inductance in Henries
2π	= 6.28, or 2 x π (pi = 3.14)

Inductors (connected in parallel)

$$L_{total} = \frac{1}{1/L_1 + 1/L_2 + 1/L_3 + \ldots 1/L_n}$$

Inductors (connected in series)

$$L_{total} = L_1 + L_2 + L_3 + \ldots\ L_n$$

Motor Slip

$$s = \frac{S_s - S_r}{S_s} \times 100$$

s	= percent of slip
Ss	= synchronous speed in RPM
Sr	= rotor speed in RPM
2π	= 6.28, or 2 x π (pi = 3.14)

Parallel circuits

Current: $I_{total} = I_1 + I_2 + I_3 + \ldots E_n$

Power: $P_{total} = P_1 + P_2 + P_3 + \ldots P_n$

Resistance (general formula) $R_t = \dfrac{1}{1/R_1 + 1/R_2 + 1/R_3 + \ldots 1/R_n}$

Resistance (two resistors) $R_t = \dfrac{R_1 \times R_2}{R_1 + R_2}$

Resistance (equal resistors) $R_t = \dfrac{R}{N}$ (N = number of resistors)

Voltage: $E_{total} = E_1 = E_2 = E_3 = \ldots E_n$

Power factor

Power factor $= \dfrac{\text{Actual power}}{\text{Apparent power}} = \dfrac{P}{E \times I}$ or PF = R/Z

Series circuits

Current: $I_{total} = I_1 = I_2 = I_3 = \ldots I_n$

Resistance: $R_{total} = R_1 + R_2 + R_3 + \ldots R_n$

Power: $P_{total} = P_1 + P_2 + P_3 + \ldots P_n$

Voltage: $E_{total} = E_1 + E_2 + E_3 + \ldots E_n$

Speed of an AC motor

$$S_s = \frac{120f}{p}$$

S_s = synchronous speed in RPM
120f = 120 x the frequency of the power supply
p = number of poles

Transformer Ratios

$$\frac{N_p}{E_s} = \frac{E_p}{N_s}$$

$$\frac{N_p}{N_s} = \frac{I_s}{I_p}$$

N_p = number of turns in primary coil
N_s = number of turns in secondary coil
E_p = voltage applied to primary coil
E_s = voltage applied to secondary coil
I_p = current in primary coil
I_s = current in secondary coil

Voltage unbalance in single phase circuits

$$\% \text{ voltage imbalance} = \frac{E_{NL} - E_{FL}}{E_{NL}} \times 100$$

E_{NL} = no-load voltage
E_{FL} = full-load voltage

Voltage imbalance in three-phase circuits

$Deviation_{MAX}$ = maximum deviation from average voltage

E_{AVG} = average voltage

$$\% \text{ voltage imbalance} = \frac{Deviation_{MAX}}{E_{AVG}} \times 100$$

Conversions

	For AC current	For direct current
Amperes when the value for kilowatts is known	$\frac{\text{kW x 1,000}}{\text{E x power factor}}$	$\frac{\text{kW x 1,000}}{\text{E}}$
Amperes when the value for kilovolt-amperes is known	$\frac{\text{kVA x 1,000}}{\text{E}}$	
Amperes when the value for horsepower is known	$\frac{\text{hp x 746 x \% efficiency}}{\text{E x power factor}}$	$\frac{\text{hp x 746 x \% efficiency}}{\text{E}}$
Horsepower	$\frac{\text{I x E x power factor}}{\text{746 x \% efficiency}}$	$\frac{\text{I x E}}{\text{746 x \% efficiency}}$
Kilovolt-amperes	$\frac{\text{I x E}}{\text{1,000}}$ (single phase) $\frac{\text{I x E x 1.7321}}{\text{1,000}}$ (three phase)	
Kilowatts	$\frac{\text{I x E x power factor}}{\text{1,000}}$ (single phase)	$\frac{\text{I x E}}{\text{1,000}}$
Kilowatts	$\frac{\text{I x E x 1.7321 x power factor}}{\text{1,000}}$ (three phase)	
Power factor	$\frac{\text{Actual power (in watts)}}{\text{I x E}}$	

Conversions continued

Voltage tolerance

$$\% = \frac{\text{no-load voltage - full-load voltage}}{\text{2 x rated voltage}} \text{ x } 100$$

Voltage regulation

$$\% = \frac{\text{no-load voltage - full-load voltage}}{\text{full-load voltage}} \text{ x } 100$$

Voltage drop	E = I x R	E = I x R
Watts	I x E x power factor (single phase)	I x E
	I x E x 1.7321 x power factor (three phase)	

Fuel

Fuel	Specific Gravity	Btu/ft^3, gross	Ultimate CO_2 %
Natural gas	0.6	~1000	11.82
Propane	1.5	~2500	13.75
Butane gas	2.0	~3200	14.00
#2 Fuel oil	n/a	140,000/gallon	15.30

Orifice Drill Size	Natural Gas 3.5" wc	Natural Gas 4.0" wc	Propane 11" wc	Butane 11" wc
80	0.52	0.55	1480	1678
78	0.76	0.78	2080	2358
76	1.13	1.21	3249	3684
74	1.44	1.55	4119	4669
72	1.77	1.90	5081	5760
70	2.22	2.39	6375	7227
68	2.72	2.93	7813	8858
66	3.09	3.32	8848	10,031
64	3.68	4.14	10,535	11,943
62	4.08	4.39	11,735	13,304
60	4.55	4.89	13,008	14,747
58	5.03	5.39	14,333	16,249
56	6.13	6.58	17,572	19,921
54	8.59	9.23	24,630	27,922
52	11.46	12.31	32,805	37,190
50	13.92	14.94	39,842	45,168
48	16.36	17.62	46,983	53,263

Galvanized Carbon Steel (sheet metal) Sheets

Galvanized Sheet gauge#	Approx. # per sq. ft.	Thickness equivalent	Thickness range in inches
18	2.15625	0.0516	.0545-.0486
19	1.90625	0.0456	.0485-.0426
20	1.65625	0.0396	.0425-.0382
21	1.53125	0.0366	.0381-.0352
22	1.40625	0.0336	.0351-.0322
23	1.28125	0.0306	.0321-.0292
24	1.15625	0.0276	.0291-.0262
25	1.03125	0.0247	.0261-.0232
26	0.90625	0.0217	.0231-.0210
27	0.84375	0.0202	.0209-.0195
28	0.78125	0.0187	.0194-.0180
29	0.71875	0.0172	.0179-.0165
30	0.65625	0.0157	.0164-.0150
31	0.59375	0.0142	.0149-.0138
32	0.56250	0.0134	.0137-.0131

Humidity

Indoor relative humidity is directly related to the wet bulb and dry bulb temperature. This table lists relative humidity in percent for various dry bulb temperatures that are listed across the top of columns and the wet bulb depression in the first column. The wet bulb depression can be found from readings taken from a sling psychrometer or hydrometer. The difference between the dry bulb and the wet bulb temperatures readings is the wet bulb depression. As the depression value decreases, the percentage of relative humidity increases and conversely as the depression value increases, the percentage of relative humidity decreases. For instance, if the sling psychrometer reads 70° F DB and 63° F WB, the wet bulb depression is 70 minus 63 or 7 degrees. From the column headed 70° F, move down the column corresponding to 7 degrees and read the relative humidity of 68%.

Degrees of Wet Bulb Depression	**DRY BULB TEMPERATURE**										
	68	69	**70**	71	72	73	74	75	76	77	78
1	95	95	95	95	95	95	95	96	96	96	96
2	90	90	90	90	91	91	91	91	91	91	91
3	85	85	86	86	86	86	86	86	87	87	87
4	80	81	81	81	82	82	82	82	82	83	83
5	76	76	77	77	77	78	78	78	78	79	79
6	71	72	72	72	73	73	74	74	74	74	75
7	67	67	**68**	68	69	69	69	70	70	71	71
8	62	63	64	64	65	65	65	66	66	67	67
9	58	59	59	60	61	61	61	62	62	63	63
10	54	55	55	56	57	57	58	58	59	59	60
11	50	51	51	52	53	53	54	54	55	56	56
12	46	47	48	48	49	50	50	51	51	52	53
13	42	43	44	45	45	46	47	47	48	48	49
14	38	39	40	41	42	42	43	44	44	45	46
15	34	35	36	37	38	39	39	40	41	42	43
16	31	32	33	33	34	35	36	37	38	39	39
17	27	28	29	30	31	32	33	34	34	35	36
18	23	24	25	27	28	29	29	30	31	32	33
19	20	21	22	23	24	25	26	27	28	29	30
20	16	18	19	20	21	22	23	24	25	26	27
21	13	14	15	17	18	19	20	21	22	23	24
22	10	11	12	13	15	16	17	18	19	20	21
23	6	8	9	10	12	13	14	15	16	17	18
24	3	5	6	7	9	10	11	12	13	14	15

Indoor/Outdoor Relative Humidity Conversion Chart

Indoor Relative Humidity % at 72° F

Outdoor Relative Humidity	-20	-10	-5	0	+5	+10	+15	+20	+25	+30	+35
100	4	6	7	9	11	14	17	21	26	31	38
95	4	5	7	8	10	13	16	20	24	30	36
90	4	5	6	8	10	12	15	19	23	28	34
85	4	5	6	8	9	12	15	18	22	27	32
80	4	5	6	7	9	11	14	17	20	25	30
75	3	4	5	7	8	10	13	16	19	23	28
70	3	4	5	6	8	10	12	15	18	22	26
65	3	4	5	6	7	8	11	14	17	20	25
60	3	3	4	5	7	8	10	13	15	19	23
55	2	3	4	5	6	8	9	12	14	17	21
50	2	3	4	4	6	7	9	10	13	16	19
45	2	3	3	4	5	6	8	9	12	14	17
40	2	2	3	4	4	6	7	8	10	12	15
35	2	2	3	3	4	5	6	7	9	11	13
30	1	2	2	3	3	4	5	6	8	9	11
25	1	1	2	2	3	3	4	5	6	8	10
20	1	1	1	2	2	3	3	4	5	6	8
15	1	1	1	1	2	2	3	3	4	5	6
10	0	1	1	1	1	1	2	2	3	3	4
5	0	0	0	0	1	1	1	1	1	1	2
0	0	0	0	0	0	0	0	0	0	0	0

Outdoor Temperature Value, ° F

Low Voltage Wiring Diagram Codes*

Color	Terminal	Function (mode of operation)
R (red)	V	Hot switched leg of 24 volt AC power used on heating only thermostats and heat/cool thermostats with a common power supply.
W (white)	H	Heating -- single stage.
Y (yellow)	C	Cooling -- single stage.
G (green)	F	Indoor fan.
B	Z	Heating circuit -- constantly energized through a manual subbase switch.
O	D	Cooling circuit -- constantly energized through a manual subbase switch.
R_H	M	Isolated power terminal for a heating circuit used on heat/cool thermostats with isolated circuits -- jumper supplied.
R_C	V	Isolated power terminal for a cooling circuit used on heat/cool thermostats with isolated circuits -- jumper supplied.
W_1	H_1	Heating -- first stage of two stage units.
W_2	H_2	Heating -- second stage of two stage units.
Y_1	C_1	Cooling -- first stage of two stage units.
Y_2	C_2	Cooling -- second stage of two stage units.
X	L	Warning light (dirty filter, electric heat, service light, etc.
4	R	Same as R.
*As with any wiring situation, always consult appropriate wiring diagrams.		

Motors

Type of Motor	Starting Torque	Running Efficiency	Application
Shaded-Pole 1Ø	Low	Good	Small fans
Split-Phase	Medium	Good	Appliances and belt drive blowers
Permanent Split Capacitor	Medium	Excellent	Small compressors and fans
Capacitor Start	Medium	Good	Belt drive fans
Capacitor Start, Capacitor Run	High	Excellent	Compressors and pumps
Three Phase	High	Excellent	All compressors 3 hp and up

Percent Heat Loss In Flue Gas -- Natural Gas											
Difference Between Flue Gas and Room Temperature In Degrees F.											
% CO_2	300	350	400	450	500	550	600	650	700	750	800
4.0	25.1	27.7	30.4	33.1	35.8	38.3	40.9	43.5	46.2	48.8	N/A
4.5	23.6	25.9	28.3	30.7	33.0	35.4	37.8	40.1	42.6	44.8	47.2
5.0	22.2	24.4	26.8	28.7	30.9	33.0	35.7	37.3	39.7	41.8	43.8
5.5	21.2	23.4	25.2	27.3	29.2	31.3	33.2	35.3	37.3	39.2	41.0
6.0	20.4	22.3	24.1	25.8	27.8	29.6	31.5	33.3	35.2	36.8	38.8
6.5	19.8	21.4	23.2	24.8	26.5	28.3	30.0	31.7	33.5	34.6	36.8
7.0	19.1	20.7	22.3	23.9	25.5	27.1	28.8	30.4	32.0	33.8	35.3
7.5	18.5	20.0	21.5	23.0	24.6	26.1	27.7	29.1	30.8	32.2	33.8
8.0	18.0	19.5	20.9	22.4	23.8	25.2	26.7	28.1	29.5	31.0	32.4
8.5	17.6	19.0	20.4	21.7	23.1	24.5	25.8	27.1	28.6	29.9	31.3
9.0	17.2	18.5	19.9	20.7	22.5	23.8	25.2	26.4	27.8	29.0	30.3
9.5	16.9	18.1	19.5	20.2	21.9	23.1	24.4	25.6	26.9	28.2	29.4
10.0	16.6	17.8	19.0	20.1	21.4	22.6	23.8	25.0	26.2	27.4	28.6
11.0	16.1	17.1	18.4	19.4	20.5	21.6	22.8	23.9	25.0	26.2	27.2

Percent Heat Loss In Flue Gas -- # 2 Fuel Oil											
Difference Between Flue Gas and Room Temperature In Degrees F.											
% CO_2	300	350	400	450	500	550	600	650	700	750	800
5.0	22.7	25.4	28.2	30.9	33.8	36.3	39.2	42.0	44.7	47.4	50.1
5.5	21.3	23.8	26.3	28.9	31.4	34.0	36.4	39.0	41.7	44.0	46.5
6.0	20.4	22.5	24.9	27.2	29.5	31.9	34.2	36.4	38.9	41.0	43.5
6.5	19.3	21.3	23.6	25.7	27.8	30.1	32.3	34.4	36.5	38.7	40.8
7.0	18.4	20.5	22.4	24.5	26.5	28.5	30.5	32.6	34.6	36.5	38.6
7.5	17.7	19.6	21.3	23.4	25.2	27.1	29.0	30.9	32.9	34.8	36.5
8.0	17.1	18.9	20.7	22.4	24.2	26.0	27.8	29.6	31.5	33.2	35.0
8.5	16.5	18.2	20.0	21.6	23.3	24.9	26.7	28.4	30.1	31.8	33.5
9.0	16.0	17.6	19.3	20.9	22.4	24.1	25.7	27.3	28.9	30.5	32.1
9.5	15.7	17.1	18.6	20.2	21.7	23.2	24.8	26.3	27.9	29.4	31.0
10.0	15.2	16.6	18.1	19.6	21.0	22.5	24.0	25.4	27.0	28.3	29.9
11.0	14.5	15.8	17.2	18.5	20.0	21.2	22.6	23.9	25.3	26.7	28.0
12.0	13.9	15.1	16.4	17.6	18.9	20.2	21.4	22.6	24.0	25.2	26.5
13.0	13.4	14.5	15.8	16.9	18.1	19.3	20.5	21.5	22.8	24.0	25.2
14.0	13.0	14.0	15.3	16.3	17.4	18.5	19.7	20.8	21.8	22.9	24.1

Pipe Thread Data

Size of pipe	Thread make-up	# of threads per inch	Diameter of Drill
1/8"	¼"	27	11/32"
¼"	3/8"	18	7/16"
3/8"	3/8"	18	37/64"
½"	½"	14	23/32"
¾"	9/16"	14	59/64"
1"	5/8"	11 ½"	1 5/32"
1 ¼"	11/16"	11 ½"	1.5"
1 ½"	11/16"	11 ½"	1 49/64"
2"	¾"	11 ½"	2 3/16"
2 ½"	15/16"	8	2 9/16"
3"	1"	8	3 3/16"
3 ½"	1 1/16"	8	3 11/16"
4"	1 1/8"	8	4 3/16"
5"	1 ¼"	8	5 5/16"
6"	1 5/16"	8	6 5/16"
8"	1 7/16"	8	8 5/16"
10"	1 5/8"	8	10 5/16"
12"	1 ¾"	8	12 5/16"

Pressure/Vacuum Equivalents

Prior to charging with refrigerant, good practice calls for evacuation to deep vacuum levels to remove non-condensable gases and also to evaporate and remove any trapped moisture from within the system. The chart below shows the equivalence of low pressure readings in microns with other familiar pressure/vacuum units. It also lists the temperature that water will vaporize at each pressure.

- Absolute pressure above zero base
- Vacuum below one atmosphere

Microns	psia	mm Hg	Inches of mercury	Approximate fraction of one atmosphere	Vaporization temperature of H_2O in °F @ each pressure
0	0.000	760.00	29.921	-----	-----
50	.001	759.95	29.92	1/15000	-50
100	.002	759.90	29.92	1/7600	-40
150	.003	759.85	29.92	1/5100	-33
200	.004	759.80	29.91	1/3800	-28
300	.006	759.70	29.91	1/2500	-21
500	.010	759.50	29.90	1/1500	-12
1000	.019	759.00	29.88	1/760	1
2000	.039	758.00	29.84	1/380	15
4000	.077	756.00	29.76	1/190	29
6000	.116	754.00	29.69	1/127	39
10000	.193	750.00	29.53	1/76	52
15000	.290	745.00	29.33	1/50	63
20000	.387	740.00	29.13	1/38	72
30000	.580	730.00	28.74	1/25	84
50000	.967	710.00	27.95	1/15	101
100000	1.930	660.00	25.98	2/15	125
200000	3.870	560.00	22.05	1/4	152
500000	9.670	260.00	10.24	2/3	192
760000	14.696	0.00	0.00	1 atmosphere	212

Natural Gas -- Products of Combustion				
% CO_2	% O_2	% Excess Air	SCFM	Weight#/Hour
1	19.2	1164.8	2050.9	8028
1.5	18.3	740.8	1367.3	5388
2.0	17.4	528.8	1025.5	4068
2.5	16.5	401.6	820.4	3276
3.0	15.7	316.8	683.7	2748
3.5	14.8	256.2	586.0	2371
4.0	14.0	210.8	512.8	2088
4.5	13.0	175.4	455.8	1868
5.0	12.2	147.2	410.2	1692
5.5	11.4	124.0	372.9	1548
6.0	10.5	104.8	341.9	1428
6.5	9.4	88.4	315.5	1326
7.0	8.75	64.5	293.0	1239
7.5	7.8	62.4	273.5	1164
8.0	7.0	51.8	256.4	1098
8.5	6.0	42.4	241.3	1040
9.0	5.1	34.1	227.9	988
9.5	4.2	26.6	215.8	942
10.0	3.4	19.9	205.0	900
10.5	2.3	13.9	195.3	862
11.0	1.6	8.4	186.4	828
11.5	.8	3.4	178.4	797
11.86	0.0	0.0	172.9	776

Propane -- Products of Combustion				
% CO_2	% O_2	% Excess Air	SCFM	Weight#/Hour
1.5	18.6	885.0	1562.9	6252
2.0	17.8	636.8	1172.2	4716
2.5	17.1	487.8	937.8	3794
3.0	16.3	388.4	781.5	3180
3.5	15.8	317.4	669.8	2741
4.0	15.0	264.2	586.1	2412
4.5	14.2	222.8	521.0	2156
5.0	13.4	189.7	468.9	1951
5.5	12.6	162.6	426.3	1784
6.0	12.0	140.0	390.7	1644
6.5	10.9	120.9	360.7	1526
7.0	10.5	104.5	334.9	1425
7.5	9.5	90.3	312.6	1337
8.0	8.9	77.9	293.0	1260
8.5	8.2	67.0	275.9	1192
9.0	7.7	57.2	260.5	1132
9.5	6.7	48.5	246.8	1078
10.0	5.9	40.6	234.4	1030
10.5	5.1	33.6	223.5	986
11.0	4.3	27.1	205.2	946
11.5	3.7	21.2	203.8	909
12.0	2.6	15.8	195.4	876
12.5	1.9	10.8	187.5	845
13.0	1.3	6.3	180.4	817
13.5	.6	2.0	173.7	791
13.75	0.0	0.0	170.7	778

Combustion Efficiency Calculator Chart					
For Category IV High Efficiency Gas Furnaces Only					
	% Excess Air				
Actual Flue Gas Temp.	**100%**	**0%**	**60%**	**40%**	**20%**
80	**97.5**	**X**	**X**	**X**	**X**
85	**96.5**	**97**	**97.5**	**98**	**X**
90	**95.5**	**96**	**96.5**	**97**	**98**
95	**94.5**	**96.25**	**95.75**	**96.25**	**97**
100	**93**	**94**	**95**	**95.5**	**96**
105	**92**	**93**	**94**	**94.5**	**95.5**
110	**90**	**91**	**92**	**93**	**94**
115	**X**	**89**	**90.5**	**92**	**93**
120	**X**	**X**	**89**	**90.5**	**92**
125	**X**	**X**	**X**	**89**	**90.5**

Refrigeration

Compression ratio

$$\text{Compression ratio} = \frac{\text{Absolute discharge pressure}}{\text{Absolute suction pressure}}$$

Volumetric efficiency

$$\%\ \text{efficiency} = \frac{\text{Actual volume}}{\text{Total calculated volume}} \times 100$$

Unit capacity

$$\text{Required capacity} = \frac{\text{Btuh load} \times 24\ \text{hours}}{\text{Hours of operation per day}}$$

Compressor speed

The speed at which a belt-drive compressor turns is determined by the following ratio:

$$\frac{\text{Compression RPM}}{\text{Motor RPM}} = \frac{\text{Pulley diameter}}{\text{Flywheel or drive diameter}}$$

General Gas Law

PV = mRT
P = absolute pressure in pounds per square foot
V = volume in cubic feet
m = mass in pounds
R = gas constant
T = absolute temperature in degrees Rankine

Charles' Laws

If the pressure remains constant, the volume of a given mass of gas is directly proportional to the absolute temperature.

$$\frac{P_1}{P_2} = \frac{T_1}{T_2}$$

If the volume remains constant, the pressure is directly proportional to the absolute temperature.

$$\frac{V_1}{V_2} = \frac{T_1}{T_2}$$

Boyle's Law

If the temperature remains constant, the volume of a given mass of gas is inversely proportional to the pressure.

$$\frac{V_1}{V_2} = \frac{P_2}{P_1}$$

Combination Gas Law

$$\frac{P_1 \text{ x } V_1}{T_1} = \frac{P_2 \text{ x } V_2}{T_2}$$

Dalton's Law of Partial Pressure

The total pressure of a mixture of gases is the sum of the pressures of the individual gases.

$$\text{Ptotal} = P_1 + P_2 + P_3 + \ldots P_n$$

Fan Laws

Fan laws

All fans operate according to a set of mathematical relationships:

$$\frac{cfm_2}{cfm_1} = \frac{rpm_2}{rpm_1}$$

$$\frac{SP_2}{SP_1} = \left(\frac{rpm_2}{rpm_1}\right)^2$$

$$\frac{bhp_2}{bhp_1} = \left(\frac{rpm_2}{rpm_1}\right)^3$$

$$\frac{D_2}{D_1} = \left(\frac{rpm_2}{rpm_1}\right)^2$$

where:

cfm = air volume in cubic feet per minute

rpm = fan speed in revolutions per minute

SP = static pressure in inches, water gauge

D = density in pounds per cubic foot

Belt Drive Fan/Motor Relationship

Blower RPM = motor RPM x motor drive pulley size/blower drive pulley size

Motor RPM = blower RPM x blower drive pulley size/motor drive pulley size

Motor drive pulley size = blower RPM x blower drive pulley size/motor RPM

Blower drive pulley size = motor RPM x motor drive pulley size/blower RPM

Heat Transfer

K = thermal conductivity.

C = thermal conductance.

R = thermal resistance. The total resistance is equal to the sum of the individual resistances $R_{total} = R_1 + R_2 + R_3 + \ldots$..

U = overall coefficient of heat transfer. The U-value is the reciprocal of the total resistance. $U = 1/R_{total}$

Rate of Heat Transfer

$Q = U \times A \times \Delta T$
Q = rate (or quantity) of heat transfer in Btu/h
U = overall heat transfer coefficient
A = surface area in square feet
▲T = temperature difference in degrees Fahrenheit

Air Flow

When blower speed is increased by:	Current requirement is increased by:
5%	16%
10%	33%
15%	52%
20%	73%
25%	95%
30%	120%
35%	146%
40%	174%
45%	205%
50%	238%

When blower speed is decreased by:	Current requirement is decreased by:
5%	14%
10%	27%
15%	39%
20%	49%
25%	58%
30%	66%
35%	72%
40%	78%
45%	83%
50%	87%

When total static is increased by 125% blower rpm must be increased by 50%.
Current requirement is increased by 238%.

When total static is increased by 100% blower rpm must be increased by 42%.
Current requirement is increased by 186%.

When total static is increased by 80% blower rpm must be increased by 34%.
Current requirement is increased by 141%.

When CFM is increased by 50% blower rpm must be increased by 50%.
Duct static (pressure drop) is increased by 125%.

When CFM is increased by 40% blower rpm must be increased by 40%.
Duct static (pressure drop) is increased by 96%.

When CFM is increased by 30% blower rpm must be increased by 30%.
Duct static (pressure drop) is increased by 69%.

When CFM is increased by 20% blower rpm must be increased by 20%.
Duct static (pressure drop) is increased by 44%.

When CFM is increased by 10% blower rpm must be increased by 10%.
Duct static (pressure drop) is increased by 21%.

Total pressure = velocity pressure + static pressure

Velocity pressure = total pressure - static pressure

Static pressure = total pressure - velocity pressure

Feet per minute = 4005 x square root of the velocity pressure

CFM (Q) = (V) velocity of air (fpm) x (A) area of duct (sq. ft.)

CFM (Q) = Area (in square inches)/144 x velocity in FPM

Area (square inches) = CFM x 144/FPM

Q = A x V **or** V = Q/A **or** A = Q/V

Sine Wave Voltages -- Conversion Chart

Convert from	Convert to	RMS (Effective)	Average	Peak	Peak-to-peak
RMS (Effective)		n/a	0.900	1.414	2.828
Average		1.110	n/a	1.570	3.141
Peak		0.707	0.637	n/a	2.000
Peak-to-peak		0.354	0.318	0.500	n/a

Specific Gravities of Gases

(Based on 68° F and 14.7 pounds absolute)

Acetylene	C_2H_2	.897
Air	---	1.000
Ammonia	NH_3	.587
Argon	A	1.378
Butane	C_4H_{10}	2.00
Carbon dioxide	CO_2	1.517
Carbon monoxide	CO	.996
Chlorine	CL_2	2.452
Ethane	C_2H_6	1.035
Helium	He	.138
Hydrogen	H_2	.070
Methane	CH_4	.553
Natural gas	CH_4	.665 (approx.)
Nitric oxide	NO	1.035
Nitrogen	N_2	.966
Nitrous oxide	N_2O	1.518
Oxygen	O_2	1.103
Propane	C_3H_8	1.550
Sulphur dioxide	SO_2	2.209

Standard Conversions

Multiple	by	to obtain
Acres	0.405	Hectares
Acres	43.560	Square feet
Acres	4047	Square meters
Acres	1.562×10^{-3}	Square miles
Acres	5645.38	Square varas
Acres	4840	Square yards
Amperes	0.11	Abamperes
Amperes	3 X 109	Statamperes
Angstrom units	1×10^{-10}	Meters
Angstrom units	3.9370×10^{-9}	Inches
Angstrom units	1×10^{-4}	Microns
Angstrom units	1×10^{-8}	Centimeters
Angstrom units	0.1	Millimicrons
Atmospheres	1.0133	Bars
Atmospheres	76.0	Centimeters of mercury
Atmospheres	33.93	Feet of water
Atmospheres	29.921	Inches of mercury
Atmospheres	1.0332	Kg./cm^2
Atmospheres	10332	Kg./m^2
Atmospheres	10.333	Kilograms per square meter
Atmospheres	760	Millimeters of mercury
Atmospheres	2216.2	Lbs./sq. ft.
Atmospheres	14.696	Pounds per square inch
Atmospheres	1.058	Tons per square foot
Barrels	42	Gallons (oil)
Barrels	31.5	Gallons
Barrels per hour	0.0936	Cubic feet per minute
Barrels per hour	5.615	Cubic feet per hour
Barrels per hour	0.700	Gallons per minute (US)
Barrels per hour	42.00	Gallons per hour (US)
Barrels per hour	2.650	Liters per minute
Barrels per hour	158.984	Liters per hour
British thermal units	0.2520	Kilograms-calories
British thermal units	252	Calories
British thermal units	777.6	Foot-pounds
British thermal units	3.927×10^{-4}	Horsepower-hours
British thermal units	1055.06	Joules
British thermal units	107.5	Kilograms-meters
British thermal units	0.0002928	Kilowatt-hours
Btu per hour	0.2929	Watts
Btu per minute	12.96	Foot-pounds per second

Btu per minute	0.02356	Horsepower
Btu per minute	0.01757	Kilowatts
Btu per minute	17.57	Watts
Btu per ft^3 per minute	0.1220	Watts per square inch
Btu per Hr/sq.ft./°F	4.88	Kg.Cal per Hr/Sq. Meter/ °C
Bushels	1.244	Cubic feet
Bushels	2150	Cubic inches
Bushels	0.03524	Cubic meters
Bushels	4	Pecks
Bushels	64	Pints (dry)
Bushels	32	Quarts (dry)
Calories	0.003968	British thermal units
Celsius (C) +273.16	1	° C Absolute, or Kelvin (K)
Celsius (C) +17.78	1.8	° Fahrenheit
Centimeters	1×10^8	Angstrom units
Centimeters	0.39371	Inches
Centimeters	0.01	Meters
Centimeters	1×10^4	Microns
Centimeters	393.7	Mils
Centimeters	10	Millimeters
Centimeters	1×10^7	Millimicrons
Centimeter-grams	980.7	Centimeters-dynes
Centimeter-grams	10^{-5}	Meter-kilograms
Centimeter-grams	7.233×10^{-5}	Pound feet
Centimeters of mercury	0.01316	Atmosphere
Centimeters of mercury	0.4460	Feet of water
Centimeters of mercury	5.3524	Inches of water
Centimeters of mercury	135.95	Kilograms per square meter
Centimeters of mercury	27.854	Pounds per square foot
Centimeters of mercury	0.1934	Pounds per square inch
Centimeters per second	1.969	Feet per minute
Centimeters per second	0.03281	Feet per second
Centimeters per second	0.036	Kilometers per hour
Centimeters per second	0.6	Meters per minute
Centimeters per second	0.02237	Miles per hour
Centimeters per second	3.728×10^{-4}	Miles per minute
Centipoise	0.000672	Lb./ft. sec.
Centipoise	0.0000209	Lb. sec./ft^2
Centipoise	2.42	Lb./ft. hr.
Cubic centimeters	3.531×10^{-5}	Cubic feet
Cubic centimeters	0.0610	Cubic inches
Cubic centimeters	10	Cubic meters
Cubic centimeters	1.308×10^{-6}	Cubic yards
Cubic centimeters	2.642×10^{-4}	Gallons
Cubic centimeters	10^{-3}	Liters
Cubic centimeters	2.113×10^{-3}	Pints (liquid)

Cubic centimeters	1.057×10^{-3}	Quarts (liquid)
Cubic centimeters/sec	0.001	Liters per second
Cubic centimeters/sec	0.060	Liters per minute
Cubic centimeters/sec	0.00212	Cubic feet per minute
Cubic feet	62.43	Pounds of water
Cubic feet	28317	Cubic centimeters
Cubic feet	1728	Cubic inches
Cubic feet	0.02832	Cubic meters
Cubic feet	0.03704	Cubic yards
Cubic feet	7.481	Gallons
Cubic feet	28.32	Liters
Cubic feet	59.84	Pints (liquid)
Cubic feet	29.92	Quarts (liquid)
Cubic feet per minute	471.70	Cubic centimeters per second
Cubic feet per minute	1.699	Cubic meters per hour
Cubic feet per minute	0.000472	Cubic meters per second
Cubic feet per minute	7.48	Gallons per minute
Cubic feet per minute	0.12468	Gallons per second
Cubic feet per minute	28.316	Liters per minute
Cubic feet per minute	0.4719	Liters per second
Cubic feet per minute	62.43	Pounds of water per minute
Cubic inches	16.3872	Cubic centimeters
Cubic inches	0.0005787	Cubic feet
Cubic inches	1.639×10^{-5}	Cubic meters
Cubic inches	2.143×10^{-5}	Cubic yards
Cubic inches	0.004329	Gallons
Cubic inches	0.01639	Liters
Cubic inches	0.03463	Pints (liquid)
Cubic inches	0.01732	Quarts (liquid)
Cubic meters	1,000,000	Cubic centimeters
Cubic meters	35.314	Cubic feet
Cubic meters	1.31	Cubic yards
Cubic meters	1,000	Liters
Cubic meters per hour	0.5886	Cubic feet per minute
Cubic meters per hour	16.666	Liters per minute
Cubic meters per hour	4.404	Gallons per minute (US)
Cubic yards	7.646×10^{5}	Cubic centimeters
Cubic yards	27	Cubic feet
Cubic yards	46.656	Cubic inches
Cubic yards	0.7646	Cubic meters
Cubic yards	202	Gallons
Cubic yards	764.6	Liters
Cubic yards	1616	Pints (liquid)
Cubic yards	807.9	Quarts (liquid)
Cubic yards per minute	0.45	Cubic yard per second
Cubic yards per minute	3.367	Gallons per second

Cubic yards per minute	12.74	Liters per second
Degrees (angle)	60	Minutes
Degrees (angle)	0.01745	Radians
Degrees (angle)	3600	Seconds
Dynes	1.020×10^{-3}	Grams
Dynes	7.233×10^{-5}	Poundals
Dynes	2.248×10^{-6}	Pounds
Ergs	9.486×10^{-11}	Kilograms
Ergs	1	Dyne-centimeters
Ergs	7.376×10^{-8}	Foot-pounds
Ergs	1.020×10^{-3}	Gram-centimeters
Ergs	10^{-7}	Joules
Ergs	2.390×10^{-11}	Kilogram-calories
Ergs	1.020×10^{-8}	Kilogram-meters
Fahrenheit (F) + 459.6	1	˚F Absolute, or Rankin
Fahrenheit (F) -32	5/9	Celsius
Feet	30.480	Centimeters
Feet	12	Inches
Feet	0.3048	Meters
Feet	0.36	Varas
Feet	0.3333	Yards
Feet of water	0.02947	Atmospheres
Feet of water	0.8826	Inches of mercury
Feet of water	304.8	Kilograms per square meter
Feet of water	62.424	Pounds per square foot
Feet of water	0.4335	Pounds per square inch
Feet of water	0.305	Meters per minute
Feet per minute	0.01136	Miles per hour
Feet per minute	0.01829	Kilometers per hour
Feet per minute	0.5080	Centimeters per second
Feet per minute	0.01667	Feet per second
Feet per second	0.305	Meters per second
Feet per second	0.6818	Miles per hour
Feet per second	1.097	Kilometers per hour
Feet per second	30.48	Centimeters per second
Feet per second	0.3048	Meters per second
Feet per second	0.5921	Knots
Foot-pounds	1.286×10^{-3}	British thermal units
Foot-pounds	1.356×10^{7}	Ergs
Foot-pounds	5.050×10^{-7}	Horsepower hours
Foot-pounds	1.356	Joules
Foot-pounds	3.241×10^{-4}	Kilogram-calories
Foot-pounds	0.1383	Kilogram-meters
Foot-pounds	3.766×10^{-7}	Kilowatt-hours
Foot-pounds per minute	1.286×10^{-3}	Btu per minute
Foot-pounds per minute	0.01667	Foot-pounds per second

Foot-pounds per minute	0.0000303	Horsepower
Foot-pounds per minute	3.241×10^{-4}	Kilogram-calories per minute
Foot-pounds per minute	0.0000226	Kilowatts
Foot-pounds per minute	7.717×10^{-2}	Btu per minute
Foot-pounds per minute	1.818×10^{-3}	Horsepower
Foot-pounds per minute	1.945×10^{-2}	Kilogram-calories per minute
Foot-pounds per second	0.001356	Kilowatts
Gallons (Imperial)	277.4	Cubic inches
Gallons (Imperial)	1.201	U.S. gallons
Gallons (Imperial)	4.546	Liters
Gallons	0.0238	Barrels
Gallons	3785	Cubic centimeters
Gallons	0.1337	Cubic feet
Gallons	231	Cubic inches
Gallons	0.003785	Cubic meters
Gallons	0.00495	Cubic yards
Gallons	128	Fluid ounces
Gallons	0.8327	Imperial gallons
Gallons	3.7854	Liters
Gallons	8	Pints (liquid)
Gallons	8.345	Pounds of water
Gallons	4	Quarts (liquid)
Gallons per minute	1.4286	Barrels per hour
Gallons per minute	2.228×10^{-3}	Cubic feet per second
Gallons per minute	0.06308	Liters per second
Gallons per minute	0.2271	Cubic meters per hour
Gallons per minute	0.1337	Cubic feet per minute
Gallons per second	8.02	Cubic feet per minute
Grains (troy)	1	Grains (av)
Grains (troy)	0.06480	Grams
Grains	64.799	Milligrams
Grains	0.00229	Ounces
Grains (troy)	0.04167	Pennyweights (troy)
Grains	0.00014286	Pounds
Grams	980.7	Dynes
Grams	15.432	Grains (troy)
Grams	0.001	Kilograms
Grams	1×10^{-6}	Micrograms
Grams	1,000	Milligrams
Grams	0.03527	Ounces
Grams	0.03215	Ounces (troy)
Grams	0.07093	Poundals
Grams	0.002205	Pounds
Grams/cubic centimeter	62.428	Pounds per cubic feet
Grams/cubic centimeter	0.03613	Pounds per cubic inch
Grams/cubic centimeter	8.345	Pounds per U.S. gallon

Hextares	10,000	Square meters
Horsepower	42.44	Btu per minute
Horsepower	33,000	Foot-pounds per minute
Horsepower	550	Foot-pounds per second
Horsepower	1.014	Horsepower (metric)
Horsepower	10.70	Kilogram-calorie per minute
Horsepower	0.7457	Kilowatts
Horsepower	745.7	Watts
Horsepower (boiler)	33,520	Btu per hour
Horsepower (boiler)	9,804	Kilowatts
Horsepower-hours	2546	British thermal units
Horsepower-hours	1.98×10^6	Foot-pounds
Horsepower-hours	2.684×10^6	Joules
Horsepower-hours	641.7	Kilogram-calorie
Horsepower-hours	2.737×10^5	Kilogram-meters
Horsepower-hours	0.7457	Kilowatt-hours
Imperial gallons	1.2009	US gallons
Inches	2.5400×10^8	Angstrom units
Inches	2.54	Centimeters
Inches	25.4	Millimeters
Inches	0.08333	Feet
Inches	0.0254	Meters
Inches	1,000	Mils
Inches	0.03	Varas
Inches	0.02778	Yards
Inches of mercury	0.03342	Atmospheres
Inches of mercury	1.133	Feet of water
Inches of mercury	13.60	Inches of water
Inches of mercury	345.32	Kilograms per square meter
Inches of mercury	70.727	Pounds per square foot
Inches of mercury	0.4912	Pounds per square inches
Inches of water	0.07355	Inches of mercury
Inches of water	25.38	Kilograms per square meter
Inches of water	0.5781	Ounces per square inches
Inches of water	5.1981	Pounds per square foot
Inches of water	0.03612	Pounds per square inches
Kelvin (K) -273.16	1	°Celsius
Kilograms	980,665	Dynes
Kilograms	1,000	Grams
Kilograms	1.5432×10^4	Grains
Kilograms	35.274	Ounces
Kilograms	70.93	Poundals
Kilograms	2.20462	Pounds
Kilograms	0.00110	Tons (short)
Kilogram-calories	3.968	British thermal units
Kilogram-calories	3,086	Foot-pounds

Kilogram-calories	1.558×10^{-3}	Horsepower-hours
Kilogram-calories	4,183	Joules
Kilogram-calories	426.6	Kilogram-meters
Kilogram-calories	0.00116	Kilowatt-hours
Kilogram-calories/minute	51.43	Foot pounds per second
Kilogram-calories/minute	0.09351	Horsepower
Kilogram-calories/minute	0.06972	Kilowatts
Kilograms per cu. meter	0.06243	Pounds per cubic foot
Kilograms per sq. cm	0.9678	Atmospheres
Kilograms per sq. cm	14.22	Lbs./sq. in. (psi)
Kilograms per sq. meter	0.00142	Lbs./sq. in. (psi)
Kilograms per sq. meter	0.20482	Lbs./sq. ft.
Kilograms per sq. meter	0.00328	Feet of water
Kilograms per sq. meter	0.1	M/cm^2
Kilograms per sq. meter	9.80665	Pascals
Kilometers	100,000	Centimeters
Kilometers	2,280.8	Feet
Kilometers	1,000	Meters
Kilometers	0.53961	Miles (nautical)
Kilometers	0.62137	Miles (statute)
Kilometers	1093.6	Yards
Kilopascals	0.00987	Atmospheres
Kilopascals	0.29613	Inches of mercury (60˚F)
Kilopascals	0.33456	Feet of water (39.2˚F)
Kilopascals	101.97162	$Kg./m^2$
Kilopascals	0.14504	Lbs./sq. in. (psi)
Kilopascals	20.88543	Lbs./sq. ft.
Kilopascals	0.01000	Bars
Kilopascals	0.01020	$Kg./cm^2$
Kilowatts	56.92	Btu per minute
Kilowatts	44,254	Foot-pounds per minute
Kilowatts	737.56	Foot-pounds per second
Kilowatts	1.3404	Horsepower
Kilowatts	14.34	Kilogram-calories per minute
Kilowatts	1,000	Watts
Kilowatt-hours	3,415	British thermal units
Kilowatt-hours	2.655×10^6	Joules
Kilowatt-hours	1.341	Horsepower-hours
Kilowatt-hours	3.6×10^6	Joules
Kilowatt-hours	860.5	Kilogram-calories
Kilowatt-hours	3.671×10^5	Kilogram-meters
Knots	1.0	Nautical miles per hour
Knots	1.6889	Feet per second
Knots	1.1515	Miles per hour
Knots	1.8532	Kilometers per hour
Knots	0.5148	Meters per second

Liters	0.03531	Cubic feet
Liters	61.025	Cubic inches
Liters	0.001	Cubic meters
Liters	0.00131	Cubic Yards
Liters	1.0567	Quarts
Liters	0.26417	US gallons
Liters per minute	2.119	Cubic feet per hour
Liters per minute	0.0600	Cubic meters per hour
Liters per minute	0.2642	Gallons per minute (US)
Log 10 n	2.303	Log en or in n
Log en or in n	0.4343	Log 10 n
Meters	100	Centimeters
Meters	3.2808	Feet
Meters	39.371	Inches
Meters	0.001	Kilometers
Meters	6.2137×10^{-4}	Miles (statute)
Meters	1,000	Millimeters
Meters	1×10^{9}	Millimicrons
Meters	1.0936	Yards
Meters per second	3.281	Feet per second
Meters per second	2.237	Miles per hour
Meters per second	3.600	Kilometers per hour
Microns	1×190^{4}	Angstrom units
Microns	1×10^{-4}	Centimeters
Microns	3.9370×10^{-5}	Inches
Microns	1×10^{-6}	Meters
Micrograms	1×10^{-6}	Grains
Miles (nautical)	1.1516	Statute miles
Miles (nautical)	2026.8	Yards
Miles (nautical)	1.8533	Kilometers
Miles (US statute)	1.609×10^{5}	Centimeters
Miles (US statute)	5,280	Feet
Miles (US statute)	1.609347	Kilometers
Miles (US statute)	1609.4	Meters
Miles (US statute)	0.86836	Nautical miles
Miles (US statute)	320	Rods
Miles	1760	Yards
Miles	1900.8	Varas
Miles per hour	44.70	Centimeters per second
Miles per hour	88	Feet per minute
Miles per hour	1.467	Feet per second
Miles per hour	1.6093	Kilometers per hour
Miles per hour	0.8684	Knots per hour
Miles per hour	0.4470	Meters per second
Milligrams	2.2046×10^{-6}	Pounds (avoirdupois)
Milligrams	3.5274×10^{-5}	Ounces (avoirdupois)

Milligrams	0.01543	Grams
Milligrams	1×10^{-6}	Kilograms
Millimeters	0.1	Centimeters
Millimeters	0.001	Meters
Millimeters	0.03937	Inches
Millimeters	1000	Microns
Millimicrons	1×10^{-9}	Meters
Millimicrons	1×10^{-7}	Centimeters
Millimicrons	10	Angstrom units
Mils	0.001	Inches
Months	30.42	Days
Months	730	Hours
Months	43,800	Minutes
Months	2.628×10^{6}	Seconds
Ounces	29.57	Cubic centimeters
Ounces	1.805	Cubic Inches
Ounces	8	Drams
Ounces	437.50	Grains
Ounces	28.3495	Grams
Ounces	0.06250	Pounds
Ounces	3.1250×10^{-5}	Tons (short)
Ounces per square inch	0.0625	Pounds per square inch
Pints (dry)	33.60	Cubic inches
Pints (liquid)	28.87	Cubic inches
Poise	1	Gm/cm. Sec.
Poise	1	Dyne sec./cm^2
Poise	100	Centipoise
Pounds	444,823	Dynes
Pounds	7,000	Grains
Pounds	453.59	Grams
Pounds	0.45359	Kilograms
Pounds	16	Ounces
Pounds	32.17	Poundals
Pounds	5×10^{-4}	Tons (short)
Pounds of water	0.01602	Cubic feet of water
Pounds of water	27.68	Cubic inches of water
Pounds of water	0.1198	Gallons of water
Pounds of water/minute	2.669×10^{-4}	Cubic feet per second
Pounds per cubic foot	0.01602	Grams per cubic centimeter
Pounds per cubic foot	16.0184	Kilograms per cubic meter
Pounds per cubic foot	5.786×10^{-4}	Pounds per cubic inch
Pounds per cubic foot	5.456×10^{-9}	Pounds per mil foot
Pounds per cubic inch	1728	Pounds per cubic foot
Pounds per cubic inch	27.68	Grams per cubic centimeter
Pounds per cubic yard	0.593	Kilograms per cubic meter
Pounds per square foot	0.01602	Feet of water

Pounds per square foot	0.014139	Inches of mercury
Pounds per square foot	4.887	Kilograms per square meter
Pounds per square foot	0.006944	Pounds per square inch
Pounds per square inch	0.06804	Atmospheres
Pounds per square inch	2.307	Feet of water
Pounds per square inch	70.31	$Gm./cm^2$
Pounds per square inch	27.685	Inches of water
Pounds per square inch	2.036	Inches of mercury
Pounds per square inch	703.1	Kilograms per square meter
Pounds per square inch	144	Pounds per square foot
Pounds per square inch	0.0703	Kilograms per square cm
Pounds per square inch	6.8948	Kilopascals
Quarts	32	Fluid ounces
Quarts	0.0334	Cubic feet
Quarts (dry)	67.20	Cubic inches
Quarts (liquid)	57.749	Cubic inches
Quarts	0.9463	Liters
Quintals	100	Kilograms
Rankin (R) -459.72	1	° Fahrenheit
Rods	16.5	Feet
Square centimeters	1.973×10^5	Circular mils
Square centimeters	0.00108	Square feet
Square centimeters	0.155	Square inches
Square centimeters	10^{-4}	Square meters
Square centimeters	100	Square millimeters
Square centimeters	1.1960×10^{-4}	Square yards
Square feet	2.2957×10^{-5}	Acres
Square feet	929.03	Square centimeters
Square feet	144	Square inches
Square feet	0.0929	Square meters
Square feet	3.587×10^{-8}	Square miles
Square feet	0.1296	Square varas
Square feet	0.11111	Square yards
Square inches	1.273×10^6	Circular mils
Square inches	6.4516	Square centimeters
Square inches	0.006944	Square feet
Square inches	6.4516×10^{-4}	Square meters
Square inches	10^6	Square mils
Square inches	645.2	Square millimeters
Square inches	0.00077	Square yards
Square kilometers	247.10	Acres
Square kilometers	1.0764×10^7	Square feet
Square kilometers	1×10^6	Square meters
Square kilometers	0.38610	Square miles (US)
Square kilometers	1.1960×10^6	Square yards
Square meters	10,000	Square centimeters

Square meters	10.7649	Square feet
Square meters	1,196	Square yards
Square miles	640	Acres
Square miles	2.7878×10^7	Square feet
Square miles	2.590	Square kilometers
Square miles	2.5900×10^8	Square meters
Square miles	3,613,040.45	Square varas
Square miles	3.0967×10^6	Square yards
Square millimeters	0.00155	Square inches
Square millimeters	1×10^{-6}	Square meters
Square millimeters	0.01	Square centimeters
Square millimeters	1.2732	Circular millimeters
Square yards	2.066×10^{-4}	Acres
Square yards	9	Square feet
Square yards	0.8361	Square meters
Square yards	3.228×10^{-7}	Square miles
Square yards	1.1664	Square varas
Tons, long	2240	Pounds
Tons, long	1.016	Tons, metric
Tons, long	1.12	Short tons
Tons, metric	1,000	Kilograms
Tons, metric	2205	Pounds
Tons, metric	1.102	Tons (short)
Tons, short	907.19	Kilograms
Tons, short	3.200×10^4	Ounces
Tons, short	2000	Pounds
Tons, short	0.907	Tons, metric
Torr	0.001316	Atmospheres
Torr	0.001333	Bar
Torr	1000	Millitorr
Torr	1.333	Millibar
Torr	1000	Micron (Hg)
Watts	3.414	Btu per hour
Watts	0.0569	Btu per minute
Watts	0.001341	Horsepower
Watts	0.001	Kilowatts
Yards	91.440	Centimeters
Yards	3	Feet
Yards	0.9144	Meters
Yards	5.6818×10^{-4}	Miles

Density

1 lb. per cu. ft.	= 16.018 kg. per cu. meter
1 lb. per cu. ft.	= 0.0005787 lb. per cu. in.
1 kg. per cu. meter	= 0.06243 lb. per cu. ft.
1 gm. per cu. cm.	= 0.03613 lb. per cu. in.

Density of Water (@ 62° F)

1 cubic foot	= 62.5 pounds
1 pound	= 0.01604 cubic foot
1 gallon	= 8.33 pounds

Energy

1 Btu	= 777.97 ft. lbs.
1 erg	= 9.4805 x 10-11 Btu
1 erg	= 7.3756 x 10-8 ft. lbs.
1 gallon #2 fuel oil	= 140,000 Btu
1 gallon #5 fuel oil	= 148,000 Btu
1 gallon #6 fuel oil	= 150,000 Btu
1 kilowatt	= 3,413 Btu
1 kilowatt hour	= 2.655 x 106 ft. lbs.
1 kilowatt hour	= 1.3410 hp.
1 kg. calorie	= 3.968 Btu
1 pound steam @ 212	= 970 Btu/lb
1 therm	= 100,000 Btu

Flow

1 cu. cm. per sec.	= 0.0021186 cu. ft. per sec.
1 cu. ft. per sec.	= 448.83 gal. per min. = 1699.3 liters per min.
1 cubic foot per minute	= 7.48 gallons per minute = 28.32 liters per minute = 449 gallons per hour
1 cubic foot per hour	= 0.0167 cubic feet per minute = 0.1247 gallons per minute = 7.48 gallons per hour = 0.472 liters per minute = 28.317 liters per hour
1 gallon per minute	= 0.1337 cubic feet per minute = 8.022 cubic feet per hour = 0.002228 cu. ft. per sec. = 0.06308 liters per sec. = 3.79 liters per minute
1 liter per minute	= 0.0353 cubic feet per minute = 2.118 cubic feet per hour = 0.2642 gallons per minute = 15.852 gallons per hour

Linear Measure

1 acre	= 43,560 square feet = 4,840 square yards = 160 square rods = 0.4047 hectare
1 centimeter	= 0.3937 inch = 10 millimeters
1 decimeter	= 3.937 inches
1 dekameter	= 1.9884 rods
1 foot	= 12 inches = 3.048 decimeters = 30.48 centimeters = 0.3048 meters = 3,048 millimeters
1 hectare	= 2.47 acres = 1.853 kilometers
1 inch	= 2.54 centimeters = 25.4×10^{-2} meters = 25.4 millimeters
1 kilometer	= 1,000 meters = 0.62137 mile
1 mile	= 5,280 feet = 1,760 yards = 1.6093 kilometers
1 meter	= 100 centimeters = 1,000 millimeters = 1,000,000 microns = 1.094 yards = 3.28083 feet = 39.37 inches
1 millimeter	= 1,000 microns = 1,000 micrometers = .03937 inches
1 nautical mile	= 6,080 feet
1 rod	= 0.5029 decameter
1 square centimeter	= 0.155 square inch = 0.001076 square foot
1 square decimeter	= 0.1076 square foot
1 square inch	= 6.452 square centimeters = 0.0065 square meters
1 square foot	= 144 square inches = 9.2903 square decimeters = 0.0929 square meter
1 square meter	= 1.196 square yards = 1,550 square inch = 10.76 square foot

Linear Measure con't

1 square yard	= 9 square feet = 0.8361 square meter = 1,296 square inches
1 square mile	= 640 acres = 2.59 square kilometers
1 square kilometer	= 0.386 square mile
1 square rod	= 0.00625% of an acre
1 square meter	= 10,000 square centimeters = 11.196 square yards = 10.76 square feet
1 square centimeter	= 100 square millimeters = 0.155 square inches
1 yard	= 3 feet = 0.9144 meters = 91.44 centimeters

Power and Heat

1 Boiler horsepower (bhp)	= 10 square feet of boiler heating surface (H^20 tube boiler) = 12 square feet of boiler heating surface (fire-tube boiler) = 34.5 pounds per hour of evaporation from and at 212° F. = 33,472 Btu/hr = 139.5 square feet EDR steam = 223 square feet EDR water
1 British thermal unit (Btu)	= 777.6 foot-pounds = 0.252 calories = 1,054 joules
1 Btu heat	= 0.2929 watts
1 calorie	= 3,088 foot-pounds = 3.968 Btu
1 horsepower	= 33,000 foot pounds per minute = 0.7457 kilowatts = 42.44 Btu per minute = 550 ft. lb. per sec = 2,546.5 Btu per hr. = 745.7 watts
1 horsepower hour (hp•h)	= 2544 Btu
1 kilowatt (kW)	= 1,000 watts = 738 foot-pounds per second = 1.341 horsepower
1 kilowatt hour (kWh)	= 3,413 Btu = 1,000 watt-hours = 860 calorie
1 watt	= 0.00134 horsepower = 44.26 ft. lbs. per minute = 3.414 Btu

Pressure

1 atmosphere	= 14.696 psi = 29.92 in. Hg. = 33.9 feet of water = 2,116.35 lb/ft^2
1 inch of $H^2 0$ (gauge) @ 62ºF	= 0.0361 psi = 5.20 lb/ft^2 = 0.07355 in. of Hg.
1 foot head of water at 62ºF	= 0.4335 psi = 0.0295 atmospheres = 0.88265 in. of Hg. = 62.424 lb/ft^2
1 inch of mercury (hg.)	= 0.49116 psi = 13.596 in. of water = 1.13299 ft. of water = 70.726 lb/ft^2 = 0.03342 atmospheres
1 atmosphere	= 14.696 psi = 760 mm of Hg. = 29.921 in. of Hg. = 33.899 ft. of water
1 psi	= 27.73 in. of water = 2.036 in. of Hg = 0.0703066 kg. per sq. cm.
1 psia	= 144 pounds per square foot = 2.036 inches of mercury = 2.307 feet of water = 27.7 inches of water = 0.068 atmospheres
0 psig	= 14.7 psia
1 psig	= 15.7 psia
1 kg. per sq. cm.	= 14.223 lb. per sq. in.
1 dyne per sq. cm.	= 0.0000145 lb. per sq. in.
1 micron	= 0.00001943 lb. per sq. in.

Velocity

1 ft. per sec.	= 30.48 cm. per sec.
1 cm. per sec.	= 0.032808 ft. per sec.

Viscosity

1 centipoise	= 0.000672 lb. per ft. sec.
1 centistoke	= 0.00001076 sq. ft. per sec.

Volume

1 cubic centimeter	= 0.06102 cubic inch
1 cubic decimeter	= 0.0353 cubic foot
1 cubic foot	= 28.347 cubic decimeters = 1,728 cubic inches = 7.48 gallons = 28.32 liters
1 cubic inch	= 16.39 cubic centimeters = 0.016 liters
1 cubic meter	= 1,000 liters = 1.308 cubic yards = 35.31 cubic feet
1 cubic yard	= 27 cubic feet = 0.7646 cubic meter = 46,656 cubic inches
1 dekaliter	= 2.6417 gallons
1 imperial gallon	= 277.4 cubic inches = 4.55 liters
1 liter	= 1,000 cubic centimeters = 0.22 imperial gallon = 0.2642 U.S. gallon = 61.03 cubic inches = 0.908 quart (dry) = 1.0567 quarts (liquid)
1 quart (dry)	= 1.101 liters
1 quart (liquid)	= 0.9463 liter
1 U.S. barrel	= 42 U. S. gallons = 35 imperial gallons
1 U.S. gallon	= 0.833 imperial gallon = 3.785 liters = 3,785 cubic centimeters = 0.1337 cubic feet = 0.3785 dekaliter = 231 cubic inches

Temperature Conversions

°Fahrenheit	= (1.8 x °C) + 32° = (9/5 x °C) + 32°
$°F_{absolute}$	= °F + 460
°Celsius	= (°F - 32°)/1.8 or 5/9 (°F - 32)
$°C_{absolute}$	= °C + 273
°Rankine	= °F + 460
°Kelvin	= °C + 273

Weight

1 cubic foot of air	= 0.075 @ 70º F and 29.92" Hg.
1 cubic foot of brick (building)	= 112-120 lb.
1 cubic foot of brick (fire)	= 145-150 lb.
1 cubic foot of coal	= 80-100 lb.
1 cubic foot of coke	= 24-30 lb.
1 cubic foot of concrete	= 120-140 lb.
1 cubic foot of earth	= 70-120 lb.
1 cubic foot of gravel	= 90-120 lb.
1 cubic foot of water	= 62.425 lb. @ 32º F
1 cubic foot of wood	= 30-60 lb.
1 cubic inch of steel	= 0.284 lb.
1 cubic inch of water	= 0.0361 lb. @ 32º F
1 English ton	= 0.9027 metric tons
1 gallon of water	= 8.33 lb. @ 32º F
1 grain	= 64.8 milligrams = 0.0648 grams = 0.00014 pounds = 0.00229 ounces
1 gram	= 1,000 milligrams = 0.03527 ounces = 0.002205 pounds = 0.001 kilograms = 15.43 grains
1 kilogram	= 1,000 grams = 2.2046 pounds
1 metric ton	= 1,000 kilograms = 0.984 U. S. long tons = 1.1023 U. S. short tons = 2,205 pounds = 1.1023 English tons
1 ounce	= 28.35 grams = 0.0625 pounds = 0.028 kilograms = 437.5 grains
1 pound	= 16 ounces = 7,000 grains = 453.59 grams = 0.4536 kilograms
1 U. S. short ton	= 2,000 pounds = 907.2 kilograms
1 U. S. long ton	= 2240 pounds = 1,016 kilograms

Fractional Equivalent

Fraction	Decimal	Millimeter
1/64	0.015625	0.397
1/32	0.03125	0.794
3/64	0.046875	1.191
1/16	0.06251	1.588
5/64	0.078125	1.984
3/32	0.09375	2.381
7/64	0.109375	2.778
1/8	0.1250	3.175
9/64	0.140625	3.572
5/32	0.15625	3.969
11/64	0.171875	4.366
3/16	0.1875	4.763
13/64	0.203125	5.159
7/32	0.21875	5.556
15/64	0.234375	5.953
1/4	0.2500	6.350
17/64	0.265625	6.747
9/32	0.28125	7.144
19/64	0.296875	7.541
5/16	0.3125	7.938
21/64	0.328125	8.334
11/32	0.34375	8.731
23/64	0.359375	9.128
3/8	0.3750	9.525
25/64	0.390625	9.922
13/32	0.40625	10.319
27/64	0.421875	10.716
7/16	0.4375	11.113
29/64	0.453125	11.509
15/32	0.46875	11.906
31/64	0.484375	12.303
1/2	0.50000	12.700
33/64	0.515625	13.097
17/32	0.53125	13.494
35/64	0.546875	13.891
9/16	0.5625	14.288
37/64	0.578125	14.684
19/32	0.59375	15.081
39/64	0.609375	15.478
5/8	0.6250	15.875
41/64	0.640625	16.272
21/32	0.65625	16.669
43/64	0.671875	17.066
11/16	0.6875	17.463

Fractional Equivalent con't

Fraction	Decimal	Millimeter
45/64	0.703125	17.859
23/64	0.71875	18.256
47/64	0.734375	18.853
3/4	0.7500	19.050
49/64	0.765625	19.447
25/32	0.78125	19.844
51/64	0.796875	20.241
13/16	0.8125	20.638
53/64	0.828125	21.034
27/32	0.84375	21.431
55/64	0.859375	21.828
7/8	0.8750	22.225
57/64	0.890625	22.622
29/32	0.90625	23.019
59/64	0.921875	23.416
15/16	0.9375	23.813
61/64	0.953125	24.209
31/32	0.96875	24.606
63/64	0.984375	25.003
64/64	1.00000	25.400

Millimeters to Inches

0.1 = 0.0039	0.2 = 0.0079	0.3 = 0.0118	0.4 = 0.0157
0.5 = 0.0197	0.6 = 0.0236	0.7 = 0.0276	0.8 = 0.0315
0.9 = 0.0354	1.0 = 0.0394	2.0 = 0.0787	3.0 = 0.1181
4.0 = 0.1575	5.0 = 0.1969	6.0 = 0.2362	7.0 = 0.2756
8.0 = 0.3150	9.0 = 0.3543	10 = 0.3937	11 = 0.4331
12 = 0.4724	13 = 0.5118	14 = 0.5512	15 = 0.5906
16 = 0.6299	17 = 0.6693	18 = 0.7087	19 = 0.7480
20 = 0.7874	21 = 0.8268	22 = 0.8661	23 = 0.9055
24 = 0.9449	25 = 0.9843	26 = 1.0236	27 = 1.0630
28 = 1.1024	29 = 1.1417	30 = 1.1811	31 = 1.2205
32 = 1.2598	33 = 1.2992	34 = 1.3386	35 = 1.3780
36 = 1.4173	37 = 1.4567	38 = 1.4961	39 = 1.5354
40 = 1.5748	41 = 1.6142	42 = 1.6535	43 = 1.6929
44 = 1.7323	45 = 1.7717	46 = 1.8110	47 = 1.8504
48 = 1.8898	49 = 1.9291	50 = 1.9685	51 = 2.0079
52 = 2.0472	53 = 2.0866	54 = 2.1260	55 = 2.1654
56 = 2.2047	57 = 2.2441	58 = 2.2835	59 = 2.3228
60 = 2.3622	61 = 2.4016	62 = 2.4409	63 = 2.4803
64 = 2.5197	65 = 2.5591	66 = 2.5984	67 = 2.6378
68 = 2.6772	69 = 2.7165	70 = 2.7559	71 = 2.7595
72 = 2.8346	73 = 2.8740	74 = 2.9134	75 = 2.9528
76 = 2.9921	77 = 3.0315	78 = 3.0709	79 = 3.1102
80 = 3.1496	81 = 3.1890	82 = 3.2283	83 = 3.2677
84 = 3.3071	85 = 3.3465	86 = 3.3858	87 = 3.4252
88 = 3.4646	89 = 3.5039	90 = 3.5433	91 = 3.5827
92 = 3.6220	93 = 3.6614	94 = 3.7008	95 = 3.7402
96 = 3.7795	97 = 3.8189	98 = 3.8583	99 = 3.8976
100 = 3.9370			

Volume

Rectangular solid: Volume = length x width x height

Circumference of a circle: C = 3.1416 x diameter

Surface of a cylinder: $$S = 3.1416D \times H + \frac{2 \times 3.1416D^2}{4}$$

Volume of a cylinder: $V = \pi r^2 \times \text{Height}$

Area

Area of a circle: **Area** = $\pi \times r^2$

Area of a ring: $$\text{Area} = \frac{3.1416D^2}{4} - \frac{3.1416d^2}{4} \text{ or } \frac{3.1416\,(D^2 - d^2)}{4}$$

where D = diameter of outer ring d = diameter of inner ring

Area of a rectangle: Area = length x height

Area of a triangle: Base x ½ perpendicular height

Area of a trapezoid: ½ sum of parallel sides x perpendicular height

Area of a regular polygon: ½ sum of sides x inside radius

Area of a parallelogram: Base x perpendicular height

Square foot area of a circle: $$A = \frac{\pi \times D^2}{144 \times 4}$$

or $\pi r^2 /4$
or $0.78540 \times D^2 /144$
or $0.07958 \times C^2 /144$

Where C = circumference
D = diameter
r = radius
$\pi = 3.1416$

Wind Chill Index

What the thermometer reads, ° F

Wind Speed MPH	50	40	30	20	10	0	-10	-20	-30	-40	-50	-60
Calm	50	40	30	20	10	0	-10	-20	-30	-40	-50	-60
5	48	37	27	16	6	-5	-15	-26	-36	-47	-57	-68
10	40	28	16	4	-9	-21	-33	-46	-58	-70	-83	-95
15	36	22	9	-5	-18	-36	-45	-58	-72	-85	-99	-112
20	32	18	4	-10	-25	-39	-53	-67	-82	-96	-110	-121
25	30	16	0	-15	-29	-44	-59	-74	-88	-104	-118	-133
30	28	13	-2	-18	-33	-48	-63	-79	-94	-109	-125	-140
35	27	11	-4	-20	-35	-49	-67	-82	-98	-113	-129	-145
40	26	10	-6	-21	-37	-53	-69	-85	-100	-116	-132	-148

Little danger if properly clothed **Danger of freezing exposed flesh** **Great danger of freezing exposed flesh**

Copper Wire Table

AWG	Metric Size	Diameter in mils	Circular mils	Ohms per 1000 ft @ 68° F	Pounds per. 1000 ft	Ft. per #
0000	11.8	460.0	211,600	0.04901	640.5	1.561
000	11.0	409.6	167,800	0.06180	507.9	1.968
00	9.0	364.8	133,100	0.07793	402.8	2.482
0	8.0	324.9	105,500	0.09827	319.5	3.130
1	7.1	289.3	83,690	0.1239	253.3	3.947
2	6.3	257.6	66,370	0.1563	200.9	4.977
4	5.0	204.3	41,740	0.2485	126.4	7.914
6	4.0	162.0	26,250	0.3951	79.46	12.58
8	3.15	128.5	16,510	0.6282	49.98	20.01
10	2.50	101.9	10,380	0.9989	31.43	31.82
12	2.00	80.81	6,530	1.588	19.77	50.59
14	1.60	64.08	4,107	2.525	12.43	80.44
16	1.25	50.82	2,583	4.016	7.818	127.9
18	1.00	40.30	1,624	6.385	4.917	203.4
20	0.80	31.96	1,022	10.51	3.092	323.4
22	0.63	25.35	642.4	16.14	1.945	514.2

Wire Size	Ampacity, amperes
14 AWG	15
12 AWG	20
10 AWG	30
8 AWG	40
6 AWG	55
4 AWG	70
2 AWG	95

Useful Information

- To find circumference of a circle, multiple diameter by 3.1416 or $2\pi r$
- To find diameter of a circle, multiple circumference by .31831.
- To find the side of an inscribed square, multiple diameter by 0.7071 or multiply circumference by 0.2251 or divide circumference by 4.4428.
- Side of inscribed cube equals radius of sphere by 1.1547.
- To find the side of an equal square, multiple diameter by .8862.
- For a square, a side multiplied by 1.14142 equals diameter of its circumscribing circle.
- A side multiplied by 4.443 equals circumference of its circumscribing circle.
- A side multiplied by 1.128 equals diameter of an equal circle.
- A side multiplied by 3.547 equals circumference of an equal circle.
- To find cubic inches in a ball, multiple cube of diameter by .5236.
- To find cubic contents of a cone, multiple area of base by 1/3 the altitude.
- Surface of frustrum of cone or pyramid: sum of circumference of both ends times 2 slant height plus area of both ends.
- Contents of frustrum of cone or pyramid: multiply area of two ends and get square root. Add the two areas and multiply 1/3 altitude.
- Doubling the diameter of a pipe increases its capacity four times.
- 1 pound steam, from and at 212° F = 970 Btu/lb
- 1 square foot EDR (steam) = 240 Btu/hr
- 1 square foot EDR (water) = 150 Btu/hr
- Work = force x distance
- Power = work/time
- Velocity = distance/time
- Pressure = force/area
- Pressure of a column of liquid = density of liquid x height of column
- Density = weight/volume
- Specific density = mass/volume
- Specific volume = volume/mass
- Centrifugal force = mass x velocity^2/radius
- Torque = force (lb) x radius (ft)
- Absolute pressure (psia) = gauge pressure (psig) + 14.7